T0237443

Lectures in Classical Thermodynamics with an Introduction to Statistical Mechanics

Daniel Blankschtein

Lectures in Classical Thermodynamics with an Introduction to Statistical Mechanics

 Springer

Daniel Blankschtein
Department of Chemical Engineering
Massachusetts Institution of Technology
Cambridge, MA, USA

ISBN 978-3-030-49200-7   ISBN 978-3-030-49198-7   (eBook)
https://doi.org/10.1007/978-3-030-49198-7

© Springer Nature Switzerland AG 2020
This work is subject to copyright. All rights are reserved by the Publisher, whether the whole or part of the material is concerned, specifically the rights of translation, reprinting, reuse of illustrations, recitation, broadcasting, reproduction on microfilms or in any other physical way, and transmission or information storage and retrieval, electronic adaptation, computer software, or by similar or dissimilar methodology now known or hereafter developed.
The use of general descriptive names, registered names, trademarks, service marks, etc. in this publication does not imply, even in the absence of a specific statement, that such names are exempt from the relevant protective laws and regulations and therefore free for general use.
The publisher, the authors, and the editors are safe to assume that the advice and information in this book are believed to be true and accurate at the date of publication. Neither the publisher nor the authors or the editors give a warranty, expressed or implied, with respect to the material contained herein or for any errors or omissions that may have been made. The publisher remains neutral with regard to jurisdictional claims in published maps and institutional affiliations.

Cover illustration: Schematic pressure (P) - molar volume (V) phase diagram of a pure substance depicting the Binodal, the Spinodal, and the Critical Point (CP). The sigmoidal dashed curve corresponds to an isotherm at a temperature (T) which is lower than the critical temperature Tc. Outside the Binodal, the system is thermodynamically Stable. Between the Binodal and the Spinodal, the system is thermodynamically Metastable. Inside the Spinodal, the system is thermodynamically Unstable. For complete details, please refer to Lecture 17.

This Springer imprint is published by the registered company Springer Nature Switzerland AG
The registered company address is: Gewerbestrasse 11, 6330 Cham, Switzerland

To my wife Anna, for her unconditional love, loyalty, and support

Contents

Part III Introduction to Statistical Mechanics

Lecture 1

Introduction to the Book

1.1 Motivation and Scope of the Book

This book is based on lectures that I delivered in the one semester, graduate-level course *Chemical Engineering Thermodynamics* (10.40) in the Department of Chemical Engineering at the Massachusetts Institute of Technology (MIT). For my teaching of 10.40, I was awarded the Outstanding Teaching Award by the graduate students nine times. Encouraged and motivated by repeated requests from my 10.40 students, in 2018, I finally decided to write this book which is based on my lecture notes, supplemented by many solved sample problems which help crystallize the material taught.

Including this lecture, the book consists of 50 lectures and is primarily designed for graduate students and senior undergraduate students in Chemical Engineering. The book is also suitable for graduate students in Mechanical Engineering, Chemistry, and Materials Science. It focuses on developing the ability of the reader to solve a broad range of challenging problems in Classical Thermodynamics, by applying the fundamental principles and concepts taught to new and often unusual scenarios. In addition, the book introduces readers to the fundamentals of Statistical Mechanics, including demonstrating how the microscopic properties of atoms and molecules, as well as their interactions, can be accounted for to calculate various practically relevant average thermodynamic properties of macroscopic systems. Again, solved sample problems are presented to help the reader better understand the material taught.

My lecture notes, and therefore this book, are inspired by concepts, principles, methods, and applications that I distilled and adapted from a number of books (see below), as well as by my own in-depth understanding of the material taught. The book provides a pedagogical presentation of the fundamentals of Classical Thermodynamics, with an introduction to Statistical Mechanics, including applying these fundamentals to the solution of illuminating and practically relevant sample problems which are dispersed throughout the lectures.

© Springer Nature Switzerland AG 2020

D. Blankschtein, *Lectures in Classical Thermodynamics with an Introduction to Statistical Mechanics*, https://doi.org/10.1007/978-3-030-49198-7_1

1.2   Organization of the Book

The book is organized as follows: Lecture 1 provides an introduction to the book. Part I discusses *Fundamental Principles and Properties of Pure Fluids* and consists of Lectures 2–20. Part II discusses *Mixtures: Models and Applications to Phase and Chemical Reaction Equilibria* and consists of Lectures 21–37. Part III presents an *Introduction to Statistical Mechanics* and consists of Lectures 38–50. Illuminating solved sample problems are presented throughout the book. In addition, following Lecture 50, the book contains solved problems pertaining to Part I (10 solved problems), Part II (10 solved problems), and Part III (15 solved problems). These problems are challenging and will assist readers to crystalize the material taught. For complete details about the organization of the book, readers are referred to the comprehensive Table of Contents.

The material on Classical Thermodynamics presented in Lectures 2–20 of Part I, as well as in Lectures 21–37 of Part II, is adapted from *Thermodynamics and Its Applications*, third edition, by Jefferson W. Tester and Michael Modell, Prentice Hall International Series in the Physical and Chemical Sciences, Upper Saddle River, NJ (1996). Hereafter, the names of the authors will be abbreviated as T&M. In addition, some of the material presented in Lectures 9 and 12 of Part I is adapted from *The Principles of Chemical Thermodynamics*, fourth edition, by Kenneth Denbigh, Cambridge University Press, London (1981). Hereafter, the name of the author will be abbreviated as Denbigh.

The material on Introduction to Statistical Mechanics presented in Lectures 38–43 of Part III is adapted from *Molecular Thermodynamics*, by Donald A. McQuarrie and John D. Simon, University Science Books, Sausalito CA (1999). Hereafter, the names of the authors will be abbreviated as M&S. The material on Introduction to Statistical Mechanics presented in Lectures 44–48 of Part III is adapted from *Statistical Mechanics*, by Donald A. McQuarrie, Harper & Row, New York (1976). Hereafter, the name of the author will be abbreviated as McQuarrie. The material presented in Lectures 49 and 50 of Part III is adapted from my lecture notes.

Starting with Lecture 2, each lecture begins with an introduction section which summarizes the material that will be discussed in the lecture, including solved sample problems. These introductions will serve as useful road maps for the 49 lectures and, as such, will help readers to more readily navigate the material taught.

1.3   Acknowledgments

I am grateful to the students who attended my 10.40 lectures over the years for challenging me with insightful questions which helped me improve the quality of my lectures. Many thanks to all the teaching assistants who worked with me over the years, and who helped the students crystallize the material taught in class through fruitful interactions during office hours and one-on-one meetings, including preparing outstanding solutions to homework and exam problems. In particular, I am indebted to Nancy Zoeller, Daniel Kamei, Ahmed Ismail, Henry Lam, Srinivas Moorkanikkara, Fei Chen, Amanda Engler, Jaisree Iyer, Michael Stern, Bomy Lee Chung, Sven Schlumpberger, Jennifer Lewis, Manish Shetty, Ananth Govind Rajan, Ran Chen, Tzyy-Shyang Lin, and Dimitrios Fraggedakis for their efforts and intellectual contributions to 10.40. Many thanks to Vishnu Sresht and Rahul Prasanna Misra for their valuable insights on various aspects of Classical Thermodynamics and Statistical Mechanics. I am also grateful to my colleagues, Jefferson Tester, who introduced me to the teaching of Chemical Engineering Thermodynamics, and Jonathan Harris, Bernhardt Trout, and Bradley Olsen, who co-taught 10.40 with me and with whom I discussed many challenging aspects of the material taught. I am indebted to Dimitrios Fraggedakis for creating all the beautiful graphical figures, and to Tzyy-Shyang Lin for creating all the tables. I am also indebted to my assistant, Cindy Welch, for her immense dedication and skill in typing the majority of the lectures in the book. Many thanks to our department head, Paula Hammond, and to our Executive Officer, Martin Bazant, for facilitating writing of the book. I am most grateful to Steven Elliot, Senior Publishing Editor, Cambridge University Press, for his insightful advise and encouragement about publishing engineering books, and for introducing me to Michael Luby, Senior Publishing Editor at Springer, who eventually oversaw the publication of my book. In fact, I am indebted to Michael Luby, Brian Halm, Sathya Stephen, Brinda Megasyamalan, and Mario Gabriele from Springer for their dedication and guidance with all aspects of my book project. Finally, my greatest gratitude and deepest love go to my parents, Samuel and Julia, for instilling in me the importance of education and hard work, to my wife, Anna, for her love, patience, and understanding, and to my daughters, Suzan and Dana, for their love and continued support.

Part I
Fundamental Principles and Properties of Pure Fluids

Lecture 2
Useful Definitions, Postulates, Nomenclature, and Sample Problems

2.1 Introduction

The material presented in this lecture is adapted from Chapter 2 in T&M. First, we will introduce useful definitions which will be utilized throughout Parts I, II, and III of this book. Second, we will discuss two (I and II) out of the four (I, II, III, and IV) postulates which serve as the pillars on which the edifice of thermodynamics is erected. These postulates can be viewed as truisms which cannot be proven from first principles, but that are consistent with our experimental observations, and have withstood the test of time. Third, we will present the nomenclature which will be utilized throughout this book. Fourth, we will solve Sample Problem 2.1 to determine if a system is simple or composite. Finally, we will solve Sample Problem 2.2 to shed light on the liquid-vapor monovariant line in a pressure (P)-temperature (T) phase diagram.

2.2 Useful Definitions

System – Subject of the experiment that we carry out. Well-defined region in terms of spatial coordinates.

Boundary – Surface enclosing the system. Can be real or imaginary.

Environment – Region of space external to the system and sharing a common boundary with the system. Work and heat interactions occur at the boundary.

Primitive Property – Property of the system which can be measured at a particular time without perturbing the system and whose value does not depend on the history of the system. Examples include temperature, pressure, and volume.

Event – An occurrence where at least one primitive property changes.

Interaction – A simultaneous event which occurs in the system and its environment and would change if the environment (or the system) was changed.

© Springer Nature Switzerland AG 2020
D. Blankschtein, *Lectures in Classical Thermodynamics with an Introduction to Statistical Mechanics*, https://doi.org/10.1007/978-3-030-49198-7_2

Closed System – System whose boundary is impermeable to mass flow.

Open System – System whose boundary is permeable to mass flow, through at least one point, of at least one component of the system.

Adiabatic Boundary – Prevents the transfer of heat. Prevents temperature equilibration.

Diathermal Boundary – Allows the transfer of heat. Leads to instantaneous temperature equilibration.

Rigid Boundary – Prevents changes in the volume of the system.

Movable Boundary – Permits changes in the volume of the system.

External Constraints – Set of boundaries: permeable versus impermeable to mass flow, adiabatic versus diathermal, rigid versus movable.

Isolated System – System enclosed by impermeable, adiabatic, and rigid boundaries.

Simple System – System devoid of any internal impermeable, rigid, and adiabatic boundaries, and not acted upon by external force fields (e.g., gravitational, electric, or inertial forces), which can change the energy (e.g., gravitational, electric, kinetic) of the system.

Composite System – System composed of two or more simple systems. No restrictions apply to the type of internal boundaries separating the various simple systems.

Phase – Region within a simple system having uniform properties.

Restraints – Barriers within a system (simple or composite) that prevent some changes from occurring within the time span of interest. Internal boundaries in a composite system which are impermeable, adiabatic, or rigid are considered restraints.

State of the System – Identified by the values of the properties needed to reproduce the system.

Stable Equilibrium State – State whose properties do not vary with time. Postulate I postulates the existence of stable equilibrium states and indicates the number of properties that need to be specified to unambiguously characterize such states (see below). Postulate II postulates what conditions are required to attain a stable equilibrium state (see below).

Change of State – Identified by a change in the value of at least one property.

Path – Describes all the states that the system traverses during a change of state.

Quasi-static Path – A path for which all the intermediate states are equilibrium states.

Derived Property – Property which exists only for stable equilibrium states, and is not measurable. A derived property is defined in terms of changes in the state of the system between initial and final stable equilibrium states. As such, a derived property is a state function (e.g., energy, enthalpy, entropy).

Extensive Property – Property which depends on the size (mass) of the system (e.g., volume, energy, number of molecules).

Intensive Property – Property which is independent of the size (mass) of the system (e.g., pressure, temperature, chemical potential).

2.3   Postulates I and II (Adapted from Appendix A in T&M)

I. For closed, simple systems with given internal restraints, there exist stable equilibrium states that can be characterized completely by two independently variable properties in addition to the masses of the particular chemical species initially charged.

   – Postulates the existence of stable equilibrium states, but does not indicate when they exist
   – As we will show in Part II, is consistent with the Gibbs Phase Rule

II. In processes for which there is no net effect on the environment, all systems (simple and composite) with given internal restraints will change in such a way that they approach one and only one stable equilibrium state for each simple subsystem. In the limiting condition, the entire system is said to be at equilibrium.

   – Describes the natural tendency of an isolated system to approach a state of stable equilibrium characterized by minimal internal energy and maximal entropy
   – If the set of internal restraints changes, then so will the state of stable equilibrium to which the system tends

2.4   Sample Problem 2.1

A closed container is filled with water coexisting with its vapor at room temperature

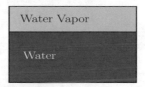

Fig. 2.1

(see Fig. 2.1). Is the system simple or composite?

2.4.1 Solution

The system in Fig. 2.1 consists of two equilibrated phases (water and water vapor), separated by an interface which is diathermal, movable, and open. In addition, the change in the gravitational potential energy of the molecules in the vapor phase is negligible and is zero for the fixed center of mass of the liquid phase. As a result, according to the requirements imposed by Postulate I, the system in Fig. 2.1 is indeed simple.

2.5 Sample Problem 2.2

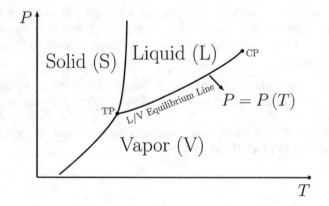

Fig. 2.2

Discuss the liquid/vapor equilibrium line in the pressure (P)-temperature (T) phase diagram shown in Fig. 2.2, where TP and CP denote the Triple Point and the Critical Point, respectively.

2.5.1 Solution

The variation of pressure (P) with temperature (T) along the L/V equilibrium line in Fig. 2.2 is governed by the Clapeyron equation, which we will derive in Part II, and is given by:

$$\left(\frac{dP}{dT}\right)_{L/V} = \frac{H^V - H^L}{T(V^V - V^L)} = \frac{\Delta H^{vap}}{T\Delta V^{vap}} \qquad (2.1)$$

In the last term in Eq. (2.1), the numerator is equal to the molar enthalpy of vaporization, and the denominator is equal to the absolute temperature times the molar volume of vaporization. Figure 2.2 and Eq. (2.1) show that along the liquid-vapor (L/V) equilibrium line, pressure and temperature are not independent. As a result, only one intensive variable can be specified, for example, the temperature, which uniquely determines the pressure. As we will show in Part II, this result is consistent with the celebrated Gibbs Phase Rule.

2.6 Nomenclature

Variable	Extensive	Intensive
General case, B	$\underline{B}$	B
Energy	$\underline{E}$	E
Internal energy	$\underline{U}$	U
Enthalpy	$\underline{H}$	H
Entropy	$\underline{S}$	S
Gibbs free energy	$\underline{G}$	G
Volume	$\underline{V}$	V
Mole number	N	—
Temperature	—	T
Pressure	—	P
Chemical potential	—	μ

In the table above and throughout this book, we will utilize the underbar to denote extensive variables like $\underline{V}$, $\underline{U}$, $\underline{H}$, and $\underline{S}$. The corresponding molar (intensive) variables like V, U, H, and S will not carry the underbar.

Lecture 3

The First Law of Thermodynamics for Closed Systems: Derivation and Sample Problems

3.1 Introduction

The material presented in this lecture is adapted from Chapter 3 in T&M. First, we will discuss mechanical work done on a rigid body. We will also introduce other types of work, including surface, electric, and magnetic. Second, we will extend the concept of mechanical work done on a rigid body to that associated with a thermodynamic system, for example, work done by a gas expanding against the atmosphere in a cylinder-piston assembly in the presence of friction. Third, we will discuss a key result that the work done by a system on the environment is equal to minus the work done by the environment on the system. We will stress that work is a mode of energy transfer which exists only at the boundary between a system and the environment. In the system or in the environment, there is only energy, which we will introduce through Postulate III as work done on the system under adiabatic conditions. Fourth, we will introduce heat absorbed by the system as the difference between the work done on the system under adiabatic conditions and the actual work done on the system. Like work, heat is a form of energy transfer which exists solely at the boundary between the system and the environment. Fifth, after we introduce work, energy, and heat, the First Law of Thermodynamics for a closed system will emerge naturally. Finally, we will solve Sample Problems 3.1, 3.2, and 3.3 to help crystallize the material taught.

3.2 Work Interactions

The differential mechanical work associated with the movement of a rigid body is defined as follows:

© Springer Nature Switzerland AG 2020

D. Blankschtein, *Lectures in Classical Thermodynamics with an Introduction to Statistical Mechanics*, https://doi.org/10.1007/978-3-030-49198-7_3

$$\underbrace{\delta W}_{\substack{\text{Differential} \\ \text{mechanical} \\ \text{work}}} = \underbrace{\left(\sum \vec{F}_s\right)}_{\substack{\text{Sum of all the forces acting on the} \\ \text{surface or boundary of the rigid body} \\ \text{at a point where there is a differential} \\ \text{displacement of the boundary}}} \cdot \underbrace{d\vec{r}}_{\substack{\text{Differential} \\ \text{displacement} \\ \text{of the boundary}}} \qquad (3.1)$$

The symbol δ denotes path-dependent differentials of functions which are not state variables, such as work. The integral of such functions yields the value of the function. For example:

$$\int_{\text{State } 1}^{\text{State } 2} \delta W = W \qquad (3.2)$$

and depends on the path connecting states 1 and 2.

3.3   Sample Problem 3.1

Calculate the work done by a gas expanding isothermally (see Fig. 3.1).

3.3.1   Solution

Given N moles of gas expanding isothermally, the gas pressure, P_g, is a function of the gas volume, $\underline{V}_g$ (see Fig. 3.1).

The work done by the expanding gas is given by (see Fig. 3.1):

$$W_g = \int_{\text{State } 1}^{\text{State } 2} \delta W_g = \int_{\underline{V}_g^1}^{\underline{V}_g^2} P_g d\underline{V}_g = \left\{ \begin{array}{l} \text{Area beneath the path;} \\ \text{depends on the path!} \end{array} \right\} \qquad (3.3)$$

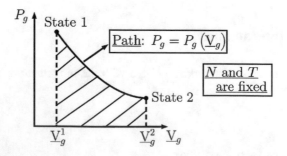

Fig. 3.1

The symbol d denotes total differentials of functions which are state variables. The integral of such functions yields the change in the value of the function. For example:

$$\int_{\text{State 1}}^{\text{State 2}} d\vec{r} = \vec{r}_2 - \vec{r}_1 \tag{3.4}$$

and is independent of the path connecting states 1 and 2.

The total mechanical work is given by (see Eq. (3.1)):

$$W = \int_{\vec{r}_1}^{\vec{r}_2} \delta W = \int_{\vec{r}_1}^{\vec{r}_2} \left(\sum \vec{F}_s \right) \cdot d\vec{r} \tag{3.5}$$

For a list of specific types of work interactions, see below.

Body (b) forces, or forces associated with external fields, act on molecules in the system. They are denoted by $\vec{F}_b$. Examples include gravitational, inertial, centrifugal, and Coulombic forces.

3.4   Specific Types of Work Interactions

As discussed above, all work interactions are path-dependent and are defined to occur at the boundary of a system. We saw that the symbol δ is used here to designate a path-dependent property. Further, we saw that, in general, differential work can be represented by the dot product of a boundary (surface, s) force, $\vec{F}_s$, and a differential displacement, $d\vec{r}$. Specifically,

$$\delta W = \vec{F}_s \cdot d\vec{r} \tag{3.6}$$

The following specific types of work are encountered in nature:

Pressure: $- Pd\underline{V}$ (P = pressure, $\underline{V}$ = volume)
Surface: $\sigma d\underline{a}$ (σ = surface tension, $\underline{a}$ = area)
Electric: $\vec{E} \cdot d\vec{D}$ ($\vec{E}$ = electric field strength, $\vec{D}$ = electric flux density)
Magnetic: $\vec{H} \cdot d\vec{B}$ ($\vec{H}$ = magnetic field strength, $\vec{B}$ = magnetic flux density)
Frictional: $\vec{F}_f \cdot d\vec{r}$ ($\vec{F}_f$ = frictional force, $\vec{r}$ = displacement)

3.5   Sample Problem 3.2

Calculate the balance of forces on a weight suspended by a string rising in the $+\hat{z}$ direction in a gravitational field in the absence of viscous forces (see Fig. 3.2).

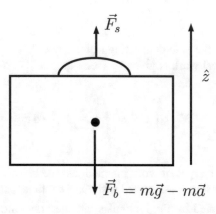

Fig. 3.2

3.5.1   Solution

Newton's Second Law of Motion states that:

$$\sum \vec{F}_s + \sum \vec{F}_b = m\vec{a} \qquad (3.7)$$

Considering the inertial force, $- m\vec{a}$, as a body force, it follows that:

$$\sum \vec{F}_s + \left(\sum \vec{F}_b - m\vec{a} \right) = 0 \left\{ \begin{array}{l} \text{The sum of all surface and body} \\ \text{forces acting on a rigid body is equal to 0} \end{array} \right\}$$

$$(3.8)$$

3.6   Sample Problem 3.3

Calculate the work done by a gas on its environment as it expands against the atmosphere in a cylinder-piston assembly in the presence of friction (see Fig. 3.3).

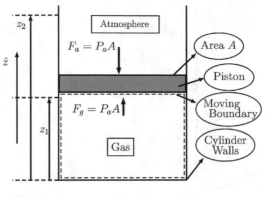

Fig. 3.3

3.6.1 Solution

Figure 3.3 shows a gas expanding against the atmosphere in a cylinder-piston assembly:

- The gas is a simple, closed system with one moving boundary
- Assume a slow, quasi-static gas expansion

According to our definition of mechanical work, the work done by the gas on its environment is given by:

$$\delta W_g = \vec{F}_g \cdot d\vec{z} = F_g dz \tag{3.9}$$

$$F_g = P_g A \Rightarrow \delta W_g = P_g (A dz) \tag{3.10}$$

$$d\underline{V}_g = A dz \Rightarrow \delta W_g = P_g d\underline{V}_g \tag{3.11}$$

$$W_g = \int_{z_1}^{z_2} \delta W_g = \int_{z_1}^{z_2} P_g A dz = \int_{\underline{V}_g^1}^{\underline{V}_g^2} P_g d\underline{V}_g \left\{ \begin{array}{l} \text{To calculate } W_g, \text{ we must know} \\ \text{the path, } P_g = P_g(z) \text{ or } P_g = P_g \\ \left(\underline{V}_g\right), \text{connecting states 1 and 2} \end{array} \right\}$$

$$\tag{3.12}$$

Note that all the forces depicted in Fig. 3.4 are colinear (along $+\hat{z}$ or $-\hat{z}$). As a result, Newton's Second Law of Motion implies that:

$$F_g - P_a A - F_f - mg - ma = 0 \tag{3.13}$$

or

$$F_g = P_a A + F_f + mg + ma \tag{3.14}$$

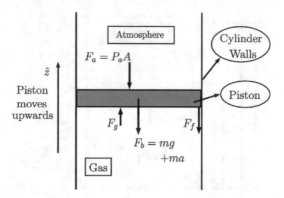

Fig. 3.4

$$\text{Recall that: a } = \frac{dv}{dt} = \frac{dv}{dz}\frac{dz}{dt} = (dv/dz)v \tag{3.15}$$

Accordingly,

$$\delta W_g = F_g dz = \left(P_a A + F_f + mg + mv\frac{dv}{dz}\right)dz \tag{3.16}$$

$$\delta W_g = P_a(Adz) + F_f dz + mgdz + mvdv \tag{3.17}$$

$$\delta W_g = \underbrace{\left(P_a d\underline{V}_g\right)}_{\substack{\text{Work done by} \\ \text{the gas to push} \\ \text{back the} \\ \text{atmosphere}}} + \underbrace{d(mgz)}_{\substack{\text{Work done by the} \\ \text{gas to increase the} \\ \text{potential energy} \\ \text{of the piston}}} + \underbrace{d\left(\frac{1}{2}mv^2\right)}_{\substack{\text{Work done by the} \\ \text{gas to increase} \\ \text{the kinetic energy} \\ \text{of the piston}}} + \underbrace{F_f dz}_{\substack{\text{Work done by} \\ \text{the gas on the} \\ \text{cylinder walls to} \\ \text{overcome friction}}} \tag{3.18}$$

Equation (3.18) shows that work done by the gas can be computed through its effect on the gas environment (the atmosphere, the piston, and the cylinder walls).

To calculate W_g, we simply integrate δW_g in Eq. (3.18) along a path connecting states 1 and 2. For simplicity, if F_f is assumed to be constant, then:

$$W_g = \int_{z_1}^{z_2} \delta W_g \tag{3.19}$$

$$W_g = P_a A(z_2 - z_1) + mg(z_2 - z_1) + \frac{1}{2}m\left(v_2^2 - v_1^2\right) + F_f(z_2 - z_1) \tag{3.20}$$

Equation (3.20) shows that if we know F_f, we can measure (P_a, A, m, z_1, z_2, v_1, and v_2) and calculate W_g in terms of its effects on the gas environment. Following a similar procedure, we can also calculate:

- W_a – Work done by the atmosphere
- W_p – Work done by the piston
- W_w – Work done by the cylinder walls

on their respective environments.

We can then show that $W_g + W_a + W_p + W_w = 0$, or that:

$$W_{\substack{\text{System} \\ \text{on} \\ \text{Environment}}} = - W_{\substack{\text{Environment} \\ \text{on} \\ \text{System}}} \tag{3.21}$$

Equation (3.21) is a key result that we will utilize to solve many problems involving the calculation of mechanical work done by a system on its environment.

It is always possible to measure mechanical work in terms of the change in the potential energy of a mass in a gravitational field. Mechanical work is a form of energy transfer. For example, work done by the gas was transferred into potential and kinetic energies of the piston. By carrying out mechanical work, a system can transfer energy to its environment.

There is a different (nonmechanical) mode of energy transfer known as heat (scc Fig. 3.5).

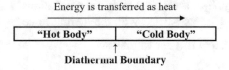

Energy is transferred as heat

| "Hot Body" | "Cold Body" |

↑
Diathermal Boundary

Fig. 3.5

To introduce heat in a rigorous manner, we invoke the notion of an adiabatic work interaction associated with a purely mechanical process (no heat can be transferred across adiabatic boundaries). Simply put, work and heat are two modes of energy transfer. Once in the system, only energy exists. Later in Part I, we will see that there is a distinction in the quality and efficiency of work and heat.

The concept of the adiabatic work interaction which is always possible between stable equilibrium states is introduced through Postulate III (see below). Because the adiabatic (a) work is only a function of the end states, it is a derived property of the system, which we will refer to as the Energy, $\underline{E}$, of the system. By convention, the energy of the system increases when work is done on the system by its environment, that is:

$$\underline{E}_f - \underline{E}_i = +W^a_{i \to f} \tag{3.22}$$

The adiabatic work, $W^a_{i \to f}$, is independent of the path connecting states i and f and is therefore a state function.

3.7   Postulate III (Adapted from Appendix A in T&M)

For any states, (1) and (2), in which a closed system is at equilibrium, the change of state represented by (1) → (2) and/or the reverse change (2) → (1) can occur by at least one adiabatic process, and the adiabatic work interaction between this system and its surroundings is determined uniquely by specifying the end states (1) and (2).

* States that adiabatic work interactions of any type in closed systems are state functions
* Implies that the First Law of Thermodynamics is, in fact, a restatement of the Law of Energy Conservation, because the adiabatic form of the First Law of Thermodynamics states that $d\underline{E} = \delta Q + \delta W$, with $\delta Q = 0$

3.8   Energy Decomposition

The total energy, $\underline{E}$, can be decomposed into three main contributions:

(i)   Internal energy, $\underline{U}$, associated with microscopic energy storage at the molecular level
(ii) Potential energy, $\underline{E}_{PE}$, associated with the gravitational force
(iii) Kinetic energy, $\underline{E}_{KE}$, associated with the inertial force

Adding up (i), (ii), and (iii) above yields:

$$\underline{E} = \underline{U} + \underline{E}_{PE} + \underline{E}_{KE} \tag{3.23}$$

For a simple system:

$$\underline{E}_{PE} = 0, \underline{E}_{KE} = 0 \Rightarrow \underline{E} = \underline{U} \tag{3.24}$$

3.9   Heat Interactions

The energy difference between two states can always be determined by measuring the work in an adiabatic process connecting the two states (Postulate III, see above). With the same initial and final states, we can visualize any process (adiabatic or nonadiabatic) connecting the two states. Because energy is a state function, the energy difference is the same as that found for the adiabatic process. However, if the process is not adiabatic, the work interaction will be different than that for the adiabatic process. Nevertheless, it is always possible to measure work as discussed above.

One can then define heat, Q, as the difference between the energy change and the actual work carried out (see Fig. 3.6).

$$\left\{ \begin{array}{l} Q = (\underline{E}_f - \underline{E}_i) - W_{i \to f} \\ \underline{E}_f - \underline{E}_i = +W^a_{i \to f} \end{array} \right\} \tag{3.25}$$

$$\therefore Q = W^a_{i \to f} - W_{i \to f} \tag{3.26}$$

3.10   The First Law of Thermodynamics for Closed Systems

The First Law of Thermodynamics for closed systems is a restatement of the Law of Energy Conservation and the manner in which energy (associated with the bulk of the system) and work and heat (associated with the boundaries of the system) are interconverted.

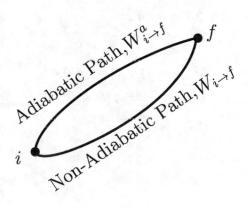

Fig. 3.6

In integral form, the First Law of Thermodynamics for closed systems is written as follows:

$$\Delta \underline{E} - \underline{E}_f - \underline{E}_i - Q + W \tag{3.27}$$

where Q is the total amount of heat absorbed by the system and W is the total amount of work done on the system.

In differential form, the First Law of Thermodynamics for closed systems is written as follows:

$$dE = \delta Q + \delta W \tag{3.28}$$

For a closed system interacting with its environment, the composite of [system (S) + environment (E)] can always be considered as a "new system" of constant volume surrounded by an adiabatic and impermeable boundary (see Fig. 3.7):

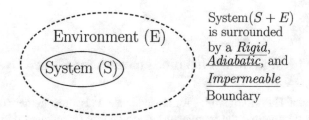

System $(S + E)$ is surrounded by a *Rigid*, *Adiabatic*, and *Impermeable* Boundary

Fig. 3.7

An examination of Fig. 3.7 shows that:

$$\Delta \underline{E}_{S+E} = \Delta \underline{E}_S + \Delta \underline{E}_E \tag{3.29}$$

$$\Delta \underline{E}_{S+E} = Q + W = 0 + 0 = 0 \tag{3.30}$$

$$\{\Delta \underline{E}_S = -\Delta \underline{E}_E; W_S = -W_E\} \tag{3.31}$$

$$\therefore Q_S = -Q_E \tag{3.32}$$

Lecture 4

The First Law of Thermodynamics for Closed Systems: Thermal Equilibrium, the Ideal Gas, and Sample Problem

4.1 Introduction

The material presented in this lecture is adapted from Chapter 3 in T&M. First, we will discuss thermal equilibrium, Postulate IV, and the directionality of heat flow. Second, we will discuss the thermodynamic properties of an ideal gas. Third, we will utilize the ideal gas as a model fluid which will allow us to obtain mathematically simple solutions when we solve problems involving the First Law of Thermodynamics. Fourth, we will begin solving Sample Problem 4.1, an illuminating problem which will allow us to select various possible systems, including ascertaining which one leads to the most challenging, or to the simplest, solution. It will become apparent that the engineer, or the scientist, needs to develop a facility to select the optimal system which will lead to the simplest solution. This, of course, will require experiense and practice. Finally, we will present a four-step strategy that can be used to solve problems involving the First Law of Thermodynamics.

4.2 Thermal Equilibrium and the Directionality of Heat Interactions

Subsystems A and B which are at equilibrium across a diathermal boundary are in thermal equilibrium. If there is a heat interaction between subsystems A and B of the isolated, composite system, it follows from Postulate IV (see below) that this intereaction must eventually cease because each subsystem, as well as the composite system, will approach equilibrium.

If a heat interaction occurs, it follows that $Q_A + Q_B = 0$ (see Fig. 4.1).

The equations below describe the isolated composite system (A + B) shown in Fig. 4.1.

© Springer Nature Switzerland AG 2020

D. Blankschtein, *Lectures in Classical Thermodynamics with an Introduction to Statistical Mechanics*, https://doi.org/10.1007/978-3-030-49198-7_4

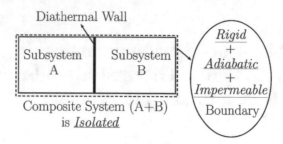

Fig. 4.1

$$\Delta \underline{E}_{A+B} = Q_{A+B} + W_{A+B} \quad \text{(First Law of Thermodynamics)} \qquad (4.1)$$

$$\Delta \underline{E}_{A+B} = 0 \quad \text{(Isolated System)} \qquad (4.2)$$

$$W_{A+B} = 0 \quad \text{(Rigid Boundary)} \qquad (4.3)$$

$$\therefore Q_{A+B} = Q_A + Q_B = 0 \quad \text{(Adiabatic Boundary)} \qquad (4.4)$$

What is the direction of a heat interaction? Postulate IV (see below) helps us answer this question. Postulate IV is often referred to as the Zeroth Law of Thermodynamics.

4.3   Postulate IV (Adapted from Appendix A in T&M)

If the sets of systems A,B and A,C each have no heat interaction when connected across nonadiabatic walls, then, there will be no heat interaction if systems B and C are also so connected.

- This postulate is sometimes referred to as the "Zeroth Law of Thermodynamics"
- States that temperature differences are required for heat transfer to occur
- Introduces the concept of thermal equilibrium in the absence of heat interactions

When sytems A and B undergo a purely heat interaction, that is, $W_A = 0$ and $W_B = 0$, such that:

$$\Delta \underline{E}_A = Q_A + W_A = Q_A = -(\Delta \underline{E}_B = Q_B + W_B = Q_B) \qquad (4.5)$$

or

$$Q_A = -Q_B \qquad (4.6)$$

then, heat is transferred from system A to system B. We can also state that $Q_{A \to B} > 0$.

The thermometric temperature, θ, can be used to rank systems with respect to the direction of heat interactions. If for the three systems, A, B, and C, $Q_{A \to B} > 0$ (heat is transferred from A to B) and $Q_{B \to C} > 0$ (heat is transferred from B to C), then,

$$\theta_A > \theta_B > \theta_C \qquad (4.7)$$

By convention, when $\theta_A > \theta_B$, the heat interaction is such that $\underline{E}_A$ decreases and $\underline{E}_B$ increases or:

$$\frac{d\underline{E}}{d\theta} > 0 \qquad (4.8)$$

4.4   Ideal Gas Properties

4.4.1   Equation of State (EOS)

$P\underline{V} = NRT$ (Extensive form)
$PV = RT$ (Intensive form)

4.4.2   Internal Energy

U is only a function of temperature: $dU = C_v^o dT$
$U = \int C_v^o dT + U_0; \underline{U} = NU$
$U_0 = $ Reference-state constant
$C_v^o = $ Ideal gas heat capacity at constant volume
$C_v^o = (\partial U / \partial T)_V$
$C_v^o = g(T) = a + bT + cT^2 + \ldots$

where a, b, c, etc. are fitted empirical constants.

4.4.3   Enthalpy

H is only a function of temperature: $dH = C_p^o dT$
$H = U + PV = \int C_p^o dT + H_0; \underline{H} = NH = \underline{U} + P\underline{V}$
$H_0 = $ Reference-state constant
$C_p^o = $ Ideal gas heat capacity at constant pressure

$$C_p^o = (\partial H / \partial T)_P$$
$$C_p^o = f(T) = a^* + b^* \, T + c^* \, T^2 + \ldots$$

where a*, b*, c*, etc. are fitted empirical constants and a* $-$ a $=$ R, b $=$ b*, c $=$ c*, etc. (see Section 4.4.2).

4.4.4   Other Useful Relationships

$$C_p^o - C_v^o = R$$

$$dV/V = dT/T - dP/P$$

$$d\underline{V}/\underline{V} + dP/P = dT/T + dN/N$$

4.5   Sample Problem 4.1: Problem 3.1 in T&M

A small well-insulated cylinder and piston assembly (see Fig. 4.2) contains an ideal gas at 10.13 bar and 294.3 K. A mechanical lock prevents the piston from moving. The length of the cylinder containing the gas is 0.305 m, and the piston cross-sectional area is 1.858×10^{-2} m^2.

The piston, which weighs 226 kg, is tightly fitted, and when allowed to move, there are indications that considerable friction is present. When the mechanical lock is released, the piston moves in the cylinder until it impacts and is engaged by another mechanical stop; at this point, the gas volume has just doubled. The heat capacity of the ideal gas is 20.93 J/mol K, independent of temperature and pressure. Consider the heat capacities of the piston and the cylinder walls to be negligible.

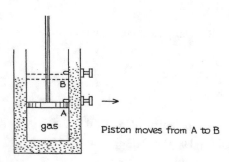

Piston moves from A to B

Fig. 4.2

(a) As an engineer, can you estimate the temperature and pressure of the gas after such an expansion? Clearly state any assumptions.
(b) Repeat the calculations if the cylinder was rotated 90° and 180° before tripping the mechanical lock.

4.5.1 Solution

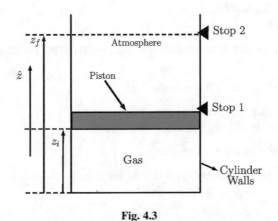

Fig. 4.3

To solve this problem, we will make use of the following four steps:

1. *Sketch Given Configuration* (see Fig. 4.3)

2. *Summarize Given Data and Information*

- Gas: Ideal

    Initial Condition:

    $T_i = 294.3$ K
    $P_i = 10.13$ bar
    $z_i = 0.305$ m
    $C_v = 20.93$ J/mol K

    Final Condition:

    $T_f = ?; P_f = ?$
    $z_f = 2z_i = 0.610$ m

- Atmosphere:

$P_a = 1.013$ bar
Infinite reservoir

- Piston:

  Area – $A_p = 1.858 \times 10^{-2}$ m^2
  Mass – $M_p = 226$ kg
  Negligible heat capacity, $C_v = 0$
  Well insulated

- Cylinder Walls:

  Negligible heat capacity, $C_v = 0$
  Well insulated

- Considerable friction is present between the tightly fitted piston and the cylinder walls!

Find T_f and P_f of the gas if the cylinder is:

   (i) Upright, as in the sketch in Fig. 4.3.
  (ii) Tilted 90° or 180° before removing Stop 1.

3. *Identify Critical Issues for Solution*

   (i) What system and boundaries should we select?
  (ii) How do we deal with the friction?

4. *Make Physically Reasonable Approximations*

   (i) Due to the friction, the gas expansion is slow and the path is quasi static.

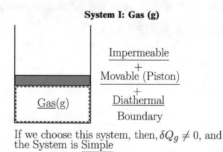

System I: Gas (g)

Impermeable
+
Movable (Piston)
+
Diathermal
Boundary

Gas(g)

If we choose this system, then, $\delta Q_g \neq 0$, and
the System is <u>Simple</u>

Fig. 4.4

  (ii) Gravitational effects on the gas and the atmosphere are negligible ⇒ both
       systems are simple (Figs. 4.4, 4.5, and 4.6).

System II: Gas + Piston + Cylinder (x)

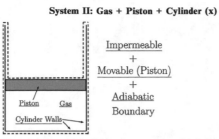

Impermeable
+
Movable (Piston)
+
Adiabatic
Boundary

If we choose this System, composed of the Gas, the Piston, and the Cylinder Walls denoted as System x, then, $\delta Q_x = 0$, and the System is <u>Simple</u>

Fig. 4.5

System III: Atmosphere (a)

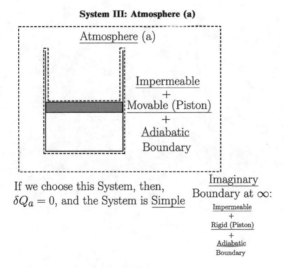

If we choose this System, then, $\delta Q_a = 0$, and the System is <u>Simple</u>

Imaginary
Boundary at ∞:
Impermeable
+
Rigid (Piston)
+
<u>Adiabatic</u>
Boundary

Fig. 4.6

Lecture 5

The First Law of Thermodynamics for Closed Systems: Sample Problem 4.1, Continued

5.1 Introduction

The material presented in this lecture is adapted from Chapter 3 in T&M. First, we will choose one of the systems that we introduced in Lecture 4, where friction does not need to be accounted for, and show that it will lead to a relatively simple solution of Sample Problem 4.1 (denoted as Solution 1). Second, we will choose another system where friction will need to be accounted for. Recall that in the Statement of Sample Problem 4.1, no information is provided about the friction! Third, the need to deal explicitly with the friction will lead to a more challenging solution (denoted as Solution 2). Nevertheless, through a creative derivation, we will show that the mechanical work done by the gas to overcome friction at the cylinder walls is recovered 100% as heat absorbed by the gas, and therefore, it cancels out in the context of the First Law of Thermodynamics for the chosen system. Finally, as expected, we will show that the simpler Solution 1 and the more challenging Solution 2 yield identical results.

5.2 Sample Problem 4.1: Problem 3.1 in T&M, Continued

For completeness, below, we again present the Statement of Sample Problem 4.1.

A small well-insulated cylinder and piston assembly (see Fig. 5.1) contain an ideal gas at 10.13 bar and 294.3 K. A mechanical lock prevents the piston from moving. The length of the cylinder containing the gas is 0.305 m, and the piston cross-sectional area is 1.858×10^{-2} m^2.

The piston, which weighs 226 kg, is tightly fitted, and when allowed to move, there are indications that considerable friction is present. When the mechanical lock is released, the piston moves in the cylinder until it impacts and is engaged by another mechanical stop; at this point, the gas volume has just doubled. The heat

© Springer Nature Switzerland AG 2020
D. Blankschtein, *Lectures in Classical Thermodynamics with an Introduction to Statistical Mechanics*, https://doi.org/10.1007/978-3-030-49198-7_5

capacity of the ideal gas is 20.93 J/mol K, independent of temperature and pressure. Consider the heat capacity of the piston and cylinder walls to be negligible.

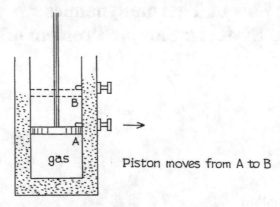

Fig. 5.1

(a) As an engineer, can you estimate the temperature and pressure of the gas after such an expansion? Clearly state any assumptions.
(b) Repeat the calculations if the cylinder were rotated 90° and 180° before tripping the mechanical lock.

5.3   Solution 1: System III-Atmosphere (a)

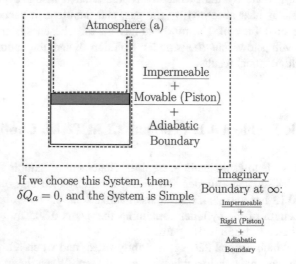

Fig. 5.2

We begin by solving System III (the Atmosphere, a) introduced in Lecture 4, and for completeness, shown again in Fig. 5.2. System III is a simple, closed system, with an adiabatic boundary that has one movable part (the piston). Accordingly,

$$\delta Q_a = 0 \tag{5.1}$$

A First Law of Thermodynamics analysis of the closed system yields:

$$d\underline{E}_a = \delta Q_a + \delta W_a \tag{5.2}$$

Combining Eqs. (5.1) and (5.2) yields:

$$d\underline{E}_a = \delta W_a \tag{5.3}$$

We also know that:

$$\underbrace{\delta W_a}_{\substack{\text{Work done on} \\ \text{the atmosphere}}} = \underbrace{-P_a d\underline{V}_a}_{\substack{\text{Work done by} \\ \text{the atmosphere}}} = P_a d\underline{V}_g \tag{5.4}$$

where we have used the fact that $d\underline{V}_a = -d\underline{V}_g$.

To compute $d\underline{E}_a$ in Eq. (5.2), we consider the composite system $(x + a)$ where system x consists of the gas + piston + cylinder (see Fig. 5.2). Note that system $(x + a)$ is isolated by the imaginary boundary at ∞ (see the outer dashed boundary in Fig. 5.2). The imaginary boundary at ∞ is impermeable, adiabatic, and rigid by choice. In that case, a First Law of Thermodynamics analysis of system $(x + a)$ yields:

$$d\underline{E}_{x+a} = d\underline{E}_x + d\underline{E}_a$$

$$d\underline{E}_{x+a} = \underbrace{\delta Q_{x+a}}_{\text{Adiabatic boundary} \to 0} + \underbrace{\delta W_{x+a}}_{\text{Rigid boundary} \to 0} = 0$$

$$\therefore \; d\underline{E}_a = -d\underline{E}_x \tag{5.5}$$

Because system $x = \text{gas (g)} + \text{piston (p)} + \text{cylinder walls (w)}$, it follows that its total energy is given by:

$$\underline{E}_x = \underline{E}_g + \underline{E}_p + \underline{E}_w \tag{5.6}$$

We also know that:

$$\underline{E}_g = \underline{U}_g \, (\text{The gas is a simple system}) \tag{5.7}$$

$$\underline{E}_p = \underline{U}_p + \underbrace{M_p g z}_{\substack{\text{Piston} \\ \text{Potential Energy}}} + \underbrace{\frac{1}{2} M_p v_p^2}_{\substack{\text{Piston} \\ \text{Kinetic Energy}}} \tag{5.8}$$

where $\underline{U}_p$ is the internal energy of the piston. Because $C_{vp} = 0$ by choice, $\underline{U}_p = 0$.

$$\underline{E}_w = \underline{U}_w \text{ (The walls are a simple system)} \tag{5.9}$$

Because $C_{vw} = 0$ by choice, it follows that:

$$\underline{E}_w = 0 \tag{5.10}$$

Using Eqs. (5.7), (5.8), and (5.10) in Eq. (5.6) yields:

$$\underline{E}_x = \underline{U}_g + M_p g z + \frac{1}{2} M_p v_p^2 \tag{5.11}$$

Taking the differential of Eq. (5.11), including using it in Eq. (5.5), yields:

$$d\underline{E}_a = -d\underline{E}_x = -d\underline{U}_g - d(M_p g z) - d\left(\frac{1}{2} M_p v_p^2\right) \tag{5.12}$$

Because the gas is ideal and N_g is constant, it follows that:

$$d\underline{U}_g = N_g C_{vg} dT \tag{5.13a}$$

Combining Eqs. (5.12) and (5.13a) yields:

$$d\underline{E}_a = -N_g C_{vg} dT - d(M_p g z) - d\left(\frac{1}{2} M_p v_p{}^2\right) \tag{5.13b}$$

Using Eq. (5.13b) in Eq. (5.3), with δW_a given in Eq. (5.4), yields:

$$\underbrace{-N_g C_{vg} dT - d(M_p g z) - d\left(\frac{1}{2} M_p V_p^2\right)}_{d\underline{E}_a} = \underbrace{P_a d\underline{V}_g}_{\delta W_a} \tag{5.14}$$

or

$$-N_g C_{vg} dT = P_a d\underline{V}_g + d(M_p g z) + d\left(\frac{1}{2} M_p v_p^2\right) \tag{5.15}$$

In Eq. (5.15), N_g, C_{vg}, P_a, M_p, and g are all known. Therefore, Eq. (5.15) can be integrated directly from $[T_i, N_{gi}, \underline{V}_{gi}, z_i, v_{pi} = 0$ (initially, the piston is at rest) to $T_f, N_{gf} = N_{gi}, \underline{V}_{gf} = 2\underline{V}_{gi}$ (chosen), $z_f, v_{pf} = 0$ (finally, the piston is at rest)].

We will integrate Eq. (5.15) later after we again derive Eq. (5.15) using System I (The Gas) introduced in Lecture 4.

We encourage the readers to undertake the solution of System II (System x – Gas + Piston + Cylinder Walls), which is similar to that of System III discussed above.

System I (Gas, g), although the most natural to choose, will lead to the most challenging solution. Of course, we will again obtain Eq. (5.15). For completeness, Fig. 5.3 depicting System I is shown again below. As discussed in Lecture 4, System I (Gas, g) is a simple, closed system, with one moving boundary (the piston), and heat is generated internally due to the friction of the piston with the cylinder walls (as a result, the dashed boundary surrounding the gas in Fig. 5.3 is diathermal).

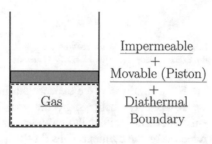

Fig. 5.3

5.4   Solution 2: System I-Gas (g)

If we choose System I, then, $\delta Q_g \neq 0$, and the system is simple.

Unlike System III and System II, first introduced in Lecture 4, in which there was no friction, the challenge with System I involves dealing with the friction, about which we have no information. The first step is to apply the First Law of Thermodynamics to System I (Gas, g). Specifically,

$$\underbrace{dE_g = dU_g}_{\text{Simple system}} = \underbrace{\delta Q_g}_{\substack{\text{Heat absorbed} \\ \text{by the gas due} \\ \text{to the friction}}} + \underbrace{\delta W_g}_{\substack{\text{Work done on} \\ \text{the gas by the} \\ \text{environment}}} \qquad (5.16)$$

We know that the source of the heat, δQ_g, is the friction generated at the cylinder walls by the work done on the cylinder walls by the moving piston against the friction. This suggests that to calculate δQ_g, we should focus on the cylinder walls as a system and carry out a First Law of Thermodynamics analysis on the cylinder walls. Note that the boundary of the cylinder walls is impermeable, rigid, and diathermal. Nevertheless, work is done on the cylinder walls by the moving piston against the friction. In other words, although there is no $Pd\underline{V}$-type work, there is friction work. It then follows that:

$$\underbrace{d\underline{E}_w = d\underline{U}_w}_{\text{Simple system}} = \underbrace{\delta Q_w}_{\substack{\text{Heat absorbed by} \\ \text{the cylinder walls}}} + \underbrace{\delta W_w}_{\substack{\text{Work done on the} \\ \text{cylinder walls to} \\ \text{overcome friction}}} \qquad (5.17)$$

As discussed in Lecture 4, because the heat capacity of the cylinder walls, C_{vw}, is zero according to the Problem Statement, the cylinder walls have no capacity to change their internal energy. In other words:

$$d\underline{U}_w = 0 \qquad (5.18)$$

Using Eq. (5.18) in Eq. (5.17) yields:

$$\underbrace{-\delta Q_w}_{\substack{\text{Heat released by} \\ \text{the cylinder walls}}} = \underbrace{\delta W_w}_{\substack{\text{Work done on the cylinder} \\ \text{walls to overcome friction}}} \qquad (5.19)$$

Next, we relate δQ_w to δW_g. Intuitively, we expect that the heat released by the cylinder walls should be completely absorbed by the gas. This follows because according to the Problem Statement, neither the cylinder walls (w) nor the piston (p) can absorb the heat released by the cylinder walls (both C_{vw} and C_{vp} are zero). In addition, the heat released by the cylinder walls, $-\delta Q_w$, cannot be released to the atmosphere, because the gas is adiabatically enclosed by the cylinder walls and the piston.

To prove that, we can choose as our system the (Gas + Cylinder Walls). Because this system is adiabatically enclosed, it follows that:

$$\delta Q_g + \delta Q_w = 0 \qquad (5.20)$$

or that

$$\underbrace{\delta Q_g}_{\substack{\text{Heat absorbed} \\ \text{by the gas}}} = \underbrace{-\delta Q_w}_{\substack{\text{Heat released by} \\ \text{the cylinder walls}}} \qquad (5.21)$$

Using Eq. (5.21) in Eq. (5.19) yields:

$$\underbrace{\delta Q_g}_{\substack{\text{Heat absorbed} \\ \text{by the gas}}} = \underbrace{\delta W_w}_{\substack{\text{Work done on} \\ \text{the cylinder walls}}} \qquad (5.22)$$

Equation (5.22) is a key result. It shows that the heat generated by the frictional work at the cylinder walls, δW_w, is transmitted 100% back to the gas which absorbs it (δQ_g). Otherwise, the problem would be much more challenging, and we would require additional information and assumptions in order to solve it.

Next, we go back to the First Law of Thermodynamics and deal with the work term δW_g. In Lecture 4, we saw that:

$$W \text{ (System on Environment)} = -W \text{ (Environment on System)} \qquad (5.23)$$

Using Eq. (5.23) where System = Gas, and Environment = (Cylinder Walls + Atmosphere + Piston) yields:

$$\underbrace{-\delta W_g}_{} \;=\; \underbrace{\delta W_w}_{} \;+\; \underbrace{\delta W_a}_{} \;+\; \underbrace{\delta W_p}_{} \qquad (5.24)$$

Work done by the gas on the three elements of its environment	Work done on the cylinder walls to overcome friction	Work done on the atmosphere to push it back	Work done on the piston to increase its potential and kinetic energies

where

$$\delta W_a = -P_a d\underline{V}_a = P_a d\underline{V}_g \qquad (5.25)$$

and

$$\delta W_p = d\left(M_p g z\right) + d\left(\tfrac{1}{2} M_p v_p^2\right) \qquad (5.26)$$

Combining Eqs. (5.24), (5.25), and (5.26) yields:

$$\delta W_g = -P_a d\underline{V}_g - \delta W_w - d\left(M_p g z\right) - d\left(\tfrac{1}{2} M_p v_p^2\right) \qquad (5.27)$$

Next, we can return to the First Law of Thermodynamics for the Gas ((Eq. 5.16)), along with Eq. (5.22), which yields:

$$d\underline{U}_g = \delta Q_g + \delta W_g = \delta W_w + \delta W_g \qquad (5.28)$$

Using Eq. (5.27) for δW_g in Eq. (5.28) yields:

$$d\underline{U}_g = \delta W_w - P_a d\underline{V}_g - \delta W_w - d\left(M_p g z\right) - d\left(\tfrac{1}{2} M_p v_p^2\right) \qquad (5.29)$$

or

$$dU_g = -P_a d\underline{V}_g - d(M_p g z) - d\left(\frac{1}{2} M_p v_p^2\right) \tag{5.30}$$

Equation (5.30) is a very interesting result which shows that the gas expands in a manner where the work done on the cylinder walls to overcome friction (δW_w) and the heat absorbed by the gas ($\delta Q_g = \delta W_w$) cancel each other out! Indeed, the internal energy of the gas (see Eq. (5.30)) decreases because, as it expands, the gas works against the atmosphere as well as to increase the potential and the kinetic energies of the piston. The frictional work done on the cylinder walls is recovered 100% in the form of heat reabsorbed by the gas!

Note that if the cylinder walls where such that $d\underline{U}_w \neq 0$, contrary to what was assumed in Eq. (5.18) based on the Problem Statement, and if the cylinder walls as well as the piston could absorb some heat (an effect neglected here based on the Problem Statement), then, only part of the frictional work lost on the cylinder walls would be recovered! In that case, the internal energy of the gas would decrease to a greater extent, and we can anticipate a lower final temperature of the gas!

Because the gas is ideal, it follows that (recall that $N_g = $ constant):

$$dU_g = N_g C_{vg} dT \tag{5.31}$$

Using Eq. (5.31) in Eq. (5.30) yields the desired result:

$$-N_g C_{vg} dT = P_a d\underline{V}_g + d(M_p g z) + d\left(\frac{1}{2} M_p v_p^2\right) \tag{5.32}$$

Note that, as expected, Eq. (5.32) is identical to Eq. (5.15)!

Finally, we can solve Eq. (5.32) to find T_f as a function of $\underline{V}_g$. Integrating Eq. (5.32) from $\left(T_i, \underline{V}_{gi}, z_i, \text{ and } v_{pi}\right)$ to $\left(T_f, \underline{V}_{gf}, z_f, \text{ and } v_{pf}\right)$, we obtain:

$$
\begin{aligned}
-N_g C_{vg}(T_f - T_i) &= P_a\left(\underline{V}_{gf} - \underline{V}_{gi}\right) + M_p g(z_f - z_i) \\
&+ \frac{1}{2} M_p \bullet \underbrace{\left[\left(v_{pf}\right)^2 - \left(v_{pi}\right)^2\right]}_{\substack{\text{Note that } v_{pi} \text{ and } v_{pf} \\ \text{are both zero, because} \\ \text{the piston starts at rest} \\ \text{and stops at rest!}}}
\end{aligned}
\tag{5.33}
$$

Setting the last term in Eq. (5.33) to zero, including rearranging, we obtain:

$$T_f = T_i - \left\{ \frac{M_p g(z_f - z_i) + P_a\left(\underline{V}_{gf} - \underline{V}_{gi}\right)}{N_g C_{vg}} \right\} \tag{5.34}$$

Because the gas is ideal, it follows that (recall that $N_g = $ constant):

$$\frac{P_f \underline{V}_{gf}}{T_f} = \frac{P_i \underline{V}_{gi}}{T_i} \tag{5.35}$$

Rearranging Eq. (5.35), P_f is given by:

$$P_f = \left(\frac{T_f}{T_i}\right) \left(\frac{\underline{V}_{gi}}{\underline{V}_{gf}}\right) P_i \tag{5.36}$$

Because $\underline{V}_{gf} = 2\underline{V}_{gi}$, as per the Problem Statement, Eq. (5.36) yields:

$$P_f = \frac{1}{2} \left(\frac{T_f}{T_i}\right) P_i \tag{5.37}$$

According to the Problem Statement, all the inputs in Eqs. (5.34) and (5.37) are known. Specifically,

$$\underline{V}_{gf} - \underline{V}_{gi} = \underline{V}_{gi} = A_p z_i = 5.67 \times 10^{-3} m^3$$

$$z_f - z_i = 0.305 m$$

$$P_a = 1 \text{ bar} = 1 \times 10^5 \frac{N}{m^2}, \ P_i = 10.13 \times 10^5 \frac{N}{m^2}$$

$$T_i = 294.3 K$$

$$N_g = \frac{P_i \underline{V}_{gi}}{R T_i} = 2.35 \text{ mol}$$

In order to address the three scenarios of the piston moving up (considered here), and the piston moving down and laterally, including obtaining a compact expression for T_f, it is convenient to introduce the variable I as follows:

$$I = \left\{ \begin{array}{ll} +1, & \text{Up} \\ 0, & \text{Sideways} \\ -1, & \text{Down} \end{array} \right\} \tag{5.38}$$

Using Eq. (5.38) in Eq. (5.34) for T_f, including accounting for the direction of the moving piston, Eq. (5.34) can be generalized as follows:

$$T_f = T_i - \left\{ \frac{I \cdot M_p g(z_f - z_i) + P_a\left(\underline{V}_{gf} - \underline{V}_{gi}\right)}{N_g C_{vg}} \right\} \qquad (5.39)$$

Using the inputs given above in Eq. (5.39) for T_f, and in Eq. (5.37) for P_f, yields the desired final results:

$$T_f = 294.3K - \left\{ \frac{675.5 \cdot I + 574.1}{49.2} \right\} K$$

$$P_f = (T_f/58.1)\,bar \qquad (5.40)$$

Table 5.1 summarizes the results for T_f and P_f.

Table 5.1

I	T_f (K)	P_f (bar)
+1 (Up)	268.9	4.64
0 (Sideways)	282.6	4.87
−1 (Down)	296.4	5.11

Readers are encouraged to rationalize the rankings of T_f and P_f as the piston moves up, sideways, and down.

5.5   Food for Thought

Repeat the calculations when a fraction α of the frictional heat goes into the gas and a fraction $(1-\alpha)$ of the frictional heat goes into the atmosphere. Recall that the two solutions presented in this lecture assumed that $\alpha = 1$!

Lecture 6

The First Law of Thermodynamics for Open Systems: Derivation and Sample Problem

6.1 Introduction

The material presented in this lecture is adapted from Chapter 3 in T&M. First, we will derive the First Law of Thermodynamics for Open Systems. Second, we will use this law to solve Sample Problem 6.1, an illuminating problem where the gas in the ullage volume of a fuel storage tank is first pressurized by gas entering from an external gas storage tank until its pressure reaches a desired value. Subsequently, the pressurized gas is used to partially empty the fuel which flows into a missile to be launched for strategic purposes. Third, we will discuss the selection of the optimal model systems when modeling the pressurization and the emptying steps. Fourth, we will describe the pressurization step using the well-mixed gas model. Fifth, we will describe the fueling step using the two-layer (or stratified) gas model. Finally, we will show how to solve the First Law of Thermodynamics for an Open System during the pressurization and the fueling steps.

6.2 The First Law of Thermodynamics for Open Systems

6.2.1 Notation

The notation below refers to Fig. 6.1:

In: Entering stream
State 1:   Initial condition
State 2:   Final condition
δn_{in}:   Number of moles entering the system
P_{in}:   Pressure of entering stream
E_{in}, V_{in}:   Molar energy and molar volume of entering stream
Q_{σ}:   Heat crossing σ-surface into σ-system

© Springer Nature Switzerland AG 2020
D. Blankschtein, *Lectures in Classical Thermodynamics with an Introduction to Statistical Mechanics*, https://doi.org/10.1007/978-3-030-49198-7_6

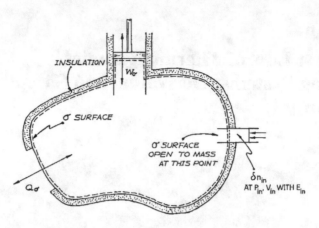

Fig. 6.1

W_σ:	Work at σ-surface done on σ-system
$P_{in}V_{in}\,\delta n_{in}$:	Work to push δn_{in} moles
H_{in}:	Molar enthalpy of entering stream

6.2.2   Derivation

Although system S, bounded by the σ-surface is open, the composite system $(S + \delta n_{in})$ is closed (see Fig. 6.1). If $\underline{E}$ is the total energy of system S, then, by applying the First Law of Thermodynamics to the closed, composite system, we obtain:

$$\underline{E}_2 - (\underline{E}_1 + E_{in}\delta n_{in}) \; = \; \delta Q_\sigma + \delta W_\sigma + P_{in}V_{in}\delta n_{in}$$

If the composite system is simple, $\underline{E} = \underline{U}$ in system S, and $E_{in} = U_{in}$ for the entering stream, where $\underline{U}$ and U are the total and molar internal energies, respectively. In that case:

$$E_{in} + P_{in}V_{in} = \; U_{in} + P_{in}V_{in} = \; H_{in}$$

Therefore, for an open, simple system, the differential form of the First Law of Thermodynamics can be written as follows:

$$d\underline{U} = \delta Q_\sigma + \delta W_\sigma + H_{in}\delta n_{in}$$

Generalizing to multiple entering (in) and leaving (out) streams, we obtain:

$$dU = \delta Q_\sigma + \delta W_\sigma + \sum_{in} H_{in}\delta n_{in} - \sum_{out} H_{out}\delta n_{out}$$

In the derivation above, we have retained the path-dependent differential operator δ for n_{in} and n_{out} because, strictly speaking, only dN (where N is the total number of moles in system S) of the system is a state variable, where:

$$dN = \sum_{in} \delta n_{in} - \sum_{out} \delta n_{out}$$

6.3 Sample Problem 6.1: Problem 3.9 in T&M

During an emergency launch operation to fill a missile with RP-4 (a kerosene-based fuel), the ullage volume of the fuel storage tank is first pressurized with air from atmospheric pressure (1.01 bar) to a pressure of 10.34 bar (see Fig. 6.2).

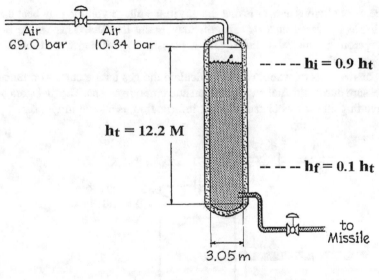

Fig. 6.2

The air is available from large external storage tanks at high pressure (69.0 bar). This operation is to be completed as rapidly as possible. After the 10.34-bar pressure level is reached, the main transfer valve is opened, and fuel flows at a steady rate until the missile is loaded. It is necessary to maintain a constant gas pressure of 10.34 bar inside the fuel tank during transfer.

The fuel storage tank can be approximated as a right circular cylinder 12.2 m tall and 3.05 m in diameter and is originally filled to 90% of capacity. Transfer of fuel to a residual volume of 10% must be completed in 18 min. Assume ideal gases and that the operation is adiabatic and all hardware has negligible heat capacity. Initial temperatures may range from 242 K (arctic sites) to 333 K (equatorial sites), but for the purposes of a first estimate, use 294 K as an initial temperature.

(a) Comment on any safety hazards that might be encountered.
(b) What problems would you anticipate if the inlet gas control valve were to malfunction and the gas space above the fuel were to reach full storage tank pressure (69.0 bar)? (The fuel tank has been hydrostatically tested to 276 bar.)
(c) What is your estimate of the time-temperature history of the gas above the fuel during the entire operation? These data are needed to size the inlet airlines.

6.3.1   Solution Strategy

(a) The most obvious danger is that the gas in the ullage volume may heat up while being compressed and, as a result, may ignite the fuel. This would be most dangerous if the valve failed and a pressure of 69 bar was reached in the storage tank.
(b) To answer this question, we will calculate the gas temperature as a function of pressure during the fuel pressurization and emptying steps. The fuel storage tank, including all the associated relevant information, is shown in Fig. 6.3.

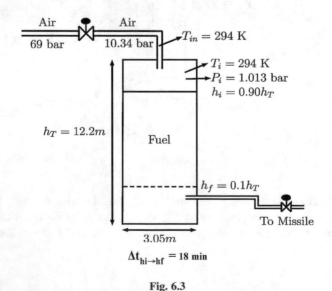

Fig. 6.3

Based on the Problem Statement, we can make the following reasonable approximations:

- The gases are ideal
- The operation is adiabatic
- The heat capacity of the operating parts (metal) $= 0$

To model the storage tank fueling step, we will assume that the entering gas and the original gas are well mixed at all times (see Fig. 6.4).

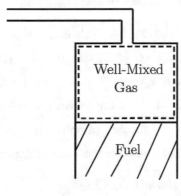

Fig. 6.4

6.3.2   Well-Mixed Gas Model System

We will make the following assumptions to solve this gas model system:

- The original and entering gases are well-mixed during the pressurization step (see Fig. 6.4)
- The gas solubility in the fuel is negligible
- The fuel is incompressible
- There is no heat transfer at the fuel/gas interface

The boundary of the Well-Mixed Gas Model System is shown in Fig. 6.5 (see the dashed line) and, based on the assumptions above, is open, rigid, and adiabatic.

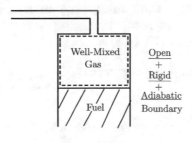

Fig. 6.5

6.3.3   Pressurization Step: Well-Mixed Gas Model System

The system is the well-mixed gas. It is a simple, open system surrounded by an adiabatic, rigid boundary. We would like to find the relation between the gas temperature and its pressure to ascertain if it can exceed the fuel ignition temperature: ~ 800 K.

A First Law of Thermodynamics analysis for this open system yields:

$$dU = \delta Q + \delta W + H_{in}\,\delta n_{in} - H_{out}\,\delta n_{out} \tag{6.1}$$

$$\delta Q = 0 \text{ (Adiabatic boundary)}$$

$$\delta W = 0 \text{ (Rigid boundary, where } V = \text{constant for the well-mixed gas)}$$

$$\delta n_{out} = 0 \text{ (Nothing leaves the volume)}$$

$$\delta n_{in} = dN \text{ (N is the number of moles of gas in the system)}$$

$$H_{in} = f(T_{in}), \text{because the entering gas is ideal}$$

Using the information above in Eq. (6.1) yields:

$$dU = H_{in}\,dN \tag{6.2}$$

Because the system is open:

$$dU = d(NU) = NdU + UdN \tag{6.3}$$

Using Eq. (6.3) in Eq. (6.2) and rearranging yields:

$$\frac{dN}{N} = \frac{dU}{H_{in} - U} \tag{6.4}$$

Recall that for an ideal gas:

$$*U = U(T); \quad dU = C_v dT \;\Rightarrow\; U(T) = U_o + C_v\,(T - T_o) \tag{6.5}$$

$$*H_{in} = H_{in}(T); \quad dH_{in} = C_p dT \;\Rightarrow\; H_{in}(T_{in}) = H_o + C_p\,(T_{in} - T_o) \tag{6.6}$$

Subtracting Eq. (6.5) from Eq. (6.6) yields:

$$H_{in}(T_{in}) - U(T) = (C_p T_{in} - C_v T) + H_o - \underbrace{\left[U_o + (C_p - C_v)T_o\right]}_{H_o} \tag{6.7}$$

where the relation $C_p - C_v = R$ was used. Accordingly,

$$H_{in}(T_{in}) - U(T) = C_p T_{in} - C_v T \tag{6.8}$$

In addition:

$$dU = C_v dT \tag{6.9}$$

Combining Eqs. (6.4), (6.8), and (6.9) yields:

$$\frac{dN}{N} = \frac{C_v dT}{C_p T_{in} - C_v T} \tag{6.10}$$

Next, it is useful to express dN/N in Eq. (6.10) in terms of changes in T and P. To this end, we use the ideal gas equation of state, which yields:

$$N = \frac{PV}{RT} \Rightarrow \frac{dN}{N} = \frac{dP}{P} + \frac{dV}{V} - \frac{dT}{T} \tag{6.11}$$

Because $\underline{V}$ is constant, $d\underline{V} = 0$ in Eq. (6.11), and therefore:

$$\frac{dN}{N} = \frac{dP}{P} - \frac{dT}{T} \tag{6.12}$$

Using Eq. (6.12) in Eq. (6.10) and rearranging Yields:

$$\frac{dP}{P} = dT \left\{ \frac{1}{T} + \frac{C_v}{C_p T_{in} - C_v T} \right\} \tag{6.13}$$

Integrating Eq. (6.13) from (P_i, T_i) to (P,T) yields the desired result:

$$T = \frac{\kappa \, T_{in}}{1 + \frac{P_i}{P} \left\{ \kappa \frac{T_{in}}{T_i} - 1 \right\}} \tag{6.14}$$

Equation (6.14), where $\kappa = C_p/C_v$, predicts how T of the gas varies with P of the gas. Using the data given in the Problem Statement, that is:

$$C_v = 20.9 \text{ J/mol K}; \ R = 8.314 \text{ J/mol K}$$

$$C_p = C_v + R = 29.21 \text{ J/mol K}; \ \kappa = C_p/C_v = 1.4$$

$$T_{in} = T_i = 294 \text{ K}$$

$$P_i = 1\,013 \text{ bar}$$

in Eq. (6.14) yields:

(i) T (at P = 10.34 bar) = 395 K
(ii) T (at P = 69 bar) = 409 K

We can conclude that temperatures (i) and (ii) above are both too low to lead to ignition of the fuel!

6.3.4   Layered or Stratified Gas Model System

The original and the entering gases do not mix, and the interface between the two gas volumes is adiabatic during pressurization (see Fig. 6.6).

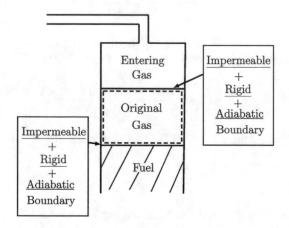

Fig. 6.6

6.3.5   Pressurization Step: Stratified Gas Model System

After predicting $T = f(P)$ for the well-mixed model system, readers are encouraged to predict $T = f(P)$ for the stratified gas model system. The following relation is found:

$$T = T_i \, (P/P_i)^{\frac{\kappa - 1}{\kappa}} \tag{6.15}$$

Using the data in the Problem Statement, Eq. (6.15) yields:

(i) T (at $P = 10.34$ bar) $= 568.7$ K
(ii) T (at $P = 69$ bar) $= 954$ K

Clearly, if the valve fails, T_{ii} could lead to fuel ignition!

6.3.6   Emptying Step: Well-Mixed Gas Model System

Once again, we choose the well-mixed gas above the fuel as our simple, open system. The operation is adiabatic, and there is one moving boundary (the gas/fuel interface). Further, according to the Problem Statement, the gas pressure is constant at 10.34 bar during the emptying step.

A First Law of Thermodynamics analysis yields:

$$dU \;=\; \delta Q + \delta W + H_{in}\,\delta n_{in} - H_{out}\,\delta n_{out} \tag{6.16}$$

$\delta Q = 0$ (Adiabatic process)
$\delta W =$ Work done on the gas $= -PdV \;=\; -Pd(NV) = -P(NdV + VdN)$
$\delta n_{in} \;=\; dN$ (N is the number of moles of gas)
$\delta n_{out} = 0$ (Nothing leaves)

$$dU = d(NU) = NdU + UdN$$

Using the information above in Eq. (6.16) yields:

$$NdU + UdN = H_{in}dN - PNdV - PVdN \tag{6.17}$$

According to the Problem Statement, P is constant, and therefore, Eq. (6.17) can be rewritten as follows:

$$Nd\,\overbrace{(U + PV)}^{H} = \left[H_{in} - \overbrace{(U + PV)}^{H} \right] dN \tag{6.18}$$

or

$$\frac{dN}{N} \;=\; \frac{dH}{H_{in} - H} \tag{6.19}$$

Because the gas is ideal:

$$*H = H(T) \;\Rightarrow\; dH = C_p dT$$
$$*H_{in}\,(T_{in}) \;=\; H_o + C_p(T_{in} - T_o)$$
$$*H(T) = H_o + C_p(T - T_o)$$
$$\therefore\; H_{in} - H \;=\; C_p(T_{in} - T)$$

Using the dH and ($H_{in} - H$) expressions above in Eq. (6.19) yields:

$$\frac{dN}{N} \;=\; \frac{dT}{T_{in} - T} \tag{6.20}$$

In Part (c) of the problem, we are asked to calculate the variation of the gas temperature with time. Because the gas volume increases with time (as the fuel exits the tank) in a known manner (see below), it is most useful to find the relation between the gas temperature and its volume using Eq. (6.20). As before, we make use of the fact that for an ideal gas:

$$N = P\underline{V}/RT \tag{6.21}$$

The differential of ln of Eq. (6.21) is given by:

$$\frac{dN}{N} = \frac{dP}{P} + \frac{d\underline{V}}{\underline{V}} - \frac{dT}{T} \tag{6.22}$$

Using Eq. (6.22), along with the fact that P is constant, in Eq. (6.20) and rearranging yields the desired relation between $\underline{V}$ and T:

$$\frac{d\underline{V}}{\underline{V}} = dT \left\{ \frac{1}{T} + \frac{1}{T_{in} - T} \right\} \tag{6.23}$$

Integrating Eq. (6.23) from $(\underline{V}_e, T_e)$ to $(\underline{V}, T)$, where $\underline{V}_e = 0.1\,\underline{V}_{tank}$ and $T_e = 395$ K, the $\underline{V}$ and T values at the end of the pressurization step, yields the desired result:

$$T(t) = \frac{T_{in}}{1 + \frac{\underline{V}_e}{\underline{V}(t)} \left\{ \frac{T_{in}}{T_e} - 1 \right\}} \tag{6.24}$$

We know that:

$$\underline{V}(t) = \underline{V}_e + [\text{Volume evacuated by the fuel}]\,(t)$$

Volume evacuated by the fuel $= 90\%\underline{V}_{tank} - 10\%\underline{V}_{tank} = 0.8\,\underline{V}_{tank}$.
According to the Problem Statement, this fuel volume leaves the tank in 18 min at a steady rate. Hence, the flow rate is $(0.8\,\underline{V}_{tank}/18)$. It then follows that:

$$[\text{Volume evacuated by fuel}]\,(t) = \left[\frac{0.8}{18}\underline{V}_{tank} \right] t \tag{6.25}$$

Because $V_e = 0.1\,V_{tank}$, it follows that:

$$\underline{V}(t) = \left(1 + \frac{4t}{9} \right) 0.1\,\underline{V}_{tank} \tag{6.26}$$

Using Eq. (6.26), along with $V_e = 0.1\,\underline{V}_{tank}$, in Eq. (6.24) yields the desired result:

$$T(t) = \frac{T_{in}}{1 + \left\{ \frac{T_{in}}{T_e} - 1 \right\} \left(\frac{1}{1 + \frac{4t}{9}} \right)} \tag{6.27}$$

Using $T_{in} = 294$ K and $T_e = 395$ K in Eq. (6.27) yields the desired temperature-time history of the gas during the emptying step. Specifically,

$$T(t) = 294\left(\frac{1 + 0.44\,t}{0.744 + 0.44\,t}\right) K \qquad (6.28)$$

Equation (6.28) indicates that the gas temperature decreases from 395 K at $t = 0$ to 303 K at $t = 18$ min. Clearly, because the gas cools down during the emptying step, there is no danger of fuel ignition!

Lecture 7

The Second Law of Thermodynamics: Fundamental Concepts and Sample Problem

7.1 Introduction

The material presented in this lecture is adapted from Chapter 4 in T&M. As we saw in Lectures 3 and 4, the First Law of Thermodynamics is simply a restatement of the "Principle of Energy Conservation" in a given process, where heat, Q, and work, W, can be converted into Energy, $\underline{E}$. Recall that if the system is open, then, enthalpic contributions need to be included in the formulation of the First Law of Thermodynamics (see Lectures 5 and 6). In this lecture, first, we will show that the First Law of Thermodynamics does not provide any information about the directionality of the process! Second, we will stress that nevertheless, from experience, there are processes which always occur spontaneously in a unique, well-defined direction and are referred to as natural processes. Third, to rationalize the directionality of natural processes, which are inherently irreversible, we will need to introduce a new thermodynamic function of state called entropy, which will provide a mathematical criterion in the context of what is known as the Second Law of Thermodynamics to ascertain if a process can occur in a particular direction. Fourth, we will introduce a heat engine, define its efficiency, and discuss under what conditions a process is reversible. Finally, we will solve Sample Problem 7.1, which will help us crystallize the material taught.

7.2 Natural Processes

The flow of heat from a "hot" body to a "cold" body. The opposite process does not occur spontaneously (see Fig. 7.1).

© Springer Nature Switzerland AG 2020
D. Blankschtein, *Lectures in Classical Thermodynamics with an Introduction to Statistical Mechanics*, https://doi.org/10.1007/978-3-030-49198-7_7

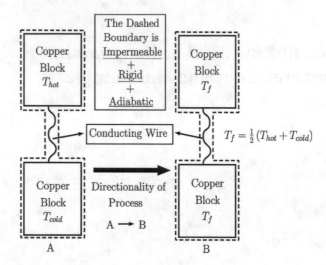

Fig. 7.1

Going from (A) to (B) is possible under isolation (spontaneously), but we have never observed that the reverse occurs spontaneously, that is, one copper block becomes "hot," while the other copper block becomes "cold." In other words, going from (B) to (A) is not possible under isolation!

Yet, processes (A)→(B) and (B)→(A) are both fully consistent with the First Law of Thermodynamics! Indeed, if we take the two copper blocks as the system, neglecting any heat capacity of the conducting wire, so that $\Delta U_{wire} = 0$, then, the flow of heat between the two copper blocks in any direction occurs at constant total internal energy, because the system is:

- Simple
- Isolated [Adiabatic $(Q = 0)$, Rigid $(W = 0)$, and Impermeable $(N = \text{constant})$ boundary]

    As a result,

$$\overbrace{\underline{E} = \underline{U}}^{\text{Simple}} = \ Q + W = 0 \tag{7.1}$$

or

$$\underline{U}_A = \underline{U}_B \tag{7.2}$$

irrespective of the directionality of the heat flow. In addition,

$$\underline{U}_A = MC_v(T_{hot} - T_o) + MC_v(T_{cold} - T_o) \tag{7.3}$$

$$\underline{U}_B = 2MC_v(T_f - T_o) \tag{7.4}$$

In Eqs. (7.3) and (7.4), M is the mass of each copper block and C_v is the copper heat capacity at constant volume. Using Eqs. (7.3) and (7.4) in Eq. (7.2), and rearranging, yields:

$$T_f = \frac{1}{2}(T_{hot} + T_{cold}) \tag{7.5}$$

Accordingly, process (B)→(A) is also consistent with the First Law of Thermodynamics!

The spontaneous transfer of heat from T_{hot} to T_{cold}, and other natural processes such as the spontaneous expansion of a gas into an evacuated volume (the reverse does not occur spontaneously), requires going beyond the First Law of Thermodynamics to elucidate their directionality.

We will see that there is a new thermodynamic function of state, called entropy, $\underline{S}$, that will provide a mathematical criterion, in the context of what is known as the "Second Law of Thermodynamics," to determine if a process can occur in a particular direction. Specifically, we will see that a process can occur spontaneously only in a direction that will increase the entropy of the system.

For example, when gas expands spontaneously into a larger available volume, the number of states available to the gas molecules increases (more positions are available to them). This, in turn, results in an increase in the entropy of the gas, as follows from Statistical Mechanics, where in Part III we will show that:

$$\underline{S} = k_B \ln \Omega \tag{7.6}$$

where

k_B – Boltzmann constant: 1.38×10^{-23} J/K
Ω – Number of distinct available states
$\underline{S}$ – System entropy

The reverse process of the gas spontaneously "compressing itself" into a smaller volume would necessarily decrease the number of available states, Ω, and, therefore, would result in a decrease in entropy, which is in violation of the Second Law of Thermodynamics. Note that we can compress the gas by "working on it," but this would not be a spontaneous (isolated) process, so that the Second Law of Thermodynamics is not violated in that case!

In the First Law of Thermodynamics statement for a closed system, heat and work are treated symmetrically as two different forms of energy transfer ($\Delta \underline{E} = Q + W$). However, experience shows that there is a difference in quality between work and heat. Indeed, as we already saw numerous times, work can be transformed into other forms of energy (e.g., potential, kinetic), and if one minimizes friction (a dissipative process that transforms work into heat), the efficiency of the conversion of work into other forms of energy can be made to approach 100%! This fact was first demonstrated in a series of celebrated experiments carried out by Joule.

Joule showed that various types of work (shaft work done by a paddle wheel, $Pd\underline{V}$-type work done by compressing gas, frictional work done by rubbing two metal parts, and electrical work done by passing current through a metal wire), all done under adiabatic conditions on a fixed mass of water (N = constant, closed system), led to an increase in the temperature of the water in direct proportion to the amount of work done. Joule viewed incorrectly the change of state of the water (i.e., its temperature increase) as resulting from the conversion of work into heat, irrespective of the type of work done, with the well-known conversion factor 4.184 J = 1 cal. Because in all cases work was done under adiabatic conditions, Joule really showed that $W = \Delta \underline{U}_{water}$, irrespective of the type of work done! Note that Joule's W is really what we referred to in Lecture 3 as the adiabatic (a) work, and because water is a simple system:

$$W^a_{i \to j} = \Delta \underline{E}^a_{i \to j} = \Delta \underline{U}^a_{i \to j} \tag{7.7}$$

On the other hand, all efforts to devise a process for the continuous conversion of heat completely (100%) into other forms of energy (work, potential energy, and kinetic energy) have failed! Regardless of the engineering improvements, conversion efficiencies of heat into work do not exceed 40%.

One can summarize this observation as follows (Statement (1) of the Second Law of Thermodynamics):

7.3   Statement (1) of the Second Law of Thermodynamics

"No apparatus can operate in such a way that its only effect (in system and environment) is to convert heat absorbed by a system completely into work" (conversion efficiency of heat into work < 100%). Statement (1) indicates that 100% conversion of heat into work is not possible if both the system and its environment do not change. To better understand Statement (1), consider the following illuminating problem.

7.4   Sample Problem 7.1

An ideal gas undergoes an isothermal expansion, without friction, in a cylinder-piston assembly. Is Statement (1) of the Second Law of Thermodynamics satisfied?

7.4.1   Solution

Because the gas is ideal, its molar internal energy $U = f(T)$ only. Because the gas is a closed system, N is constant. Therefore, in this isothermal expansion:

$$\underline{U}_f(T, N) = \underline{U}_i(T, N) \Rightarrow \Delta\underline{U} = 0 \tag{7.8}$$

Because the gas is a simple system, it follows that $\Delta\underline{E} = \Delta\underline{U} = 0$ in its isothermal expansion. A First Law of Thermodynamics analysis of the gas then yields:

$$\Delta\underline{E} = \Delta\underline{U} = 0 = Q + W \tag{7.9}$$

or

$$\underbrace{Q}_{\substack{\text{Heat absorbed by the gas from} \\ \text{the environment to maintain} \\ \text{constant temperature}}} = \underbrace{-W}_{\substack{\text{Work done by the} \\ \text{gas as it expands}}} \tag{7.10}$$

At first glance, Eq. (7.10) appears to violate Statement (1) of the Second Law of Thermodynamics, because we have converted 100% of heat absorbed by the gas from the environment (to maintain constant temperature) into work produced by the (reversible) expansion of the gas. However, a closer examination reveals that there is no violation of Statement (1) of the Second Law of Thermodynamics because the system changed its state as it expanded from initial state i to final state f, that is, the state of the gas changed from $(N, T, \underline{V}_i)$ to $(N, T, \underline{V}_f)$, with $\underline{V}_f > \underline{V}_i$!

If the original state of the system $(N, T, \underline{V}_i)$ is restored, so that the system does not change its state, as required by Statement (1) of the Second Law of Thermodynamics, then energy from the environment in the form of work is needed to compress the gas back from $(N, T, \underline{V}_f)$ to $(N, T, \underline{V}_i)$. At the same time, energy in the form of heat will be transferred from the system to the environment to maintain constant temperature. Therefore, the reverse gas compression process requires exactly the same amount of work produced by the gas expansion, and therefore, no net work is produced!

The solution to sample Problem 7.1 clearly shows that Statement (1) of the Second Law of Thermodynamics can be expressed in the following alternative way.

7.5   Statement (1a) of the Second Law of Thermodynamics

"It is impossible by a cyclic process to convert the heat absorbed by a system completely (100%) into work." The word cyclic in Statement (1a) of the Second Law of Thermodynamics requires that the system be restored periodically to its original state. In the gas example discussed in Sample Problem 7.1, the expansion and compression of the gas back to its original state constitute a complete cycle. If the process is repeated, it becomes a cyclic process. The restriction to a "cyclic process" in Statement (1a) of the Second Law of Thermodynamics provides the same limitations as the restriction "only effect" in Statement (1) of the Second Law of Thermodynamics.

The Second Law of Thermodynamics does not prohibit the production of work from heat, but it does place a limit on the fraction of heat that may be converted into work in a cyclic process [Statement (1a) of the Second Law of Thermodynamics]. The partial conversion of heat into work is the basis for most commercial generation of power. Next, we introduce a heat engine and derive a quantitative expression for the conversion efficiency.

7.6   Heat Engine

A heat engine is a device (or a machine) which produces work from heat in a cyclic process. It is useful to illustrate the operation of a heat engine with a steam power cycle (Rankine cycle, see Fig. 7.2).

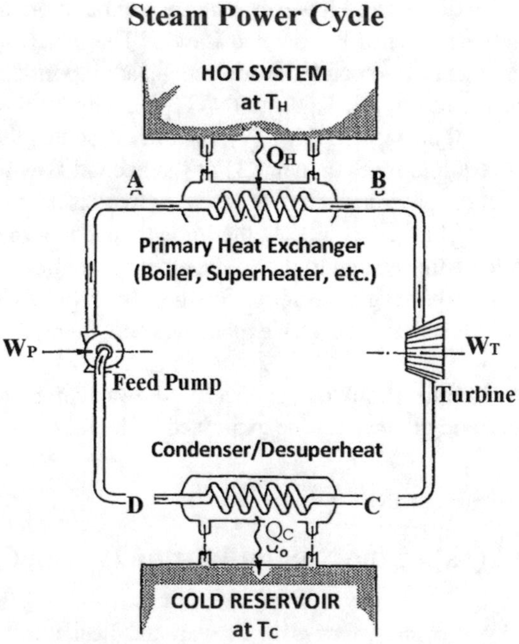

Fig. 7.2

The Rankine cycle consists of the following four steps:

1. Liquid water at ambient temperature is pumped, W_P, into a boiler
2. Heat, Q_H, from a fuel is transferred in the boiler to the water, converting it to steam at high temperature and pressure
3. Useful work is obtained by expanding the steam to a low pressure in a turbine, W_T
4. Exhaust steam from the turbine is condensed by the transfer of heat, Q_C, to cooling water, thus completing the cycle

7.7 Efficiency of a Heat Engine

Essential to the operation of a heat engine are (see Fig. 7.3):

– Absorbing heat from a hot reservoir
– Doing work
– Rejecting heat to a cold reservoir

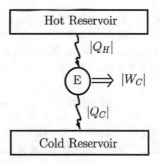

Fig. 7.3

In Fig. 7.3, $|Q_H|$, $|Q_C|$, and $|W_E|$ denote the magnitudes of the "vectors," while the arrows ⤳ for Q and ⟹ for W denote directions. In Lecture 8, we will use the vectorial representation of heat and work to model an ideal engine, known as the Carnot engine.

The efficiency of a heat engine is defined as follows:

$$\eta_E = \frac{\text{Work done by the engine}}{\text{Heat absorbed by the engine from the hot reservoir}} \tag{7.11}$$

Using our work and heat conventions, Eq. (7.11) can be expressed as follows:

$$\eta_E = \frac{-W_E}{Q_H} \tag{7.12}$$

where W_E is the work done on the engine and Q_H is the heat absorbed by the engine from the hot reservoir. In all allowable processes, either W_E or Q_H will be < 0, but not both, so that $\eta_E > 0$ in Eq. (7.12).

To make the equations independent of the sign conventions for W and Q, we can use $|W_E|$ and $|Q_H|$, with the directions indicated by the arrows $\rightsquigarrow$ for Q and $\Rightarrow$ for W. In that case:

$$\eta_E = \frac{|W_E|}{|Q_H|} \tag{7.13}$$

As stressed above, $\eta_E < 1$! If this is the case, what is it that limits the engine efficiency? From a practical viewpoint, factors such as friction and other resistances that dissipate energy will certainly decrease the value of η_E. However, even if one could construct an ideal engine, "free of friction," which could operate fully reversibly, one would still find that there is a Second Law of Thermodynamics limit of the efficiency of a heat engine, so that:

$$\eta_E^{ideal} < 1 \tag{7.14}$$

Because every real engine will certainly have an efficiency which is less than η_E^{ideal}, it is very important to compute this limiting value of the efficiency. For this purpose, we will introduce an ideal engine which operates fully reversibly. This engine is known as the Carnot engine, for which, by definition:

$$\eta_C = \eta_E^{ideal} \tag{7.15}$$

where the subscript C in Eq. (7.15) denotes Carnot. Before we examine the Carnot engine, however, let us define more precisely what we mean by reversibility.

7.8   Reversible Process

"A process is reversible if a second process could be carried out in at least one way so that the system and all elements of its environment can be restored to their respective initial states, except for differential changes of second order (e.g., $dPd\underline{V}$)." A reversible process is:

(a) Frictionless
(b) Only differentially removed from equilibrium and traverses a quasi-static path
(c) The driving forces for the process are differential in magnitude (see Fig. 7.4)

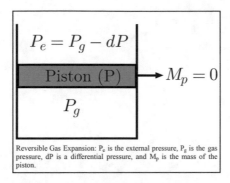

Reversible Gas Expansion: P_e is the external pressure, P_g is the gas pressure, dP is a differential pressure, and M_p is the mass of the piston.

Fig. 7.4

(d) Its direction can be reversed at any point by a differential change in external conditions, causing the process to retrace its path, leading to restoration of the initial state of the system and its environment

Lecture 8

Heat Engine, Carnot Efficiency, and Sample Problem

8.1 Introduction

The material presented in this lecture is adapted from Chapter 4 in T&M. First, we will continue our discussion of the heat engine introduced in Lecture 7, including deriving an expression for the efficiency of a heat engine. We will then show that the efficiency of a heat engine can never be 100% and will discuss what determines the upper limit. Second, we will introduce an ideal heat engine which operates fully reversibly, known as the Carnot engine. We will show that even for the Carnot engine, the efficiency is less than 100%. Third, we will discuss the Theorem of Carnot and the Corollary to this theorem. Fourth, we will solve Sample Problem 8.1 to analyze the Carnot cycle for an ideal gas, including using this cycle to calculate the efficiency of a Carnot engine. Finally, we will present the Theorem of Clausius.

8.2 Heat Engine

As discussed in Lecture 7, essential to the operation of a heat engine is the absorption of heat at a high temperature, the production of work, and the rejection of heat at a lower temperature. The two temperature levels which characterize the operation of the heat engine are maintained by heat reservoirs, bodies assumed to be capable of absorbing or rejecting an infinite amount of heat without a change in temperature.

In the derivation of the efficiency of a heat engine, η_E, presented below, we will use a vectorial description of Q and W, first introduced in Lecture 7, so that the result is independent of our sign convention for Q and W. Specifically,

(i) The magnitudes of Q and W are denoted by |Q| and |W|, respectively, both positive quantities

© Springer Nature Switzerland AG 2020
D. Blankschtein, *Lectures in Classical Thermodynamics with an Introduction to Statistical Mechanics*, https://doi.org/10.1007/978-3-030-49198-7_8

(ii) A vector ～～➤ entering the engine denotes heat absorbed by the engine. A vector ～～➤ exiting the engine denotes heat rejected by the engine. Vectors are added accordingly

(iii) A vector ⇒ entering the engine denotes work done on the engine. A vector ⇒ exiting the engine denotes work done by the engine. Vectors are added accordingly

In operation, the working fluid of a heat engine absorbs heat, $|Q_H|$, from a hot reservoir, produces work, $|W_E|$, and rejects heat, $|Q_C|$, to a cold reservoir, returning to its initial state (see Fig. 8.1). In this cyclic process, if we consider the working fluid of the heat engine as a closed, simple system, it follows that $\underline{U}_f = \underline{U}_i$.

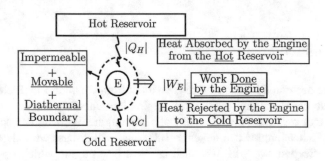

Fig. 8.1

Because the system (the heat engine) is simple and the process is cyclic, the First Law of Thermodynamics can be written as follows:

$$\Delta \underline{E} = \Delta U = 0 = (|Q_H| - |Q_C|) - |W_E| \tag{8.1}$$

or

$$|W_E| = |Q_H| - |Q_C| \tag{8.2}$$

where $|W_E|$ is the work done by the heat engine and $(|Q_H| - |Q_C|)$ is the net heat absorbed by the heat engine (see Fig. 8.1).

Recall that:

$$\eta_E = \frac{\text{Work done by the engine}}{\begin{array}{c}\text{Heat absorbed by the engine}\\\text{from the hot reservoir}\end{array}} \Rightarrow \eta_E = \frac{|W_E|}{|Q_H|} = \frac{|Q_H| - |Q_C|}{|Q_H|} \tag{8.3}$$

or

$$\eta_E = 1 - \frac{|Q_C|}{|Q_H|} \tag{8.4}$$

Because $\eta_E = 1 - |Q_C|/|Q_H|$, for η_E to be unity (100% conversion of heat into work), $|Q_C|$ must be zero, or $|Q_H|$ must be infinite! No heat engine has ever been built for which this is possible. In other words, some heat is always rejected to the cold reservoir, and there is no heat reservoir that can supply infinite heat.

If η_E cannot be unity, what determines the upper limit? Practically, friction and other dissipative processes do, because they affect the degree of reversibility of the operation. Clearly, η_E is expected to attain its upper limit for a reversible cyclic operation. Because this is impossible to achieve in practice, the French Engineer Carnot introduced the concept of an ideal engine that operates in a fully reversible manner, which is called the Carnot engine.

This hypothetical heat engine will be used to calculate the maximum efficiency that a heat engine can attain. Any real heat engine will necessarily have $\eta_E < \eta_{Carnot}$. Note that because the Carnot engine operates reversibly, it must absorb heat from the hot reservoir at the constant temperature of the hot reservoir, θ_H, and it also must reject heat to the cold reservoir at the constant temperature of the cold reservoir, θ_C. A real heat engine must necessarily transfer heat across finite temperature differences and, therefore, cannot be reversible!

Because a Carnot engine is reversible, it can be operated in reverse to produce a reversible refrigerator cycle in which $|Q_H|$, $|Q_C|$, and $|W_E|$ are the same as before but are reversed in direction.

8.3 Theorem of Carnot

"For two given heat reservoirs, no heat engine can have a higher thermal efficiency than a Carnot engine."

$$\eta_E < \eta_C \tag{8.5}$$

The proof is based on a violation of Postulate II (left as an exercise).

8.4 Corollary to Theorem of Carnot

"All Carnot engines operating between heat reservoirs at the same two temperatures have the same thermal efficiency."

Therefore, η_C depends only on the temperature levels, θ_H and θ_C, of the hot (H) and the cold (C) reservoirs, and not on the properties of the heat engine's working fluid (see Fig. 8.2).

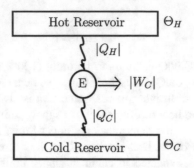

Fig. 8.2

Using the Corollary to the Theorem of Carnot, one can prove that:

$$\frac{|Q_H|}{|Q_C|} = f(\theta_H, \theta_C) = \frac{\psi(\theta_H)}{\psi(\theta_C)} \tag{8.6}$$

We may define the right-hand side of Eq. (8.6) as the ratio of two new thermodynamic temperatures, $\psi(\theta_H)$ and $\psi(\theta_C)$. To determine ψ, we must choose a temperature scale, for example, Kelvin (see below).

Because for any heat engine we showed that $\eta_E = 1 - |Q_C|/|Q_H|$, if E = Carnot engine, we can use Eqs. (8.4) and (8.6) to obtain:

$$\eta_C = 1 - \frac{|Q_C|}{|Q_H|} = 1 - \frac{\psi(\theta_C)}{\psi(\theta_H)} \tag{8.7}$$

or

$$\eta_C = \frac{\psi(\theta_H) - \psi(\theta_C)}{\psi(\theta_H)} \tag{8.8}$$

To calculate ψ, we utilize an ideal gas as the working fluid and the Kelvin temperature scale. In other words, the Kelvin temperature scale is chosen for the ideal gas.

8.5   Sample Problem 8.1

Analyze the Carnot cycle for an ideal gas depicted in Fig. 8.3.

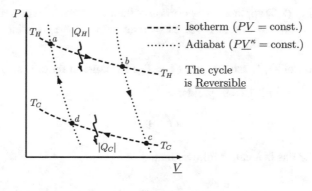

Fig. 8.3

8.5.1 Solution

As shown in Fig. 8.3, the Carnot cycle consists of the following four steps:

1. a→b: Isothermal expansion with heat absorption, $|Q_H|$
2. b→c: Adiabatic expansion from $T_H→T_C$
3. c→d: Isothermal compression with heat rejection, $|Q_C|$
4. d→a: Adiabatic compression from $T_C→T_H$

Note that other thermodynamic plots can be used to describe the Carnot cycle, that is, not necessarily the (P-$\underline{V}$) phase diagram shown in Fig. 8.3. For example, one can also use an entropy vs. temperature ($\underline{S}$-T) phase diagram. Although an ideal gas is used here, the result that we will obtain is general for any working fluid, as long as the thermodynamic temperature scale is the Kelvin (T) scale! Using this temperature scale, it follows that $|Q_H|/|Q_C| = \psi(T_H)/\psi(T_C)$ (see Eq. (8.6)). To find $\psi(T)$, we will proceed to calculate $|Q_H|/|Q_C|$, because this is equal to $\psi(T_H)/\psi(T_C)$.

The process (Carnot cycle) is reversible along any portion of the cycle. It is convenient to follow a fixed differential amount of gas as a closed, simple system traversing the cycle from a to b to c to d to a. One can then apply the First Law of Thermodynamics to this system working on a per mole basis. For any portion of the Carnot cycle, we can write:

$dU = \delta Q + \delta W$
$dU = C_V dT$ (Ideal gas)
$-\delta W = P d V$ (Work done by the gas)
$\therefore \delta Q = dU - \delta W$

or

$$\delta Q = C_V dT + P dV \qquad (8.9)$$

Note that Eq. (8.9) applies to any portion of the Carnot cycle, where along the two isothermal portions, a→b and c→d, $dT = 0$ and along the two adiabatic portions, b→c and d→a, $\delta Q = 0$!

1. **Calculation of $|Q_H|$: Isothermal ($T = T_H$) Expansion from a to b**

Setting $dT = 0$ in Eq. (8.9) yields:

$$\delta Q = PdV \tag{8.10}$$

Because the gas is ideal, it follows that:

$$P = \frac{RT_H}{V} \text{ (along the a→b isotherm)} \tag{8.11}$$

Using Eq. (8.11) in Eq. (8.10) yields:

$$\delta Q = \frac{RT_H}{V} dV \tag{8.12}$$

Integrating Eq. (8.12) from V_a to V_b yields:

$$\int_{V_a}^{V_b} \delta Q = Q_H = \int_{V_a}^{V_b} \frac{RT_H}{V} dV = RT_H \ln\left(\frac{V_b}{V_a}\right) \tag{8.13}$$

Note that because $V_b/V_a > 1 \Rightarrow Q_H > 0$, heat is absorbed by the Carnot engine from the hot reservoir.

2. **Calculation of $|Q_C|$: Isothermal ($T = T_C$) Compression from c→d**

Once again, along this isotherm:

$$\delta Q = \frac{RT_C}{V} dV \tag{8.14}$$

Integrating Eq. (8.14) from V_c to V_d yields:

$$\int_{V_c}^{V_d} \delta Q = Q_C = \int_{V_c}^{V_d} \frac{RT_C}{V} dV = -RT_C \ln\left(\frac{V_c}{V_d}\right) \tag{8.15}$$

Note that because $V_c/V_d > 1 \Rightarrow Q_C < 0$, heat is rejected by the Carnot engine to the cold reservoir.

Note that Q_H being positive in Eq. (8.13) and Q_C being negative in Eq. (8.15) are consistent with the directions of the arrows (〰➤) in the Carnot cycle (see Fig. 8.3).

Using Eqs. (8.13) and (8.15), it follows that:

$$\frac{|Q_H|}{|Q_C|} = \frac{T_H \ln(V_b/V_a)}{T_C \ln(V_c/V_d)} \tag{8.16}$$

We can show (left as an exercise) that:

$$\ln(V_b/V_a) = \ln(V_c/V_d) \tag{8.17}$$

by analyzing the two adiabatic portions ($\delta Q = 0$), b→c and d→a, using Eq. (8.9). Combining Eqs. (8.16) and (8.17) yields:

$$\frac{|Q_H|}{|Q_C|} = \frac{|T_H|}{|T_C|} \tag{8.18}$$

Equation (8.18) can be rewritten without the absolute magnitude signs as follows:

$$\frac{Q_H}{T_H} + \frac{Q_C}{T_C} = 0 \tag{8.19}$$

However, because

$$\frac{|Q_H|}{|Q_C|} = \frac{\psi(T_H)}{\psi(T_C)} \tag{8.20}$$

a comparison of Eqs. (8.20) and (8.18) clearly shows that:

$$\psi(T) = T \tag{8.21}$$

Because

$$\eta_C = \frac{\psi(T_H) - \psi(T_C)}{\psi(T_H)} \tag{8.22}$$

using Eq. (8.21) for T_H and T_C in Eq. (8.22) yields:

$$\eta_C = \frac{T_H - T_C}{T_H} \tag{8.23}$$

Equation (8.23) is a central result. It clearly shows that only if $T_C = 0$ K (-273 °C) or if $T_H \rightarrow$ infinity can $\eta_E = 1$! Cold reservoirs (e.g., the atmosphere, lakes, rivers, oceans) have T_C values of ~300 K. Hot reservoirs (e.g., fuel combustion, nuclear reactors) have T_H values of ~600 K. As a result, $\eta_C \approx 0.5$. Real irreversible heat engines have η_E values which rarely exceed 0.35!

8.6   Theorem of Clausius

"Given any reversible process in which the temperature changes in any prescribed manner, it is always possible to find a reversible zigzag process consisting of adiabatic-isothermal-adiabatic steps such that the heat interaction in the isothermal step is equal to the heat interaction in the original process" (see Fig. 8.4):

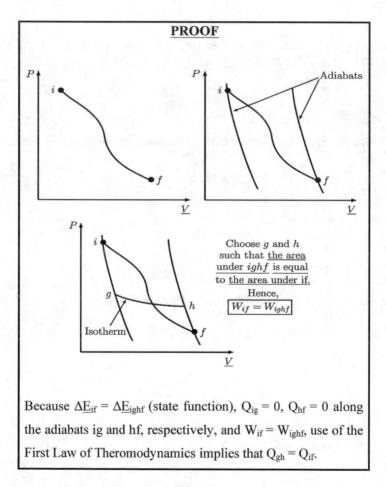

Fig. 8.4

Lecture 9
Entropy and Reversibility

9.1 Introduction

The material presented in this lecture is adapted from Chapter 4 in T&M, as well as from Chapter 1 in Denbigh. First, we will introduce a new thermodynamic function, the entropy $\underline{S}$, which strictly is only defined for a reversible process. Second, we will show that the entropy $\underline{S}$ is a function of state, first for a reversible Carnot cycle and then for a reversible arbitrary cycle. Fourth, after showing that the entropy is a function of state, we will show that if two states can be bridged both along reversible and irreversible paths, then, the entropy change along the irreversible path, although not defined, can nevertheless be equated to that along the reversible path. This will allow us to calculate entropy changes for irreversible processes. Interestingly, we will show that some states cannot be bridged both along reversible and irreversible paths. Finally, we will show that the entropy change for a closed and adiabatic system is always positive for an irreversible (natural) process and is zero for a reversible one. This statement is often referred to as the Second Law of Thermodynamics.

9.2 Entropy

The differential of the new thermodynamic function, the Entropy, denoted as $\underline{S}$, is defined as follows:

$$d\underline{S} = \left(\frac{\delta Q}{T}\right)_{rev} \tag{9.1}$$

© Springer Nature Switzerland AG 2020
D. Blankschtein, *Lectures in Classical Thermodynamics with an Introduction to Statistical Mechanics*, https://doi.org/10.1007/978-3-030-49198-7_9

where δQ is the differential amount of heat absorbed by the system at temperature, T, as it undergoes a reversible change of state. First, we would like to show that $\underline{S}$ is a function of state. To this end, we need to show that:

$$\oint d\underline{S} = 0 \tag{9.2}$$

where the circle symbol in the integral indicates that the integration is carried out around the reversible cycle, such that the system begins and ends at the same state. Figure 9.1 depicts going reversibly from state i to state f along Path α and then returning reversibly from state f to state i along Path β.

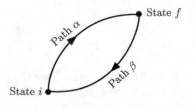

Fig. 9.1

Indeed, if we can prove that:

$$\oint d\underline{S} = \int_{i}^{f} (d\underline{S})_{\text{Path } \alpha} + \int_{f}^{i} (d\underline{S})_{\text{Path } \beta} = 0 \tag{9.3}$$

it would then follow that:

$$\int_{i}^{f} (d\underline{S})_{\text{Path } \alpha} = - \int_{f}^{i} (d\underline{S})_{\text{Path } \beta} = + \int_{i}^{f} (d\underline{S})_{\text{Path } \beta} \tag{9.4}$$

or that:

$$(\Delta \underline{S}_{i \rightarrow f})_{\text{Path } \alpha} = (\Delta \underline{S}_{i \rightarrow f})_{\text{Path } \beta} \tag{9.5}$$

Equation (9.5) clearly shows that the change in the entropy of the system is the same along any path connecting states i and f. The change is determined entirely by i and f, and therefore, $\underline{S}$ is a function of state.

With the above in mind, we would like to prove that for a reversible cyclic process:

$$\oint d\underline{S} = 0 \tag{9.6}$$

To this end, we will first prove this result for a special reversible cyclic process, the Carnot cycle, introduced in Lecture 8, which consists of two isotherms (one at a constant hot temperature, T_H, and the other at a constant cold temperature, T_C) and two adiabats. We will use a (P-$\underline{V}$) representation of the Carnot cycle (see Fig. 9.2):

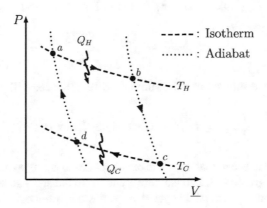

Fig. 9.2

Using the (P-$\underline{V}$) phase diagram in Fig. 9.2, it follows that:

$$\oint d\underline{S} = \Delta\underline{S}_{a\rightarrow b\rightarrow c\rightarrow d} = \Delta\underline{S}_{a\rightarrow b} + \Delta\underline{S}_{b\rightarrow c} + \Delta\underline{S}_{c\rightarrow d} + \Delta\underline{S}_{d\rightarrow a} \tag{9.7}$$

where again the circle symbol in the integral indicates that the integration is carried out around the reversible Carnot cycle.

Next, we evaluate separately each of the four entropy contributions in Eq. (9.7). Specifically,

$$\Delta\underline{S}_{a\rightarrow b} = \int_a^b \left(\frac{\delta Q_H}{T_H}\right)_{rev} = \frac{1}{T_H}\int_a^b (\delta Q_H)_{rev} = \frac{Q_H}{T_H} \text{ (Isotherm)} \tag{9.8}$$

$$\Delta\underline{S}_{b\rightarrow c} = \int_b^c \left(\frac{\delta Q_{b\rightarrow c}}{T}\right)_{rev} = 0 \text{ (Adiabat)} \tag{9.9}$$

$$\Delta \underline{S}_{c \to d} = \int_c^d \left(\frac{\delta Q_C}{T_C} \right)_{rev} = \frac{1}{T_C} \int_c^d (\delta Q_c)_{rev} = \frac{Q_C}{T_C} \text{ (Isotherm)} \qquad (9.10)$$

$$\Delta \underline{S}_{d \to a} = \int_d^a \left(\frac{\delta Q_{d \to a}}{T} \right)_{rev} = 0 \text{ (Adiabat)} \qquad (9.11)$$

Using Eqs. (9.8), (9.9), (9.10), and (9.11) in Eq. (9.7) it follows that:

$$\oint d\underline{S} = \frac{Q_H}{T_H} + \frac{Q_C}{T_C} \qquad (9.12)$$

In Lecture 8, we showed that the sum in Eq. (9.12) is zero! Therefore,

$$\oint d\underline{S} = 0 \text{ (For a reversible Carnot cycle)} \qquad (9.13)$$

Next, we need to prove that the same result applies to any type of reversible cycle, not necessarily consisting of two isotherms and two adiabats. The key to the proof is to replace the arbitrary reversible cycle by a series of Carnot cycles (this is possible due to the Theorem of Clausius that we proved in Lecture 8) and then to use the results just obtained for each Carnot cycle. One can then show that:

$$\oint d\underline{S}_s = 0 \text{ (For a reversible arbitrary cycle)} \qquad (9.14)$$

Note that in Eq. (9.14), the subscript s in $\underline{S}_s$ has been added to stress the fact that $d\underline{S}_s$ is the entropy change of the system which absorbed heat, δQ, in a reversible process at temperature, T, and it does not include the entropy change of the heat reservoir that supplied the heat.

Because the entropy, $\underline{S}$, is a function of state, if two states, A and B, can be bridged by both reversible and irreversible paths (see Fig. 9.3):

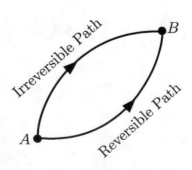

Fig. 9.3

it follows that:

$$(\Delta \underline{S}_{A \to B})_{rev} = (\Delta \underline{S}_{A \to B})_{irrev} \qquad (9.15)$$

This implies that although we do not know how to directly compute $(\Delta \underline{S})_{irrev}$, because $d\underline{S}$ was only defined for a reversible process, we can nevertheless compute this quantity by equating it with $(\Delta \underline{S})_{rev}$, because $\underline{S}$ is a function of state! We will use this important result to solve various problems involving irreversible processes, where one is asked to compute $(\Delta \underline{S})_{irrev}$.

It is important to recognize that for some processes, it will be impossible to bridge the same two states along irreversible and reversible paths! For example, if states A and B can be bridged by an adiabatic, irreversible path, then, they cannot be bridged by an adiabatic, reversible path (see Fig. 9.4):

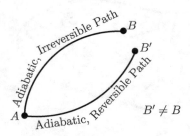

Fig. 9.4

To prove this, one assumes that $B' = B$ and then shows that this leads to an inconsistency (see Fig. 9.5):

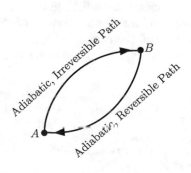

Fig. 9.5

Because the adiabatic path is reversible, one can come back from B to A! Clearly, the cyclic process A→B→A would be irreversible. However, the system returned to its original state (A), and no heat interaction occurred (the process A→B→A is

adiabatic). This implies that the environment did not change its state either. Therefore, the system and its environment were restored to their original states. This would
correspond to the process A→B→A being reversible. However, this is inconsistent
with the original statement that the process A→B→A is irreversible. Hence, B' ≠ B!

If a process is such that B' = B, then:

$$(\Delta \underline{S}_{A \to B})_{\text{irrev}} = (\Delta \underline{S}_{A \to B})_{\text{rev}} \tag{9.16}$$

However, we will soon see that the heat and work interactions along the reversible and the irreversible paths are different.

If one has a collection of subsystems 1, 2, ..., n, the total entropy of the composite
system is the sum of the entropies of each subsystem, because the entropy, $\underline{S}$, is
extensive (see Fig. 9.6).

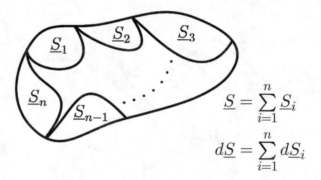

Fig. 9.6

"The entropy change of a closed and adiabatic system is always positive for an
irreversible (natural) process and zero for a reversible one." This statement is often
referred to as the Second Law of Thermodynamics. Let us examine this statement,
first for a reversible process (see Fig. 9.7) and then for an irreversible one.

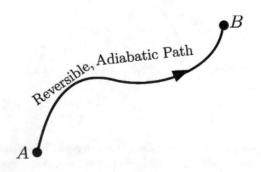

Fig. 9.7

9.3   Reversible Process

$$\bullet \, d\underline{S} = \left(\frac{\delta Q}{T}\right)_{rev} \tag{9.17}$$

$$\bullet \, (\Delta \underline{S}_{A\to B})_{rev} = (\underline{S}_B - \underline{S}_A)_{rev} \tag{9.18}$$

$$= \int_{A}^{B} \left(\frac{\delta Q}{T}\right)_{rev} = 0 \ \text{(Because } \delta Q = 0 \text{ along the AB adiabatic path)} \tag{9.19}$$

We have therefore shown that the entropy change for a closed and adiabatic system is zero for a reversible process.

9.4   Irreversible Process

Because the process A→B is irreversible (see Fig. 9.8), the defining equation, $d\underline{S} = (\delta Q/T)_{rev}$, cannot be used to calculate $(\Delta \underline{S}_{A\to B})_{irrev}$. Therefore, we will assume that after the original change of state, A→B, has taken place irreversibly and adiabatically, the reverse change of state, B→A, is carried out reversibly. Note that, in general, the return path, B→A, cannot be carried out adiabatically. Indeed, because the overall cyclic process from A to B to A is irreversible, it cannot be completed without leaving some change in the environment, because the system itself returns to its original state at A. If the process could be completed adiabatically, then, the entire process, A→B→A, would be reversible, which is not the case.

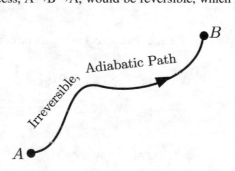

Fig. 9.8

We can design the reversible return path, B→A, in such a way that any heat interaction occurring along the path, $Q_{B\to A}$, occurs isothermally. Recall that this is possible according to the Theorem of Clausius that we proved in Lecture 8. Specifically, we choose a reversible return path consisting of the following three steps:

(i) The adiabatic step, B→C

(ii) The isothermal step, C→D

(iii) The adiabatic step, D→A

The chosen return path B→C→D→A is such that:

$$Q_{B\to C\to D\to A} = Q_{B\to A} = Q_{C\to D} \qquad (9.20)$$

because $Q_{B\to C}$ and $Q_{D\to A}$ are both zero along the adiabats BC and DA, respectively. Figure 9.9 shows the (P-$\underline{V}$) phase diagram corresponding to the entire process.

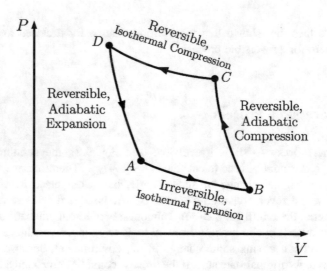

Fig. 9.9

We seek to calculate $(\underline{S}_B - \underline{S}_A)_{irrev}$. Because $\underline{S}$ is a function of state, it follows that:

$$0 = \oint_{A\to A} d\underline{S} = (\underline{S}_B - \underline{S}_A)_{irrev} + (\underline{S}_A - \underline{S}_B)_{rev} \qquad (9.21)$$

or

$$(\underline{S}_B - \underline{S}_A)_{irrev} = -(\underline{S}_A - \underline{S}_B)_{rev} \qquad (9.22)$$

We know that along the reversible return path B→A = B→C→D→A, one has:

$$d\underline{S}_{B\to A} = d\underline{S}_{B\to C} + d\underline{S}_{C\to D} + d\underline{S}_{D\to A} \qquad (9.23)$$

Because along the two adiabatic paths, $\delta Q_{B \to C} = 0$ and $\delta Q_{D \to A} = 0$ and, therefore, $d\underline{S}_{B \to C} = 0$ and $d\underline{S}_{D \to A} = 0$, it follows that:

$$d\underline{S}_{B \to A} = d\underline{S}_{C \to D} = \left(\frac{\delta Q_{C \to D}}{T}\right)_{rev} = \left(\frac{\delta Q_{B \to A}}{T}\right)_{rev} \tag{9.24}$$

In Eq. (9.24), T is the same along the isothermal step (C→D), and $\delta Q_{C \to D} = \delta Q_{B \to A}$ according to the Theorem of Clausius.

Integrating Eq. (9.24) from B to A, we obtain:

$$\int_B^A d\underline{S}_{B \to A} = \underline{S}_A - \underline{S}_B = \int_B^A \left(\frac{\delta Q_{B \to A}}{T}\right)_{rev} \tag{9.25}$$

Because T is constant, it follows that:

$$\underline{S}_A - \underline{S}_B = \frac{1}{T} \int_B^A (\delta Q_{B \to A})_{rev} = \frac{(Q_{B \to A})_{rev}}{T} \tag{9.26}$$

or

$$(\underline{S}_A - \underline{S}_B)_{rev} = \frac{(Q_{B \to A})_{rev}}{T} \tag{9.27}$$

Next, we need to show that $(Q_{B \to A})_{rev} < 0$, that is, that the system releases heat along the reversible return path from B to A.

To do this, we consider the closed system undergoing the cyclic process A→B→A and apply the First Law of Thermodynamics. Specifically,

$$\Delta \underline{U}_{A \to B \to A} = 0 = Q_{A \to B} + Q_{B \to A} + W_{A \to B} + W_{B \to A} \tag{9.28}$$

Because $Q_{A \to B}$ is zero along the adiabat AB, rearranging Eq. (9.28), we obtain:

$$Q_{B \to A} = -(W_{A \to B} + W_{B \to A}) \tag{9.29}$$

Note that $-(W_{A \to B} + W_{B \to A})$ is the total work done by the system during the cyclic process A→B→A.

Next, we will show that $Q_{B \to A}$ must be negative. Indeed,

(i) If $Q_{B \to A} = 0$, then, the environment would not change its state, and because the system does not change its state either, the cyclic process A→B→A would be reversible, contrary to what we know (the process is irreversible)

(ii) If $Q_{B \to A} > 0$, then, this would correspond to the complete conversion of heat, $Q_{B \to A}$, absorbed by the system into work in a cyclic process, which is not possible according to Statement 1(a) of the Second Law of Thermodynamics

In view of (i) and (ii) above, it follows that:

$$Q_{B \to A} < 0 \tag{9.30}$$

Equation (9.30) reveals that the system releases heat to the environment during its reversible return path from B to A.

Because we showed that:

$$(\underline{S}_B - \underline{S}_A)_{irrev} = -(\underline{S}_A - \underline{S}_B)_{rev} = -\frac{Q_{B \to A}}{T} \tag{9.31}$$

and that $Q_{B \to A} < 0$, it follows that:

$$(\underline{S}_B - \underline{S}_A)_{irrev} > 0 \tag{9.32}$$

or that:

$$(\Delta \underline{S}_{A \to B})_{irrev} = (\underline{S}_B - \underline{S}_A)_{irrev} > 0 \tag{9.33}$$

For the system to return to A, the entropy created along the irreversible path A→B must be decreased by releasing heat to the environment along the reversible return path B→A!

Lecture 10

The Second Law of Thermodynamics, Maximum Work, and Sample Problems

10.1 Introduction

The material presented in this lecture is adapted from Chapter 4 in T&M. First, we will present a more general statement of the Second Law of Thermodynamics than the one presented in Lecture 9. To this end, we will introduce the concept of entropy created in a process (zero for a reversible process and greater than zero for an irreversible one). Second, we will show that, as discussed in Lecture 9, although the entropy changes along reversible and irreversible paths are equal (because entropy is a function of state), more heat is produced along the irreversible path and more work is produced along the reversible path. Third, we will solve Sample Problem 10.1 to illustrate the calculation of maximum work for a particular reversible process. Fourth, we will solve Sample Problem 10.2 to prove that heat always flows from a hot body to a cold body, including showing that as long as heat transfer takes place (an irreversible process), entropy is created. Fifth, we will derive a criterion of equilibrium based on the entropy, which states that the entropy attains its maximum value at thermodynamic equilibrium. Finally, we will solve Sample Problem 10.3 to calculate the entropy change of a closed, simple system undergoing a reversible process, including when the process is isobaric.

10.2 A More General Statement of the Second Law of Thermodynamics

Continuing with the Second Law of Thermodynamics discussed in Lecture 9, we will present a more general statement of this law. Indeed, because the entire "universe" = (system + environment) is an isolated (adiabatic + closed + rigid) system, the Second Law of Thermodynamics can also be stated as follows:

© Springer Nature Switzerland AG 2020

D. Blankschtein, *Lectures in Classical Thermodynamics with an Introduction to Statistical Mechanics*, https://doi.org/10.1007/978-3-030-49198-7_10

"The entropy of the universe must increase in any irreversible (natural) process, and is conserved (remains constant) in any reversible process".

In other words:

$dS_{universe} > 0$, for an irreversible process

$dS_{universe} = 0$, for a reversible process

It is useful to convert the inequality in the Second Law of Thermodynamics for an irreversible process into an equality. To this end, we introduce the quantity, $\underline{\sigma}$, the total entropy created in the irreversible process. One can then rewrite the Second Law of Thermodynamics as follows:

$$dS_S = d\underline{\sigma} \tag{10.1}$$

where the subscript S in $\underline{S}$ denotes system, and:

$$d\underline{\sigma} = 0, \text{For a reversible process} \tag{10.2}$$

$$d\underline{\sigma} > 0, \text{For an irreversible process} \tag{10.3}$$

10.3   Heat Interactions Along Reversible and Irreversible Paths (Closed System)

In Lecture 9, we showed that $(\Delta \underline{S}_{A \to B})_{irrev} = (\Delta \underline{S}_{A \to B})_{rev}$. Next, we will show that the heat interaction is different for each process. Consider system (S) undergoing a purely heat interaction with a heat reservoir (R). A heat reservoir is an ideal body which acts solely as a donor or an acceptor of heat while maintaining constant volume (PdV- type work $= 0$). The heat reservoir can absorb or reject an infinite amount of heat while maintaining constant temperature. This absorption or rejection of heat results in a unique change of state of the reservoir. Indeed, a First Law of Thermodynamics analysis of the heat reservoir as a closed, simple, rigid, and diathermal system (see Fig. 10.1) yields:

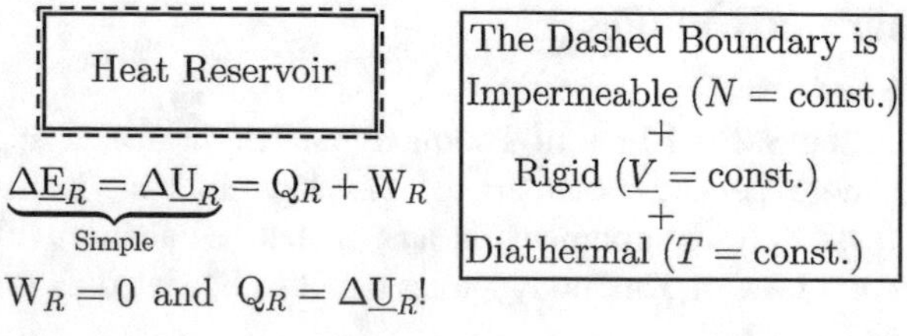

$$\underbrace{\Delta E_R = \Delta U_R}_{Simple} = Q_R + W_R$$

$W_R = 0$ and $Q_R = \Delta U_R!$

The Dashed Boundary is
Impermeable ($N =$ const.)
$+$
Rigid ($\underline{V} =$ const.)
$+$
Diathermal ($T =$ const.)

Fig. 10.1

In other words, the heat, Q_R, absorbed ($Q_R > 0$) or released ($Q_R < 0$) by the heat reservoir is equal to the change of the state function, $\Delta \underline{U}_R$. Therefore, the change of state of the heat reservoir is uniquely determined by the amount of heat transferred, δQ_R, irrespective of whether δQ_R is transferred reversibly or irreversibly. As a result, in defining $d\underline{S}_R$, there is no need to indicate $(\delta Q_R)_{rev}$ or $(\delta Q_R)_{irrev}$, because they lead to the same change of state of the heat reservoir. That is, $d\underline{S}_R = \delta Q_R / T_R$.

If the heat reservoir (R) releases heat, $-\delta Q_R$, to system (S) at temperature, T_R, then, the following entropy changes occur (see Fig. 10.2):

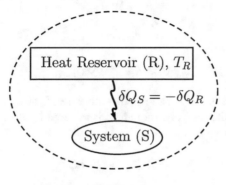

Fig. 10.2

In Fig. 10.2, the composite (t) of [heat reservoir (R) + system (S)] is a closed + adiabatic system surrounded by the dashed boundary, for which the Second Law of Thermodynamics applies. Specifically,

$$d\underline{S}_t = d\underline{S}_S + d\underline{S}_R = d\underline{\sigma} \tag{10.4}$$

For the heat reservoir:

$$d\underline{S}_R = \frac{\delta Q_R}{T_R} = -\frac{\delta Q_S}{T_R} \tag{10.5}$$

In Eq. (10.5), δQ_R is the heat absorbed by the heat reservoir (R), and δQ_S is the heat absorbed by the system (S), where $\delta Q_S > 0$ and $\delta Q_R > 0$.

Combining Eqs. (10.4) and (10.5) yields:

$$d\underline{S}_S = \frac{\delta Q_S}{T_R} + d\underline{\sigma} \tag{10.6}$$

However,

$$d\underline{S}_S = \left(\frac{\delta Q_S}{T_R}\right)_{rev} \tag{10.7}$$

Accordingly, Eqs. (10.6) and (10.7) show that:

$$\left(\frac{\delta Q_S}{T_R}\right)_{rev} = \frac{\delta Q_S}{T_R} + d\underline{\sigma} \tag{10.8}$$

Equation (10.8) shows that:

$$\left(\frac{\delta Q_S}{T_R}\right)_{rev} > \left(\frac{\delta Q_S}{T_R}\right)_{irrev}, \text{ for } d\underline{\sigma} > 0 \tag{10.9}$$

or that:

$$(\delta Q_S)_{rev} > (\delta Q_S)_{irrev} \tag{10.10}$$

The inequality in Eq. (10.10) shows that more heat is absorbed by the system in a reversible process. Alternatively, more heat is released by the system in an irreversible process. In other words,

$$(-\delta Q_S)_{irrev} > (-\delta Q_S)_{rev} \tag{10.11}$$

10.4   Work Interactions Along Reversible and Irreversible Paths (Closed System)

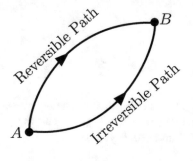

Fig. 10.3

Given states A and B that can be bridged both along reversible and irreversible paths (see Fig. 10.3), the First Law of Thermodynamics for a closed system states that:

$$\Delta \underline{E}_{A\to B} = Q_{A\to B} + W_{A\to B} \tag{10.12}$$

Because $\underline{E}$ is a function of state, it follows that:

$$(\Delta \underline{E}_{A\to B})_{rev} = (\Delta \underline{E}_{A\to B})_{irrev} \tag{10.13}$$

We have just shown that:

$$(Q_{A\to B})_{rev} > (Q_{A\to B})_{irrev} \tag{10.14}$$

Equations (10.12), (10.13), and (10.14) show that:

$$\therefore (-W_{A\to B})_{rev} > (-W_{A\to B})_{irrev} \text{ (Work done by the system)} \tag{10.15}$$

The inequalities in Eqs. (10.15) and (10.11) indicate that in any reversible change of state, A→B, of a closed system, a greater amount of work and a smaller amount of heat are produced relative to the corresponding irreversible change of state. This, of course, is consistent with the expected dissipative nature of an irreversible process. In a reversible process, one can obtain maximum work.

10.5 Sample Problem 10.1

In a given process, a closed system absorbs heat, $|\delta Q|$, from a heat reservoir at the constant temperature, T_R, of the hot reservoir and performs work (see Fig. 10.4). Calculate the maximum work.

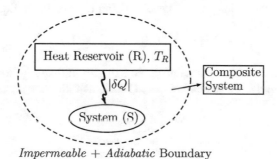

Impermeable + Adiabatic Boundary

Fig. 10.4

10.5.1   Solution

The composite system (t) of system (S) + heat reservoir (R) is closed + adiabatic, and therefore, the Second Law of Thermodynamics applies. Specifically,

$$d\underline{S}_t = d\underline{S}_S + d\underline{S}_R = d\underline{\sigma} \tag{10.16}$$

In Eq. (10.16),

$$d\underline{S}_R = \frac{\delta Q_R}{T_R} = -\frac{|\delta Q|}{T_R} \tag{10.17}$$

where $- |\delta Q|$ is the differential heat released by the heat reservoir (see Fig. 10.4). Combining Eqs. (10.16) and (10.17) yields:

$$|\delta Q| = T_R d\underline{S}_S - T_R d\underline{\sigma} \tag{10.18}$$

where $\underline{\sigma}$ is the entropy created in the process, with $d\underline{\sigma} > 0$ for an irreversible process and $d\underline{\sigma} = 0$ for a reversible process.

Next, we focus on system (S) and carry out a First Law of Thermodynamics analysis on it. The system is closed and simple and has a diathermal, movable boundary. Accordingly:

$$d\underline{U}_S = |\delta Q| + \delta W \tag{10.19}$$

or

$$-\delta W = |\delta Q| - d\underline{U}_S \tag{10.20}$$

where $-\delta W$ is the work done by the system. Using Eq. (10.18) in Eq. (10.20) yields:

$$-\delta W \leq d\left[T_R \underline{S}_S - \underline{U}_S\right] - T_R d\underline{\sigma} \tag{10.21}$$

where $d\underline{\sigma}$ is either zero for a reversible process or greater than zero for an irreversible process. Accordingly, we can express Eq. (10.21) as follows:

$$-\delta W \leq d\underbrace{\left[T_R \underline{S}_S - \underline{U}_S\right]}_{\text{State Function}} \tag{10.22}$$

Integrating Eq. (10.22) from state A to state B, we obtain:

$$-W_{A \to B} \leq T_R \left(\underline{S}_S^B - \underline{S}_S^A\right) - \left(\underline{U}_S^B - \underline{U}_S^A\right) \tag{10.23}$$

where the two terms on the right-hand side of Eq. (10.23) are independent of the path connecting states A and B. Equation (10.23) shows that $(-W_{A \to B})$ cannot be larger than the quantity on the right-hand side, which we will denote as W_{max}, where:

$$W_{max} = T_R \left(\underline{S}_S^B - \underline{S}_S^A \right) - \left(\underline{U}_S^B - \underline{U}_S^A \right) \tag{10.24}$$

is the maximum work done by the system for the process considered and corresponds to the reversible case ($d\underline{\sigma} = 0$). If the process considered were irreversible ($d\underline{\sigma} > 0$), then:

$$(-W_{A \to B})_{irrev} < W_{max} \tag{10.25}$$

Note that if a different process is considered, a different expression for W_{max} will result.

10.6   Sample Problem 10.2

Calculate the differential total entropy change for a process involving heat transfer from a heat reservoir at T_2 to another heat reservoir at T_1. Figure 10.5 depicts the composite system (t) under consideration.

10.6.1   Solution

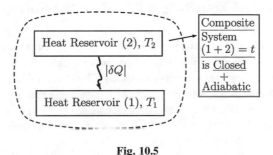

Fig. 10.5

Applying the Second Law of Thermodynamics to the composite system (t) in Fig. 10.5 yields:

$$d\underline{S}_t = d\underline{S}_1 + d\underline{S}_2 = d\underline{\sigma} \tag{10.26}$$

$$d\underline{S}_1 = \frac{|\delta Q|}{T_1}, \quad d\underline{S}_2 = -\frac{|\delta Q|}{T_2} \tag{10.27}$$

Combining Eqs. (10.26) and (10.27) yields:

$$\therefore \; d\underline{S}_t = |\delta Q| \, (T_2 - T_1)/T_1 T_2 = d\underline{\sigma} \tag{10.28}$$

Because for any allowable process $d\underline{\sigma} \geq 0$, Eq. (10.28) indicates that $T_2 \geq T_1$. Therefore, heat can only be transferred from the "hotter" to the "colder" body, as observed in nature.

Note that if $T_2 < T_1$, it would follow that $d\underline{\sigma} < 0$, which would violate the Second Law of Thermodynamics!

As long as $T_2 > T_1$, entropy is created ($d\underline{\sigma} > 0$) as heat is transferred from "hot" body 2 to "cold" body 1. Thermal equilibrium is eventually attained when $T_2 = T_1$ and $d\underline{S}_t = 0$. Note that as long as $T_2 > T_1$, the heat transfer process is irreversible. When $T_2 - T_1 = dT$, heat can be transferred reversibly, because:

$$d\underline{S}_t = d\underline{\sigma} = \frac{|\delta Q| dT}{T_2^2} \approx 0 \; \text{(Differential of the second order)} \tag{10.29}$$

This is, in fact, what is assumed for the reversible heat transfer process in a Carnot engine.

If a system is open and not adiabatically enclosed, the Second Law of Thermodynamics does not have to apply.

10.7   Criterion of Equilibrium Based on the Entropy

As we already saw, the only changes of state that can occur within a closed + adiabatic boundary are those for which the entropy either increases (irreversible process) or remains constant (reversible process). The same holds true for an isolated system, which in addition to being closed ($N = $ constant) and adiabatic ($Q = 0$) is also surrounded by a rigid ($\underline{V} = $ constant) boundary.

For an isolated system, the First Law of Thermodynamics indicates that $\Delta \underline{E} = Q + W = 0$ or that $\underline{E} = $ constant.

It then follows that whenever an isolated system ($\underline{E}$, $\underline{V}$, and N are constant) can change from a state of lower entropy to one of higher entropy, it is possible for this change of state to occur according to the Second Law of Thermodynamics. Because the isolated system must eventually reach an equilibrium state whose properties (in particular, the entropy) are constant, it follows that $\underline{S}$ must be a maximum at equilibrium. Mathematically, this implies that (see Fig. 10.6):

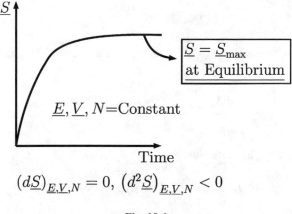

$$\left(d\underline{S}\right)_{E,V,N} = 0, \; \left(d^2\underline{S}\right)_{E,V,N} < 0$$

Fig. 10.6

10.8   Sample Problem 10.3

Calculate the entropy of a closed, simple system undergoing a reversible process.

10.8.1   Solution

Consider states A and B connected along a reversible path (see Fig. 10.7):

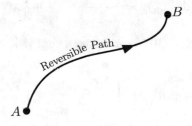

Fig. 10.7

Figure 10.7 indicates that:

$$\left(\Delta\underline{S}_{A\to B}\right)_{\text{rev}} = \int_{A}^{B} \left(d\underline{S}_{A\to B}\right)_{\text{rev}} \tag{10.30}$$

According to the definition of entropy:

$$\left(d\underline{S}_{A\to B}\right)_{rev} = \frac{\left(\delta Q_{A\to B}\right)_{rev}}{T} \tag{10.31}$$

We can compute $\left(\delta Q_{A\to B}\right)_{rev}$ by invoking the First Law of Thermodynamics, that is:

$$\underbrace{\left(d\underline{E}_{A\to B}\right)_{rev} = \left(d\underline{U}_{A\to B}\right)_{rev}}_{\text{Simple}} = \left(\delta Q_{A\to B}\right)_{rev} + \left(\delta W_{A\to B}\right)_{rev} \tag{10.32}$$

Rearranging Eq. (10.32) yields:

$$\left(\delta Q_{A\to B}\right)_{rev} = \left(d\underline{U}_{A\to B}\right)_{rev} - \left(\delta W_{A\to B}\right)_{rev} \tag{10.33}$$

Using Eq. (10.33) in Eq. (10.31) yields:

$$\left(d\underline{S}_{A\to B}\right)_{rev} = \frac{\left(d\underline{U}_{A\to B}\right)_{rev}}{T} - \frac{\left(dW_{A\to B}\right)_{rev}}{T} \tag{10.34}$$

If only $Pd\underline{V}$- type work is involved, it follows that:

$$\left(\delta W_{A\to B}\right)_{rev} = -Pd\underline{V} \tag{10.35}$$

where P varies with $\underline{V}$ along path A to B.
    Combining Eqs. (10.35) and (10.34) yields:

$$\left(d\underline{S}_{A\to B}\right)_{rev} = \frac{\left(d\underline{U}_{A\to B}\right)_{rev}}{T} + \frac{P}{T}\, d\underline{V} \tag{10.36}$$

Using Eq. (10.36) in Eq. (10.30), and integrating, yields:

$$\left(\Delta\underline{S}_{A\to B}\right)_{rev} = \int_A^B \frac{\left(d\underline{U}_{A\to B}\right)_{rev}}{T} + \int_A^B \frac{P}{T}\, d\underline{V} \tag{10.37}$$

If the process is irreversible and one can bridge states A and B along a reversible path, Eq. (10.37) can be used to compute $\left(\Delta\underline{S}_{A\to B}\right)_{irrev}$, because $\underline{S}$ is a function of state.
    If in addition to being reversible the process is isobaric (P is constant), Eq. (10.36) can be expressed as follows:

$$\left(d\underline{S}_{A\to B}\right)_{rev,P} = \frac{1}{T}\, d\left(\underline{U}_{A\to B} + P\underline{V}\right)_{rev,P} \tag{10.38}$$

Recalling that $\underline{U} + P\underline{V} = \underline{H}$, Eq. (10.38) can be expressed as follows:

$$(d\underline{S}_{A \to B})_{rev,P} = \frac{(d\underline{H}_{A \to B})_{rev,P}}{T} \tag{10.39}$$

Equation (10.39) will be particularly useful to compute entropy changes in the case of reversible processes which occur isobarically.

The product, $T\underline{S}$, has units of energy. Typically, if we choose $[T] =$ Kelvin (K) and $[E] =$ Joule (J), it follows that $[S] =$ J/K.

Lecture 11

The Combined First and Second Law of Thermodynamics, Availability, and Sample Problems

11.1 Introduction

The material presented in this lecture is adapted from Chapter 4 in T&M. First, we will derive the Combined First and Second Law of Thermodynamics for both a closed and an open, single-phase, simple system. Second, we will solve Sample Problem 11.1 which will help us crystallize the material presented in Lecture 10, including calculating changes in entropy and changes in energy, heat, and work when a gas expands isothermally in a cylinder-piston assembly, without friction, in one case, reversibly, and in another case, irreversibly. Finally, we will solve Sample Problem 11.2 to calculate the maximum work done by an open system, consisting of shaft work and Carnot work, including introducing a new thermodynamic function of state, the Availability or Exergy.

11.2 Closed, Single-Phase, Simple System

For this system, the First Law of Thermodynamics states that:

$$d\underline{E} = d\underline{U} = \delta Q + \delta W \tag{11.1}$$

For an internally reversible, quasi-static process with only PdV-type work, it follows that:

$$\delta Q = \delta Q_{rev} = Td\underline{S} \tag{11.2}$$

$$\delta W = \delta W_{rev} = -Pd\underline{V} \tag{11.3}$$

© Springer Nature Switzerland AG 2020
D. Blankschtein, *Lectures in Classical Thermodynamics with an Introduction to Statistical Mechanics*, https://doi.org/10.1007/978-3-030-49198-7_11

Using Eqs. (11.2) and (11.3) in Eq. (11.1) yields:

$$d\underline{U} = Td\underline{S} - Pd\underline{V} \tag{11.4}$$

11.3   Open, Single-Phase, Simple System

For an internally reversible, quasi-static process with a one-component stream entering (in) and leaving (out) the system (see Fig. 11.1), all the intensive properties must remain the same. In other words:

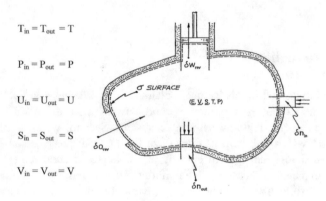

$T_{in} = T_{out} = T$

$P_{in} = P_{out} = P$

$U_{in} = U_{out} = U$

$S_{in} = S_{out} = S$

$V_{in} = V_{out} = V$

Fig. 11.1

A First Law of Thermodynamics analysis of the system bounded by the σ-surface in Fig. 11.1 yields:

$$d\underline{E} = d\underline{U} = \delta Q_{rev} + \delta W_{rev} + (U_{in} + P_{in}V_{in})\,\delta n_{in} - (U_{out} + P_{out}V_{out})\,\delta n_{out} \tag{11.5}$$

Because the intensive properties of the entering (in) and leaving (out) streams are the same and $\delta n_{in} - \delta n_{out} = dN$, where N is the total number of moles in the system bounded by the σ-surface, Eq. (11.5) can be rewritten as follows:

$$d\underline{E} = d\underline{U} = \delta Q_{rev} + \delta W_{rev} + (U + PV)dN \tag{11.6}$$

A Second Law of Thermodynamics entropy balance on the system bounded by the σ-surface (see Fig. 11.1) yields:

$$d\underline{S} = \delta Q_{rev}/T + S_{in}\,\delta n_{in} - S_{out}\,\delta n_{out} \tag{11.7}$$

$$= \delta Q_{rev}/T + S(\delta n_{in} - \delta n_{out}) \tag{11.8}$$

$$dS = \delta Q_{rev}/T + SdN \tag{11.9}$$

Rearranging Eq. (11.9), we obtain:

$$\delta Q_{rev} = TdS - TSdN \tag{11.10}$$

For PdV-type work:

$$\delta W_{rev} = -PdV \tag{11.11}$$

Combining Eqs. (11.6), (11.10), and (11.11) yields the Combined First and Second Law of Thermodynamics, that is:

$$dU = TdS - PdV + (U + PV - TS)dN = TdS - PdV + \mu dN \tag{11.12}$$

In Eq. (11.12), μ is the molar Gibbs free energy, or the chemical potential, defined as follows:

$$\mu = G = U + PV - TS - H - TS \tag{11.13}$$

Equation (11.12) can be generalized for a multi-component, single-phase system that traverses a quasi-static path. Specifically:

$$dU = TdS - PdV + \sum_{i=1}^{n} \mu_i dN_i \tag{11.14}$$

where U is a continuous function of its $(n + 2)$ independent extensive variables, that is, $U = f(S, V, N_1, N_2, \ldots, N_n)$, whose differential can be expressed as follows:

$$dU = (\partial U/\partial S)_{V,N} \, dS + (\partial U/\partial V)_{S,N} \, dV + \sum_{i=1}^{n} (\partial U/\partial N_i)_{S,V,N_{j(i)}} \, dN_i \tag{11.15}$$

In Eq. (11.15), the subscript N is a short-hand notation for $N_1, N_2, \ldots, N_n$ are kept constant, and the subscript Nj(i) indicates that all the mole numbers j except i are kept constant. A comparison of Eqs. (11.14) and (11.15) shows that:

$$(\partial U/\partial S)_{V,N} = T; \ (\partial U/\partial V)_{S,N} = -P; \text{ and } (\partial U/\partial N_i)_{S,V,N_{j(i)}} = \mu_i \tag{11.16}$$

11.4   Sample Problem 11.1

Two moles of an ideal gas expand isothermally and without friction in a cylinder-piston assembly which is in contact with a heat reservoir maintained at a constant temperature of 300 K. The initial state of the gas is $(0.5 \ m^3, 300 \ K)$, and the final state of the gas is $(5 \ m^3, 300 \ K)$. Assuming that the piston is massless, calculate:

(1) The entropy change of the gas (g), (2) the entropy change of the heat reservoir (R), and (3) the entropy change of the universe (U). Carry out the calculation when (a) the gas expansion is reversible and (b) the gas expansion is irreversible.

11.4.1   Solution

(a) The gas expansion is reversible and quasi-static, such that $P_g = P_{ext} + dP$ at all times. Figure 11.2 illustrates the various elements of this problem.

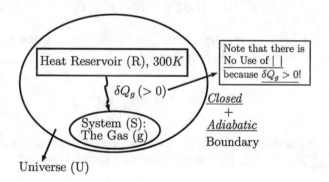

Fig. 11.2

$\Delta U_g = 0$ (Ideal gas, closed + isothermal)
$0 = \Delta U_g = Q_g + W_g$ (First Law of Thermodynamics analysis of the gas)
$Q_g = -W_g$ (Heat absorbed by the gas from the heat reservoir = work done by the gas)

The entropy change of the gas results from the heat interaction with the heat reservoir and is given by:

$$\Delta \underline{S}_g = \frac{Q_g}{T_R} = -\frac{W_g}{T_R} \qquad (11.17)$$

where T_R is the temperature of the heat reservoir (300 K).

The work done by the gas, $-W_g$, is given by:

$$-W_g = \int_{\underline{V}_i}^{\underline{V}_f} P_g d\underline{V}_g = \int_{\underline{V}_i}^{\underline{V}_f} N_g R T_g \frac{d\underline{V}_g}{\underline{V}_g} \qquad (11.18)$$

or

$$-W_g = N_g R T_g \ln\left(\frac{\underline{V}_f}{\underline{V}_i}\right) \qquad (11.19)$$

The values of the gas properties are given in the Problem Statement and are summarized below for completeness:

$N_g = 2$ mol
$R = 8.314$ J/mol K
$T_g = 300$ K
$\underline{V}_i = 0.5$ m^3
$\underline{V}_f = 5$ m^3

Using the gas property values above in Eq. (11.19) yields:

$$(-W_g)_{rev} = (Q_g) = 1.15 \times 10^4 J \qquad (11.20)$$

Accordingly:

$$\Delta \underline{S}_g = -\frac{W_g}{T_R} = \frac{1.15 \times 10^4 J}{300 K} \qquad (11.21)$$

or

$$\left(\Delta \underline{S}_g\right)_{rev} = +38.3 \text{ J/K} \qquad (11.22)$$

Because the heat reservoir lost precisely an amount of heat, $Q_R = -Q_g$, it follows that:

$$(\Delta \underline{S}_R)_{rev} = -38.3 \text{ J/K} \qquad (11.23)$$

Consistent with the Second Law of Thermodynamics for a reversible process, it follows that the change in the entropy of the universe (U) is given by (see Eqs. (11.22) and (11.23)):

$$(\Delta \underline{S}_U)_{rev} = \Delta \underline{S}_g + \Delta \underline{S}_R = 0 \qquad (11.24)$$

(b) Because $P_g > P_{ext}$, the gas expansion is irreversible and not quasi-static. Let us consider the most extreme irreversibility which corresponds to $P_{ext} = 0$, where the gas expands against vacuum. Because P_g is no longer uniform, we cannot use the expression $P_g d\underline{V}_g$ to compute $(-\delta W_g)$! Instead, we can use the central result, presented in Lecture 3, that the work done by the gas (g) is equal to the work done on the three elements of its environment (in this problem, the vacuum, the friction (f) of the cylinder walls, and the piston (p)). Specifically:

$$-W_g = W_{vacuum} + W_f + W_p \qquad (11.25)$$

As per the Problem Statement, there is no friction with the cylinder walls ($W_f = 0$), and the piston is massless ($W_p = 0$). Therefore, Eq. (11.25) reduces to:

$$-W_g = W_{vacuum} = 0 \qquad (11.26)$$

or

$$\left(-W_g\right)_{irrev} = 0 \qquad (11.27)$$

Because we already saw that, in general:

$$Q_g = -W_g \qquad (11.28)$$

it follows that:

$$\left(Q_g\right)_{irrev} = 0 \qquad (11.29)$$

Comparing Eqs. (11.20) and (11.29), we obtain:

$$\left(Q_g\right)_{rev} = 1.15 \times 10^4 J > \left(Q_g\right)_{irrev} = 0 \qquad (11.30)$$

Further, comparing Eqs. (11.20) and (11.27), we obtain:

$$\left(-W_g\right)_{rev} = 1.15 \times 10^4 J > \left(-W_g\right)_{irrev} = 0 \qquad (11.31)$$

Equations (11.30) and (11.31) are consistent with the observations that we made in Lecture 10 about heat and work for reversible and irreversible processes.

Because $\underline{S}$ is a function of state, it follows that:

$$\left(\Delta \underline{S}_g\right)_{irrev} = \left(\Delta \underline{S}_g\right)_{rev} = +38.3 \text{ J/K} \qquad (11.32)$$

Because the heat reservoir did not deliver any heat, that is, $Q_R = -Q_g = 0$, it did not change its state, and therefore:

$$(\Delta \underline{S}_R)_{irrev} = 0 \tag{11.33}$$

Consistent with the Second Law of Thermodynamics for an irreversible process and using Eqs. (11.32) and (11.33), it then follows that:

$$(\Delta \underline{S}_U)_{irrev} = \left(\Delta \underline{S}_g\right)_{irrev} + (\Delta \underline{S}_R)_{irrev} = +38.3 \, \text{J/K} \tag{11.34}$$

11.5 Sample Problem 11.2

Calculate the maximum work done by an open system undergoing the processes, depicted in Fig. 11.3.

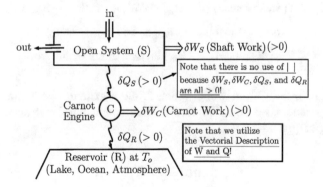

Fig. 11.3

The following assumptions can be made: (1) reversible, quasi-static processes, (2) steady-state operation, indicating that:

$$dN - \delta n_{in} - \delta n_{out} - 0 \rightarrow \delta n_{in} - \delta n_{out} = \delta n \tag{11.35}$$

$$d\underline{E} = d(NE) = d\underline{U} = d(NU) = 0 \tag{11.36}$$

$$d\underline{S} = d(NS) = 0 \tag{11.37}$$

,and (3) the heats δQ_S and δQ_R are absorbed and rejected isothermally by the Carnot engine. The Carnot engine is used to produce Carnot work out of these heat interactions.

11.5.1 Solution

(1) To calculate the maximum work, we will assume that all the processes considered are reversible. First, we will calculate the maximum value of the shaft work, δW_S, by doing a First Law of Thermodynamics analysis of the open system (see Fig. 11.4):

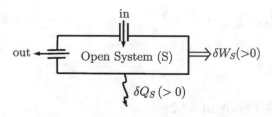

Fig. 11.4

$$d\underline{E} = d\underline{U} = 0 \,(\text{Steady state}) = -\delta Q_S - \delta W_S + H_{in}\delta n_{in} - H_{out}\delta n_{out} \quad (11.38)$$

$$dN = 0 \Rightarrow \delta n_{in} = \delta n_{out} \equiv \delta n \,(\text{Steady state}) \quad (11.39)$$

$$\therefore - \delta W_S^{max} = \delta Q_S + (H_{out} - H_{in})\,\delta n \,(\text{Maximum shaft work}) \quad (11.40)$$

(2) To calculate the Carnot work, which is the maximum work that the Carnot engine produces, we do a First Law of Thermodynamics analysis of the Carnot engine (see Fig. 11.5):

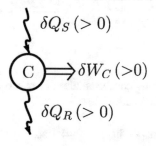

Fig. 11.5

$$d\underline{U} = 0 \,(\text{Steady state}) = \delta Q_S - \delta Q_R - \delta W_C \quad (11.41)$$

or

$$-\delta W_C^{max} = -\delta Q_S + \delta Q_R \,(\text{Maximum Carnot work}) \quad (11.42)$$

Therefore, the total maximum work is given by the sum of the maximum shaft work (Eq. 11.40) and the maximum Carnot work (Eq. 11.42), that is, by:

$$-(\delta W_S + \delta W_C)_{max} = \delta Q_R + (H_{out} - H_{in})\delta n \qquad (11.43)$$

(3) It is convenient to express δQ_R in Eq. (11.43) in terms of a function of state. To this end, we carry out a Second Law of Thermodynamics entropy balance on the composite system of (system + Carnot engine; see Fig. 11.6):

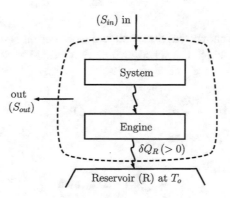

Fig. 11.6

$$d\underline{S} = 0 \text{ (Steady state)} \; = \; -\frac{\delta Q_R}{T_o} + S_{in}\,\delta n_{in} - S_{out}\,\delta n_{out} \qquad (11.44)$$

Because, at steady state, $\delta n_{in} = \delta n_{out} \equiv \delta n$, Eq. (11.44) can be expressed as follows:

$$\delta Q_R = -T_o\,(S_{out} - S_{in})\,\delta n \qquad (11.45)$$

Denoting

$$(\delta W_S + \delta W_C) = \delta W_{net} \text{ [Total (net) work obtained]} \qquad (11.46)$$

where the three differential work contributions in Eq. (11.46) are positive. Using Eq. (11.45) in Eq. (11.43) yields:

$$\delta W_{max}^{net} = -\{(H_{out} - H_{in}) - T_o\,(S_{out} - S_{in})\}\,\delta n \qquad (11.47)$$

It is convenient to define a new derived thermodynamic property, the Availability or Exergy, $\underline{B}$, as follows:

$$\underline{B} = \underline{H} - T_o \underline{S} \tag{11.48}$$

where T_o is the constant temperature of the cold reservoir. It then follows that:

$$\delta W_{max}^{net} = -(B_{out} - B_{in})\,\delta n = -\Delta B_{in \to out}\,\delta n \tag{11.49}$$

or that:

$$\frac{\delta W_{max}^{net}}{\delta n} = -(B_{out} - B_{in}) = -\Delta B_{in \to out} \tag{11.50}$$

Equation (11.50) shows that the maximum work per mole done by the system is equal to $-\Delta B_{in \to out}$, which is a state function. Further, the maximum power is given by:

$$\frac{\delta W_{max}^{net}}{\delta t} = \dot{W}_{net}^{max} = -\Delta B_{in \to out}\frac{\delta n}{\delta t} \tag{11.51}$$

or

$$\dot{W}_{max}^{net} = -\Delta B_{in \to out}\dot{n} \tag{11.52}$$

Lecture 12

Flow Work and Sample Problems

12.1 Introduction

In this lecture, we will solve Sample Problem 12.1 to calculate the reversible work of expansion or compression in flow systems (adapted from Chapter 2 in Denbigh). Second, we will solve Sample Problem 12.2 to analyze the operation of a Hilsch vortex tube (see below), which will allow us to crystallize material presented in Lectures 10, 11, and 12.

12.2 Sample Problem 12.1

Reversible work of expansion or compression in flow systems.

Figure 12.1 depicts the process under consideration:

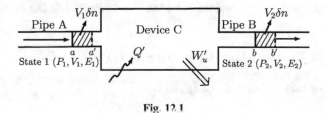

Fig. 12.1

12.2.1 Solution

First, we will carry out a First Law of Thermodynamics analysis on δn moles of fluid as they flow from state 1 to state 2 (denoted hereafter as $1{\rightarrow}2$). Specifically,

© Springer Nature Switzerland AG 2020
D. Blankschtein, *Lectures in Classical Thermodynamics with an Introduction to Statistical Mechanics*, https://doi.org/10.1007/978-3-030-49198-7_12

$$\Delta \underline{E}_{1\to2} = Q_{1\to2} - W_{1\to2} \tag{12.1}$$

In Eq. (12.1), $Q_{1\to2}$ is the heat absorbed by the δn moles of fluid as they flow from $1\to2$, and $W_{1\to2}$ is the work done by the δn moles of fluid as they flow from $1\to2$, where:

$$\Delta \underline{E}_{1\to2} = (E_2 - E_1)\,\delta n \tag{12.2}$$

$$Q_{1\to2} = Q'\delta n \tag{12.3}$$

and

$$W_{1\to2} = W'_u\,\delta n + \underbrace{P_2\,(V_2\,\delta n)}_{\substack{\text{Work done by the }\delta n\text{ moles} \\ \text{on the fluid which lies ahead} \\ \text{to change its volume by }V_2\delta n}} - \underbrace{P_1\,(V_1\,\delta n)}_{\substack{\text{Work done on the }\delta n\text{ moles by} \\ \text{the fluid which lies behind to} \\ \text{change its volume by }V_1\,\delta n}} \tag{12.4}$$

where $W_{1\to2}$ is the total work done by the δn moles of fluid as they flow from $1\to2$, and W'_u is the useful (shaft) work done by one mole of fluid as it flows from $1\to2$.
Using Eqs. (12.2), (12.3), and (12.4) in Eq. (12.4), we obtain:

$$(E_2 - E_1)\,\delta n = \left[Q' - W'_u + P_1V_1 - P_2V_2\right]\delta n \tag{12.5}$$

or

$$E_2 - E_1 = Q' - W'_u + (P_1V_1 - P_2V_2) \tag{12.6}$$

where Q' is the heat absorbed per mole, W'_u is the useful (shaft) work done per mole, and $(P_1V_1 - P_2V_2)$ represents the flow work done on a mole of fluid.
In the absence of external fields, e.g., gravitational and inertial, the system is simple. In that case, $E = U$, and the left-hand side in Eq. (12.6) can be expressed as follows:

$$E_2 - E_1 = U_2 - U_1 = (H_2 - P_2V_2) - (H_1 - P_1V_1) \tag{12.7}$$

Combining Eqs. (12.6) and (12.7) yields:

$$H_2 - H_1 + P_1V_1 - P_2V_2 = Q' - W'_u + P_1V_1 - P_2V_2 \tag{12.8}$$

Cancelling the P_1V_1 and P_2V_2 terms in Eq. (12.8) yields:

$$H_2 - H_1 = \Delta H_{1 \to 2} = Q' - W'_u \tag{12.9}$$

or, in differential form:

$$dH = \delta Q' - \delta W'_u \tag{12.10}$$

Because the process is reversible, it follows that:

$$\delta Q' = TdS \tag{12.11}$$

In addition, the following well-known thermodynamic relation applies:

$$dH = TdS + VdP \tag{12.12}$$

Combining Eqs. (12.10), (12.11), and (12.12) yields:

$$TdS + VdP = TdS - \delta W'_u \tag{12.13}$$

or

$$\delta W'_u = -VdP \tag{12.14}$$

Integrating Eq. (12.14) from P_1 to P_2 yields:

$$W'_u = - \int_{P_1}^{P_2} VdP \tag{12.15}$$

In general,

$$dH = d(U + PV) = dU + PdV + VdP \tag{12.16}$$

If the process is reversible, such that the δn moles of fluid constitute a closed, simple system traversing a quasi-static path, then, the Combined First and Second Law of Thermodynamics applies. Specifically,

$$dU = TdS - PdV \tag{12.17}$$

Equation (12.15) indicates that:

$$W'_u = \frac{\delta W_u}{\delta n} = - \int_{P_1}^{P_2} VdP \tag{12.18}$$

Integrating Eq. (12.18) with respect to n yields:

$$W_u = - \int_n \int_{P_1}^{P_2} V dP \, \delta n \tag{12.19}$$

where W_u is the useful (shaft) work done by the δn moles of fluid as they flow from $1 \rightarrow 2$. Note that the limits of integration with respect to pressure may vary with the number of moles of fluid (n) flowing through the device.

Dividing Eq. (12.4) by δn yields:

$$W'_{1 \rightarrow 2} = W'_u + [P_2 V_2 - P_1 V_1] \tag{12.20}$$

In addition, the pressure integral in Eq. (12.18) can be rewritten as follows:

$$W'_u = - \int_{P_1}^{P_2} V dP = - \int_{P_1}^{P_2} [d(PV) - PdV] \tag{12.21}$$

or

$$W'_u = - [P_2 V_2 - P_1 V_1] + \int_{V_1}^{V_2} PdV \tag{12.22}$$

Using Eq. (12.22) in Eq. (12.20), and then cancelling the $P_1 V_1$ and $P_2 V_2$ terms, yields:

$$W'_{1 \rightarrow 2} = \int_{V_1}^{V_2} PdV \tag{12.23}$$

12.3   Sample Problem 12.2: Problem 4.3 in T&M

A Hilsch vortex tube (Hvt) for sale commercially is fed with air at 300 K and 5 bar into a tangential slot near the center (point A in Fig. 12.2). Stream B leaves at the left end at 1 bar and 250 K, and stream C leaves at the right end at 1 bar and 310 K. These two streams then act as a sink and a source for a Carnot engine, and both streams leave the engine at 1 bar and T_D. Assume ideal gases that have a constant heat capacity $C_p = 29.3$ J/mol K.

(a) If stream A flows at 1 mol/s, what are the flow rates of streams B and C?
(b) What is T_D?
(c) What is the Carnot power output per mole of stream A?
(d) What is the entropy change of the overall process per mole of stream A?
(e) What is the entropy change of the Hilsch vortex tube (Hvt) per mole of stream A?
(f) What is the maximum power that one could obtain by any process per mole of stream A if all the heat were rejected or absorbed from an isothermal reservoir at T_D?

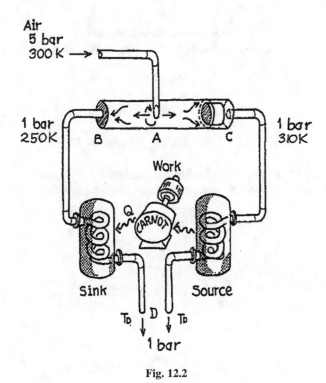

Fig. 12.2

12.3.1   Solution: Assumptions

- Ideal gases, constant C_p
- The Hilsch vortex tube operates adiabatically
- Steady-state operation
- The gases are well mixed in the Hilsch vortex tube

(a) If stream A flows at $\dot{n}_A = 1$ mol/s, what are the flow rates of streams B and C?

Choose the Hilsch vortex tube as the system (see Fig. 12.3):

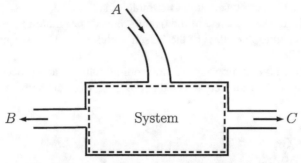

Boundary (dashed line): Open + Adiabatic + Rigid

Fig. 12.3

A First Law of Thermodynamics analysis of the system in Fig. 12.3 yields:

$$d\underline{U} = \delta Q + \delta W + H_A \delta n_A - H_B \delta n_B - H_C \delta n_C \qquad (12.24)$$

Dividing Eq. (12.24) by δt, and using the fact that $d\underline{U} = 0$ (Steady state), $\delta Q = 0$ (Adiabatic boundary), and $\delta W = 0$ (Rigid boundary), we obtain:

$$H_A \dot{n}_A - H_B \dot{n}_B - H_C \dot{n}_C = 0 \qquad (12.25)$$

where

$$\frac{\delta n_i}{\delta t} = \dot{n}_i \ (i = A, B, C) \qquad (12.26)$$

and

$$\frac{dN}{dt} = \frac{\delta n_A}{\delta t} - \frac{\delta n_B}{\delta t} - \frac{\delta n_C}{\delta t} = 0 \text{ (Steady state)} \qquad (12.27)$$

Combining Eq. (12.27) with Eq. (12.26) for $i = A$, B, and C yields:

$$\dot{n}_A = \dot{n}_B + \dot{n}_C \qquad (12.28)$$

Combining Eqs. (12.25) and (12.28) yields:

$$(H_A - H_B)\dot{n}_B = (H_C - H_A)\dot{n}_C \qquad (12.29)$$

Because the gases are ideal, it follows that:

$$H_A - H_B = C_p(T_A - T_B) \tag{12.30}$$

and

$$H_C - H_A = C_p(T_C - T_A) \tag{12.31}$$

Therefore, combining Eqs. (12.29), (12.30), and (12.31) yields:

$$C_p(T_A - T_B)\dot{n}_B = C_p(T_C - T_A)\dot{n}_C \tag{12.32}$$

or

$$\dot{n}_B = \left(\frac{T_C - T_A}{T_A - T_B}\right)\dot{n}_C \tag{12.33}$$

As per the Problem Statement, we have:

$$T_A = 300 \text{ K}$$
$$T_B = 250 \text{ K}$$
$$T_C = 310 \text{ K}$$

Using these three temperature values in Eq. (12.33) yields:

$$\therefore \dot{n}_B = \frac{1}{5}\dot{n}_C \tag{12.34}$$

At steady state, we know that:

$$\dot{n}_B + \dot{n}_C = \dot{n}_A = 1 \text{ mol/sec} \tag{12.35}$$

Solving Eqs. (12.34) and (12.35) yields:

$$\dot{n}_B = \frac{1}{6}\dot{n}_A = 0.167 \text{ mol/sec} \quad (a1) \tag{12.36}$$

and

$$\dot{n}_B = \frac{5}{6}\dot{n}_A = 0.833 \text{ mol/sec} \quad (a2) \tag{12.37}$$

(b) What is T_D?

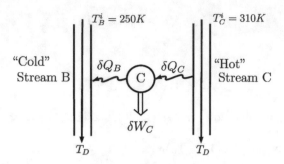

Fig. 12.4

Consider the reversible Carnot engine. It absorbs heat from the "Hot" stream C and rejects heat to the "Cold" stream B at temperatures which are changing (see Fig. 12.4).

For the reversible Carnot engine in Fig. 12.4, we have:

$$\frac{\delta Q_B}{T_B} + \frac{\delta Q_C}{T_C} = 0 \tag{12.38}$$

Dividing Eq. (12.38) by δt, we obtain:

$$\frac{(\delta Q_B/\delta t)}{T_B} + \frac{(\delta Q_C/\delta t)}{T_C} = 0 \tag{12.39}$$

or

$$\frac{\dot{Q}_B}{T_B} + \frac{\dot{Q}_C}{T_C} = 0 \tag{12.40}$$

Because the flows in streams B and C are isobaric, we can express $\dot{Q}$ in terms of $d\underline{H}$ as follows:

$$\delta Q_{St} = d\underline{H}_{St} = \delta n\, C_p dT \tag{12.41}$$

where the subscript St denotes stream. Dividing Eq. (12.41) by δt, we obtain:

$$(\delta Q_{St}/\delta t) = \dot{Q}_{St} = (\delta n/\delta t)\, C_p dT \tag{12.42}$$

or

$$\dot{Q}_{St} = \dot{n}\, C_p dT \text{ (Isobaric flow)} \tag{12.43}$$

However, $\dot{Q}_{Engine} = -\dot{Q}_{Stream}$, and therefore:

$$\dot{Q}_C = -\dot{n}_C C_p dT \tag{12.44}$$

and

$$\dot{Q}_B = -\dot{n}_B C_p dT \tag{12.45}$$

Using Eqs. (12.44) and (12.45) in Eq. (12.40), we obtain:

$$-\dot{n}_B \frac{dT_B}{T_B} - \dot{n}_C \frac{dT_C}{T_C} = 0 \tag{12.46}$$

Integrating Eq. (12.46) from $T_B^i \rightarrow T_D$ and from $T_C^i \rightarrow T_D$ yields:

$$-\dot{n}_B \int_{T_B^i}^{T_D} \frac{dT_B}{T_B} - \dot{n}_C \int_{T_C^i}^{T_D} \frac{dT_C}{T_C} = 0 \tag{12.47}$$

where we expect that $T_C^i > T_D$ and $T_B^i < T_D$ (see Fig. 12.4). Carrying out the two integrations in Eq. (12.47), we obtain:

$$-\dot{n}_B \ln \left(\frac{T_D}{T_B^i} \right) + \dot{n}_C \ln \left(\frac{T_C^i}{T_D} \right) = 0 \tag{12.48}$$

In Eq. (12.48), the first term represents the heat rejected (<0) by the Carnot engine to stream B, and the second term represents the heat absorbed (>0) by the Carnot engine from stream C.

Defining $x = \dot{n}_B / \dot{n}_C$, and rearranging Eq. (12.48), we obtain:

$$T_D = \left[T_C^i \left(T_B^i \right)^x \right]^{\frac{1}{1+x}} \tag{12.49}$$

Using $T_C^i = 310$ K, $T_B^i = 250$ K, and $x = \dot{n}_B / \dot{n}_C = 1/5$ in Eq. (12.49) yields:

$$T_D = 299.1 \text{K (b)} \tag{12.50}$$

(c) What is the Carnot power output per mole of stream A?

Because no work is done in the Hilsch vortex tube, and no work is done by the flows after they exit the Hilsch vortex tube ($dP = 0$), we expect that:

$$W_{total} = W_C \tag{12.51}$$

where W_{total} (>0) is the total work done by the system, and W_C (>0) is the work done by the Carnot engine (see Fig. 12.4).

We can therefore choose a system consisting of the Carnot engine and the Hilsch vortex tube and compute the work done by this system, W_{total}, which should be equal to the Carnot work, W_C.

Alternatively, we can calculate W_C directly by choosing the Carnot engine as the system

Here, we will utilize the first approach to calculate W_C.

Figure 12.5 illustrates the system in question.

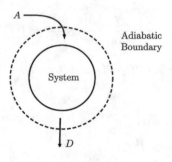

Fig. 12.5

At steady state, a First Law of Thermodynamics analysis of the system depicted in Fig. 12.5 yields:

$$\frac{dU}{dt} = 0 = \frac{\delta Q}{\delta t} - \frac{\delta W_{total}}{\delta t} + H_A \frac{\delta \dot{n}_A}{\delta t} - H_D \frac{\delta \dot{n}_D}{\delta t} \tag{12.52}$$

where $\delta Q = 0$ (Adiabatic boundary). Rearranging Eq. (12.52) yields:

$$\dot{W}_{total} = H_A \dot{n}_A - H_D \dot{n}_D \tag{12.53}$$

At steady state,

$$\dot{n}_A = \dot{n}_D \tag{12.54}$$

Combining Eqs. (12.53) and (12.54) yields:

$$\dot{W}_{total} = \dot{n}_A (H_A - H_D) = \dot{n}_A C_p (T_A - T_D) \tag{12.55}$$

Using the data provided in the Problem Statement in Eq. (12.55) yields:

$$\dot{W}_{total} = \dot{W}_C = 26.4 \text{ J/sec (c)} \tag{12.56}$$

(d) What is the entropy change of the overall process per mole of stream A?

To answer this question, it is convenient to consider a small element of fluid flowing from the initial condition of (5 bar, 300 K) to the final condition of (1 bar, 299.1 K) along a reversible path. We can then use the Combined First and Second Law of Thermodynamics for a closed system (the element of fluid). Specifically,

$$d\underline{S} = \frac{d\underline{U}}{T} + \frac{P}{T} d\underline{V} \tag{12.57}$$

Because δn is constant, Eq. (12.57) can be expressed as follows:

$$dS = \frac{dU}{T} + \frac{P}{T} dV \tag{12.58}$$

For an ideal gas:

$$\frac{P}{T} dV = R \frac{dT}{T} - \frac{R}{P} dP, \text{ and } dU = C_v dT \tag{12.59}$$

Using the results in Eq. (12.59) in the dS relation in Eq. (12.58), and rearranging, we obtain:

$$dS = (C_v + R) \frac{dT}{T} - R \frac{dP}{P} \tag{12.60}$$

or

$$dS = C_p \frac{dT}{T} - R \frac{dP}{P} \tag{12.61}$$

Integrating Eq. (12.61) from $(T_A, P_A) \rightarrow (T_D, P_D)$ yields:

$$\Delta S_{A \rightarrow D} = C_p \ln\left(\frac{T_D}{T_A}\right) - R \ln\left(\frac{P_D}{P_A}\right) \tag{12.62}$$

Using the data provided in the Problem Statement in Eq. (12.62) yields:

$$\Delta S_{A \rightarrow D} - 13.3 \text{ J/mol K (d)} \tag{12.63}$$

Note that because $\Delta S_R = 0$, it follows that:

$$\Delta S_U = \Delta S_{A \rightarrow D} + \Delta S_R = 13.3 \text{ J/mol K (Irreversible process)} \tag{12.64}$$

(e) What is the entropy change of the Hilsch vortex tube (Hvt) per mole of stream A?

Because $\Delta S_C = 0$, it follows that:

$$\Delta S_{Hvt} = \Delta S_{A \rightarrow D} = 13.3 \, \text{J/mol K (e)} \tag{12.65}$$

We can obtain the result in (e) above in an alternative manner by passing $\dot{n}_B$ moles of stream B from (T_A, P_A) to (T_B, P_B) reversibly, and $\dot{n}_C$ moles of stream C from (T_A, P_A) to (T_C, P_C) reversibly, calculating the entropy changes associated with each stream per mole of stream A and then adding up these two changes. This yields:

$$\Delta S_B = \frac{\dot{n}_B}{\dot{n}_A} \left\{ C_p \ln\left(\frac{T_B}{T_A}\right) - R\ln\left(\frac{P_B}{P_A}\right) \right\} = 1.35 \text{J/mol} K \tag{12.66}$$

and

$$\Delta S_C = \frac{\dot{n}_C}{\dot{n}_A} \left\{ C_p \ln\left(\frac{T_C}{T_A}\right) - R\ln\left(\frac{P_C}{P_A}\right) \right\} = 11.95 \text{J/mol} K \tag{12.67}$$

Adding Eqs. (12.66) and (12.67), we obtain the desired result:

$$\Delta S_{Hvt} = \Delta S_B + \Delta S_C = 13.3 \text{ J/mol K} = \Delta S_{A \rightarrow D} \text{ (e)} \tag{12.68}$$

(f) What is the maximum power that one could obtain by any process per mole of stream A if all the heat were rejected or absorbed from an isothermal reservoir at T_D?

Figure 12.6 depicts the system of interest:

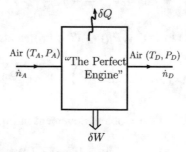

Fig. 12.6

where

$$\dot{n}_A = \dot{n}_D = 1 \, \text{mol/sec}$$

$$T_A = 300 \, \text{K}; \ T_D = 299.1 \, \text{K}$$

$$P_A = 5 \, \text{bar}; \ P_D = 1 \, \text{bar}$$

First, we carry out a First Law of Thermodynamics analysis of "The Perfect Engine":

$$d\underline{U} = 0 \text{ (Steady state)} = -\delta Q - \delta W + H_A \delta n_A - H_D \delta n_D \qquad (12.69)$$

where $\delta n_D = \delta n_A$, and therefore, Eq. (12.69) can be expressed as follows:

$$\delta W = (H_A - H_D)\delta n - \delta Q \qquad (12.70)$$

Next, for this reversible process, it is convenient to express δQ in terms of entropies. To this end, we carry out a Second Law of Thermodynamics entropy balance on "The Perfect Engine." This yields (see Fig. 12.6):

$$d\underline{S} = 0 \text{ (Steady state)} = -\frac{\delta Q}{T_D} + S_A \delta n_A - S_D \delta n_D \qquad (12.71)$$

Because, at steady state, $\delta n_A = \delta n_D = \delta n$, rearranging Eq. (12.71) yields:

$$\delta Q = T_D (S_A - S_D)\delta n \qquad (12.72)$$

where T_D is the temperature of the heat sink.

Using Eq. (12.72) for δQ in Eq. (12.70) for δW yields the following expression for the differential maximum work done by "the Perfect Engine":

$$\delta W_{max} = \{(H_A - T_D S_A) - (H_D - T_D S_D)\} \delta n \qquad (12.73)$$

Using the availability introduced in Lecture 11, we can express Eq. (12.73) as follows:

$$\delta W_{max} = -(B_D - B_A)\delta n = -\Delta B_{A \to D} \delta n \qquad (12.74)$$

Dividing Eq. (12.74) by δt, we obtain:

$$\frac{\delta W_{max}}{\delta t} = -\Delta B_{A \to D} \frac{\delta n}{\delta t} \qquad (12.75)$$

or

$$\delta \dot{W}_{max} = -\Delta B_{A \to D} \dot{n} \qquad (12.76)$$

where $\dot{W}_{max}$ is the maximum power produced by "The Perfect Engine" per mole of stream A.

Using the data provided in the Problem Statement in Eq. (12.76), we obtain:

$$\dot{W}_{max} = 4000 \text{ W (f)} \qquad (12.77)$$

Lecture 13

Fundamental Equations and Sample Problems

13.1 Introduction

The material presented in this lecture is adapted from Chapter 5 in T&M. First, we will motivate the need to calculate partial derivatives of thermodynamic properties. Second, we will discuss the internal energy and the entropy fundamental equations. Third, we will introduce the celebrated Theorem of Euler in the context of thermodynamics, including solving Sample Problems 13.1 and 13.2 to illustrate its use. We will also discuss a method that we refer to as "the Euler integration," including demonstrating its usefulness. Fourth, we will discuss how to systematically transform from one set of independent thermodynamic variables to another, each associated with a different fundamental equation, in a manner that ensures that the original and the new fundamental equations possess the same thermodynamic information content. Fifth, we will solve Sample Problems 13.3 and 13.4 to illustrate how to carry out these variable transformations. Finally, we will derive the celebrated Gibbs-Duhem equation which relates the differential changes in temperature, pressure, and n chemical potentials and shows that these (n + 2) intensive variables are not independent. This conclusion is consistent with the Corollary to Postulate I that we will discuss in Lecture 14.

13.2 Thermodynamic Relations for Simple Systems

In this lecture, and in Lecture 14, we will discuss material that is somewhat mathematical. Nevertheless, the results presented are essential for the calculation of changes in thermodynamic properties of systems which are not necessarily ideal. For example, suppose that we are given a one-component system (n = 1), whose extensive equilibrium thermodynamic state is characterized by the two independent intensive variables, T and V, plus the number of moles, N, initially charged,

© Springer Nature Switzerland AG 2020
D. Blankschtein, *Lectures in Classical Thermodynamics with an Introduction to Statistical Mechanics*, https://doi.org/10.1007/978-3-030-49198-7_13

consistent with Postulate I, first introduced in Lecture 2, which requires that $n + 2 = 1 + 2 = 3$.

We are then asked to calculate the change in the molar entropy of the system as it evolves from state 1 (characterized by T_1, V_1) to state 2 (characterized by T_2, V_2). Having chosen T and V as the two independent intensive variables, it follows that the molar entropy, S, can be expressed as $S = S(T,V)$. Therefore, the differential of S is given by:

$$dS = \left(\frac{\partial S}{\partial T}\right)_V dT + \left(\frac{\partial S}{\partial V}\right)_T dV \tag{13.1}$$

Because S in Eq. (13.1) is a function of state, we can use a convenient constant volume (isochoric) + constant temperature (isothermal) two-step path to carry out the integration from state 1 to state 2. The (V-T) phase diagram in Fig. 13.1 shows the real path and the two-step path involved in connecting state 1 and state 2:

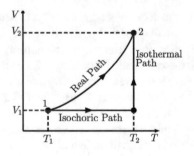

Fig. 13.1

Integrating Eq. (13.1) from state 1 to state 2 along the isochoric-isothermal path shown in Fig. 13.1 yields:

$$\Delta S_{1\rightarrow 2} = \int_1^2 dS = \int_{T_1}^{T_2} \left(\frac{\partial S}{\partial T}\right)_{V|V_1} dT + \int_{V_1}^{V_2} \left(\frac{\partial S}{\partial V}\right)_{T|T_2} dV \tag{13.2}$$

In Eq. (13.2), the notation $|V_1$ indicates that the temperature partial derivative of the molar entropy is evaluated at $V = V_1$, and the notation $|T_2$ indicates that the molar volume partial derivative of the molar entropy is evaluated at $T = T_2$.

In order to carry out the temperature integration in Eq. (13.2), we need to know the partial derivative, $(\partial S/\partial T)_V$. In order to carry out the molar volume integration in Eq. (13.2), we need to know the partial derivative, $(\partial S/\partial V)_T$. We will soon show that $(\partial S/\partial T)_V = C_v/T$ and that $(\partial S/\partial V)_T = (\partial P/\partial T)_V$.

With the above in mind, in this lecture, we will discuss various methods to calculate the required partial derivatives. To this end, we will relate partial derivatives to other partial derivatives or to quantities that can be evaluated in terms of measurable properties. As stated above and as we will show, $(\partial S/\partial T)_V$ can be determined using the heat capacity at constant volume, C_V. In addition, we will show that $(\partial S/\partial V)_T$ can be determined using an appropriate equation of state, $P = f(T, V)$.

13.3 Fundamental Equation

We have seen that the properties $\{\underline{S}, \underline{V}, N_1, N_2, \ldots, N_n\}$ form a set of $(n + 2)$ natural independent extensive variables for the internal energy, $\underline{U}$. This is consistent with Postulate I first introduced in Lecture 2. Specifically,

$$\underline{U} = \underline{U}\,(\underline{S}, \underline{V}, N_1, N_2, \ldots, N_n) \tag{13.3}$$

Equation (13.3) is referred to as the Internal Energy Fundamental Equation (FE).

By solving the $\underline{U}$ expression in Eq. (13.3) for $\underline{S}$, we can express $\underline{S}$ in terms of the $(n + 2)$ independent extensive variables $(\underline{U}, \underline{V}, N_1, N_2, \ldots, N_n)$, specifically,

$$\underline{S} = \underline{S}\,(\underline{U}, \underline{V}, N_1, N_2, \ldots, N_n) \tag{13.4}$$

Equation (13.4) is referred to as the Entropy Fundamental Equation (FE). Geometrically, the FE is a surface in $(n + 3)$ dimensional space. For example, the $\underline{U}$ surface as a function of the $(n + 2)$ independent extensive variables $(\underline{S}, \underline{V}, N_1, N_2, \ldots, N_n)$ is plotted schematically in Fig. 13.2:

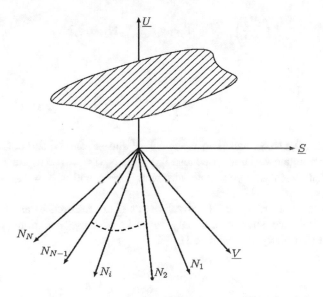

Fig. 13.2

The points on the (n + 3) FE surface represent stable equilibrium states of the simple system. Quasi-static processes can be represented by curves on the surface. Processes that are not quasi-static are not identified with curves on the surface.

Given $\underline{U} = \underline{U}\ (\underline{S}, \underline{V}, N_1, \ldots, N_n)$, the differential of $\underline{U}$ is given by:

$$dU = \left(\frac{\partial \underline{U}}{\partial \underline{S}}\right)_{\underline{V},N} d\underline{S} + \left(\frac{\partial \underline{U}}{\partial \underline{V}}\right)_{\underline{S},N} d\underline{V} + \sum_{i=1}^{n} \left(\frac{\partial \underline{U}}{\partial N_i}\right)_{\underline{S},\underline{V},N_{j[i]}} dN_i \qquad (13.5)$$

In Eq. (13.5), N is a shorthand notation for $\{N_1, \ldots, N_n\}$ are constant, and $N_{j[i]}$ is a shorthand notation for all N_js except N_i are constant.

In Lecture 11, we saw that the Combined First and Second Law of Thermodynamics for an open, simple, n-component system undergoing a reversible process is given by:

$$dU = TdS - PdV + \sum_{i=1}^{n} \mu_i dN_i \qquad (13.6)$$

Recall that Eq. (13.6) accounts solely for the reversible PdV-type work done on the system.

A comparison of Eqs. (13.5) and (13.6) for dU shows that:

$$\left(\frac{\partial \underline{U}}{\partial \underline{S}}\right)_{\underline{V},N} = T = g_T\ (\underline{S}, \underline{V}, N_1, \ldots, N_n) \qquad (13.7)$$

$$\left(\frac{\partial \underline{U}}{\partial \underline{V}}\right)_{\underline{S},N} = -P = g_P\ (\underline{S}, \underline{V}, N_1, \ldots, N_n) \qquad (13.8)$$

$$\left(\frac{\partial \underline{U}}{\partial N_i}\right)_{\underline{S},\underline{V},N_{j[i]}} = \mu_i = g_i\ (\underline{S}, \underline{V}, N_1, \ldots, N_n), i = 1, \ldots, n \qquad (13.9)$$

In Eqs. (13.7), (13.8), and (13.9), $\underline{S}$ and T, $\underline{V}$ and $-P$, and N_i and μ_i $(i = 1, \ldots, n)$ are (n + 2) conjugate variables. Because $\underline{S}$, $\underline{V}$, and N_i $(i = 1, \ldots, n)$ are all extensive, it follows that the (n + 2) conjugate variables, T, $-P$, and μ_i $(i = 1, \ldots, n)$, are all intensive.

The (n + 2) first-order partial derivatives of $\underline{U}$ are referred to as "Equations of State." We will see shortly that only (n + 1) of these equations of state are independent (Corollary to Postulate I).

The following results apply to an n-component system:

1. $\underline{U} = \underline{U}\ (\underline{S}, \underline{V}, N_1, \ldots, N_n), d\underline{U} = Td\underline{S} - Pd\underline{V} + \sum_{i=1}^{n} \mu_i dN_i$ (13.10)

where in the $d\underline{U}$ expression above, only $Pd\underline{V}$-type work is accounted for.

2. There are $(n + 1)$ independent first-order partial derivatives of the FE, $\underline{U}$. When $n = 1$, there are two: $(\partial \underline{U}/\partial \underline{S})_{\underline{V},N} = T$ and $(\partial \underline{U}/\partial \underline{V})_{\underline{S},N} = -P$. Note that, as we will show later, the first-order partial derivative of $\underline{U}$ with respect to N is given by $(\partial \underline{U}/\partial N)_{\underline{S},\underline{V}} = \mu\ (T,P)$ and, therefore, depends on the other two first-order partial derivatives of $\underline{U}$.

3. There are $(n + 1)(n + 2)/2$ independent second-order partial derivatives of $\underline{U}$. When $n = 1$, there are three: $(\partial^2 \underline{U}/\partial \underline{S}^2)_{\underline{V},N}$, $(\partial^2 \underline{U}/\partial \underline{V}^2)_{\underline{S},N}$, and $(\partial^2 \underline{U}/\partial \underline{S} \partial \underline{V})_N = (\partial^2 \underline{U}/\partial \underline{V} \partial \underline{S})_N$, because the order of differentiation in the mixed second-order partial derivative is immaterial.

The set of $(n + 1)$ independent first-order partial derivatives of $\underline{U}$, and the set of $(n + 1)(n + 2)/2$ independent second-order partial derivatives of $\underline{U}$, is very important, because it is possible to prove that any other partial derivative of $\underline{U}$ can be expressed in terms of this set of partial derivatives.

If the FEs ($\underline{U}$, $\underline{S}$, etc.) were known, they would allow us to obtain all the $(n + 2)$ equations of state by differentiation with respect to their independent variables. Unfortunately, FEs depend explicitly on the material that they describe. In other words, there is no single, universal FE governing the properties of all materials. As such, the calculation of a FE is outside the scope of Thermodynamics. Instead, given information about the molecules comprising a material, and the intermolecular forces operating in the material, we can use Statistical Mechanics to calculate a FE of interest. In Part III, we will use Statistical Mechanics to illustrate the calculation of several FEs.

Thermodynamics does not specify the complete form of the FE, but it imposes some restrictions on its functional form. We know that $\underline{U}$ and $\underline{S}$ are extensive properties which are first-order in mass or (moles). A well-known mathematical theorem due to Euler can be implemented in the context of Thermodynamics to obtain a very powerful result which we present next.

13.4   The Theorem of Euler in the Context of Thermodynamics (Adapted from Appendix C in T&M)

The Theorem of Euler applies to all smoothly varying, homogeneous functions f (a, b, ..., x, y, ...), where a, b, ... are intensive variables homogeneous to zero-order in mass (or moles), and x, y, ... are extensive variables homogeneous to first-order in mass (or moles). Further, df (total differential) can be integrated directly, where df is an exact differential, that is, it is not path dependent.

It then follows that if $Y = ky$ and $X = kx$, we obtain:

$$f(a, b, \ldots, X, Y, \ldots) = kf(a, b, \ldots, x, y, \ldots) \tag{13.11}$$

and

$$x\left(\partial f/\partial x\right)_{a,b,\ldots,y\ldots} + y\left(\partial f/\partial y\right)_{a,b,\ldots,x\ldots} + \ldots = (1)f(a, b, \ldots, x, y, \ldots) \tag{13.12}$$

In Eq. (13.12), only the extensive variables x, y, ... appear on the left-hand side of the equality.

13.5   Sample Problem 13.1

$\underline{U} = \underline{U}(\underline{S}, \underline{V}, N_1, \ldots, N_n)$ is homogeneous to first-order in all its $(n + 2)$ independent extensive variables $\{\underline{S}, \underline{V}, N_1, \ldots, N_n\}$. Apply the Theorem of Euler to $\underline{U}$.

13.5.1   Solution

According to the Theorem of Euler, it follows that:

$$\underline{U}(k\underline{S}, k\underline{V}, kN_1, \ldots, kN_n) = k\underline{U}(\underline{S}, \underline{V}, N_1, \ldots, N_n) \tag{13.13}$$

and

$$\left(\frac{\partial \underline{U}}{\partial \underline{S}}\right)_{\underline{V},N} \underline{S} + \left(\frac{\partial \underline{U}}{\partial \underline{V}}\right)_{\underline{S},N} \underline{V} + \sum_{i=1}^{n} \left(\frac{\partial \underline{U}}{\partial N_i}\right)_{\underline{S},\underline{V},N_{j[i]}} N_i = (1)\underline{U} \tag{13.14}$$

Note that because the $(n + 2)$ independent variables in $\underline{U}$ are all extensive, they all appear on the left-hand side of Eq. (13.14). In Eq. (13.14), the first-order partial derivatives multiplying $\underline{S}$, $\underline{V}$, and N_i are equal to $g_T = T$, $g_P = -P$, and $g_i = \mu_i$, respectively (see Eqs. (13.7), (13.8), and (13.9)). Therefore, Eq. (13.14) can be expressed as follows:

$$T\underline{S} - P\underline{V} + \sum_{i=1}^{n} \mu_i N_i = \underline{U} \tag{13.15}$$

The Theorem of Euler enables us to reconstruct $\underline{U}$, if we know the set of $(n + 2)$ intensive first-order partial derivatives $\{g_T, g_P, \text{ and } g_i = \mu_i \ (i = 1, \ldots, n)\}$. In fact, below, we will show that only $(n + 1)$ of these $(n + 2)$ intensive variables are independent.

As shown in Lecture 11, the Combined First and Second Law of Thermodynamics states that:

$$T d\underline{S} - P d\underline{V} + \sum_{i=1}^{n} \mu_i dN_i = d\underline{U} \qquad (13.16)$$

If we "Euler Integrate" (EI) the Combined First and Second Law of Thermodynamics in Eq. (13.16), that is, if we replace:

$$d \, (\text{Extensive Property}) \ EI \rightarrow \ \text{Extensive Property} \qquad (13.17)$$

and

$$d \, (\text{Intensive Property}) \ EI \rightarrow \ 0 \qquad (13.18)$$

we recover:

$$T\underline{S} - P\underline{V} + \sum_{i=1}^{n} \mu_i N_i = \underline{U} \qquad (13.19)$$

because all the differentials in Eq. (13.16) are of extensive properties.

The "Euler Integration" can be carried out on other differential forms of fundamental equations, as illustrated in the problem below.

13.6   Sample Problem 13.2

Apply the Theorem of Euler to the enthalpy, $\underline{H}$.

13.6.1   Solution

The enthalpy, $\underline{H}$, is given by:

$$\underline{H} = \underline{H} \, (\underline{S}, P, N_1, \ldots, N_n) \qquad (13.20)$$

Because the variable P in $\underline{H}$ is intensive, Eq. (13.20) shows that only $(n + 1)$ variables out of the $(n + 2)$ independent variables in $\underline{H}$ are extensive. Therefore, use of the Theorem of Euler yields:

$$\underline{H}(k\underline{S}, P, kN_1, \ldots, kN_n) = k\underline{H}(\underline{S}, P, N_1, \ldots, N_n) \tag{13.21}$$

and

$$\left(\frac{\partial \underline{H}}{\partial \underline{S}}\right)_{P,N} \underline{S} + \sum_{i=1}^{n} \left(\frac{\partial \underline{H}}{\partial N_i}\right)_{\underline{S},P,N_{j[i]}} N_i = (1)\,\underline{H} \tag{13.22}$$

In Eq. (13.22), the first-order partial derivatives multiplying $\underline{S}$ and N_i are equal to $g_T = T$ and $g_i = \mu_i$, respectively. Note that the intensive variable, P, does not appear on the left-hand side of Eq. (13.22).

13.7 Variable Transformations and New Fundamental Equations

From the Combined First and Second Law of Thermodynamics emerged naturally the FE, $\underline{U}$, whose natural variables $\{\underline{S}, \underline{V}, N_1, \ldots, N_n\}$ are all extensive. However, practically, $\underline{S}$ (the entropy) is not a convenient variable to work with. Indeed, we do not know how to precisely control the entropy $\underline{S}$, and no devices are available to measure entropy.

On the other hand, the variable conjugate to the entropy, $T = (\partial \underline{U}/\partial \underline{S})_{\underline{V},N}$, the temperature, is simple to control using a water bath and to measure using a thermometer. Therefore, for practical reasons, it is often convenient to work with other variables. When we change a variable to its conjugate variable, we introduce a new FE. Further, when we change variables, we need to ensure that the new FE obtained contains the same thermodynamic information content as the FE from which it was derived. Typically, Legendre transforms (LTs) are used to ensure that this is the case. However, here, we will not discuss LTs. Readers who are interested in learning more about LTs, please refer to Chapter 5 in T&M.

Instead, we will explore a simpler procedure that can be used to ensure that the thermodynamic information content of each FE is preserved. As a first example, consider the FE, $\underline{U}$ $\{\underline{S}, \underline{V}, N_1, \ldots, N_n\}$, and replace the variable, $\underline{V}$, by its conjugate variable, $-P$. We know that:

$$d\underline{U} = Td\underline{S} - Pd\underline{V} + \sum_{i=1}^{n} \mu_i dN_i \tag{13.23}$$

Let us add the quantity $d(P\underline{V})$ to both sides of the equality in Eq. (13.23) to create the new FE, $\underline{H}$:

$$d\underline{U} + d(P\underline{V}) = d(\underline{U} + P\underline{V}) = d\underline{H}$$

$$= Td\underline{S} - Pd\underline{V} + \sum_{i=1}^{n} \mu_i dN_i + Pd\underline{V} + \underline{V}dP \qquad (13.24)$$

Cancelling the two $Pd\underline{V}$ terms on the right-hand side of the equality in Eq. (13.24) yields:

$$d\underline{H} = Td\underline{S} + \underline{V}dP + \sum_{i=1}^{n} \mu_i dN_i \qquad (13.25)$$

If we EI Eq. (13.25), recalling that $\underline{H}$, $\underline{S}$, and N_i ($i = 1, \ldots, n$) are all extensive variables, and that P is an intensive variable, we obtain:

$$\underline{H} = T\underline{S} + 0 + \sum_{i=1}^{n} \mu_i N_i \qquad (13.26)$$

Subtracting and adding $P\underline{V}$ to the right-hand side of the equality in Eq. (13.26), and rearranging, we obtain:

$$\underline{H} = \left(T\underline{S} - P\underline{V} + \sum_{i=1}^{n} \mu_i dN_i \right) + P\underline{V} \qquad (13.27)$$

Recognizing that the expression in parenthesis in Eq. (13.27) is the internal energy, $\underline{U}$, it follows that:

$$\underline{H} = \underline{U} + P\underline{V} \qquad (13.28)$$

Recall that $\underline{H} = \underline{H}(\underline{S}, P, N_1, \ldots, N_n)$. Note also that, strictly, the intensive variable appearing in $\underline{H}$ in Eq. (13.28) should be $-P$, which is the conjugate variable of $\underline{V}$. However, for all practical purposes, not showing explicitly the $-$ sign is fine.

Taking the differential of $\underline{H} = \underline{H}(\underline{S}, P, N_1, \ldots, N_n)$, we obtain:

$$d\underline{H} = \left(\frac{\partial \underline{H}}{\partial \underline{S}} \right)_{P,N} d\underline{S} + \left(\frac{\partial \underline{H}}{\partial P} \right)_{\underline{S},N} dP + \sum_{i=1}^{n} \left(\frac{\partial \underline{H}}{\partial N_i} \right)_{\underline{S},P,N_{j[i]}} dN_i \qquad (13.29)$$

Comparing Eqs. (13.25) and (13.29) for d$\underline{H}$, we obtain:

$$\left(\frac{\partial \underline{H}}{\partial \underline{S}}\right)_{P,N} = T, \quad \left(\frac{\partial \underline{H}}{\partial P}\right)_{\underline{S},N} = \underline{V}, \quad \left(\frac{\partial \underline{H}}{\partial N_i}\right)_{\underline{S},P,N_{j[i]}} = \mu_i \ (i = 1, \ldots, n) \quad (13.30)$$

We also know that:

$$\left(\frac{\partial \underline{U}}{\partial \underline{S}}\right)_{\underline{V},N} = T, \quad \left(\frac{\partial \underline{U}}{\partial N_i}\right)_{\underline{S},\underline{V},N_{j[i]}} = \mu_i \ (i = 1, \ldots, n) \quad (13.31)$$

Because T, derived from $\underline{U}$ in Eq. (13.31) and from $\underline{H}$ in Eq. (13.30), are the same, and μ_i, derived from $\underline{U}$ in Eq. (13.31) and from $\underline{H}$ in Eq. (13.30), are the same, it follows that $\underline{U}$ ($\underline{S}$, $\underline{V}$, N_1, ..., N_n) and $\underline{H}$ ($\underline{S}$, P, N_1, ..., N_n) possess the same thermodynamic information content, although they depend on different sets of (n + 2) independent variables: {$\underline{S}$, $\underline{V}$, N_1, ..., N_n} for $\underline{U}$, and {$\underline{S}$, P, N_1, ..., N_n} for $\underline{H}$.

To further crystallize the material taught, below, we will solve Sample Problems 13.3 and 13.4.

13.8 Sample Problem 13.3

Transform $\underline{U}$ to $\underline{A}$ (Helmholtz Free Energy).

13.8.1 Solution

We first replace $\underline{S}$ in $\underline{U}(\underline{S}, \underline{V}, N_1, \ldots, N_n)$ by its conjugate variable:

$$T = \left(\frac{\partial \underline{U}}{\partial \underline{S}}\right)_{\underline{V},N} \quad (13.32)$$

When we do, we will find that:

$$\underline{A} = \underline{A}(T, \underline{V}, N_1, \ldots, N_n) \quad (13.33)$$

and

$$d\underline{A} = -\underline{S}dT - Pd\underline{V} + \sum_{i=1}^{n} \mu_i dN_i \quad (13.34)$$

13.9 Sample Problem 13.4

Transform $\underline{U}$ to $\underline{G}$ (Gibbs Free Energy).

13.9.1 Solution

We first replace $\underline{S}$ by its conjugate variable:

$$T = \left(\frac{\partial \underline{U}}{\partial \underline{S}}\right)_{\underline{V},N} \tag{13.35}$$

and then $\underline{V}$ by its conjugate variable:

$$-P = \left(\frac{\partial \underline{U}}{\partial \underline{V}}\right)_{\underline{S},N} \tag{13.36}$$

When we do, we will find that $\underline{G} = \underline{G}\,(T, P, N_1, \ldots, N_n)$ and that:

$$d\underline{G} = -\underline{S}dT + \underline{V}dP + \sum_{i=1}^{n} \mu_i dN_i \tag{13.37}$$

The transformation of variables presented above suggests the following "Rule of Thumb": To transform a variable to its conjugate variable in the differential form of a fundamental equation, we simply:

1. Flip the variable with its conjugate variable
2. Multiply the result by -1

For example:

$$T d\underline{S} \text{ in } d\underline{U} \rightarrow -\underline{S}dT \text{ in } d\underline{A}$$

$$-Pd\underline{V} \text{ in } d\underline{U} \rightarrow +\underline{V}dP \text{ in } d\underline{H} \text{ and } d\underline{G}$$

If we transform all the $(n + 2)$ original extensive variables $\{\underline{S}, \underline{V}, N_1, \ldots, N_n\}$ in the FE, $\underline{U}$, to their $(n + 2)$ conjugate intensive variables, that is,

$$T = \left(\frac{\partial \underline{U}}{\partial \underline{S}}\right)_{\underline{V},N}, \quad -P = \left(\frac{\partial \underline{U}}{\partial \underline{V}}\right)_{\underline{S},N}, \quad \mu_i = \left(\frac{\partial \underline{U}}{\partial N_i}\right)_{\underline{S},\underline{V},N_{j[i]}} \quad (i = 1, \ldots, n) \tag{13.38}$$

we obtain a very useful relation between the $(n + 2)$ intensive conjugate variables known as the Gibbs-Duhem Equation.

To derive the Gibbs-Duhem Equation, we begin with the differential form of the Combined First and Second Law of Thermodynamics, that is:

$$d\underline{U} = Td\underline{S} - Pd\underline{V} + \sum_{i=1}^{n} \mu_i dN_i \tag{13.39}$$

If we Euler Integrate Eq. (13.39), we obtain:

$$\underline{U} = T\underline{S} - P\underline{V} + \sum_{i=1}^{n} \mu_i N_i \tag{13.40}$$

If we then differentiate the Euler Integrated form of $\underline{U}$ in Eq. (13.40), we obtain:

$$d\underline{U} = Td\underline{S} + \underline{S}dT - Pd\underline{V} - \underline{V}dP + \sum_{i=1}^{n} \mu_i dN_i + \sum_{i=1}^{n} N_i d\mu_i \tag{13.41}$$

Because $d\underline{U}$ in Eq. (13.39) and $d\underline{U}$ in Eq. (13.41) are the same, it follows that:

$$Td\underline{S} - Pd\underline{V} + \sum_{i=1}^{n} \mu_i dN_i = Td\underline{S} + \underline{S}dT - Pd\underline{V} - \underline{V}dP + \sum_{i=1}^{n} \mu_i dN_i$$

$$+ \sum_{i=1}^{n} N_i d\mu_i \tag{13.42}$$

Cancelling the equal terms on both sides of the equality in Eq. (13.42), we obtain:

$$\underline{S}dT - \underline{V}dP + \sum_{i=1}^{n} N_i d\mu_i = 0 \ \text{(The Gibbs-Duhem Equation)} \tag{13.43}$$

The Gibbs-Duhem Equation (GDE) is a relation between the $(n + 2)$ intensive variables $\{T, P, \mu_1, \ldots, \mu_n\}$ and shows that only $(n + 1)$ of these are independent, a statement that was made earlier in this lecture, and that we have now proven.

Lecture 14

Manipulation of Partial Derivatives and Sample Problems

14.1 Introduction

The material presented in this lecture is adapted from Chapter 5 in T&M. Continuing with the material introduced in Lecture 13, first, we will discuss two additional restrictions which need to be imposed on the internal energy fundamental equation, $\underline{U}$, including presenting the Corollary to Postulate I. Second, we will discuss how to reconstruct the fundamental equation, $\underline{U}$, if we know the (n + 2) first-order partial derivatives of $\underline{U}$. Third, we will motivate the need to manipulate partial derivatives of thermodynamic functions in order to calculate changes in these thermodynamic functions when the system evolves from an initial state i to a final state f. To this end, we will discuss how to devise useful integration paths which connect state i with state f. Fourth, we will discuss various useful methods to relate an unknown partial derivative to other partial derivatives which can be calculated or determined experimentally. Fifth, among the useful methods to relate a desired partial derivative to others, we will discuss the triple product rule, the add another variable rule, the derivative inversion rule, and Maxwell's reciprocity rule. In addition, we will discuss Jacobian transformations. Finally, we will solve Sample Problems 14.1 and 14.2 to illustrate the use of some of the methods presented to calculate partial derivatives.

14.2 Two Additional Restrictions on the Internal Energy Fundamental Equation

In addition to the restrictions imposed by the Theorem of Euler on the form of the fundamental equation, $\underline{U}$, the following two restrictions need to be imposed:

(i) $\underline{U}$ is a single-valued function of its (n + 2) independent variables $\underline{S}$, $\underline{V}$, N_1, ..., N_n

(ii) $(\partial \underline{U}/\partial \underline{S})_{\underline{V},N} = T$ is nonnegative (≥ 0)

© Springer Nature Switzerland AG 2020
D. Blankschtein, *Lectures in Classical Thermodynamics with an Introduction to Statistical Mechanics*, https://doi.org/10.1007/978-3-030-49198-7_14

14.3   Corollary to Postulate I

"For a single-phase, simple system of n components, any intensive property can be defined by the values of any other (n + 1) independent intensive properties."

14.4   Reconstruction of the Internal Energy Fundamental Equation

We saw that given $\underline{U} = \underline{U}\,(\underline{S},\,\underline{V},\,N_1,\,\ldots,\,N_n)$, it follows that:

$$\left(\frac{\partial \underline{U}}{\partial \underline{S}}\right)_{\underline{V},N} = T = g_T\,(\underline{S},\underline{V},N_1,\,\ldots,N_n) \tag{14.1}$$

$$\left(\frac{\partial \underline{U}}{\partial \underline{V}}\right)_{\underline{S},N} = -P = g_P\,(\underline{S},\underline{V},N_1,\,\ldots,N_n) \tag{14.2}$$

$$\left(\frac{\partial \underline{U}}{\partial N_i}\right)_{\underline{S},\underline{V},N_{j[i]}} = \mu_i = g_i\,(\underline{S},\underline{V},N_1,\,\ldots,N_n),\ \ i = 1,2,\,\ldots,n \tag{14.3}$$

Alternatively, if we know the (n + 2) "Equations of State":

$$\{g_T = T, g_P = -P, g_1 = \mu_1,\,\ldots,g_n = \mu_n\} \tag{14.4}$$

we can reconstruct $\underline{U}$ using the Euler integrated form. Specifically,

$$\underline{U} = T\underline{S} - P\underline{V} + \sum_{i+1}^{n} \mu_i N_i \tag{14.5}$$

Because the (n + 2) variables, $g_T, g_P, g_1, \ldots, g_n$, are all intensive, according to the Corollary to Postulate I, only (n + 1) of them are independent. Therefore, given (n + 1) of them, we can determine the remaining one to within an arbitrary constant of integration. Accordingly, only (n + 1) "Equations of State" are needed to determine the Fundamental Equation, $\underline{U}$, to within an arbitrary constant of integration. Because only $\Delta \underline{U}$ has physical significance, this constant of integration is not important.

14.5   Manipulation of Partial Derivatives of Thermodynamic Functions

Suppose that we want to calculate $\Delta S_{1 \to 2}$ for a one-component ($n = 1$) system and that we choose T and P as the $n + 1 = 1 + 1 = 2$ independent intensive variables. Consistent with the Corollary to Postulate I, it then follows that another intensive variable, for example, the molar entropy, S, can be uniquely expressed as a function of T and P, that is,

$$S = S\,(T, P) \tag{14.6}$$

The differential of S in Eq. (14.6) is given by:

$$dS = \left(\frac{\partial S}{\partial T}\right)_P dT + \left(\frac{\partial S}{\partial P}\right)_T dP \tag{14.7}$$

Integrating Eq. (14.7) from state 1 to state 2 yields:

$$\Delta S_{1 \to 2} = \int_1^2 dS_{1 \to 2} \tag{14.8}$$

In order to carry out the integration in Eq. (14.8), it is useful to consider the schematic (P-T) phase diagram in Fig. 14.1.

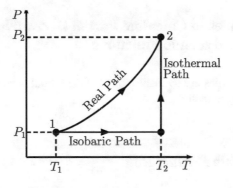

Fig. 14.1

Because S is a function of state, it is convenient to choose an isobaric (constant pressure)-isothermal (constant temperature) two-step path to go from state 1 to state 2 in order to carry out the integration (see Fig. 14.1). Specifically,

$$\Delta S_{1 \to 2} = \int_{T_1}^{T_2} \left(\frac{\partial S}{\partial T}\right)_{P|P_1} dT + \int_{P_1}^{P_2} \left(\frac{\partial S}{\partial P}\right)_{T|T_2} dP \tag{14.9}$$

To carry out the temperature integration in Eq. (14.9), we need to know $(\partial S/\partial T)_P = C_P/T$. To carry out the pressure integration in Eq. (14.9), we need to know $(\partial S/\partial P)_T = -(\partial V/\partial T)_P$.

14.6   Internal Energy and Entropy Fundamental Equations

The Fundamental Equations $\underline{U}$ and $\underline{S}$ can be obtained via an Euler integration of the Combined First and Second Law of Thermodynamics. Specifically, $\underline{U}$ can be expressed as follows:

$$\underline{U} = f_u [\underline{S}, \underline{V}, N_1, \ldots, N_n] = T\underline{S} - P\underline{V} + \sum_{i=1}^{n} \mu_i N_i \tag{14.10}$$

Further, by solving Eq. (14.10) for $\underline{S}$, we obtain:

$$\underline{S} = f_s [\underline{U}, \underline{V}, N_1, \ldots, N_n] = \underline{U}/T + (P/T)\underline{V} - \sum_{i=1}^{n} (\mu_i/T)N_i \tag{14.11}$$

14.7   Useful Rules to Calculate Partial Derivatives of Thermodynamic Functions

A number of useful rules are available to calculate partial derivatives of thermodynamic functions. These include:

14.7.1   The Triple Product Rule

$$F(x, y) \Rightarrow \left(\frac{\partial F}{\partial x}\right)_y \left(\frac{\partial x}{\partial y}\right)_F \left(\frac{\partial y}{\partial F}\right)_x = -1 \tag{14.12}$$

Example : $H(T, P) \Rightarrow (\partial H/\partial T)_P (\partial T/\partial P)_H (\partial P/\partial H)_T = -1$
$$\tag{14.13}$$

14.7.2   The Add Another Variable Rule

$$(\partial F/\partial y)_x = \frac{(\partial F/\partial \phi)_x}{(\partial y/\partial \phi)_x} \tag{14.14}$$

Example : $\phi = T \Rightarrow (\partial S/\partial H)_P = \dfrac{(\partial S/\partial T)_P}{(\partial H/\partial T)_P} = \dfrac{C_p/T}{C_p} = \dfrac{1}{T}$ (14.15)

14.7.3   The Derivative Inversion Rule

$$(\partial F/\partial y)_x = 1/(\partial y/\partial F)_x \tag{14.16}$$

Example : $(\partial T/\partial S)_P = 1/(\partial S/\partial T)_P = T/C_P$ (14.17)

14.7.4   Maxwell's Reciprocity Rule

Given a function F (x, y):

$$dF = \left(\frac{\partial F}{\partial x}\right)_y dx + \left(\frac{\partial F}{\partial y}\right)_x dy \tag{14.18}$$

$$\left[\frac{\partial}{\partial y}\left(\frac{\partial F}{\partial x}\right)_y\right]_x = \left[\frac{\partial}{\partial x}\left(\frac{\partial F}{\partial y}\right)_x\right]_y \tag{14.19}$$

or

$$F_{yx} = F_{xy} \tag{14.20}$$

For a smoothly varying function, the order of differentiation is immaterial.

14.8 Sample Problem 14.1

Derivation of a Maxwell reciprocity relation starting with U(S, V).

14.8.1 Solution

For a pure material (n = 1):

$$U = U(S, V) \tag{14.21}$$

$$dU = TdS - PdV \tag{14.22}$$

$$\left(\frac{\partial U}{\partial S}\right)_V = T \tag{14.23}$$

$$\left(\frac{\partial U}{\partial V}\right)_S = -P \tag{14.24}$$

However,

$$\left[\frac{\partial}{\partial V}\left(\frac{\partial U}{\partial S}\right)_V\right]_S = \left[\frac{\partial}{\partial S}\left(\frac{\partial U}{\partial V}\right)_S\right]_V \tag{14.25}$$

or

$$\left(\frac{\partial T}{\partial V}\right)_S = -\left(\frac{\partial P}{\partial S}\right)_V \tag{14.26}$$

Equation (14.26) is an example of a Maxwell reciprocity relation.

14.9 Jacobian Transformations

Given f(x,y) and g(x,y), the Jacobian is defined in terms of the following determinant:

$$\text{Jacobian} = \frac{\partial(f,g)}{\partial(x,y)} \equiv \begin{vmatrix} \left(\dfrac{\partial f}{\partial x}\right)_y & \left(\dfrac{\partial f}{\partial y}\right)_x \\ \left(\dfrac{\partial g}{\partial x}\right)_y & \left(\dfrac{\partial g}{\partial y}\right)_x \end{vmatrix} \tag{14.27}$$

$$\text{Jacobian} = \left(\frac{\partial f}{\partial x}\right)_y \left(\frac{\partial g}{\partial y}\right)_x - \left(\frac{\partial f}{\partial y}\right)_x \left(\frac{\partial g}{\partial x}\right)_y \tag{14.28}$$

14.9.1 Properties of Jacobians

1. Transposition

$$\frac{\partial(f,g)}{\partial(x,y)} = -\frac{\partial(g,f)}{\partial(x,y)} \tag{14.29}$$

2. Inversion

$$\frac{\partial(f,g)}{\partial(x,y)} = \frac{1}{\dfrac{\partial(x,y)}{\partial(f,g)}} \tag{14.30}$$

3. Chain Rule Expansion

$$\frac{\partial(f,g)}{\partial(x,y)} = \frac{\partial(f,g)}{\partial(z,w)} \frac{\partial(z,w)}{\partial(x,y)} \tag{14.31}$$

A simplification occurs if we need to compute $(\partial f/\partial z)_g$, where $f = f(z,g)$. First, we recognize that:

$$\left(\frac{\partial f}{\partial z}\right)_g = \frac{\partial(f,g)}{\partial(z,g)} \tag{14.32}$$

Implementing Properties 3 and 2 above in Eq. (14.32), we obtain:

$$\left(\frac{\partial f}{\partial z}\right)_g = \frac{\partial(f,g)}{\partial(x,y)} \Big/ \frac{\partial(z,g)}{\partial(x,y)} \tag{14.33}$$

14.10 Sample Problem 14.2

Calculate $(\partial T/\partial P)_H$ for a one-component fluid.

14.10.1 Solution

Choosing T and P as the two independent intensive variables, it follows that $H = H(T,P)$. We can therefore use Eq. (14.33), with $f = T$, $g = H$, and $z = P$, that is,

$$\left(\frac{\partial T}{\partial P}\right)_H = \frac{\dfrac{\partial (T,H)}{\partial (x,y)}}{\dfrac{\partial (P,H)}{\partial (x,y)}} \tag{14.34}$$

If we substitute $x = T$ and $y = P$ in Eq. (14.34), we obtain:

$$\left(\frac{\partial T}{\partial P}\right)_H = \frac{\dfrac{\partial (T,H)}{\partial (T,P)}}{\dfrac{\partial (P,H)}{\partial (T,P)}} \tag{14.35}$$

Using Property 1 in Eq. (14.35), we obtain:

$$\left(\frac{\partial T}{\partial P}\right)_H = \frac{\left(\frac{\partial (H,T)}{\partial (P,T)}\right)}{-\left(\frac{\partial (H,P)}{\partial (T,P)}\right)} = \frac{\left(\frac{\partial H}{\partial P}\right)_T}{-\left(\frac{\partial H}{\partial T}\right)_P} = \frac{\left(\frac{\partial H}{\partial P}\right)_T}{-C_P} \tag{14.36}$$

Note that we could have utilized the triple product rule directly to obtain the result in Eq. (14.36).

Lecture 15

Properties of Pure Materials and Gibbs Free Energy Formulation

15.1 Introduction

The material presented in this lecture is adapted from Chapter 5 in T&M. In this lecture, we will consider pure materials (n = 1). First, we will discuss the Gibbs Free Energy Fundamental Equation, $\underline{G}$. Second, we will again derive the Gibbs-Duhem Equation in a complementary way. Third, we will relate the Gibbs Free Energy to other thermodynamic functions. Fourth, we will calculate the first-order and second-order partial derivatives of $\underline{G}$ with respect to its independent variables T, P, and N. In particular, we will show that the three independent second-order partial derivatives of $\underline{G}$ are related to three widely used fluid properties – the heat capacity at constant pressure, the isothermal compressibility, and the coefficient of thermal expansion. Finally, we will determine which data set has the same thermodynamic information content as the Fundamental Equation, $\underline{G}$.

15.2 Gibbs Free Energy Fundamental Equation

For a single phase, simple system of a pure material (n = 1), the independent variables for the Internal Energy Fundamental Equation (FE), $\underline{U}$, are $\{\underline{S}, \underline{V}, N\}$. From a practical point of view, it is often useful to transform $\underline{S}$ and $\underline{V}$ into their conjugate variables, $T = (\partial \underline{U}/\partial \underline{S})_{\underline{V},N}$ and $-P = (\partial \underline{U}/\partial \underline{V})_{\underline{S},N}$, respectively. The new variables $\{T, -P, N\}$ correspond to the new FE, $\underline{G} = \underline{G}(T, -P, N)$. Note that the $-$ sign in P is not needed unless we use Legendre transforms, which we will not discuss. Therefore, hereafter, we will write: $\underline{G} = \underline{G}(T, P, N)$.

Beginning with the Combined First and Second Law of Thermodynamics, $d\underline{U} = Td\underline{S} - Pd\underline{V} + \mu dN$, and following the "Rule of Thumb" that we discussed in Lecture 13, we obtain:

© Springer Nature Switzerland AG 2020
D. Blankschtein, *Lectures in Classical Thermodynamics with an Introduction to Statistical Mechanics*, https://doi.org/10.1007/978-3-030-49198-7_15

$$+Td\underline{S} \text{ in } d\underline{U} \rightarrow -\underline{S}dT \text{ in } d\underline{G} \tag{15.1}$$

and

$$-Pd\underline{V} \text{ in } d\underline{U} \rightarrow +\underline{V}dP \text{ in } d\underline{G} \tag{15.2}$$

It then follows that:

$$d\underline{G} = -\underline{S}dT + \underline{V}dP + \mu dN \tag{15.3}$$

Because $\underline{G}$ and N are extensive variables, and T and P are intensive variables, if we Euler integrate Eq. (15.3), we obtain:

$$\underline{G} = 0 + 0 + \mu N \tag{15.4}$$

or

$$\underline{G} = \mu N \tag{15.5}$$

Key Result :   For $n = 1 \Rightarrow \mu = \underline{G}/N = G$ \hfill (15.6)

Equation (15.6) shows that, for a pure material ($n = 1$), the chemical potential, μ, is equal to the molar Gibbs Free Energy, $G = \underline{G}/N$. As we will see in Part II, knowing the chemical potential, μ, is essential to solve phase equilibria and chemical reaction equilibria problems.

15.3   Derivation of the Gibbs-Duhem Equation

Recalling that $\underline{S} = NS$, $\underline{V} = NV$, and $\underline{G} = NG$, and using these relations in Eq. (15.3), we obtain:

$$d\underline{G} = d(NG) = d(N\mu) = Nd\mu + \mu dN = -\underline{S}dT + \underline{V}dP + \mu dN \tag{15.7}$$

Cancelling the equal terms in Eq. (15.7), and rearranging, yields:

$$Nd\mu = -NSdT + NVdP \tag{15.8}$$

Dividing Eq. (15.8) by N, we obtain:

$$d\mu = -SdT + VdP \tag{15.9}$$

Equation (15.9) is, in fact, the Gibbs-Duhem Equation (GDE) derived in Lecture 13 for $n = 1$, and clearly shows that $\mu = \mu(T, P)$ for a pure ($n = 1$) material. The GDE is also consistent with the Corollary to Postulate I which we presented in Lecture 14. Indeed, for $n = 1$, T and P are $(n + 1) = (1 + 1) = 2$ independent intensive variables on which the intensive variable, μ, depends.

15.4 Relating the Gibbs Free Energy to Other Thermodynamic Functions

We also know that:

$$d\underline{U} = Td\underline{S} - Pd\underline{V} + \mu dN \ \underrightarrow{EI} \ \underline{U} = T\underline{S} - P\underline{V} + \mu N \tag{15.10}$$

Using $\mu N = \underline{G}$ in the expression for $\underline{U}$ in Eq. (15.10), we obtain:

$$\underline{U} = T\underline{S} - P\underline{V} + \underline{G} \tag{15.11}$$

or

$$\underline{G} = \underline{U} + P\underline{V} - T\underline{S} \tag{15.12}$$

or

$$\underline{G} = (\underline{U} + P\underline{V}) - T\underline{S} = \underline{H} - T\underline{S} \tag{15.13}$$

or

$$\underline{G} = (\underline{U} - T\underline{S}) + P\underline{V} = A + P\underline{V} \tag{15.14}$$

15.5 First-Order and Second-Order Partial Derivatives of the Gibbs Free Energy

Similar to $\underline{U} = \underline{U}(\underline{S}, \underline{V}, N)$, for $n = 1$, the FE, $\underline{G} = \underline{G}(T, P, N)$, for $n = 1$, also has three first-order partial derivatives or "Equations of State" (EOS). Specifically, beginning with:

$$d\underline{G} = -\underline{S}dT + \underline{V}dP + \mu dN \tag{15.15}$$

it follows that:

- $(\partial\underline{G}/\partial T)_{P,N} = -\underline{S} = \underline{g}_1(T,P,N) = Ng_1(T,P) \rightarrow -SN = Ng_1(T,P)$
$$\rightarrow -S = g_1(T,P) \tag{15.16}$$

- $(\partial\underline{G}/\partial P)_{T,N} = \underline{V} = \underline{g}_2(T,P,N) = Ng_2(T,P) \rightarrow VN = Ng_2(T,P)$
$$\rightarrow V = g_2(T,P) \tag{15.17}$$

- $(\partial\underline{G}/\partial N)_{T,P} = G = \mu = g_3(T,P) \rightarrow \mu = g_3(T,P) \tag{15.18}$

Note that the set of $(n + 2) = (1 + 2) = 3$ "EOS," $\{g_1(T,P), g_2(T,P), \text{ and } g_3(T,P)\}$, which are all intensive, is not independent. As discussed earlier, only $(n + 1) = (1 + 1) = 2$ are independent. This is fully consistent with the Corollary to Postulate I that we presented in Lecture 14. We will utilize this important result below.

Similar statements to those made earlier about the second-order partial derivatives of $\underline{U}$ can be made about the second-order partial derivatives of $\underline{G}$. In particular, there are $[(n + 2)(n + 1)]/2$ independent second-order partial derivatives of $\underline{G}$. Specifically, for $n = 1$, there are $[(1 + 2)(1 + 1)]/2 = 3$, given by:

(1)
$$\left(\partial^2\underline{G}/\partial T^2\right)_{P,N} \Rightarrow -(\partial\underline{S}/\partial T)_{P,N} = \underline{g}_{11}(T,P,N) \tag{15.19}$$

or

$$-N(\partial S/\partial T)_P = Ng_{11}(T,P) \tag{15.20}$$

or

$$-(\partial S/\partial T)_P = g_{11}(T,P) \tag{15.21}$$

(2)
$$\left(\partial^2\underline{G}/\partial P^2\right)_{T,N} \Rightarrow (\partial\underline{V}/\partial P)_{T,N} = \underline{g}_{22}(T,P,N) \tag{15.22}$$

or

$$N(\partial V/\partial P)_T = Ng_{22}(T,P) \tag{15.23}$$

or

$$(\partial V/\partial P)_T = g_{22}(T,P) \tag{15.24}$$

and

(3)
$$\left(\partial^2 \underline{G}/\partial T \partial P\right)_N \Rightarrow -(\partial \underline{S}/\partial P)_{T,N} = \left(\partial^2 \underline{G}/\partial P \partial T\right)_N \tag{15.25}$$

$$= (\partial \underline{V}/\partial T)_{P,N} = \underline{g}_{12}(T,P,N) = Ng_{12}(T,P) = \underline{g}_{21}(T,P,N)$$
$$= Ng_{21}(T,P) \tag{15.26}$$

Accordingly,

$$N(\partial V/\partial T)_P = -N(\partial S/\partial P)_T = Ng_{12}(T,P) = Ng_{21}(T,P) \tag{15.27}$$

or

$$-(\partial S/\partial P)_T = (\partial V/\partial T)_P = g_{12}(T,P) = g_{21}(T,P) \tag{15.28}$$

There are three additional second-order partial derivatives of $\underline{G}$ involving differentiations with respect to N. However, these partial derivatives are either zero or redundant with the previous partial derivatives of $\underline{G}$. For example:

$$\left(\partial^2 \underline{G}/\partial N^2\right)_{T,P} = (\partial \mu/\partial N)_{T,P} = 0 \text{ (Recall that } \mu = \mu(T,P)) \tag{15.29}$$

For additional practice, interested readers may want to calculate the two mixed partial derivatives of $\underline{G}$ with respect to T and N, as well as with respect to P and N, that is, to calculate:

$$\left(\partial^2 \underline{G}/\partial T \partial N\right)_P \text{ and } \left(\partial^2 \underline{G}/\partial P \partial N\right)_T \tag{15.30}$$

The three independent, second-order partial derivatives of $\underline{G}$, that is, $-(\partial S/\partial T)_P$, $(\partial V/\partial P)_T$, and $(\partial V/\partial T)_P$, are related to the following three widely used fluid properties:

(1) $\qquad C_P = T(\partial S/\partial T)_P$, Heat capacity at constant pressure $\qquad$ (15.31)

(2) $\qquad \kappa_T = -\dfrac{1}{V}(\partial V/\partial P)_T$, Isothermal compressibility $\qquad$ (15.32)

(3) $\qquad \alpha_P = \dfrac{1}{V}(\partial V/\partial T)_P$, Coefficient of thermal expansion $\qquad$ (15.33)

Recall that:

(i)
$$(\partial S/\partial T)_P = -g_{11}(T, P) \qquad (15.34)$$

$$(\partial V/\partial P)_T = g_{22}(T, P) \qquad (15.35)$$

$$(\partial V/\partial T)_P = g_{12}(T, P) = g_{21}(T, P) \qquad (15.36)$$

where g_1, g_2, and g_3 are the three "EOS" associated with $\underline{G}$.

(ii) Knowing κ_T and α_P is equivalent to knowing the P-V-T volumetric equation of state of the pure (n = 1) fluid

(iii) C_P, κ_T, and α_P are experimentally accessible

(iv) To prove that $C_P = T(\partial S/\partial T)_P$, we begin from the basic definition $C_P = (\partial H/\partial T)_P$. Because H = H (S,P) and dH = TdS + VdP, it follows that $(\partial H/\partial T)_P = T(\partial S/\partial T)_P + 0 = C_P$

15.6 Determining Which Data Set Has the Same Thermodynamic Information Content as the Gibbs Free Energy Fundamental Equation

Clearly, because:

$$\underline{G}(T, P, N) = N\mu(T, P) = Ng_3(T, P) \qquad (15.37)$$

it is sufficient to know $g_3(T,P) = \mu(T,P)$ to reconstruct $\underline{G}$ when n = 1. However, to derive $\mu(T,P) = g_3(T,P)$ to within an arbitrary constant of integration, we need to know the two independent "EOS," $g_1(T,P)$ and $g_2(T,P)$. To show this more clearly, we recall that:

$$dG = d\mu = -SdT + VdP \text{ (The GDE for n = 1)} \qquad (15.38)$$

where S = S(T,P) and V = V(T,P).
    Recall that:

$$-S(T, P) = g_1(T, P) \qquad (15.39)$$

and

$$V(T, P) = g_2(T, P) \qquad (15.40)$$

Therefore,

$$d\mu = -S(T, P)\, dT + V(T, P) dP \qquad (15.41)$$

or

$$d\mu = g_1(T, P)\, dT + g_2(T, P) dP \qquad (15.42)$$

To integrate Eq. (15.42), we recall that $\mu = G$ is a function of state, and therefore, we can choose a convenient integration path connecting the initial state (i) to the final state (f). Specifically, in a (P-T) phase diagram, we can choose a two-step isobaric (constant pressure) $-$ isothermal (constant temperature) path to go from state i (Po,To) to state f (P,T). The (P-T) phase diagram in Fig. 15.1 shows the chosen two-step path.

Carrying out the integration of $d\mu$ in Eq. (15.42) from (To,Po) to (T,P) yields:

$$\int_{\mu o}^{\mu} d\mu = \mu - \mu o = \int_{To}^{T} g_1(T, P)|_{Po}\, dT + \int_{Po}^{P} g_2(T, P)|_{T}\, dP \qquad (15.43)$$

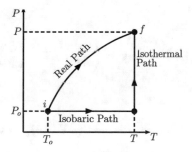

Fig. 15.1

where in Eq. (15.43) and Fig. 15.1, the temperature integration corresponds to the isobaric path, and the pressure integration corresponds to the isothermal path.
Equation (15.43) can be expressed as follows:

$$\mu(T,P) = G(T,P) = \mu o + \int_{To}^{T} g_1(T,P)|_{Po}\, dT + \int_{Po}^{P} g_2(T,P)|_{T}\, dP \qquad (15.44)$$

where μo is an arbitrary constant of integration. When calculating the difference between $\mu(T,P)$, or $G(T,P)$, in states 2 and 1, μo cancels out.

Therefore, to compute μ (or G) to within an arbitrary constant of integration, μo, we need to know the two "Equations of State," $g_1(T,P)$ and $g_2(T,P)$. Note that this is fully consistent with the Corollary to Postulate I that we discussed in Lecture 13.

Next, what do we do if we do not know $g_1(T,P)$ and $g_2(T,P)$? How can we nevertheless reconstruct the FE for G (or μ) in that case? To this end, we can first evaluate $g_1(T,P)$ and $g_2(T,P)$ from knowledge of C_P, κ_T, and α_P, or equivalently, from knowledge of C_P and the P-V-T volumetric equation of state. Recall that:

$$-g_1(T,P) = S(T,P) \Rightarrow dS = \left(\frac{\partial S}{\partial T}\right)_P dT + \left(\frac{\partial S}{\partial P}\right)_T dP \qquad (15.45)$$

where

$$(\partial S/\partial T)_P = C_P/T,\text{ and } (\partial S/\partial P)_T = -(\partial V/\partial T)_P = -V\alpha_p \qquad (15.46)$$

and

$$g_2(T,P) = V(T,P) \Rightarrow dV = \left(\frac{\partial V}{\partial T}\right)_P dT + \left(\frac{\partial V}{\partial P}\right)_T dP \qquad (15.47)$$

where

$$(\partial V/\partial T)_P = V\alpha_p,\text{ and } (\partial V/\partial P)_T = V\kappa_T \qquad (15.48)$$

Let us rewrite Eq. (15.45) for dS and Eq. (15.47) for dV in terms of C_P, κ_T, and α_P. Specifically,

$$dS = \left(\frac{C_P}{T}\right)dT - (V\alpha_P)dP \qquad (15.49)$$

and

$$dV = (V\alpha_P)dT - (V\kappa_T)dP \qquad (15.50)$$

To integrate Eqs. (15.49) and (15.50), we again utilize a convenient isobaric-isothermal two-step path. Specifically, we integrate from an initial state i at (Po,To), characterized by $S = So$ (arbitrary) and $V = Vo(To,Po)$ (not arbitrary), to a final state f at (P,T). The (P-T) phase diagram in Fig. 15.2 illustrates the various paths involved.

Integrating Eqs. (15.49) and (15.50) yields:

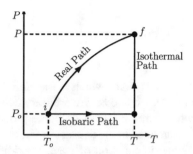

Fig. 15.2

$$S(T, P) = So + \int_{To}^{T} \left(\frac{C_P}{T}\right)\Big|_{Po} dT - \int_{Po}^{P} (V\alpha_P)\big|_T dP \qquad (15.51)$$

$$V(T, P) = Vo + \int_{To}^{T} (V\alpha_P)\big|_{Po} dT - \int_{Po}^{P} (V\kappa_T)\big|_T dP \qquad (15.52)$$

Note that in Eqs. (15.51) and (15.52), the temperature integrations correspond to the isobaric path, and the pressure integrations correspond to the isothermal path (see Fig. 15.2).

Because $d\mu = -S(T,P)dT + V(T,P)dP$, we can use the expressions for $S(T,P)$ in Eq. (15.51) and for $V(T,P)$ in Eq. (15.52) in the expression for $d\mu$ and then integrate from (Po,To) to (P,T) to calculate $\mu(T,P) = g_3(T,P)$. To this end, the integration will be carried out along the same isobaric-isothermal two-step path shown in Fig. 15.2. The result of the integration is presented below:

$$\mu(T, P) = \mu o$$
$$- \int_{To}^{T} \left[So + \int_{To}^{T} \left(\frac{C_P}{T}\right)\big|_{Po} dT - \int_{Po}^{\mu} (V\alpha_P)\big|_T dP \right]_{Po} dT \qquad (15.53)$$
$$+ \int_{Po}^{P} \left[Vo(To, Po) + \int_{To}^{T} (V\alpha_P)\big|_{Po} dT - \int_{Po}^{P} (V\kappa_T)\big|_T dP \right]_T dP$$

where in Eq. (15.53), μo and So are both arbitrary constants of integration, and the pressure integration between Po and P, when $P = Po$, is equal to zero.

Rearranging Eq. (15.53) for $\mu(T,P)$, we obtain:

$$\mu(T, P) = (\mu o - PoVo) - So(T - To) + VoP - \int\limits_{To}^{T} \left[\int\limits_{To}^{T} \left(\frac{C_P}{T}\right)\Big|_{Po} dT \right] \Big|_{Po} dP$$

$$+ \int\limits_{Po}^{P} \left[\int\limits_{To}^{T} (V\alpha_P)\Big|_{Po} dT - \int\limits_{Po}^{P} (V\kappa_T)\Big|_{T} dP \right] \Big|_{T} dP \qquad (15.54)$$

where in Eq. (15.54), $\mu o - PoVo = Ao$, which is the Helmholtz free energy at (To, Po). In addition, in Eq. (15.54), Ao and So are arbitrary constants, while Vo(To,Po) is not. Finally, the thermodynamic information content of the Fundamental Equation, $G = \mu$ (for n = 1), is contained in the data set of C_P, α_P, and κ_T, or alternatively, of C_P and the P-V-T volumetric equation of state, in addition to one value of the molar volume Vo in the reference state characterized by (Po,To). However, we can only determine $\Delta\mu = \Delta G$ to within the arbitrary constant of integration, So.

In other words:

$$\Delta G_{1\to2} = \Delta\mu_{1\to2} = -So(T_2 - T_1) + Vo(P_2 - P_1) + \text{Integrals} \qquad (15.55)$$

In the special case of an isothermal process, Eq. (15.38) shows that:

$$d\mu = dG = -SdT + VdP = VdP \quad (\text{At constant temperature}) \qquad (15.56)$$

Integrating Eq. (15.56) for dG from P_1 to P_2 at constant T yields:

$$\Delta G_{1\to2}\big|_T = \int\limits_{P_1}^{P_2} V\big|_T dP \qquad (15.57)$$

Equation (15.57) shows that $\Delta G_{1\to2}\big|_T$ does not depend on any arbitrary constant of integration and, as shown earlier in Part I, represents the negative of the shaft work associated with a flowing n = 1 fluid undergoing a reversible, isothermal process.

Lecture 16

Evaluation of Thermodynamic Data of Pure Materials and Sample Problems

16.1 Introduction

The material presented in this lecture is adapted from Chapter 8 in T&M. In this lecture, we will continue discussing pure materials (n = 1). First, we will present a summary of the differentials of S, U, and H, expressed in terms of two sets of (n + 1) = 2 intensive variables: (T, P) and (T, V). We will also consider the ideal gas limit and derive an expression relating the heat capacities at constant pressure and volume. Second, we will solve Sample Problem 16.1 to calculate the variation of the heat capacity at constant pressure with pressure at a given temperature. Third, we will solve Sample Problem 16.2 to calculate the variation of the heat capacity at constant volume with volume at a given temperature. Fourth, we will show how to calculate changes in thermodynamic properties given a mathematical expression relating P, T, and V (referred to as the equation of state, EOS) and different types of heat capacity data. In particular, we will present three strategies that can be used depending on the type of heat capacity data available to us. Fifth, we will present a three-step integration method, using either a (P-T) or a (V-T) phase diagram, which can be used when ideal gas heat capacity data is available (i.e., in the attenuated state, corresponding to the pressure approaching zero, or to the molar volume approaching infinity). Finally, we will demonstrate the use of this three-step integration method, referred to as the attenuated state approach, to calculate entropy changes when a pure material evolves from an initial state 1 to a final state 2, using either a (P-T) or a (V-T) phase diagram.

© Springer Nature Switzerland AG 2020
D. Blankschtein, *Lectures in Classical Thermodynamics with an Introduction to Statistical Mechanics*, https://doi.org/10.1007/978-3-030-49198-7_16

16.2 Summary of Changes in Entropy, Internal Energy, and Enthalpy

16.2.1 Using T and P as the Two Independent Intensive Variables

$$dS = \left(\frac{C_P}{T}\right) dT - \left(\frac{\partial V}{\partial T}\right)_P dP \tag{16.1}$$

$$dU = \left[C_P - P\left(\frac{\partial V}{\partial T}\right)_P\right] dT - \left[T\left(\frac{\partial V}{\partial T}\right)_P + P\left(\frac{\partial V}{\partial P}\right)_T\right] dP \tag{16.2}$$

$$dH = C_P \, dT + \left[V - T\left(\frac{\partial V}{\partial T}\right)_P\right] dP \tag{16.3}$$

16.2.2 Using T and V as the Two Independent Intensive Variables

$$dS = \left(\frac{C_V}{T}\right) dT + \left(\frac{\partial P}{\partial T}\right)_V dV \tag{16.4}$$

$$dU = C_V dT + \left[T\left(\frac{\partial P}{\partial T}\right)_V - P\right] dV \tag{16.5}$$

$$dH = \left[C_V + V\left(\frac{\partial P}{\partial T}\right)_V\right] dT + \left[T\left(\frac{\partial P}{\partial T}\right)_V + V\left(\frac{\partial P}{\partial V}\right)_T\right] dV \tag{16.6}$$

16.2.3 The Ideal Gas Limit

$$PV = RT \Rightarrow \left(\frac{\partial V}{\partial T}\right)_P = \frac{R}{P}, \quad \left(\frac{\partial V}{\partial P}\right)_T = -\frac{RT}{P^2} \tag{16.7}$$

$$PV = RT \Rightarrow \left(\frac{\partial P}{\partial T}\right)_V = \frac{R}{V}, \ \left(\frac{\partial P}{\partial V}\right)_T = -\frac{RT}{V^2} \qquad (16.8)$$

16.2.4   Relation between C_P and C_V

$$dS = \left(\frac{C_P}{T}\right)dT - \left(\frac{\partial V}{\partial T}\right)_P dP \qquad (16.9)$$

$$\left(\frac{\partial S}{\partial T}\right)_V = \frac{C_P}{T} - \left(\frac{\partial V}{\partial T}\right)_P \left(\frac{\partial P}{\partial T}\right)_V \qquad (16.10)$$

However, $\left(\frac{\partial S}{\partial T}\right)_V = \frac{C_V}{T}$. Using this result in Eq. (16.10), multiplying by T, and rearranging, yields:

$$C_P - C_V = T \left(\frac{\partial V}{\partial T}\right)_P \left(\frac{\partial P}{\partial T}\right)_V \qquad (16.11)$$

For an ideal gas (IG), Eqs. (16.7) and (16.8) show that:

$$\left(\frac{\partial V}{\partial T}\right)_P = \frac{R}{P}, \ \left(\frac{\partial P}{\partial T}\right)_V = \frac{R}{V} \qquad (16.12)$$

Using the results in Eq. (16.12) in Eq. (16.11) for an IG, we obtain:

$$(C_P - C_V)_{IG} = T \frac{R}{P} \frac{R}{V} = \frac{TR^2}{PV} = \frac{TR^2}{RT} = R \qquad (16.13)$$

16.3   Sample Problem 16.1

Calculate how C_P varies with P at constant T.

16.3.1   Solution

Imagine that C_P was measured as a function of temperature at a pressure, P_o, and that we would like to know $C_P(T,P)$ at a different pressure P, without having to carry out any additional measurements. For this purpose, if we could calculate:

$$\left(\frac{\partial C_P}{\partial P}\right)_T \qquad\qquad (16.14)$$

it would follow that:

$$C_P(T, P) = C_P(T, Po) + \int_{Po}^{P} \left(\frac{\partial C_P}{\partial P}\right)_T dP \qquad (16.15)$$

To compute the desired partial derivative of C_P with respect to P, at constant T, in the integrand in Eq. (16.15), we begin with:

$$C_P = T \left(\frac{\partial S}{\partial T}\right)_P \qquad\qquad (16.16)$$

Differentiating Eq. (16.16) with respect to P, at constant T, and using the fact that the order of differentiation is immaterial, we obtain:

$$\left(\frac{\partial C_P}{\partial P}\right)_T = T \frac{\partial}{\partial P}\left[\left(\frac{\partial S}{\partial T}\right)_P\right]_T = T \frac{\partial}{\partial T}\left[\left(\frac{\partial S}{\partial P}\right)_T\right]_P \qquad (16.17)$$

Equation (16.1), repeated below for completeness, shows that:

$$dS = \left(\frac{C_P}{T}\right)dT - \left(\frac{\partial V}{\partial T}\right)_P dP \Rightarrow \left(\frac{\partial S}{\partial P}\right)_T = -\left(\frac{\partial V}{\partial T}\right)_P \qquad (16.18)$$

Using the result on the right-hand side of the arrow in Eq. (16.18) in Eq. (16.17), we obtain:

$$\left(\frac{\partial C_P}{\partial P}\right)_T = -T\left(\frac{\partial^2 V}{\partial T^2}\right)_P \qquad\qquad (16.19)$$

Equation (16.19) shows that, for an ideal gas, for which $V = RT/P$, it follows that:

$$(\partial C_P/\partial P)_T = 0$$

16.4   Sample Problem 16.2

Calculate how C_V varies with V at constant T.

16.4.1 Solution

Given $C_V(T,Vo)$, in order to calculate $C_V(T,V)$ at another volume V, we begin with:

$$C_V(T, V) = C_V(T, Vo) + \int_{Vo}^{V} \left(\frac{\partial C_V}{\partial V}\right)_T dV \qquad (16.20)$$

We know that:

$$C_V = T \left(\frac{\partial S}{\partial T}\right)_V \qquad (16.21)$$

and therefore that:

$$\left(\frac{\partial C_V}{\partial V}\right)_T = T\frac{\partial}{\partial V}\left[\left(\frac{\partial S}{\partial T}\right)_V\right]_T = T\frac{\partial}{\partial T}\left[\left(\frac{\partial S}{\partial V}\right)_T\right]_V \qquad (16.22)$$

We also know that (see Eq. (16.4)):

$$dS = \left(\frac{C_V}{T}\right)dT + \left(\frac{\partial P}{\partial T}\right)_V dV \Rightarrow \left(\frac{\partial S}{\partial V}\right)_T = \left(\frac{\partial P}{\partial T}\right)_V \qquad (16.23)$$

Using the result on the right-hand side of the arrow in Eq. (16.23) in Eq. (16.22), we obtain:

$$\left(\frac{\partial C_V}{\partial V}\right)_T = T\left(\frac{\partial^2 P}{\partial T^2}\right)_V \qquad (16.24)$$

Equation (16.24) shows that, for an ideal gas, for which $P = RT/V$, it follows that:

$$(\partial C_V/\partial V)_T = 0$$

16.5 Evaluation of Changes in the Thermodynamic Properties of Pure Materials

16.5.1 Calculation of the Entropy Change

Choosing the $n + 1 = 2$ independent intensive variables, T and P, we know that:
$S = S(T, P)$, and that (see Eq. (16.1)):

$$dS = \left(\frac{C_P}{T}\right)dT - \left(\frac{\partial V}{\partial T}\right)_P dP \qquad (16.25)$$

16.5.2   Strategy I

If $C_P(T,P)$ is known, and the volumetric P-V-T equation of state (EOS) is known, as we did in Lecture 15, we can integrate Eq. (16.25) directly using an isobaric-isothermal two-step path as follows:

$$\Delta S_{1 \to 2} = \int_{T_1}^{T_2} \left(\frac{C_P}{T}\right)\bigg|_{P_1} dT - \int_{P_1}^{P_2} \left(\frac{\partial V}{\partial T}\right)_P \bigg|_{T_2} dP \qquad (16.26)$$

In Eq. (16.26), the temperature integration corresponds to the isobaric path at P_1, and the pressure integration corresponds to the isothermal path at T_2.

16.5.3   Strategy II

If $C_P(T,Po)$ is known, and the volumetric P-V-T EOS is known, we can first calculate $C_P(T,P)$ by integrating Eq. (16.19). Specifically,

$$C_P(T, P) = C_P(T, Po) - \int_{Po}^{P} T\left(\frac{\partial^2 V}{\partial T^2}\right)_P dP \qquad (16.27)$$

Following that, we can use Strategy I.

16.5.4   Strategy III

If we know the ideal gas heat capacity at constant pressure, $C_P^o = a^* + b^*T + c^*T^2 + \ldots$, (see Lecture 4) corresponding to $P \to P^* \to 0$, referred to as an attenuated state, as well as the volumetric P-V-T EOS, we can choose the following three-step path in a (P-T) phase diagram to evaluate $\Delta S_{1 \to 2}$ when the system evolves from an initial state 1 (at P_1, T_1) to a final state 2 (at P_2, T_2). The schematic (P-T) phase diagram in Fig. 16.1 illustrates the three-step path involved:

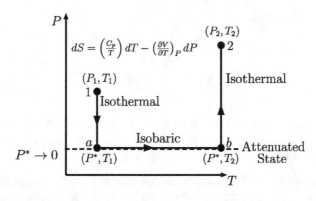

$$dS = \left(\frac{C_P}{T}\right) dT - \left(\frac{\partial V}{\partial T}\right)_P dP$$

Fig. 16.1

Following the three-step path shown in Fig. (16.1) yields:

$$\Delta S_{1\to2} = \Delta S_{1\to a} + \Delta S_{a\to b} + \Delta S_{b\to2} \tag{16.28}$$

Next, we will calculate each of the three entropy contributions on the right-hand side of the equal sign in Eq. (16.28).

1. Path 1→a: Isothermal

$$\Delta S_{1\to a} = - \int\limits_{P_1}^{P^*=0} \left(\frac{\partial V}{\partial T}\right)_P\bigg|_{T1} dP \tag{16.29}$$

In Eq. (16.29), when $P^* = 0$, $V = RT/P$, $(\partial V/\partial T)_P = R/P$, and there is a log $(P^* = 0)$ divergence. Further, the partial derivative in the integrand in Eq. (16.29) can be evaluated if a volumetric P-V-T EOS in available.

2. Path a→b: Isobaric

$$\Delta S_{a\to b} = \int\limits_{T_1}^{T_2} \left(\frac{C_P^o}{T}\right)\bigg|_{P^*} dT \tag{16.30}$$

where the temperature integration is only carried out in the attenuated state, and where C_P^o is known. Further, holding P^* constant in the C_P^o/T term in Eq. (16.30) is redundant, because C_P^o is independent of pressure.

3. Path b→2: Isothermal

$$\Delta S_{b\to 2} = -\int_{P^*=0}^{P_2} \left(\frac{\partial V}{\partial T}\right)_P \bigg|_{T_2} dP \tag{16.31}$$

In Eq. (16.31), when $P^* = 0$, $V = RT/P$, $(\partial V/\partial T)_P = R/P$, and there is a log ($P^* = 0$) divergence. Further, the partial derivative in the integrand in Eq. (16.31) can be evaluated if a volumetric P-V-T EOS is available.

4. How do we deal with the divergences that exist at $P^* = 0$ when carrying out the pressure integrals isothermally along path 1→a (see Eq. (16.19)) and path b→2 (see Eq. (16.31))? Because the system behaves ideally when $P^* = 0$, it is possible to express the singularity in analytical form as follows:

(a) In Eq. (16.29), we **add** and **subtract**:

$$\int_{P_1}^{P^*=0} \left(\frac{R}{P}\right) \bigg|_{T_1} dP \tag{16.32}$$

which yields:

$$\Delta S_{1\to a} = -\int_{P_1}^{P^*=0} \left[\left(\frac{\partial V}{\partial T}\right)_P - \frac{R}{P}\right]_{T_1} dP - R \ln\left(\frac{P^*}{P_1}\right) \tag{16.33}$$

In the limit when $P^* \to 0$:

$$V \to \frac{RT}{P} \Rightarrow \left(\frac{\partial V}{\partial T}\right)_P \to \frac{R}{P} \tag{16.34}$$

Using Eq. (16.34) in the integrand in Eq. (16.33), it follows that $\left[\left(\frac{\partial V}{\partial T}\right)_P - \frac{R}{P}\right] \to 0$ when $P^* \to 0$, thereby removing the singularity from the integral.

(b) In Eq. (16.31), we **add** and **subtract**:

$$\int_{P^*=0}^{P_2} \left(\frac{R}{P}\right) \bigg|_{T_2} dP \tag{16.35}$$

which yields:

$$\Delta S_{b\to 2} = - \int_{P^*=0}^{P_2} \left[\left(\frac{\partial V}{\partial T}\right)_P - \frac{R}{P}\right]_{T_2} dP - R\ln\left(\frac{P_2}{P^*}\right) \qquad (16.36)$$

Like in Eq. (16.33), the integrand in Eq. (16.36) $\to 0$ in the limit $P^*\to 0$ and the singularity is removed from the integral.

When we calculate $\Delta S_{1\to a} + \Delta S_{b\to 2}$ (see Eqs. (16.33) and (16.36)), the two ln $(P^*\to 0)$ contributions cancel out. Specifically, we obtain:

$$\Delta S_{1\to a} + \Delta S_{b\to 2} = - \int_{P_1}^{P^*=0} \left[\left(\frac{\partial V}{\partial T}\right)_P - \frac{R}{P}\right]_{T_1} dP - R\ln\left(\frac{P_2}{P_1}\right)$$

$$- \int_{P^*=0}^{P_2} \left[\left(\frac{\partial V}{\partial T}\right)_P - \frac{R}{P}\right]_{T_2} dP \qquad (16.37)$$

In Eq. (16.37), the first pressure integration corresponds to the real gas state to the ideal gas state transition at T_1, and the second pressure integration corresponds to the ideal gas state to the real gas state transition at T_2.

If the two independent intensive variables are V and T, we can use a (V-T) phase diagram to calculate $\Delta S_{1\to 2}$. In this case, the ideal gas attenuated state corresponds to $V^* = \infty$, and we can choose the convenient three-step path shown in Fig. 16.2:

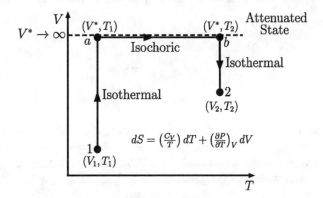

Fig. 16.2

In the attenuated state at $V^* = \infty$, the gas behaves ideally with a heat capacity at constant volume given by: $C_V^\circ = a + bT + cT^2 + \dots$ (see Lecture 4). If C_V° and the volumetric P-V-T EOS are known, we can calculate $\Delta S_{1\to 2}$ as follows (see Fig. 16.2):

$$\Delta S_{1 \to 2} = \Delta S_{1 \to a} + \Delta S_{a \to b} + \Delta S_{b \to 2} \tag{16.38}$$

Next, we calculate each of the three entropy contributions on the right-hand side of the equal sign in Eq. (16.38).

1. Path $1 \to a$: Isothermal

$$\Delta S_{1 \to a} = \int_{V_1}^{V^* = \infty} \left(\frac{\partial P}{\partial T} \right)_V \bigg|_{T_1} dV \tag{16.39}$$

In Eq. (16.39), when $V^* = \infty$, $P = \frac{RT}{V}$, $\left(\frac{\partial P}{\partial T} \right)_V = \frac{R}{V}$, and there is a log ($V^* = \infty$) divergence. In addition, the partial derivative in the integrand in Eq. (16.39) can be evaluated if a volumetric P-V-T EOS is available.

2. Path $a \to b$: Isochoric

$$\Delta S_{a \to b} = \int_{T_1}^{T_2} \left(\frac{C_V^o}{T} \right) \bigg|_{V^*} dV \tag{16.40}$$

The temperature integration in Eq. (16.40) is only carried out in the attenuated state, where C_V^o is known. In addition, holding V^* constant in the (C_V^o/T) term in Eq. (16.40) is redundant, because C_V^o is independent of the molar volume.

3. Path $b \to 2$: Isothermal

$$\Delta S_{b \to 2} = \int_{V^* = \infty}^{V_2} \left(\frac{\partial P}{\partial T} \right)_V \bigg|_{T_2} dV \tag{16.41}$$

In Eq. (16.41), when $V^* = \infty$, $P = RT/V$, $(\partial P/\partial T)_V = R/V$, and there is a log ($V^* = \infty$) divergence. In addition, the partial derivative in the integrand in Eq. (16.41) can be evaluated if a volumetric P-V-T EOS is available.

In order to deal with the divergences in $\Delta S_{1 \to a}$ and $\Delta S_{b \to 2}$ when $V^* = \infty$ (see Eqs. (16.39) and (16.40)), we **add** and **subtract**:

$$\int_{V_1}^{V^* = \infty} \left(\frac{R}{V} \right)_{T_1} dV \text{ for } \Delta S_{1 \to a} \tag{16.42}$$

and

$$\int_{V^*=\infty}^{V_2} \left(\frac{R}{V}\right)_{T_2} dV \text{ for } \Delta S_{b\to 2} \tag{16.43}$$

This yields:

$$\Delta S_{1\to a} + \Delta S_{b\to 2} = \int_{V_1}^{V^*=\infty} \left[\left(\frac{\partial P}{\partial T}\right)_V - \frac{R}{V}\right]_{T_1} dV + R\ln\left(\frac{V_2}{V_1}\right)$$

$$+ \int_{V^*=\infty}^{V_2} \left[\left(\frac{\partial P}{\partial T}\right)_V - \frac{R}{V}\right]_{T_2} dV \tag{16.44}$$

In Eq. (16.44), the first molar volume integration corresponds to the real gas state to the ideal gas state transition at T_1, and the second molar volume integration corresponds to the ideal gas state to the real gas state transition at T_2.

Lecture 17

Equations of State of a Pure Material, Binodal, Spinodal, Critical Point, and Sample Problem

17.1 Introduction

The material presented in this lecture is adapted from Chapters 8 and 7 in T&M. In this lecture, we will continue discussing pure materials ($n = 1$). First, we will discuss the mathematical relation between P, V, and T, referred to as the volumetric equation of state, or in short, the equation of state (EOS). In particular, we will discuss the ideal gas EOS and the van der Waals EOS, including providing an underlying molecular interpretation for both EOS. Second, we will solve Sample Problem 17.1 to calculate the excluded volume between two spheres of equal radius. Third, we will examine the various forms of the isotherms in a pressure (P)-volume (V) phase diagram at temperatures which are high, equal, or low relative to the critical temperature. Fourth, we will discuss the coexistence curve, the spinodal curve, and the critical point, including providing mathematical criteria to calculate them. Finally, we will discuss stability, metastability, and instability, including providing mathematical criteria to characterize these behaviors, as well as useful mechanical analogies to rationalize what each behavior entails.

17.2 Equations of State of a Pure Material

In Lecture 16, we saw that knowledge of the relation between P, V, and T is essential for the calculation of thermodynamic properties of a pure (one-component, $n = 1$) system. The P-V-T relation is known as the volumetric equation of state, or in short, the equation of state (EOS), and is available as a:

$$\text{Pressure} - \text{Explicit EOS} : P = f(T, V) \tag{17.1}$$

or

© Springer Nature Switzerland AG 2020
D. Blankschtein, *Lectures in Classical Thermodynamics with an Introduction to Statistical Mechanics*, https://doi.org/10.1007/978-3-030-49198-7_17

$$\text{Volume} - \text{Explicit EOS}: V = f(T, P) \tag{17.2}$$

Equations of state may be very simple, such as the ideal gas EOS which contains no parameters, or very complex, such as the Martin-Hou EOS which contains 21 parameters. The choice of EOS depends on (1) the desired accuracy and (2) the endurance of the user. The various parameters which appear in the EOS are evaluated by fitting the EOS to experimental P-V-T data. Therefore, an EOS can never be more accurate than the experimental P-V-T data that it describes.

17.3   Examples of Equations of State (EOS)

17.3.1   The Ideal Gas EOS

Applicable when P is low (~5 atm), or V is high, or $\rho = 1/V$ is low. The ideal gas EOS can be written as follows:

$$\text{Intensive Form}: PV = RT \tag{17.3}$$

and

$$\text{Extensive Form}: P\underline{V} = NRT \tag{17.4}$$

In Eqs. (17.3) and (17.4), the gas constant $R = 8.314$ J/mol K.

17.3.2   The van der Waals EOS

The ideal gas EOS assumes that:

 (i) Molecules have no volume, such that the entire system volume, $\underline{V}$, is available to each of the N molecules comprising the system

(ii) No interactions operate between the N molecules comprising the system

Assumptions (i) and (ii) result in Eqs. (17.4) and (17.3). van der Waals relaxed assumptions (i) and (ii) as follows:

(1) Every molecule excludes a volume, b, from every other molecule, such that the free volume available in the system is ($\underline{V}$-Nb)

17.4 Sample Problem 17.1

Evaluate the excluded volume between two hard spheres of equal radius, Ro, (Fig. 17.1).

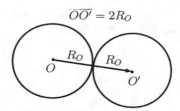

$$\overline{OO'} = 2R_O$$

Fig. 17.1

17.4.1 Solution

As Fig. 17.1 shows, the distance of closest approach between the centers, 0 *and* 0′, of the two hard spheres is equal to 2Ro. The volume excluded by the sphere of radius Ro, centered at 0, from the center of the second sphere of radius Ro, centered at 0′, is given by:

$$V_{excluded} = \frac{4\pi}{3}(2R_O)^3 = 8\left(\frac{4\pi}{3}R_O^3\right) \tag{17.5}$$

Note that: $b = 0.5 V_{excluded}$ (per sphere/molecule).

The assumption of additive excluded volumes is reasonable at low densities, where there is no overlap of excluded volumes (see Fig. 17.2):

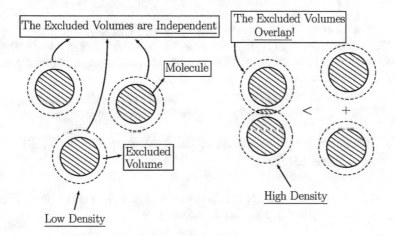

Fig. 17.2

However, as Fig. 17.2 shows, at high densities, the excluded volume calculated assuming that the excluded volumes are independent is overpredicted.

(2) Because liquids and solids exist in nature, van der Waals recognized the inevitable existence of attractive interactions between molecules, which he described as pairwise and proportional to the number density squared. If each pair of molecules experiences an attraction of strength, a, the resulting attractive contribution to the system pressure is given by:

$$a\rho^2 = a\left(\frac{N}{\underline{V}}\right)^2 \tag{17.6}$$

van der Waals (vdW) used assumption (1) to replace $\underline{V}$ in Eq. (17.4) by $(\underline{V}\text{-Nb})$, and assumption (2) to replace the pressure P in Eq. (17.4) by $(P + a\,(N/\underline{V})^2)$, and proposed the celebrated vdW EOS given by:

$$\left[P + a\left(\frac{N}{\underline{V}}\right)^2\right][\underline{V} - Nb] = NRT \tag{17.7}$$

Equation (17.7) can be expressed in the following two forms:

$$P = \frac{NRT}{\underline{V} - Nb} - \frac{aN^2}{\underline{V}^2} \text{ (Pressure-explicit extensive form of the vdW EOS)}$$

$$\tag{17.8}$$

$$P = \frac{RT}{V - b} - \frac{a}{V^2} \text{ (Pressure-explicit intensive form of the vdW EOS)} \tag{17.9}$$

Note that when $\underline{V} \to \infty$, Eq. (17.8) reduces to Eq. (17.4), and when $V \to \infty$, Eq. (17.9) reduces to Eq. (17.3). Finally, as we will show in Lecture 18, the two parameters, a and b, in the vdW EOS can be determined using the experimental critical point conditions.

17.5   Pressure-Explicit Form of the Isotherm P = f (V, T) of a Pure Material

Figure 17.3 shows schematically the family of isotherms in a (P-V) phase diagram for a pure (n = 1) material (e.g., carbon dioxide).

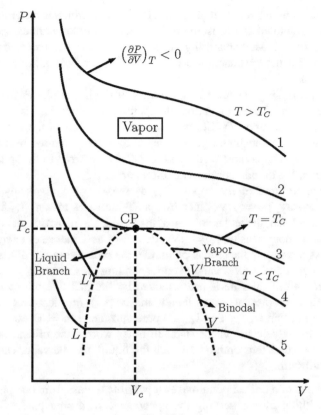

Fig. 17.3

The following observations can be made about Fig. 17.3:

(1) At sufficiently high temperatures, each isotherm is a continuous curve, where as V increases, P decreases. In other words, on isotherms such as 1 and 2, $(\partial P/\partial V)_T < 0$.

(2) At sufficiently low temperatures, each isotherm is discontinuous and consists of three sections. For example, in isotherm 5, the first section of the curve at high pressures corresponds to the liquid state, while the third section of the curve at low pressures corresponds to the vapor (gas) state.

(3) The two curves (sections) in (2) are joined by a horizontal line which corresponds to the simultaneous presence of two phases, liquid and vapor (gas), coexisting at thermodynamic equilibrium.

(4) Isotherm 3 corresponds to the transition between isotherms corresponding to the vapor phase only (like 1 and 2) and those which include a horizontal section and correspond to liquid-vapor equilibrium (like 5). In isotherm 3, the horizontal

section has contracted to a single point of inflection, which is the critical point of the system, denoted as CP. Isotherm 3 is referred to as the critical isotherm, for which $T = T_C$. Mathematically, therefore, the CP is characterized by the conditions for the existence of a point of inflection with a horizontal tangent (see below).

(5) The curve LL'CP gives the molar volumes of the liquid phases in equilibrium with the vapor phases, at various temperatures below T_C, as a function of P. Similarly, the curve VV'CP gives the molar volumes of the vapor phases in equilibrium with the liquid phases, at various temperatures below T_C, as a function of P. The complete curve LL'CPV'V is referred to as the L/V coexistence curve, the binodal, or the saturation curve.

(6) The molar volume of the vapor, V_V, decreases, or equivalently, the vapor number density, $\rho_V = 1/V_V$, increases as T increases toward T_C from below. On the other hand, the molar volume of the liquid, V_L, increases, or equivalently, the liquid number density, $\rho_L = 1/V_L$, decreases as T increases toward T_C from below. As a result, at the critical point, CP, the molar volumes of the vapor and the liquid are equal to V_C, and the corresponding pressure is P_C.

(7) In general, a critical state is characterized by the fact that the two coexisting phases (in the present case, liquid and vapor) are identical. Above the critical point, that is, for $T > T_C$, the pure substance ($n = 1$) can exist in true equilibrium only in the vapor state. In other words, the critical temperature, T_C, is the highest temperature at which the liquid and the vapor can coexist in true equilibrium.

The existence of a critical point makes it possible to pass from one physical state to another without ever observing the appearance of a new phase. Figure 17.4 illustrates this point.

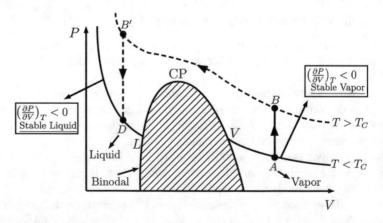

Fig. 17.4

Indeed, starting with vapor at A, increasing T, at constant V, along the isochoric path AB to $T > T_C$, then compressing the system while maintaining $T > T_C$ (T changes) from B to B', and finally decreasing T to below T_C, at constant V, from B' to D, we end with liquid at D. Therefore, we can pass in a continuous manner from the vapor state at A to the liquid state at D.

The continuity of state, shown in Fig. 17.4, which results from the existence of a critical point, indicates that, in a certain sense, the vapor and liquid states are two different manifestations of the same physical state (a fluid state). This was recognized by Kelvin who suggested that the segments DL and AV in Fig. 17.4, as well as in the (P-V) phase diagram in Fig. 17.5, are really parts of the single continuous curve DLMNVA.

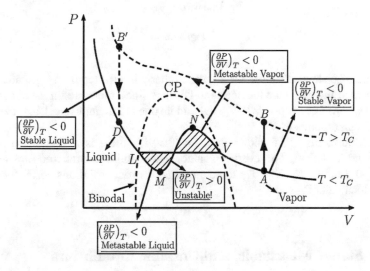

Fig. 17.5

The idea of a continuous isotherm for $T < T_C$, shown in Fig. 17.5, was subsequently picked up by van der Waals, and developed further into the celebrated vdW EOS, which is cubic in V. Specifically,

$$V^3 - \left(b + \frac{RT}{P}\right)V^2 + \left(\frac{a}{P}\right)V - \left(\frac{ab}{P}\right) = 0 \qquad (17.10)$$

The (P-V) phase diagram in Fig. 17.6 shows the form of the isotherm of the vdW EOS for $T < T_C$.

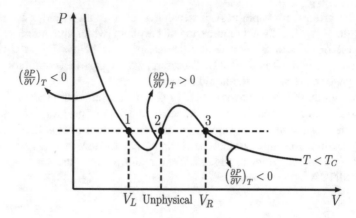

Fig. 17.6

Figure 17.6 shows that, for $T < T_C$, the isotherm has the required sigmoidal shape. At a given value of P (see the horizontal dashed line), there are three solutions: the smallest one at 1 corresponds to V_L, the middle one at 2 is unphysical, and the largest one at 3 corresponds to V_V.

In addition, Fig. 17.5 shows that, for $T > T_C$, there is a single real solution corresponding to V_V for every P value. The resulting isotherm satisfies $(\partial P/\partial V)_T < 0$. Finally, for $T = T_C$, there are three identical solutions corresponding to the CP located at P_C and V_C.

17.6   Stable, Metastable, and Unstable Equilibrium

As shown in the (P-V) phase diagram in Fig. 17.5, the various portions of the DLMNVA continuous curve have distinct physical significance. First, between M (a minimum) and N (a maximum), we have:

$$\left(\frac{\partial P}{\partial V}\right)_T > 0 \qquad (17.11)$$

which indicates that the states between M and N are mechanically unstable (that is, when P increases, V increases as well). As a result, such states are not realizable in practice.

The portion VN corresponds to an overcompressed (supersaturated) vapor, which can exist in a metastable state which will disappear spontaneously if condensation nuclei are introduced in the system. This is precisely what happens in clouds (large masses of supersaturated vapor), which upon seeding with silver halide particles, produce liquid drops which subsequently fall under gravity as artificial rain drops.

Similarly, the portion LM corresponds to an overexpanded liquid, which again is metastable. The points M and N are therefore boundary points between metastable and unstable states of the system. For every isotherm having $T < T_C$, there are pairs of points like M and N. If we join all the M points and all the N points with the curve aMCPNb (see Fig. 17.7), the resulting curve is referred to as the spinodal, a curve which separates unstable states, for which $(\partial P/\partial V)_T > 0$, from metastable states, for which $(\partial P/\partial V)_T < 0$. Clearly, the mathematical condition for the spinodal is given by:

$$\left(\frac{\partial P}{\partial V}\right)_T = 0 \qquad (17.12)$$

The CP is located at the maximum of the spinodal where it touches the binodal. Therefore, the binodal and the spinodal divide the (P-V) phase diagram into stable, metastable, and unstable regions. The (P-V) phase diagram in Fig. 17.7 illustrates the various features discussed above:

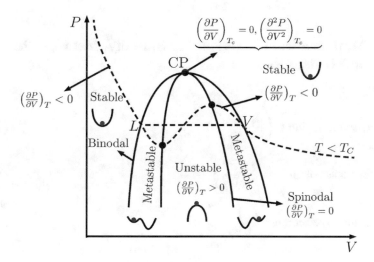

Fig. 17.7

17.7   Mechanical Analogy of Stable, Metastable, and Unstable Equilibrium States

Figure 17.8 provides mechanical analogs of a stable equilibrium state (a ball resting at the bottom of a valley), an unstable equilibrium state (a ball resting at the top of a hill), and a metastable equilibrium state (a ball resting at the bottom of a shallow local minimum separated from a deeper global minimum by an energy barrier).

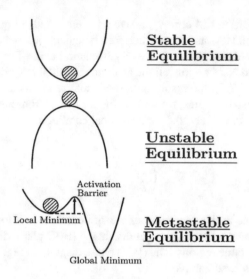

Fig. 17.8

17.8   Mathematical Conditions for Stability, Metastability, and Instability

1. In the stable region : $\left(\dfrac{\partial P}{\partial V}\right)_T < 0$ (17.13)

2. In the metastable region : $\left(\dfrac{\partial P}{\partial V}\right)_T < 0$ (17.14)

3. In the unstable region : $\left(\dfrac{\partial P}{\partial V}\right)_T > 0$ (17.15)

17.9   Mathematical Conditions for the Spinodal and the Critical Point

Recall that:

$$P = -\left(\frac{\partial A}{\partial V}\right)_T$$ (17.16)

so that:

$$\left(\frac{\partial P}{\partial V}\right)_T = -\left(\frac{\partial^2 A}{\partial V^2}\right)_T \tag{17.17}$$

(i) Spinodal Conditions:

$$\left(\frac{\partial P}{\partial V}\right)_T = 0, \ \textbf{or} \ \left(\frac{\partial^2 A}{\partial V^2}\right)_T = 0 \tag{17.18}$$

(ii) Critical Point Conditions:

$$\left(\frac{\partial P}{\partial V}\right)_{T_C} = 0 \ \textbf{and} \ \left(\frac{\partial^2 P}{\partial V^2}\right)_{T_C} = 0, \ \textbf{or} \ \left(\frac{\partial^2 A}{\partial V^2}\right)_{T_C} = 0 \ \textbf{and} \ \left(\frac{\partial^3 P}{\partial V^3}\right)_{T_C} = 0 \tag{17.19}$$

Lecture 18

The Principle of Corresponding States and Sample Problems

18.1 Introduction

The material presented in this lecture is adapted from Chapter 8 in T&M. In this lecture, we will continue discussing EOS of pure materials ($n = 1$). First, we will discuss the Redlich-Kwong EOS. Second, we will solve Sample Problem 18.1 to calculate the parameters, a and b, in the van der Waals EOS presented in Lecture 17 using the critical point conditions. We will also solve Sample Problem 18.2 to express the van der Waals EOS using reduced coordinates. Third, we will discuss the Peng-Robinson EOS and the Virial EOS, including solving Sample Problem 18.3 to calculate the second virial coefficient in the case of excluded-volume interactions. Fourth, we will introduce the two-parameter principle of corresponding states and show that it needs to be generalized to encompass all the fluids in nature. Specifically, we will introduce the three-parameter Pitzer compressibility factor correlation approach, which introduces Pitzer's acentric factor as the third correlative parameter. Fifth, we will solve Sample Problem 18.4 to show how to calculate Pitzer's acentric factor using experimental vapor pressure data. Finally, we will discuss how to use graphical representations to calculate the compressibility factor, whose knowledge is equivalent to knowing the fluid EOS.

18.2 Examples of Additional Equations of State (EOS)

18.2.1 The Redlich-Kwong (RK) EOS

Like the van der Waals EOS presented in Lecture 17, the Redlich-Kwong (RK) EOS is another two-constant cubic EOS. The pressure-explicit form of the RK EOS is given by:

© Springer Nature Switzerland AG 2020   171
D. Blankschtein, *Lectures in Classical Thermodynamics with an Introduction to Statistical Mechanics*, https://doi.org/10.1007/978-3-030-49198-7_18

$$P = \frac{RT}{V - b'} - \frac{a'}{T^{1/2}V(V + b')} \tag{18.1}$$

where

$$a' = \frac{0.42748 \, R^2 T_C^{5/2}}{P_C} \tag{18.2}$$

and

$$b' = 0.08664 \, \frac{RT_C}{P_C} \tag{18.3}$$

In addition, the critical compressibility factor corresponding to the RK EOS is given by:

$$Z_C = \frac{P_C V_C}{RT_C} = \frac{1}{3} \tag{18.4}$$

Equations (18.2), (18.3), and (18.4) show that the two EOS parameters, a' and b', in Eq. (18.1), as well as the critical compressibility factor, Z_C, can be determined using the critical point conditions.

18.2.2   Sample Problem 18.1

Calculate the parameters, a and b, in the van der Waals (vdW) EOS using the critical point conditions.

18.2.2.1   Solution

We begin with the van der Waals EOS introduced in Lecture 17, that is:

$$P = \frac{RT}{V - b} - \frac{a}{V^2} \tag{18.5}$$

We then utilize the critical point conditions presented in Lecture 17 and given by:

$$\left(\frac{\partial P}{\partial V}\right)_T = 0 \quad \text{and} \quad \left(\frac{\partial^2 P}{\partial V^2}\right)_T = 0 \tag{18.6}$$

Using Eq. (18.5) for P in the two conditions in Eq. (18.6), and solving the two resulting equations simultaneously at the critical point $(P = P_C, V = V_C,$ and $T = T_C)$, yields:

$$a = \frac{27}{64} \frac{R^2 T_C^2}{P_C} = \frac{9}{8} R T_C V_C \qquad (18.7)$$

$$b = \frac{R T_C}{8 P_C} = \frac{V_C}{3} \qquad (18.8)$$

$$Z_C = \frac{P_C V_C}{R T_C} = \frac{3}{8} \qquad (18.9)$$

18.2.3 Sample Problem 18.2

Express the van der Waals EOS in reduced coordinates.

18.2.3.1 Solution

We begin by defining:

(1) Reduced Pressure: $P_r = P/P_C$
(2) Reduced Molar Volume: $V_r = V/V_C$
(3) Reduced Temperature: $T_r = T/T_C$

In terms of P_r, V_r, and T_r above, the van der Waals EOS in Eq. (18.5) can be expressed as follows:

$$P_r = \frac{8 T_r}{3 V_r - 1} - \frac{3}{V_r^2} \qquad (18.10)$$

Equation (18.10) shows that all fluids satisfying the vdW EOS fall onto a single universal $P_r - V_r$ curve at a given T_r. This statement is an example of the Principle of Corresponding States, which we will discuss in more detail below.

18.2.4   The Peng-Robinson (PR) EOS

This is a three-constant cubic EOS, which can be applied to both the liquid and vapor phases. The pressure-explicit form of the Peng-Robinson (PR) EOS is given by:

$$P = \frac{RT}{V-b} - \frac{a(\omega, T_r)}{V(V+b) + b(V-b)} \tag{18.11}$$

where

$$a(\omega, T_r) = a(T_C)\alpha(\omega, T_r); \ T_r = T/T_C; \omega = \text{Pitzer's acentric factor} \tag{18.12}$$

$$a(T_C) = 0.45724 \frac{R^2 T_C^2}{P_C} \tag{18.13}$$

$$\alpha(\omega, T_r) = \left[1 + \kappa\left(1 - \sqrt{T_r}\right)\right]^2 \tag{18.14}$$

$$\kappa = 0.37464 + 1.54226\omega - 0.26992\omega^2 \tag{18.15}$$

$$b = 0.07780 \frac{RT_C}{P_C} \tag{18.16}$$

Below, we will derive an expression to calculate Pitzer's acentric factor which appears in Eqs. (18.12), (18.14), and (18.15). Equations (18.12)–(18.16) show that all the parameters in the PR EOS can be determined using the critical point conditions and Pitzer's acentric factor.

18.2.5   The Virial EOS

The Virial EOS has a molecular basis in Statistical Mechanics (see Part III of the book). The pressure-explicit form of the Virial EOS can be written in the two following forms:

$$P = \frac{RT}{V} + \frac{BRT}{V^2} + \frac{CRT}{V^3} + \cdots \tag{18.17}$$

or

$$P = RT \left[\rho + B\rho^2 + C\rho^3 + \ldots \right]; \quad \rho = 1/V \qquad (18.18)$$

When $V \to \infty$, or equivalently, when $\rho \to 0$, we recover the ideal gas (IG) EOS discussed in Lecture 17. In addition, the virial form of the compressibility factor, Z, as a function of V or ρ can be written as follows:

$$Z = \frac{PV}{RT} = 1 + \frac{B}{V} + \frac{C}{V^2} + \ldots = 1 + B\rho + C\rho^2 + \ldots \qquad (18.19)$$

The coefficients B, C, ... in Eqs. (18.17), (18.18), and (18.19) are referred to as the second, third, ... virial coefficients. As will be shown next, the second virial coefficient, B, can be determined from knowledge of the intermolecular potential in the context of statistical mechanics.

18.2.6 Sample Problem 18.3

Calculate the second virial coefficient in the case of excluded-volume interactions.

18.2.6.1 Solution

Given molecules i and j, modeled as identical hard spheres of radius, Ro, which are separated by a center-to-center distance, r (see Fig. 18.1),

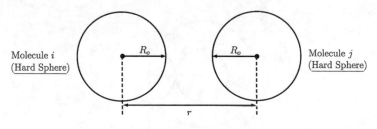

Fig. 18.1

the intermolecular potential between i and j, referred to as the hard sphere potential, is given by:

$$U_{ij} = \begin{cases} \infty, & \text{if } r \leq 2Ro \\ 0, & \text{if } r > 2Ro \end{cases} \qquad (18.20)$$

Using statistical mechanics, we can show that (see Part III):

$$B = \frac{N_A}{2} (4\pi) \int_0^\infty \left[1 - \exp\left[-\frac{U_{ij}}{k_B T} \right] \right] r^2 dr \qquad (18.21)$$

where N_A is Avogadro's number and k_B is the Boltzmann constant.

Using the expression for U_{ij} in Eq. (18.20) in the expression for B in Eq. (18.21) yields:

$$B = \frac{N_A}{2} (4\pi) \left\{ \int_0^{2Ro} [1 - 0] r^2 dr + \int_{2Ro}^\infty [1 - 1] r^2 dr \right\} \qquad (18.22)$$

In Eq. (18.22), the first integration yields $8Ro^3$, and the second integration yields zero. Using these results in Eq. (18.22) yields:

$$B = \frac{N_A}{2} \left[8 \left(\frac{4\pi}{3} \right) Ro^3 \right] \qquad (18.23)$$

or

$$B = \frac{16\pi \, Ro^3 \, N_A}{3} \qquad (18.24)$$

In Fig. 18.2, the volume of the dashed sphere corresponds to the volume excluded by a sphere of radius Ro from the center of a second sphere of radius Ro which touches the first sphere and is given by $8(4\pi/3)Ro^3$.

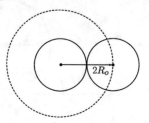

Fig. 18.2

As discussed in Lecture 17, the parameter, b, in the van der Waals EOS reflects the excluded-volume interactions between two molecules. Therefore, b should be related to B calculated above. To show this, we begin with the vdW EOS, repeated here for completeness (see Eq. (18.5)):

$$P = \frac{RT}{V-b} - \frac{a}{V^2} \tag{18.25}$$

We then assume that no attractive interactions operate between the molecules by setting $a = 0$ in Eq. (18.25). Following that, we expand Eq. (18.5), with $a = 0$, in powers of $1/V$ to second order. This yields:

$$P = \frac{RT}{V} + \frac{bRT}{V^2} \tag{18.26}$$

Comparing the virial EOS in Eq. (18.17), truncated to second order in $1/V$, with Eq. (18.26) shows that:

$$b = B = \frac{16\pi \, Ro^3 \, N_A}{3} \tag{18.27}$$

The important result in Eq. (18.27) shows that, if we know the chemical structure of the molecule, we can estimate Ro and then compute b at the molecular level.

18.3   The Principle of Corresponding States

"All fluids in corresponding states (same P_r and T_r) have the same reduced volume, V_r."

18.3.1   The Compressibility Factor

$$Z = \frac{PV}{NRT} = \frac{PV}{RT} \tag{18.28}$$

Equation (18.28) shows that: (i) for an ideal gas (IG), $Z_{IG} = 1$, (ii) knowledge of $Z = Z\,(T,P)$ or $Z = Z\,(T,V)$ is equivalent to knowing the EOS.

Recalling that $P = P_r P_C$, $V = V_r V_C$, and $T = T_r T_C$, and then using these P, V, and T expressions in Eq. (18.28), we obtain:

$$Z = Z_C \, \frac{P_r V_r}{T_r} \tag{18.29}$$

where

$$Z_C = \frac{P_C V_C}{R T_C} \tag{18.30}$$

Table 18.1 shows predictions of equations of state for the critical compressibility factor.

Based on the expression for Z in Eq. (18.29), we can ask: "Can all the fluids in nature be correlated from knowledge of their P_r and T_r values," as predicted by the Principle of Corresponding States (POCS)? According to the POCS, V_r is a universal function of P_r and T_r. As a result, for all the fluids with the same Z_C, Z should be universal too.

Table 18.1 Predictions of equations of state for Z_C

Equation of state	Z_C
Ideal	1
van der Waals (vdW)	0.375
Redlich-Kwong (RK)	0.333
Redlich-Kwong-Soave (RKS)	0.333
Peng-Robinson (PR)	0.307
"Best Fit" Martin	0.25

Table 18.2 Experimental values of Z_C for various fluids

Substance	Z_C
He	0.3141
H_2	0.3049
CO_2	0.2869
SO_2	0.2774
C_6H_6	0.2663
NH_3	0.2420
H_2O	0.2290

Table 18.2 shows that for various fluids, $0.22 < Z_C < 0.32$. Because the experimental Z_C values are not identical, the experimental $Z (P_r, T_r)$ values are not universal, contrary to the prediction made by the POCS (see Fig. 18.3).

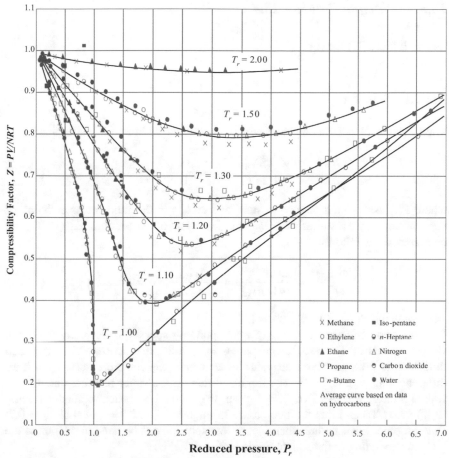

[Reprinted from Yunus A. Cengel & Michael A. Boles, 2005, McGraw Hill, Figure 3-51,
p 141, based on data from Gouq-Jen Su, 1946, Ind. Eng. Chem. 38, p 803.]

Fig. 18.3   Gas compressibilities as a function of reduced pressure and temperature

Because the original two-parameter (P_r, T_r) POCS cannot reproduce the experimental data in Fig. 18.3 for all the fluids considered, it is convenient to introduce a
third differentiating parameter. A natural first choice is Z_C.
  In terms of P_r, T_r, and Z_C, we can generalize the original POCS as follows:

$$Z = Z\,(T_r, P_r, Z_C) \tag{18.31}$$

Another choice for the third differentiating parameter is Pitzer's acentric factor,
ω, defined as follows:

$$\omega = -1.0 - \log_{10}\left[(P_r^{\;sat})_{T_r=0.7}\right] \tag{18.32}$$

where $P_r^{\;sat}$ is the reduced vapor pressure and is evaluated at $T_r = 0.7$. In Sample
Problem 18.4, we present a derivation of Eq. (18.32).

18.3.2 Sample Problem 18.4

Derive Eq. (18.32) for ω.

18.3.2.1 Solution

The Clausius-Clapeyron equation, which we will derive in Part II, states that:

$$\frac{d(\ln P_{vap})}{d(1/T)} = -\frac{\Delta H_{vap}}{R} \tag{18.33}$$

where P_{vap} is the fluid vapor pressure, and ΔH_{vap} is the fluid molar enthalpy of vaporization. If we assume that ΔH_{vap} is independent of temperature, Eq. (18.33) can be expressed as follows:

$$\frac{d(\ln P_{vap})}{d(1/T)} = \frac{d(\ln P_r^{sat})}{d(1/T_r)} = -\frac{\Delta H_{vap}}{R} = a = \text{constant} \tag{18.34}$$

where a is the slope of $\ln (P_r^{sat})$ vs. $1/T_r$.

If the two-parameter principle of corresponding states was generally valid, the slope a would be the same for all the pure $(n = 1)$ fluids. In practice, we find that each fluid has its own characteristic value of a, which can therefore serve as a third differentiating parameter. Pitzer recognized that all the vapor pressure data for the simple fluids (SF: Ar, Kr, Xe) lie on the same line when plotted as $\log_{10} (P_r^{sat})$ vs. $(1/T_r)$ and that the line passes through $\log_{10} (P_r^{sat}) = -1.0$ at $T_r = 0.7$. Data for other fluids define other lines whose location can be fixed in relation to the line corresponding to the simple fluids by the difference:

$$[\log_{10} (P_r^{sat}(SF)) - \log_{10} (P_r^{sat})]_{T_r=0.7} \tag{18.35}$$

The Pitzer acentric factor, ω, is defined as this difference, that is,

$$\omega = \log_{10} (P_r^{sat}(SF))\,|_{T_r=0.7} - \log_{10} (P_r^{sat})|_{T_r=0.7} \tag{18.36}$$

The first term on the right-hand side of the equal sign in Eq. (18.36) is equal to -1.0, and therefore, Eq. (18.36) can be expressed as follows:

$$\omega = -1.0 - \log_{10} (P_r^{sat})|_{T_r=0.7} \tag{18.37}$$

Equation (18.37) completes our derivation.

To actually determine ω, we need to know T_C, P_C, and a single vapor pressure value at $T_r = 0.7$.

By definition, $\omega = 0$ for the simple fluids (Ar, Kr, Xe). Values of ω, as well as critical point coordinates, for other fluids, are listed in Table 18.3.

Table 18.3   Critical point coordinates and Pitzer's acentric factors

	T_c/K	P_c/bar	$V_c/10^{-6}$ m^3mol^{-1}	Z_c	ω
Paraffins:					
Methane	190.6	46	99.	0.288	0.008
Ethane	305.4	48.8	148.	0.285	0.098
Propane	369.8	42.5	203.	0.281	0.152
n-butane	425.2	38	255.	0.274	0.193
Isobutane	408.1	36.5	263.	0.283	0.176
n-Pentane	469.6	33.7	304.	0.262	0.251
Isopentane	460.4	33.8	306.	0.271	0.227
Neopentane	433.8	32.0	303.	0.269	0.197
n-Hexane	507.4	29.7	370.	0.26	0.296
n-Heptane	540.2	27.4	412.	0.263	0.351
n-Octane	568.8	24.8	492.	0.259	0.394
Monoolefins:					
Ethylene	282.4	50.4	129.	0.276	0.085
Propylene	365	46.2	181.	0.273	0.148
1-Butene	419.6	40.2	240.	0.277	0.187
1-Pentene	464.7	40.5	300.	0.31	0.245
Miscellaneous organic compounds:					
Acetic acid	594.4	57.9	171.	0.2	0.454
Acetone	508.1	47	209.	0.232	0.309
Acetonitrile	547.9	48.3	173.	0.184	0.321
Benzene	562.1	48.9	259.	0.271	0.212
Elementary gases:					
Argon	150.8	48.7	74.9	0.291	0.0
Bromine	584.	103.	127.	0.270	0.132
Hydrogen	33.2	13.0	65.0	0.305	-0.22
Krypton	209.4	55.0	91.2	0.288	0.0
Neon	44.4	27.6	41.7	0.311	0.0
Nitrogen	126.2	33.9	89.5	0.290	0.040
Oxygen	154.6	50.5	73.4	0.288	0.021
Miscellaneous inorganic compounds:					
Ammonia	405.6	112.8	72.5	0.242	0.250
Carbon dioxide	304.2	73.8	94.0	0.274	0.225
Carbon disulfide	552.	79.	170.	0.293	0.115
Carbon monoxide	132.2	35.0	93.1	0.295	0.049
Carbon tetrachloride	556.1	45.6	276.	0.272	0.194
Chloroform	536.4	55.	239.	0.293	0.216
Hydrazine	653.	147.	96.1	0.260	0.328
Hydrogen chloride	324.6	83.	81.	0.249	0.12
Hydrogen cyanide	456.8	53.9	139.	0.197	0.407
Hydrogen sulfide	373.2	89.4	98.5	0.284	0.108
Nitric oxide (NO)	180.	65.	58.	0.25	0.607
Nitrous oxide (N2O)	309.6	72.4	97.4	0.274	0.168
Sulfur dioxide	430.8	78.8	122.	0.268	0.251
Sulfur trioxide	491.0	82.	130.	0.26	0.41
Water	647.3	220.5	56.	0.229	0.344

Simple Fluids $\omega = 0!$ (Argon, Krypton, Neon)

References: A.P. Kudchadker, G.H. Alani, and B.J. Zwolinski, Chem. Rev. 68: 659, 1968; J.F. Mathews, Chem. Rev. 72: 71, 1972; R.C. Reid, J.M. Prausnitz, and T.K. Sherwood, "The Properties of Gases and Liquids." 3rd ed., McGraw-Hill, N.Y., 1977; C.A. Passut, and R.P. Danner, Ind. Eng. Chem. Process Des. Dev. 12: 3, 1973.

It is noteworthy that, if we do not have access to an EOS, the ability to calculate Z at a given T and P enables us to calculate V = RTZ/P. Of course, even if an EOS is available, as shown next, we can use the generalized three-parameter Pitzer correlation approach to compute Z.

It is possible to express Z as an expansion in powers of ω truncated to linear order. Specifically,

$$Z = Z^0\,(T_r, P_r) + \omega Z^1\,(T_r, P_r) \tag{18.38}$$

Interestingly, Z^0 and Z^1 are available graphically (see Figs. 18.4, 18.5, 18.6, and 18.7). In Lecture 19, we will utilize Eq. (18.38), along with Figs. 18.4 and 18.5, to solve an interesting sample problem.

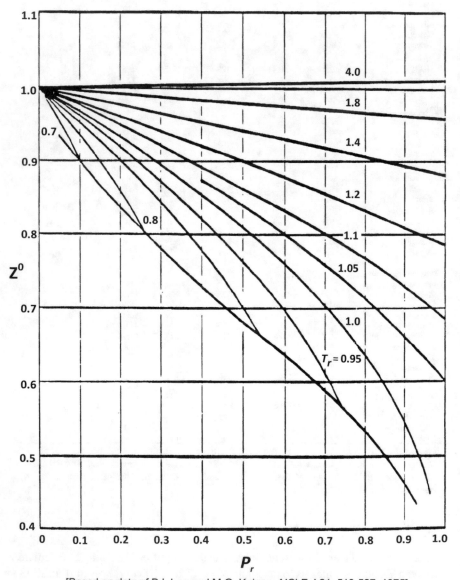

[Based on data of B.I. Lee and M.G. Kaiser. AIChE J 21: 510-527, 1975].

Fig. 18.4 Generalized correlation for Z^0, $P_r < 1.0$

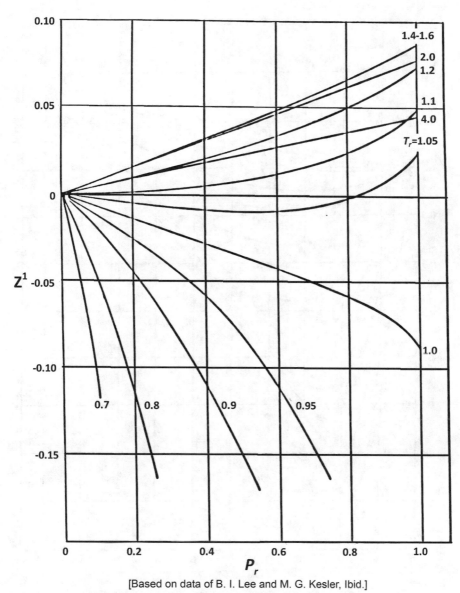

[Based on data of B. I. Lee and M. G. Kesler, Ibid.]

Fig. 18.5 Generalized correlation for Z^1, $P_r < 1.0$

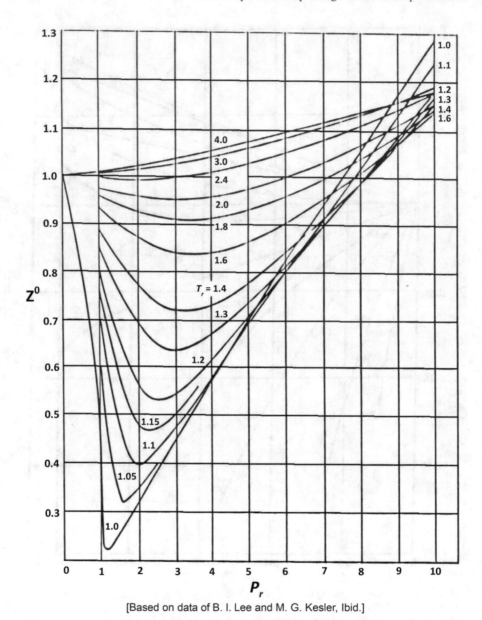

[Based on data of B. I. Lee and M. G. Kesler, Ibid.]

Fig. 18.6   Generalized correlation for Z^0, $P_r > 1.0$

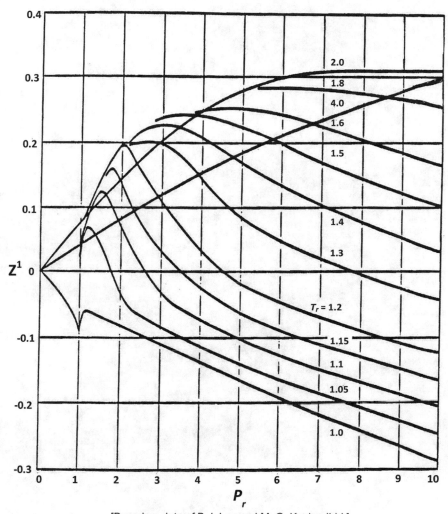

[Based on data of B. I. Lee and M. G. Kesler, Ibid.]

Fig. 18.7 Generalized correlation for Z^1, Pr > 1.0

Lecture 19

Departure Functions and Sample Problems

19.1 Introduction

The material presented in this lecture is adapted from Chapter 8 in T&M. In this lecture, we will continue discussing pure materials (n = 1). First, we will solve Sample Problem 19.1 to calculate the molar volume of n-butane at a given temperature and pressure. For this purpose, we will first use the ideal gas EOS and then can use more realistic EOS. In addition, we will use the Generalized Pitzer Correlation Approach presented in Lecture 18. Second, we will discuss the departure function (DF) approach to calculate changes in thermodynamic properties, including the entropy departure function, DS, and the Helmholtz free energy departure function, DA. Third, we will solve Sample Problem 19.2 to calculate the internal energy departure function, DU. Finally, in order to calculate isothermal variations of A or G, we will show that it is simpler to directly integrate the differentials of A or G, instead of using departure functions.

19.2 Sample Problem 19.1

Calculate the molar volume of n-butane at 510 K and 25 bar. You can assume that, for n-butane, $T_c = 425.2$ K, $P_c = 38$ bar, and $\omega = 0.193$.

19.2.1 Solution

The (P-T) phase diagram in Fig. 19.1 shows the L/V equilibrium line, the CP, and the operating conditions (the black circle).

© Springer Nature Switzerland AG 2020
D. Blankschtein, *Lectures in Classical Thermodynamics with an Introduction to Statistical Mechanics*, https://doi.org/10.1007/978-3-030-49198-7_19

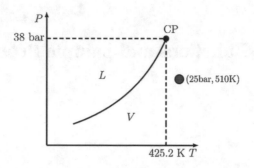

Fig. 19.1

Because the temperature and the pressure of interest are known, we can use an EOS to calculate the molar volume. For example, we can:

(a) Use the ideal gas EOS

$$V = \frac{RT}{P} = 1696.1 \text{ cm}^3/\text{mol} \tag{19.1}$$

where we have used the gas constant R, T = 510 K, and P = 25 bar.

(b) Use other EOS

For example, the vdW EOS, the RK EOS, or the PR EOS discussed in Lecture 18. To estimate the various parameters in these EOS, we can use the given values of T_C, P_C, and ω as needed. We would then need to solve a cubic equation for V, where only one solution will be physical in the gas phase (25 bar, 510 K).

Alternatively, if an EOS is not available, we can use the Generalized Pitzer Correlation Approach discussed in Lecture 18, as shown in (c) below.

(c) Generalized Pitzer Correlation Approach

We know that:

$$V = \frac{RTZ}{P} \tag{19.2}$$

where

$$Z = Z^0(T_r, P_r) + \omega Z^1(T_r, P_r) \tag{19.3}$$

To use the graphs for Z^0 and Z^1 presented in Lecture 18, we first calculate T_r and P_r corresponding to the T and P values given in the Problem Statement. Specifically,

$$T_r = T/T_C = 510/425.2 = 1.198 \tag{19.4}$$

$$P_r = P/P_C = 25/38 = 0.658 \qquad (19.5)$$

Because $P_r < 1.0$, we need to use the Generalized Correlation Graphs for Z^0 and Z^1 corresponding to $P_r < 1.0$, first introduced in Lecture 18, and shown again in Fig. 19.1 and Fig. 19.3, respectively, for completeness.

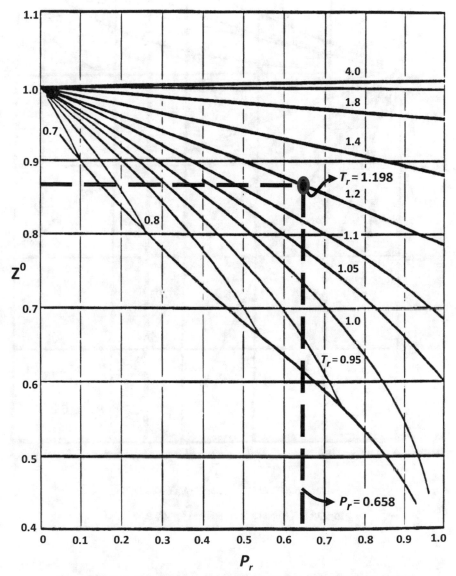

[Based on data of B. I. Lee and M.G. Kesler, AIChE J., 21: 510-527, 1975.]

Fig. 19.2 Generalized correlation for Z^0, $P_r < 1.0$

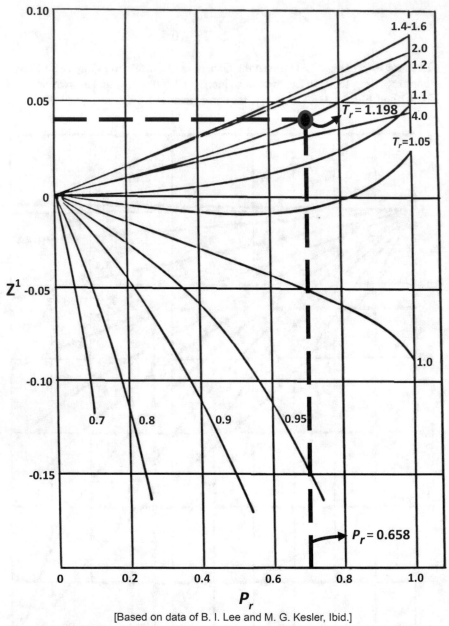

[Based on data of B. I. Lee and M. G. Kesler, Ibid.]

Fig. 19.3   Generalized correlation for Z^1, $P_r < 1.0$

- Using the graph for Z^0 in Fig. 19.2, at $T_r = 1.198$ and $P_r = 0.658$, yields:

$$Z^0 = 0.865 \Rightarrow V^{\circ} = \frac{Z^0 RT}{P} \simeq 1467 \ \frac{cm^3}{mol} \tag{19.6}$$

- Using the graph for Z^1 in Fig. 19.3, at the same T_r and P_r values, yields:

$$Z^1 \simeq 0.038 \tag{19.7}$$

Using the calculated values for Z^0 and Z^1 in Eq. (19.6) and Eq. (19.7), respectively, as well as $\omega = 0.193$, in Eq. (19.3) yields:

$$Z = Z^0 + \omega Z^1 \simeq 0.872 \tag{19.8}$$

Using the calculated Z value in Eq. (19.8), along with the T and P values given in the Problem Statement, in Eq. (19.2) for V yields the desired result:

$$V = \frac{ZRT}{P} \simeq 1479 \ \frac{cm^3}{mol} \tag{19.9}$$

Interestingly, the predicted molar volume of n-butane in Eq. (19.9) is in remarkably good agreement with the experimental value given below:

$$V_{exp} = 1480.7 \ cm^3/mol \tag{19.10}$$

19.3 Departure Functions

In Lecture 16, we discussed the three-step integration approach which can be used to calculate a thermodynamic property change. This approach combines two isothermal property changes from the real to an attenuated, ideal gas (IG) state ($P^* = 0$ or $V^* = \infty$) and a temperature variation in the attenuated state. As we will show in this lecture, the isothermal variation can be more formally defined in terms of a so-called departure function of a property, defined as follows:

"A Departure Function is the difference between the property of interest in its real state at a specified (T,P) or (T,V), and in an Ideal Gas State at the same T and P."

Therefore, there are two equivalent formulations of departure functions:

$$(1) \quad B(T, P) - B^{IG}(T, P) = B(T, P) - B^\circ(T, P) \tag{19.11}$$

and

$$(2) \quad B(T, V) - B^{IG}(T, V^\circ) = B(T, V) - B^\circ(T, V^\circ), \text{ where } V^\circ = \frac{RT}{P} \tag{19.12}$$

In Eqs. (19.11) and (19.12), B is any derived property of the system (S, H, U, A, G, etc.). Hereafter, we will abbreviate departure function as DF.

19.4   Calculation of the Entropy Departure Function, DS(T, P)

If S = S(T,P), then:

$$dS = \left(\frac{C_P}{T}\right) dT - \left(\frac{\partial V}{\partial T}\right)_P dP \tag{19.13}$$

At constant temperature, integrating Eq. (19.13) with respect to pressure, from the ideal gas state at P→0 to the real state at P, yields:

$$S(T, P) - S^\circ(T, P \to 0) = - \int_0^P \left(\frac{\partial V}{\partial T}\right)_P dP \tag{19.14}$$

Further, at constant temperature, integrating Eq. (19.13) with respect to pressure, from the ideal gas state at P→0 to an ideal gas state at a low P value, yields:

$$S^\circ(T, P) - S^\circ(T, P \to 0) = - \int_0^P \left(\frac{R}{P}\right) dP \tag{19.15}$$

In Eqs. (19.14) and (19.15), the superscript o in S denotes an ideal gas when P = 0. In Eq. (19.15), because we are connecting two ideal gas states, V = RT/P and $(\partial V/\partial T)_P = R/P$ in the integrand.

Subtracting Eq. (19.15) from Eq. (19.14), cancelling the two equal terms on the left-hand side, and combining the two integrals, yields the entropy departure function, DS(T,P), given by:

$$DS(T,P) = S(T,P) - S^\circ(T,P) = -\int_0^P \left[\left(\frac{\partial V}{\partial T} \right)_P - \frac{R}{P} \right] dP \qquad (19.16)$$

Note that, in Lecture 16, we also obtained the integral in Eq. (19.16) when we removed the apparent divergence in the limit $P^* \to 0$. In essence, we used a DF to do that.

Departure functions can always be computed using a P-V-T EOS. Because most P-V-T EOS are pressure explicit (with V and T as the independent variables), this suggests that we should first calculate the departure function of the molar Helmholtz free energy A(T,V), from which we should be able to calculate all the other departure functions (see below).

19.5 Calculation of the Helmholtz Free Energy Departure Function, DA(T, V)

Recall that:

$$A = A(T, V) \Rightarrow dA = -SdT - PdV \qquad (19.17)$$

At constant T, integrating dA in Eq. (19.17) with respect to V from the ideal gas state at $V = \infty$ to the real state at V yields:

$$A(T, V) - A^\circ(T, \infty) = -\int_\infty^V PdV \qquad (19.18)$$

In Eq. (19.18), $A^\circ(T, \infty)$ carries the superscript o to indicate that the system is an ideal gas when $V = \infty$.

Because the behavior between $V = V^\circ$ and $V = \infty$ corresponds to that of an ideal gas (IG), it follows that:

$$A^\circ(T, \infty) \quad A^\circ(T, V^\circ) \; - -\int_{V^\circ}^\infty PdV \Big|_{IG} \qquad (19.19)$$

Adding Eqs. (19.18) and (19.19), cancelling the two equal terms on the left-hand side, and combining the two integrals, yields:

$$A(T, V) - A°(T, V°) = - \int\limits_{\infty}^{V} PdV - \int\limits_{V°}^{\infty} \frac{RT}{V} \, dV \qquad (19.20)$$

where in Eq. (19.19), $PdV|_{IG}$ is equal to $(RT/V)dV$. In Eq. (19.20), there is an apparent divergence in the two integrals in the limit $V = \infty$. To remove the apparent divergence in Eq. (19.20), we add and subtract the following integral:

$$\int\limits_{\infty}^{V} \left(\frac{RT}{V}\right) dV \qquad (19.21)$$

Rearranging the left-hand side of the resulting equation, we obtain the Helmholtz free energy departure function, DA(T, V). Specifically,

$$DA(T, V) = A(T, V) - A°(T, V°) = - \int\limits_{\infty}^{V} \left[P - \frac{RT}{V}\right] dV + RT \ln\left(\frac{V°}{V}\right) \quad (19.22)$$

When $V \to \infty$, the integrand in Eq. (19.22) is proportional to $1/V^2$. Therefore, the integral over dV scales as $1/V$ and is equal to 0 when $V = \infty$, thereby eliminating the apparent divergence.

By using any pressure-explicit EOS (vdW, RK, PR, etc.), DA(T,V) can be computed using Eq. (19.22). Once DA(T,V) is known, we can readily compute the other DFs. As an illustration, below, we calculate DS(T, V).

19.6   Calculation of the Entropy Departure Function, DS(T, V)

To calculate DS(T,V), we begin from the definition of S in terms of A. Specifically,

$$S = -\left(\frac{\partial A}{\partial T}\right)_V \qquad (19.23)$$

Next, we can calculate DS(T,V) by differentiating DA(T,V) in Eq. (19.22) with respect to T, at constant V. This yields:

$$\left(\frac{\partial A}{\partial T}\right)_V - \left(\frac{\partial A^\circ}{\partial T}\right)_V = -\frac{\partial}{\partial T}\left[\int_\infty^V \left(P - \frac{RT}{V}\right)dV\right]_V + R\ln\left(\frac{V^\circ}{V}\right) + RT\left(\frac{\partial \ln V^\circ}{\partial T}\right)_V$$

$$(19.24)$$

where $(\partial A/\partial T)_V = -S$.

Using the fact that $V^\circ = (RT/P)$, the last term in Eq. (19.24) can be expressed as follows:

$$RT\left(\frac{\partial \ln V^\circ}{\partial T}\right)_V = \frac{RT}{V^\circ}\left(\frac{\partial V^\circ}{\partial T}\right)_V = P\left(\frac{\partial V^\circ}{\partial T}\right)_V \qquad (19.25)$$

In addition, $dA^\circ = -S^\circ dT - PdV^\circ$, which upon differentiation with respect to T, at constant V, yields:

$$\left(\frac{\partial A^\circ}{\partial T}\right)_V = -S^\circ - P\left(\frac{\partial V^\circ}{\partial T}\right)_V \qquad (19.26)$$

Combining Eqs. (19.24), (19.25), and (19.26), including cancelling the equal terms, yields the desired expression for $DS(T,V) = S(T,V) - S^\circ(T,V^\circ)$. Specifically,

$$DS(T, V) = S(T, V) - S^\circ(T, V^\circ) = \left[\int_\infty^V \left(P - \frac{RT}{V}\right)dV\right]_V - R\ln\left(\frac{V^\circ}{V}\right) \qquad (19.27)$$

From knowledge of $DA(T,V)$ and $DS(T,V)$, we can calculate the departure function of any other thermodynamic property, $B(T,V)$, abbreviated as $DB(T,V)$. For example, below, we provide expressions for the internal energy departure function, $DU(T,V)$, the enthalpy departure function, $DH(T,V)$, and the Gibbs free energy departure function, $DG(T,V)$. Specifically,

$$DU(T, V) = U(T, V) - U^\circ(T, V^\circ) = \underbrace{(A - A^\circ)}_{DA} + T\underbrace{\left(S - S^\circ\right)}_{DS} \qquad (19.28)$$

$$DH(T, V) = H(T, V) - H^\circ(T, V^\circ) = \underbrace{(U - U^\circ)}_{DU} + \left(PV - \underbrace{RT}_{PV^\circ}\right) \qquad (19.29)$$

$$DG(T, V) = G(T, V) - G^\circ(T, V^\circ) = \underbrace{(H - H^\circ)}_{DH} - T\underbrace{\left(S - S^\circ\right)}_{DS} \qquad (19.30)$$

19.7   Sample Problem 19.2

Use the DF approach to calculate $\Delta U_{1\to 2}$ as a function of T and V.

19.7.1   Solution

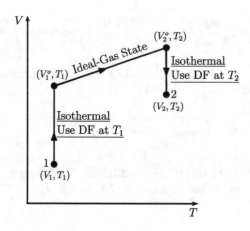

Fig. 19.4

The (V-T) phase diagram in Fig. 19.4 illustrates the initial state (1), the final state (2), and the three paths connecting states 1 and 2. Recall that the superscript o denotes properties of the ideal gas state.

In Fig. 19.4,

$$V_1^o = \frac{RT_1}{P_1} \text{ and } V_2^o = \frac{RT_2}{P_2} \tag{19.31}$$

where P_1 and P_2 are the actual pressures in states 1 and 2, respectively.

Utilizing the DF approach as it applies to the three-path process depicted in Fig. 19.4 yields:

$$U(T_2, V_2) - U(T_1, V_1) = \Delta U_{1\to 2} = \left[U(T_2, V_2) - U^o\left(T_2, V_2^o\right)\right]$$

$$+ \left[U^o\left(T_2, V_2^o\right) - U^o\left(T_1, V_1^o\right)\right] - \left[U(T_1, V_1) - U^o\left(T_1, V_1^o\right)\right] \tag{19.32}$$

In Eq. (19.32), the first bracketed term corresponds to DU at T_2, the third bracketed term corresponds to DU at T_1, and the second bracketed term is evaluated in the ideal gas state (o) as follows:

$$U^\circ(T_2, V_2^\circ) - U^\circ(T_1, V_1^\circ) = \int_{T_1}^{T_2} C_v^\circ dT \tag{19.33}$$

Recall that for an ideal gas, $U^\circ = f(T)$ only. In order to compute DU in the first and the third bracketed terms in Eq. (19.32), we only require a pressure-explicit EOS. The same procedure can be used to compute $\Delta S_{1\to2}$, $\Delta H_{1\to2}$, $\Delta A_{1\to2}$, $\Delta G_{1\to2}$, etc. In all cases, we only require a pressure-explicit EOS + C_v^0 data. Note the similarity between the DF approach, which uses the ideal gas state with $V^\circ = RT/P$, and the attenuated state approach which uses $V^* = \infty$.

19.8   Important Remark

To compute isothermal variations of G or A, it is more convenient to work directly with the equations for dG or dA derived earlier, rather than to use DFs.
    Recall that:

$$dG|_T = VdP \Rightarrow G(T, P_2) - G(T, P_1) = \int_{P_1}^{P_2} V(T, P) \, dP \tag{19.34}$$

where $V(T, P)$ is a volume-explicit EOS.
    Similarly:

$$dA|_T = -PdV \Rightarrow A(T, V_2) - A(T, V_1) = -\int_{V_1}^{V_2} P(T, V) \tag{19.35}$$

where $P(T, V)$ is a pressure-explicit EOS.

Lecture 20

Review of Part I and Sample Problem

20.1 Introduction

In this lecture, first, we will present a comprehensive review of the material discussed in Part I. Following that, we will solve Sample Problem 20.1 to demonstrate that the expression for the efficiency of a Carnot engine that we derived in Lecture 8 using an ideal gas as the engine working fluid is in fact generally valid for any working fluid.

20.2 Basic Concepts, Definitions, and Postulates

Definitions of systems and boundaries, states and paths, work and heat, and the four postulates of thermodynamics.

20.3 Ideal Gas

$$P\underline{V} = NRT \qquad (20.1)$$

$$dU = C_V dT \qquad (20.2)$$

$$dH = C_P dT \qquad (20.3)$$

$$C_P = C_V + R \qquad (20.4)$$

© Springer Nature Switzerland AG 2020
D. Blankschtein, *Lectures in Classical Thermodynamics with an Introduction to Statistical Mechanics*, https://doi.org/10.1007/978-3-030-49198-7_20

20.4   The First Law of Thermodynamics for Closed Systems

$$d\underline{E} = \delta Q + \delta W \tag{20.5}$$

$$\underline{E} = \underline{U} + \underline{E}_{PE} + \underline{E}_{KE} \tag{20.6}$$

$$\text{For a simple system}, \underline{E} = \underline{U} \tag{20.7}$$

20.5   The First Law of Thermodynamics for Open, Simple Systems

$$d\underline{U} = \delta Q + \delta W + \sum_{i} H_{in,i}\delta n_{in,i} - \sum_{j} H_{out,j}\delta n_{out,j} \tag{20.8}$$

If the system is not simple, then, $d\underline{U} \to d\underline{E} = d\underline{U} + d\underline{E}_{KE} + d\underline{E}_{PE}$. There can also be KE and PE terms in the "in/out" stream terms in Eq. (20.8).

Mass Balance for an Open System

$$dN = \sum_{i} \delta n_{in,i} - \sum_{j} \delta n_{out,j} \tag{20.9}$$

20.6   The First Law of Thermodynamics for Steady-State Flow Systems

$$\dot{Q} + \dot{W} + \sum_{i} H_{in,i}\dot{n}_{in,i} - \sum_{j} H_{out,j}\dot{n}_{out,j} = 0 \tag{20.10}$$

$$\sum_{i} \dot{n}_{in,i} - \sum_{j} \dot{n}_{out,j} = 0 \tag{20.11}$$

20.7   Carnot Engine

$$\frac{|\delta Q_H|}{T_H} = \frac{|\delta Q_C|}{T_C} \tag{20.12}$$

$$\frac{\delta Q_H}{T_H} + \frac{\delta Q_C}{T_C} = 0 \tag{20.13}$$

$$\eta_c = \frac{-W}{Q_H} = \frac{T_H - T_C}{T_H} \tag{20.14}$$

20.8   Entropy of a Closed System

$$d\underline{S} = \left(\frac{\delta Q}{T}\right)_{rev} \tag{20.15}$$

20.9   The Second Law of Thermodynamics

For a closed system, the entropy $d\underline{S}$:

(1) Is greater than 0 for an irreversible adiabatic process
(2) Is equal to 0 for a reversible adiabatic process

20.10   The Combined First and Second Law of Thermodynamics for Closed Systems

$$d\underline{U} = Td\underline{S} - Pd\underline{V} \tag{20.16}$$

In addition, choosing a convenient path to compute a change in a state function, e.g., choosing a reversible path to compute $\Delta \underline{S}$.

20.11 Entropy Balance for Open Systems

$$d\underline{S} = \left(\frac{\delta Q}{T}\right)_{rev} + \sum_i S_{in,i}\delta n_{in,i} - \sum_j S_{out,j}\delta n_{out,j} + \delta\underline{\sigma} \qquad (20.17)$$

20.12 Maximum Work, Availability

$$\Delta B = \Delta H - T_0\Delta S \quad (B = \text{Availability}) \qquad (20.18)$$

For an open system,

$$\delta W_{max} = \delta n\Delta B \qquad (20.19)$$

where δW_{max} is the maximum work done on the system.

For a reversible, steady-state process with all heat interactions at T_o,

$$\dot{W}_{max} = \dot{n}\Delta B \qquad (20.20)$$

For a non-reversible, steady-state, constant volume, adiabatic process,

$$\dot{W}_{net} = \dot{n}\Delta B + T_o\underline{\dot{\sigma}} \qquad (20.21)$$

20.13 Fundamental Equations

$$\text{Internal Energy}: \qquad d\underline{U} = Td\underline{S} - Pd\underline{V} + \sum_i \mu_i dN_i \qquad (20.22)$$

$$\text{Enthalpy}: \qquad d\underline{H} = Td\underline{S} + \underline{V}dP + \sum_i \mu_i dN_i \qquad (20.23)$$

Helmholtz Free Energy:

$$d\underline{A} = -\underline{S}dT - Pd\underline{V} + \sum_i \mu_i dN_i \tag{20.24}$$

Gibbs Free Energy:

$$d\underline{G} = -\underline{S}dT + \underline{V}dP + \sum_i \mu_i dN_i \tag{20.25}$$

For a Single-Component (Pure) System:

$$dU = TdS - PdV \tag{20.26}$$

$$dH = TdS + VdP \tag{20.27}$$

$$dA = -SdT - PdV \tag{20.28}$$

$$dG = -SdT + VdP \tag{20.29}$$

$$\text{Also useful}: \quad G = U + PV - TS = H - TS = \mu \tag{20.30}$$

In addition, the Theorem of Euler, and Euler integration to reconstruct Fundamental Equations.

20.14 Manipulation of Partial Derivatives

$$\text{Inversion Rule}: \quad \left(\frac{\partial X}{\partial Y}\right)_Z = \left(\frac{\partial Y}{\partial X}\right)_Z^{-1} \tag{20.31}$$

$$\text{Chain Rule}: \quad \left(\frac{\partial X}{\partial Y}\right)_W = \left(\frac{\partial X}{\partial Z}\right)_W \left(\frac{\partial Z}{\partial Y}\right)_W \tag{20.32}$$

$$\text{Add Another Variable Rule}: \quad \left(\frac{\partial X}{\partial Y}\right)_W = \frac{\left(\frac{\partial X}{\partial Z}\right)_W}{\left(\frac{\partial Y}{\partial Z}\right)_W} \tag{20.33}$$

Triple Product Rule : $\left(\dfrac{\partial X}{\partial Y}\right)_Z = -\left[\left(\dfrac{\partial Y}{\partial Z}\right)_X \left(\dfrac{\partial Z}{\partial X}\right)_Y\right]^{-1}$ (20.34)

20.15 Manipulation of Partial Derivatives Using Jacobian Transformations

$$\left(\frac{\partial f}{\partial z}\right)_g = \frac{\partial(f, g)}{\partial(z, g)} \tag{20.35}$$

The Jacobian : $\dfrac{\partial(f, g)}{\partial(x, y)} = \begin{vmatrix} \left(\frac{\partial f}{\partial x}\right)_y & \left(\frac{\partial f}{\partial y}\right)_x \\ \left(\frac{\partial g}{\partial x}\right)_y & \left(\frac{\partial g}{\partial y}\right)_x \end{vmatrix}$ (20.36)

Transposition : $\dfrac{\partial(f, g)}{\partial(x, y)} = -\dfrac{\partial(g, f)}{\partial(x, y)}$ (20.37)

Inversion : $\dfrac{\partial(f, g)}{\partial(x, y)} = \dfrac{1}{\dfrac{\partial(x, y)}{\partial(f, g)}}$ (20.38)

Chain Rule Expansion : $\dfrac{\partial(f, g)}{\partial(x, y)} = \dfrac{\partial(f, g)}{\partial(z, w)} \dfrac{\partial(z, w)}{\partial(x, y)}$ (20.39)

20.16 Maxwell's Reciprocity Rules

for dU $\left(\dfrac{\partial T}{\partial V}\right)_S = -\left(\dfrac{\partial P}{\partial S}\right)_V$ (20.40)

for dH
$$\left(\frac{\partial T}{\partial P}\right)_S = \left(\frac{\partial V}{\partial S}\right)_P \qquad (20.41)$$

for dA
$$\left(\frac{\partial S}{\partial V}\right)_T = \left(\frac{\partial P}{\partial T}\right)_V \qquad (20.42)$$

for dG
$$\left(\frac{\partial S}{\partial P}\right)_T = -\left(\frac{\partial V}{\partial T}\right)_P \qquad (20.43)$$

In addition, the fundamental equations and their partial derivatives.

20.17 Important Thermodynamic Relations for Pure Materials

Change in thermodynamic properties using T and P as variables:

$$dS = \frac{C_P}{T} dT - \left(\frac{\partial V}{\partial T}\right)_P dP \qquad (20.44)$$

$$dU = \left[C_P - P\left(\frac{\partial V}{\partial T}\right)_P\right] dT - \left[T\left(\frac{\partial V}{\partial T}\right)_P + P\left(\frac{\partial V}{\partial P}\right)_T\right] dP \qquad (20.45)$$

$$dH = C_P dT + \left[V - T\left(\frac{\partial V}{\partial T}\right)_P\right] dP \qquad (20.46)$$

Change in thermodynamic properties using T and V as variables:

$$dS = \frac{C_V}{T} dT + \left(\frac{\partial P}{\partial T}\right)_V dV \qquad (20.47)$$

$$dU = C_V dT + \left[T\left(\frac{\partial P}{\partial T}\right)_V - P\right] dV \qquad (20.48)$$

$$dH = \left[C_V + V\left(\frac{\partial P}{\partial T}\right)_V\right]dT + \left[T\left(\frac{\partial P}{\partial T}\right)_V + V\left(\frac{\partial P}{\partial V}\right)_T\right]dV \qquad (20.49)$$

Relation between C_P and C_V:

$$C_P - C_V = T\left(\frac{\partial V}{\partial T}\right)_P\left(\frac{\partial P}{\partial T}\right)_V \qquad (20.50)$$

Variations of C_P and C_V at constant T:

$$C_P(T, P) = C_P(T, P_0) - T\int_{P_0}^{P}\left(\frac{\partial^2 V}{\partial T^2}\right)_P dP \qquad (20.51)$$

$$C_V(T, V) = C_V(T, V_0) + T\int_{V_0}^{V}\left(\frac{\partial^2 P}{\partial T^2}\right)_V dV \qquad (20.52)$$

In addition, the attenuated state and the departure function approaches.

20.18   Gibbs-Duhem Equation for a Pure Material

$$-\underline{S}dT + \underline{V}dP - Nd\mu = 0 \qquad (20.53)$$

20.19   Equations of State (EOS)

The van der Waals EOS:

$$P = \frac{RT}{V - b} - \frac{a}{V^2} \qquad (20.54)$$

The Redlich-Kwong EOS:

$$P = \frac{RT}{V - b} - \frac{a}{\sqrt{T}V(V + b)} \tag{20.55}$$

The Peng-Robinson EOS:

$$P = \frac{RT}{V - b} - \frac{a(\omega, T)}{V(V + b) + b(V - b)} \tag{20.56}$$

The Virial EOS:

$$P = \frac{RT}{V} + \frac{BRT}{V^2} + \frac{CRT}{V^3} \tag{20.57}$$

The Pitzer correlation for the compressibility factor:

$$Z = \frac{PV}{RT} = Z^0(T_r, P_r) + \omega Z^1(T_r, P_r) \tag{20.58}$$

In addition, the principle of corresponding states and the evaluation of EOS parameters based on the critical point conditions.

20.20 Stability Criteria for a Pure Material

Spinodal Condition:

$$\left(\frac{\partial P}{\partial V}\right)_T = 0 \tag{20.59}$$

Critical Point Conditions:

$$\left(\frac{\partial P}{\partial V}\right)_{T_C} = 0; \quad \left(\frac{\partial^2 P}{\partial V^2}\right)_{T_C} = 0 \tag{20.60}$$

In addition, stable, metastable, and unstable equilibrium states and the binodal and spinodal curves in the (P-V) phase diagram of a pure material (n = 1).

20.21   Sample Problem 20.1

Demonstrate that, for any working fluid, the thermal efficiency of a Carnot engine is given by:

$$\eta_C = \frac{T_H - T_C}{T_H} = 1 - \frac{T_C}{T_H}$$

20.21.1   Solution

As discussed in Lecture 8, the Carnot cycle can be viewed as consisting of the following four steps:

(1) A reversible, isothermal addition of heat to the working fluid at a high temperature, T_H, as it expands from state 1 to state 2.

(2) A reversible, adiabatic expansion of the working fluid to a low temperature, T_C, as it flows from state 2 to state 3.

(3) A reversible, isothermal rejection of heat from the working fluid at a low temperature, T_C, as it contracts from state 3 to state 4.

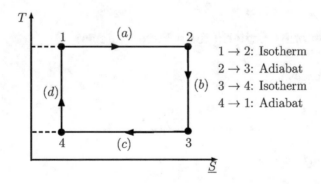

Fig. 20.1

(4) A reversible, adiabatic compression of the working fluid to a high temperature, T_H, as it flows from state 4 to state 1 in order to complete the cycle and return to its original state 1.

When we first solved this problem in Lecture 8, we used an ideal gas as the engine's working fluid. In addition, we used a (P-$\underline{V}$) phase diagram to represent the

Carnot cycle. However, this required using an equation of state to express the dependence of P on $\underline{V}$. To simplify the algebra, in Lecture 8, we chose the ideal gas equation of state (EOS) given by:

$$P = \frac{NRT}{\underline{V}}$$

and claimed, without proof, that the solution would not change if the working fluid was not ideal. As per the Sample Problem 20.1 Statement, in order to deal with any working fluid without being limited to any particular EOS, it is convenient to represent the Carnot cycle using a temperature-entropy (T-S) phase diagram, instead of using a pressure-volume (P-V) phase diagram (see Fig. 20.1).

Let us carry out a First Law of Thermodynamics analysis of the Carnot cycle by following an element of the working fluid (closed system) as it flows around the Carnot cycle from state1→state2→state3→state4→state1 (see Fig. 20.1). Specifically,

$$\Delta\underline{U}\big|_{cycle} = 0 = \underline{Q}\big|_{cycle} + \underline{W}\big|_{cycle} \tag{20.61}$$

where the first term on the right-hand side of the second equal sign in Eq. (20.61) is the total heat absorbed by the element of fluid as it completes the Carnot cycle, and the second term on the right-hand side of the second equal sign in Eq. (20.61) is the total work done on the element of fluid as it completes the Carnot cycle. Rearranging Eq. (20.61) yields:

$$-\underline{W}\big|_{cycle} = \underline{Q}\big|_{cycle} \tag{20.62}$$

The heat interactions along the Carnot cycle are as follows (see Fig. 20.1):

$$(a)\ (\delta \underline{Q})_{rev,a} = T_H(d\underline{S})_{rev,a} \Rightarrow \underline{Q}_a = T_H(\underline{S}_2 - \underline{S}_1) \tag{20.63}$$

$$(b)\ (\delta \underline{Q})_{rev,b} = 0 \Rightarrow \underline{Q}_b = 0 \tag{20.64}$$

$$(c)\ (\delta \underline{Q})_{rev,c} = T_C(d\underline{S})_{rev,c} \Rightarrow \underline{Q}_C = T_C(\underline{S}_4 - \underline{S}_3) \tag{20.65}$$

$$(d)\ (\delta \underline{Q})_{rev,d} = 0 \Rightarrow \underline{Q}_d = 0 \tag{20.66}$$

Adding up Eqs. (20.63), (20.64), (20.65), and (20.66) yields $\underline{Q}|_{cycle}$, the total amount of heat absorbed by the element of fluid as it flows from 1→2→3→4→1. Specifically,

$$Q|_{cycle} = T_H(\underline{S}_2 - \underline{S}_1) + 0 + T_C(\underline{S}_4 - \underline{S}_3) + 0 \qquad (20.67)$$

However, the (T-$\underline{S}$) phase diagram in Fig. 20.1 shows that:

$$\underline{S}_4 = \underline{S}_1, \text{and } \underline{S}_3 = \underline{S}_2 \qquad (20.68)$$

Using the two equalities in Eq. (20.68) in Eq. (20.67) yields:

$$Q|_{cycle} = (T_H - T_C)(\underline{S}_2 - \underline{S}_1) \qquad (20.69)$$

Using Eq. (20.69) in Eq. (20.62) yields:

$$W|_E = W_E = Q|_{cycle} = (T_H - T_C)(\underline{S}_2 - \underline{S}_1) \qquad (20.70)$$

Recall that the efficiency of a Carnot engine (the element of working fluid in this case) is defined as follows:

$$\eta_C = \frac{(\text{Work Done by the Engine})}{\left(\begin{array}{c}\text{Heat Absorbed by the Engine} \\ \text{from the Hot Reservoir at } T_H\end{array}\right)} \qquad (20.71)$$

or, using Eq. (20.70) and Eq. (20.63) in Eq. (20.71), yields:

$$\eta_C = \frac{-W|_{cycle}}{Q_a} = \frac{(T_H - T_C)(\underline{S}_2 - \underline{S}_1)}{T_H(\underline{S}_2 - \underline{S}_1)} \qquad (20.72)$$

Cancelling the $\underline{S}_2 - \underline{S}_1$ terms in Eq. (20.72) yields the desired result, that is,

$$\eta_C = \frac{(T_H - T_C)}{T_H} = 1 - \frac{T_C}{T_H} \qquad (20.73)$$

Part II
Mixtures: Models and Applications to Phase and Chemical Reaction Equilibria

Lecture 21

Extensive and Intensive Mixture Properties and Partial Molar Properties

21.1 Introduction

The material presented in this lecture is adapted from Chapter 9 in T&M. First, we will discuss extensive and intensive differentials of mixture thermodynamic properties, $\underline{B}$ (extensive) and B (intensive), using two sets of (n + 2) independent variables. The first set consists of the temperature, the pressure, and the n mole numbers of all the components comprising the mixture. The second set consists of the temperature, the pressure, the mixture composition, and the total number of moles of the components comprising the mixture. Second, we will introduce partial molar properties of thermodynamic properties of mixtures. Finally, we will show how to "assemble" an extensive thermodynamic property of a mixture comprising n components by adding up the products of the partial molar property of component i times the number of moles of component i.

In Part I, we considered pure materials (n = 1). However, there are many systems of fundamental and practical importance where several components (n > 1) are present, for example, mixtures of: (i) solvent (say, water) and salt (say, NaCl), water and polymer (say, polyethylene glycol), water and protein (say, ovalbumin), water and colloids (say, silica particles), or water and surfactant (say, sodium dodecyl sulfate), (ii) mixtures of gases (say, methane and carbon dioxide), or mixtures of liquids (say, water and methanol), just to mention a few. The new important feature in all these mixtures is that the composition of the mixture, in addition to its temperature (T) and pressure (P), controls the thermodynamic behavior of the mixture. In other words, while for a pure material (n − 1), the intensive (molar) properties depend only on T and P; for mixtures (n > 1), these properties depend on T, P, and the mixture composition (hereafter, referred to as composition). This will add a new dimension to the calculation of changes in the thermodynamic properties of mixtures which we did not encounter in Part I.

With the above in mind, in Lectures 21, 22, 23, 24, 25, 26, 27, 28, 29, 30, 31, 32, 33, 34, 35, 36, and 37 of Part II, in addition to introducing several new concepts,

© Springer Nature Switzerland AG 2020
D. Blankschtein, *Lectures in Classical Thermodynamics with an Introduction to Statistical Mechanics*, https://doi.org/10.1007/978-3-030-49198-7_21

including partial molar properties, mixing functions excess functions, mixture fugacities and fugacity coefficients, mixture activities and activity coefficients, an important challenge in going from n = 1 to n > 1 will involve the more complex notation of multiple indices that we need to keep track of. First, we will consider single-phase, multi-component (n > 1) mixtures. Subsequently, we will consider multi-phase, multi-component (n > 1) mixtures. Initially, we will assume that the various components comprising the mixture are all inert, postponing the treatment of chemical reactions to the conclusion of Part II (for details, please refer to the Table of Contents).

21.2   Extensive and Intensive Differentials of Mixtures

Postulate I states that any thermodynamic property can be expressed as a function of (n + 2) independent variables. For a single-phase, simple system, there are no restrictions regarding the choice of these (n + 2) variables.

On the other hand, for a mixture (n > 1), we will choose two convenient sets of (n + 2) independent variables to express an extensive mixture property, $\underline{B}$, or an intensive mixture property, B. Specifically,

(i) Set 1 : $\{T, P, N_1, N_2, \ldots, N_n\}$

In terms of Set 1, we have:

$$\underline{B} = \underline{B}(T, P, N_1, N_2, \ldots, N_n)$$

$$B = B(T, P, N_1, N_2, \ldots, N_n)$$

where N_i is the number of moles of component i (1, 2, ..., n).

(ii) Set 2 : $\{T, P, x_1, x_2, \ldots, x_{n-1}, N\}$

In terms of Set 2, we have:

$$\underline{B} = \underline{B}(T, P, x_1, x_2, \ldots, x_{n-1}, N)$$
$$B = B(T, P, x_1, x_2, \ldots, x_{n-1}, N)$$

where x_i is the mole fraction of component i (1, 2, ..., n) and is given by:

$$x_i = N_i/N, \quad N = \sum_{i=1}^{n} N_i \tag{21.1}$$

It is noteworthy that instead of using T and P, we can use the two general variables, Y_1 and Y_2, which can be extensive or intensive. However, consideration of such variables is not necessary here. Next, we will analyze separately how $\underline{B}$ and B vary as a function of the variables in Set 1 and Set 2, respectively.

21.3 Choose Set 1: {T, P, N₁, ..., Nₙ} and Analyze $\underline{B} = \underline{B} \, (T, P, N_1, ..., N_n)$

We begin with the differential of $\underline{B}$, given by:

$$d\underline{B} = \left(\frac{\partial \underline{B}}{\partial T}\right)_{P,N_i} dT + \left(\frac{\partial \underline{B}}{\partial P}\right)_{T,N_i} dP + \sum_{i=1}^{n} \left(\frac{\partial \underline{B}}{\partial N_i}\right)_{T,P,N_{j[i]}} dN_i \tag{21.2}$$

Recall that, in Eq. (21.2), the subscript N_i in the partial derivatives of $\underline{B}$ with respect to temperature and pressure indicates that all the N_is remain constant. In addition, the subscript $N_{j[i]}$ in the partial derivative of $\underline{B}$ with respect to N_i indicates that all the N_js, except for N_i, remain constant.

The partial derivative of $\underline{B}$ with respect to N_i, at constant $T,P,N_{j[i]}$, appears so frequently in the study of mixtures, that it was given its own name and symbol. Specifically,

$$\left(\frac{\partial \underline{B}}{\partial N_i}\right)_{T,P,N_{j[i]}} = \overline{B}_i(\text{Partial molar B of component i}) \tag{21.3}$$

where $\overline{B}_i$ is an intensive mixture property. Recall that we have already seen that when $B = \underline{G}$, $\overline{B}_i = \overline{G}_i = \mu_i$, the chemical potential of component i. We will discuss partial molar properties in more detail in Lecture 22, including assigning physical significance to them.

If we Euler integrate the $d\underline{B}$ relation in Eq. (21.2), recalling that T and P are intensive variables, we obtain:

$$\underline{B} = 0 + 0 + \sum_{i=1}^{n} \overline{B}_i N_i \tag{21.4}$$

or

$$\underline{B} = \sum_{i=1}^{n} \overline{B}_i N_i \tag{21.5}$$

Equation (21.5) is a central result, which indicates that any extensive property, $\underline{B}$, of a mixture can be "assembled" from the set of n partial molar properties, $\{\overline{B}_1, \overline{B}_2, \ldots, \overline{B}_n\}$, using the mole numbers, $\{N_1, N_2, \ldots, N_n\}$, as the weighting factors.

21.4   Important Remarks

(i) In a mixture $(n > 1)$, the partial molar property, $\overline{B}_i$, plays a role which is analogous to that played by the molar property, B, in a pure (one-component, $n = 1$) system. In other words:

• For $n = 1 \Rightarrow \underline{B} = NB(T, P)$ $\hspace{2cm}$ (21.6)

• For $n > 1 \Rightarrow \underline{B} = \sum_{i=1}^{n} N_i \overline{B}_i$ $\hspace{2cm}$ (21.7)

where $\overline{B}_i = \overline{B}_i (T, P, x_1, x_2, \ldots, x_{n-1})$, and $x_1, x_2, \ldots, x_{n-1}$ is the composition.

(ii) When $n = 1 \Rightarrow \overline{B}_i = B_i (T, P)$.

21.5   Choose Set 2: $\{T, P, x_1, \ldots, x_{n-1}, N\}$ and Analyze $\underline{B} = \underline{B} (T, P, x_1, \ldots, x_{n-1}, N)$

Note that in Set 2, we eliminated the mole fraction:

$$x_n = 1 - \sum_{i=1}^{n-1} x_i \tag{21.8}$$

We begin with the differential of $\underline{B}$, given by:

$$dB = \left(\frac{\partial \underline{B}}{\partial T}\right)_{P,x,N} dT + \left(\frac{\partial \underline{B}}{\partial P}\right)_{T,x,N} dP + \sum_{i=1}^{n-1} \left(\frac{\partial \underline{B}}{\partial x_i}\right)_{T,P,x[i,n],N} dx_i$$

$$+ \left(\frac{\partial \underline{B}}{\partial N}\right)_{T,P,x} dN \tag{21.9}$$

In Eq. (21.9), the following shorthand notation was used in the partial derivatives:

x – Indicates that all the x_is are kept constant

x[i,n] – Indicates that, when we change x_i, we also change the eliminated mole fraction x_n, because $dx_n = - dx_i$. Accordingly, x[i,n] indicates that all the xs, except for x_i and x_n, are kept constant

In Eq. (21.9) for $d\underline{B}$, the variables, $\underline{B}$ and N, are extensive, while the variables, T, P, and x_i, are intensive. Accordingly, if we Euler integrate Eq. (21.9), we obtain:

$$\underline{B} = 0 + 0 + 0 + \left(\frac{\partial \underline{B}}{\partial N}\right)_{T,P,x} N \qquad (21.10)$$

or

$$\left(\frac{\partial \underline{B}}{\partial N}\right)_{T,P,x} = \frac{\underline{B}}{N} = B \qquad (21.11)$$

We can relate the partial derivative, $(\partial \underline{B}/\partial x_i)_{T,P,x[i,n],N}$, in Eq. (21.9) to the partial molar properties, $\bar{B}_i$ and $\bar{B}_n$, as follows. Starting with $\underline{B} = \underline{B}(T, P, N_1, …, N_n)$, we write the differential of $\underline{B}$ as follows:

$$d\underline{B} = \left(\frac{\partial \underline{B}}{\partial T}\right)_{P,N_i} dT + \left(\frac{\partial \underline{B}}{\partial P}\right)_{T,N_i} dP + \sum_{j=1}^{n} \bar{B}_j \, dN_j \qquad (21.12)$$

Differentiating $d\underline{B}$ in Eq. (21.12) with respect to x_i, keeping T, P, x[i,n], and N constant, yields:

$$\left(\frac{\partial \underline{B}}{\partial x_i}\right)_{T,P,x[i,n],N} = 0 + 0 + \sum_{j=1}^{n} \bar{B}_j \left(\frac{\partial N_j}{\partial x_i}\right)_{T,P,x[i,n],N} \qquad (21.13)$$

Because $N_j = x_j N$, it follows that:

$$\left(\frac{\partial N_j}{\partial x_i}\right)_{T,P,x[i,n],N} = N \left(\frac{\partial x_j}{\partial x_i}\right)_{x[i,n]} = N \bullet \left\{ \begin{array}{l} 0, \text{if } j \neq i \text{ and } n \\ +1, \text{if } j = i \\ -1, \text{if } j = n \end{array} \right\} \qquad (21.14)$$

Using Eq. (21.14) on the right-hand side of Eq. (21.13), we obtain:

$$\sum_{j=1}^{n} \bar{B}_j \left(\frac{\partial N_j}{\partial x_i}\right)_{T,P,x[i,n],N} = 0 + N\bar{B}_i - N\bar{B}_n = N\left(\bar{B}_i - \bar{B}_n\right) \qquad (21.15)$$

Using Eq. (21.15) in Eq. (21.3) yields the desired result:

$$\left(\frac{\partial B}{\partial x_i}\right)_{T,P,x\,[i,j],N} = N\left(\overline{B}_i - \overline{B}_n\right) \tag{21.16}$$

21.6   Choose Set 1: $\{T, P, N_1, \ldots, N_n\}$ and Analyze $B = B(T, P, N_1, \ldots, N_n)$

We begin with the differential of B, given by:

$$dB = \left(\frac{\partial B}{\partial T}\right)_{P,N_i} dT + \left(\frac{\partial B}{\partial P}\right)_{T,N_i} dP + \sum_{i=1}^{n} \left(\frac{\partial B}{\partial N_i}\right)_{T,P,N_{j\,[i]}} dN_i \tag{21.17}$$

Because B, T, and P are intensive, and the N_is $\{i = 1, 2, \ldots, n\}$ are extensive, if we Euler integrate the dB expression in Eq. (21.17), we obtain:

$$0 = 0 + 0 + \sum_{i=1}^{n} \left(\frac{\partial B}{\partial N_i}\right)_{T,P,N_{j\,[i]}} N_i \tag{21.18}$$

or

$$\sum_{i=1}^{n} \left(\frac{\partial B}{\partial N_i}\right)_{T,P,N_{j\,[i]}} N_i = 0 \tag{21.19}$$

Equation (21.19) shows that the n N_is satisfy a constraint, and therefore, cannot be varied independently. In Eq. (21.19):

$$\left(\frac{\partial B}{\partial N_i}\right)_{T,P,N_{j\,[i]}} \neq \overline{B}_i \tag{21.20}$$

The readers are encouraged to always include the underbar to highlight an extensive property.

One can also show that (see Appendix F in T&M):

$$\left(\frac{\partial B}{\partial N_i}\right)_{T,P,N_{j\,[i]}} = \frac{1}{N}\left(\overline{B}_i - B\right) \tag{21.21}$$

or

$$\overline{B}_i = B + N\left(\frac{\partial B}{\partial N_i}\right)_{T,P,N_{j[i]}} \tag{21.22}$$

Equation (21.22) for $\overline{B}_i$ is particularly useful to compute $\overline{B}_i$ when $B = B$ (T, P, N_1, ..., N_n) is known.

21.7   Choose Set 2: {T, P, x₁, ..., x_{n-1}, N} and Analyze B = B (T, P, x₁, ..., x_{n-1}, N)

Note that B is intensive, and the $(n + 1)$ variables {T, P, x_1, ..., x_{n-1}} are also intensive. Accordingly, and consistent with the Corollary to Postulate I, we will show that B does not depend on N.

We begin with the differential of B (T, P, x_1, ..., x_{n-1}, N), given by:

$$dB = \left(\frac{\partial B}{\partial T}\right)_{P,x,N} dT + \left(\frac{\partial B}{\partial P}\right)_{T,x,N} dP + \sum_{i=1}^{n-1}\left(\frac{\partial B}{\partial x_i}\right)_{T,P,x[i,n],N} dx_i$$
$$+ \left(\frac{\partial B}{\partial N}\right)_{T,P,x} dN \tag{21.23}$$

In Eq. (21.23) for dB only N is extensive, and therefore, if we Euler integrate the dB expression, we obtain:

$$0 = 0 + 0 + 0 + \left(\frac{\partial B}{\partial N}\right)_{T,P,x} N \tag{21.24}$$

or

$$\left(\frac{\partial B}{\partial N}\right)_{T,P,x} N = 0 \tag{21.25}$$

Because $N \neq 0$, it follows that:

$$\left(\frac{\partial B}{\partial N}\right)_{T,P,x} = 0 \tag{21.26}$$

Equation (21.26) shows that the intensive property, B, depends on the $(n + 1)$ intensive properties {T, P, x_1, ..., x_{n-1}}, as required by the Corollary to Postulate I.

We can also show that (see Appendix F in T&M):

$$\left(\frac{\partial B}{\partial x_i}\right)_{T,P,x[i,n]} = \overline{B}_i - \overline{B}_n \tag{21.27}$$

where Eq. (21.27) follows from Eq. (21.16). Specifically,

$$\left(\frac{\partial \underline{\underline{B}}}{\partial x_i}\right)_{T,P,x[i,n],N} = N\left(\frac{\partial B}{\partial x_i}\right)_{T,P,x[i,n]} = N\left(\overline{B}_i - \overline{B}_n\right) \tag{21.28}$$

Cancelling the two Ns in Eq. (21.28) yields:

$$\left(\frac{\partial B}{\partial x_i}\right)_{T,P,x[i,n]} = \left(\overline{B}_i - \overline{B}_n\right) \tag{21.29}$$

We can derive another useful relation for dB as follows. We have just shown that:

$$B = B(T, P, x_1, x_2, \ldots, x_{n-1}) \tag{21.30}$$

Taking the differential of B yields:

$$dB = \left(\frac{\partial B}{\partial T}\right)_{P,x} dT + \left(\frac{\partial B}{\partial P}\right)_{T,x} dP + \sum_{i=1}^{(n-1)} \left(\frac{\partial B}{\partial x_i}\right)_{T,P,x[i,n]} dx_i \tag{21.31}$$

Using Eq. (21.29) in the last sum in Eq. (21.31) yields:

$$\sum_{i=1}^{(n-1)} \left(\overline{B}_i - \overline{B}_n\right) dx_i = \sum_{i=1}^{(n-1)} \overline{B}_i dx_i - \sum_{i=1}^{(n-1)} \overline{B}_n dx_i \tag{21.32}$$

The second sum on the right-hand side of Eq. (21.32) can be expressed as follows:

$$-\overline{B}_n \left(\sum_{i=1}^{(n-1)} dx_i\right) = -\overline{B}_n(-dx_n) = \overline{B}_n dx_n \tag{21.33}$$

Using Eq. (21.33) in the sum on the left-hand side of Eq. (21.32) yields:

$$\sum_{i=1}^{(n-1)} \left(\overline{B}_i - \overline{B}_n\right) dx_i = \sum_{i=1}^{(n-1)} \overline{B}_i \, dx_i + \overline{B}_n dx_n \tag{21.34}$$

or

$$\sum_{i=1}^{(n-1)} \left(\overline{B}_i - \overline{B}_n \right) dx_i \;=\; \sum_{i=1}^{n} \overline{B}_i \, dx_i \tag{21.35}$$

Using Eq. (21.29) in the last term in Eq. (21.31), and then using Eq. (21.35), we obtain:

$$dB = \left(\frac{\partial B}{\partial T} \right)_{P,x} dT + \left(\frac{\partial B}{\partial P} \right)_{T,x} dP + \sum_{i=1}^{n} \overline{B}_i \, dx_i \tag{21.36}$$

Lecture 22

Generalized Gibbs-Duhem Relations for Mixtures, Calculation of Partial Molar Properties, and Sample Problem

22.1 Introduction

The material presented in this lecture is adapted from Chapter 9 in T&M. First, we will provide physical insight into the concept of a partial molar property by considering the volume of the system as the property of interest. Second, we will introduce the partial molar operator and use it to derive useful relations between partial molar properties. Third, we will consider three cases to illustrate how to calculate partial molar properties of (i) an extensive thermodynamic property which depends on T, P, and the n mole numbers, (ii) an intensive (molar) thermodynamic property which depends on T, P, and the n mole numbers, and (iii) an intensive (molar) thermodynamic property which depends on T, P, and the mixture composition. Fourth, we will solve Sample Problem 22.1 to calculate the partial molar Bs of components 1 and 2 in a binary mixture, given the mixture molar property B as a function of T, P, and the mole fraction of component 2. Fifth, we will provide a useful geometrical interpretation of the results derived in item four above. Finally, we will derive the generalized Gibbs-Duhem relations for mixtures.

22.2 Partial Molar Properties

To gain physical insight into the concept of a partial molar property, including how it differs from a molar property, it is convenient to consider the volume of the system as the property of interest.

Pure Water (w) at T $= 25\,°C$ and P $= 1$ bar. What is the volume occupied by one mole of water at 25 °C and 1 bar? In other words, what is the molar volume of water, V_w (25 °C, 1 bar)? In pure water, the measured value is:

© Springer Nature Switzerland AG 2020   223
D. Blankschtein, *Lectures in Classical Thermodynamics with an Introduction to Statistical Mechanics*, https://doi.org/10.1007/978-3-030-49198-7_22

$$V_w(T = 25^\circ C, P = 1 \text{ bar}) = 18 \text{ cm}^3/\text{mol} \tag{22.1}$$

On average, in pure water, at this T and P, the water molecules are a distance d (w, pure) from each other.

A 50:50 Mixture of Water (w) and Methanol (m) at T = 25 °C and P = 1 bar.
What is the volume occupied by one mole of water at these conditions? In other words, what is the partial molar volume of water, $\overline{V}_w$ (25 °C, 1 bar, $x_w = 0.5$)? Recall that $x_w + x_m = 1$.

In the binary water (w)-methanol (m) mixture, the measured value is:

$$\overline{V}_w(T = 25^\circ C, P = 1 \text{ bar}, x_w = 0.5) = 14 \text{ cm}^3/\text{mol} \tag{22.2}$$

A comparison of Eq. (22.2) and Eq. (22.1) reveals that, in the binary water-methanol mixture, water contracts! In other words, the volume occupied by one mole of water in the binary water-methanol mixture is smaller than that occupied by one mole of water in pure water at the same T and P. That is,

$$\overline{V}_w \ (25^\circ C, 1 \text{ bar}, x_w = 05) < V_w(25^\circ C, 1 \text{ bar}) \tag{22.3}$$

Molecularly, due to the presence of the methanol molecules in the mixture with water, the average distance between the water molecules, d (w, mixture), decreases, that is, d (w, mixture) < d (w, pure). It turns out that depending on the water mixture:

$$\overline{V}_w \overset{<}{\underset{=}{>}} V_w \tag{22.4}$$

at the same T and P. Understanding the relation between $\overline{V}_w$ and V_w is not trivial and requires a deep understanding of the intermolecular forces operating between the various components comprising the water mixture.

22.3   Useful Relations Between Partial Molar Properties

In Lecture 21, we saw that:

$$\overline{B}_i = \left(\frac{\partial B}{\partial N_i}\right)_{T,P,N_{j[i]}} \tag{22.5}$$

The operator

$$\left(\frac{\partial}{\partial N_i}\right)_{T,P,N_{j[i]}} \tag{22.6}$$

is referred to as the partial molar operator. Clearly, if the partial molar operator in Eq. (22.6) operates on an extensive property $\underline{B}$ which is multiplied by either T or P, it follows that:

$$\left(\frac{\partial}{\partial N_i}\right)_{T,P,N_{j[i]}} T\underline{B} = T\bar{B}_i \tag{22.7}$$

and

$$\left(\frac{\partial}{\partial N_i}\right)_{T,P,N_{j[i]}} P\underline{B} = P\bar{B}_i \tag{22.8}$$

The relations in Eqs. (22.7) and (22.8) imply that:

$$\underline{H} = \underline{U} + P\underline{V} \implies \bar{H}_i = \bar{U}_i + P\bar{V}_i \tag{22.9}$$

and

$$\underline{A} = \underline{U} - T\underline{S} \implies \bar{A}_i = \bar{U}_i - T\bar{S}_i \tag{22.10}$$

Recall that $\bar{B}_i$ is an intensive mixture property, and according to the Corollary to Postulate I, it can be expressed in terms of $(n + 1)$ independent intensive properties. For example:

$$\bar{B}_i = \bar{B}_i (T, P, x_1, \ldots, x_{n-1}) \tag{22.11}$$

Taking the differential of $\bar{B}_i$ in Eq. (22.11) yields:

$$d\bar{B}_i = \left(\frac{\partial \bar{B}_i}{\partial T}\right)_{P,x} dT + \left(\frac{\partial \bar{B}_i}{\partial P}\right)_{T,x} dP + \sum_{j=1}^{(n-1)} \left(\frac{\partial \bar{B}_i}{\partial x_j}\right)_{T,P,x[j,n]} dx_j \tag{22.12}$$

In Eq. (22.12), x_n was eliminated in the last term.

22.4   How Do We Calculate $\bar{B}_i$? Cases 1, 2, and 3

If experimental data, or analytical expressions, of $\underline{B}$ or B are available, $\bar{B}_i$ can be calculated directly. Consider three possible cases:

Case 1: $\underline{B} = \underline{B}(T, P, N_1, \ldots, N_n)$ is known.

In this case, $\overline{B}_i$ is calculated using the definition of the partial molar B of component i, which we repeat below for completeness:

$$\overline{B}_i = \left(\frac{\partial \underline{B}}{\partial N_i}\right)_{T,P,N_{j[i]}} \tag{22.13}$$

Case 2: $B = B(T, P, N_1, \ldots, N_n)$ is known. In this case,

(i) We can first multiply B by N to obtain $\underline{B} = NB(T, P, N_1, \ldots, N_n)$ and then use Case 1.

or

(ii) We can first obtain $(\partial B/\partial N_i)_{T,P,N_{j[i]}}$ directly from the given data and then use Eq. (21.22) presented in Lecture 21, which we repeat below for completeness:

$$\left(\frac{\partial B}{\partial N_i}\right)_{T,P,N_{j[i]}} = \frac{1}{N}\left(\overline{B}_i - B\right) \Rightarrow \overline{B}_i = B + N\left(\frac{\partial B}{\partial N_i}\right)_{T,P,N_{j[i]}} \tag{22.14}$$

where the partial derivative of B with respect to N_i can be calculated directly using the given data.

Case 3: $B = B(T, P, x_1, \ldots, x_{i-1}, x_{i+1}, \ldots, x_n)$ is known.

In this case, x_i was eliminated from the set of n mole fractions, and as a result, B depends only on the (n-1) independent mole fractions:

$$\{x_1, \ldots, x_{i-1}, x_{i+1}, \ldots, x_n\} \tag{22.15}$$

In addition,

(i) Similar to Case 2 (i) above, we can first multiply the given B by N to obtain $\underline{B} = NB\ (T, P, x_1, \ldots, x_{i-1}, x_{i+1}, \ldots, x_n)$ and then use Case 1.

or

(ii) We can first obtain $(\partial B/\partial x_j)_{T,P,x[j,i]}$ directly from the given data and then use this partial derivative in one of the expressions presented above. For example, we know that:

$$\overline{B}_i = B + N\left(\frac{\partial B}{\partial N_i}\right)_{T,P,N_{j[i]}} \tag{22.16}$$

Accordingly, let us relate $(\partial B/\partial N_i)_{T,P,N_{j[i]}}$ in Eq. (22.16) to $(\partial B/\partial x_j)_{T,P,x[j,i]}$. Because $B = B(T, P, x_1, \ldots, x_{i-1}, x_{i+1}, \ldots, x_n)$, where x_i was eliminated, the differential of B is given by:

$$dB = \left(\frac{\partial B}{\partial T}\right)_{P,x} dT + \left(\frac{\partial B}{\partial P}\right)_{T,x} dP + \sum_{j \neq i} \left(\frac{\partial B}{\partial x_j}\right)_{T,P,x[j,i]} dx_j \qquad (22.17)$$

Differentiating dB in Eq. (22.17) with respect to N_i, at constant T, P, $N_{j[i]}$, yields:

$$\left(\frac{\partial B}{\partial N_i}\right)_{T,P,N_{j[i]}} = \sum_{j \neq i} \left(\frac{\partial B}{\partial x_j}\right)_{T,P,x[j,i]} \cdot \left(\frac{\partial x_j}{\partial N_i}\right)_{T,P,N_{j[i]}} \qquad (22.18)$$

where

$$\left(\frac{\partial x_j}{\partial N_i}\right) = \left(\frac{\partial \frac{N_j}{N}}{\partial N_i}\right) = \frac{N\left(\frac{\partial N_j}{\partial N_i}\right) - N_j\left(\frac{\partial N}{\partial N_i}\right)}{N^2} = \frac{0 - N_j}{N^2} = -\frac{N_j}{N^2}$$

$$= -\frac{x_j}{N} \qquad (22.19)$$

Using Eq. (22.19) in Eq. (22.18) yields:

$$\left(\frac{\partial B}{\partial N_i}\right)_{T,P,N_{j[l]}} = \sum_{j \neq i} \left(\frac{\partial B}{\partial x_j}\right)_{T,P,x[j,i]} \left(-\frac{x_j}{N}\right) \qquad (22.20)$$

Using Eq. (22.20) in Eq. (22.16) and rearranging, we obtain:

$$\overline{B}_i = B - \sum_{j \neq i} x_j \left(\frac{\partial B}{\partial x_j}\right)_{T,P,x[j,i]} \qquad (22.21)$$

It is noteworthy that Eq. (22.21) for $\overline{B}_i$ should only be used for a data set in which the mole fraction, x_i, was eliminated, that is, when B is known as a function of:

$$\{T, P, x_1, \ldots, x_{i-1}, x_{i+1}, \ldots, x_n\} \qquad (22.22)$$

If we would like to obtain a partial molar $\overline{B}_k$, where k is in the set of variables on which B depends (i.e., where x_k was not eliminated), we need to use a different relation (see T&M, Section 9.3, pages 331 and 332). Specifically,

$$\overline{B}_k = B + \left(\frac{\partial B}{\partial x_k}\right)_{T,P,x[k,i]} - \sum_{j\neq i} x_j \left(\frac{\partial B}{\partial x_j}\right)_{T,P,x[j,i]} \tag{22.23}$$

where the mole fraction, x_i, was eliminated. Note that Eq. (22.23) for $\overline{B}_k$ should only be used for all the x_ks which are in the set of variables on which B depends.

22.5   Sample Problem 22.1

Consider a binary mixture of components 1 and 2, where $B(T, P, x_2)$ is known. Calculate $\overline{B}_1(T, P, x_2)$ and $\overline{B}_2(T, P, x_2)$ and provide a geometrical interpretation of your results.

22.5.1   Solution

Clearly, $x_1 = 1 - x_2$ was eliminated from the set of the two mole fractions. In order to compute $\overline{B}_1$, we use Eq. (22.21) for a mole fraction that was eliminated (x_1 in this case). Specifically,

$$\overline{B}_1 = B - \sum_{j\neq 1} x_j \left(\frac{\partial B}{\partial x_j}\right)_{T,P,x[j,i]} \tag{22.24}$$

or, because $j = 2$,

$$\overline{B}_1 = B - x_2 \left(\frac{\partial B}{\partial x_2}\right)_{T,P} = \overline{B}_1(T,P,x_2) \tag{22.25}$$

To compute $\overline{B}_2$, we use Eq. (22.23), where $k = 2$ is in the set of mole fractions on which B depends. Specifically, for $k = 2$, $i = 1$, and $j = 2$, we obtain:

$$\overline{B}_2 = B + \left(\frac{\partial B}{\partial x_2}\right)_{T,P} - \sum_{2\neq 1}^{n} x_2 \left(\frac{\partial B}{\partial x_2}\right)_{T,P} \tag{22.26}$$

where in Eq. (22.26), the sum is redundant. In other words:

$$\overline{B}_2 = B + \left(\frac{\partial B}{\partial x_2}\right)_{T,P} - x_2 \left(\frac{\partial B}{\partial x_2}\right)_{T,P} \tag{22.27}$$

or

$$\overline{B}_2 = B + (1 - x_2)\left(\frac{\partial B}{\partial x_2}\right)_{T,P} = \overline{B}_2\,(T, P, x_2) \qquad (22.28)$$

The expressions for $\overline{B}_1$ and $\overline{B}_2$ in Eq. (22.25) and Eq. (22.28), respectively, have a very nice geometrical interpretation that can be used to graphically compute $\overline{B}_1$ and $\overline{B}_2$. Suppose that we are given B as a function of x_2, at constant T and P, in graphical format (see Fig. 22.1). In Fig. 22.1, as well as in the derivation below, we have replaced x by X to enhance visualization.

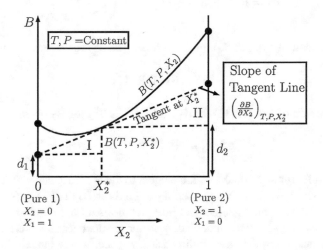

Fig. 22.1

Let us consider triangles I and II in Fig. 22.1:

Triangle I: Using Trigonometry

$$\frac{B\left(T, P, X_2^*\right) - d_1}{X_2^* - 0} = \left(\frac{\partial B}{\partial X_2}\right)_{T,P\,|\,X_2^*} \qquad (22.29)$$

where $|\ X_2^*$ indicates that the partial derivative is evaluated at $X_2 = X_2^*$. Equation (22.29) can be rearranged as follows (see Fig. 22.1):

$$d_1 = B\left(T, P, X_2^*\right) - X_2^*\left(\frac{\partial B}{\partial X_2}\right)_{T,P\,|\,X_2^*} \qquad (22.30)$$

Triangle II: Using trigonometry

$$\frac{d_2 - B\left(T, P, X_2^*\right)}{1 - X_2^*} = \left(\frac{\partial B}{\partial X_2}\right)_{T,P \mid X_2^*} \tag{22.31}$$

Rearranging Eq. (22.31) yields (see Fig. 22.1):

$$d_2 = B\left(T, P, X_2^*\right) + (1 - X_2) \left(\frac{\partial B}{\partial X_2}\right)_{T,P \mid X_2^*} \tag{22.32}$$

A comparison of Eq. (22.30) for d_1 with Eq. (22.25) for $\overline{B}_1$ reveals that:

$$d_1 = \overline{B}_1\left(T, P, X_2^*\right) \tag{22.33}$$

A comparison of Eq. (22.32) for d_2 with Eq. (22.28) for $\overline{B}_2$ reveals that:

$$d_2 = \overline{B}_2\left(T, P, X_2^*\right) \tag{22.34}$$

Accordingly, for any $0 < X_2^* < 1$, a tangent to the B vs. X_2 curve at X_2^* intercepts the $X_2 = 0$ axis at $\overline{B}_1$ (T, P, X_2^*) and the $X_2 = 1$ axis at $\overline{B}_2$ (T, P, X_2^*) (see Fig. 22.1). This provides a very nice geometrical interpretation. As the number of mole fractions increases, the geometrical depiction of our results becomes increasingly challenging, because it requires a multidimensional space, where tangent lines become tangent hyper planes in n dimensional space.

22.6   Generalized Gibbs-Duhem Relations for Mixtures

For an n-component system, there are n partial molar quantities representing any extensive property $\underline{B}$: $\overline{B}_1, \overline{B}_2, \overline{B}_3, \ldots, \overline{B}_n$. These n intensive properties, along with T and P, form a set of (n + 2) intensive properties. According to the Corollary to Postulate I, only (n + 1) of these intensive properties are independent. The mathematical equations relating these (n + 2) intensive properties are called the generalized Gibbs-Duhem relations.

To find the relation between $\overline{B}_1, \overline{B}_2, \ldots, \overline{B}_n$, T and P, we proceed as follows:

$$\underline{B} = \underline{B}(T, P, N_1, \ldots, N_n) \tag{22.35}$$

$$d\underline{B} = \left(\frac{\partial \underline{B}}{\partial T}\right)_{P,N_i} dT + \left(\frac{\partial \underline{B}}{\partial P}\right)_{T,N_i} dP + \sum_{i=1}^{n} \overline{B}_i \, dN_i \tag{22.36}$$

where N_i is a shorthand notation which indicates that $N_1, N_2, \ldots, N_n$ are all constant. Euler integrating Eq. (22.36) yields:

$$\underline{B} = 0 + 0 + \sum_{i=1}^{n} \overline{B}_i \, N_i \tag{22.37}$$

or

$$\underline{B} = \sum_{i=1}^{n} \overline{B}_i \, N_i \tag{22.38}$$

The differential of Eq. (22.38) is given by:

$$d\underline{B} = \sum_{i=1}^{n} \overline{B}_i \, dN_i + \sum_{i=1}^{n} N_i \, d\overline{B}_i \tag{22.39}$$

Equating $d\underline{B}$ in Eq. (22.36) and Eq. (22.39), and cancelling the equal terms, yields:

$$\sum_{i=1}^{n} N_i \, d\overline{B}_i = \left(\frac{\partial \underline{B}}{\partial T}\right)_{P,N_i} dT + \left(\frac{\partial \underline{B}}{\partial P}\right)_{T,N_i} dP \tag{22.40}$$

Dividing Eq. (22.40) by $N = \sum_{i=1}^{n} N_i$ yields:

$$\sum_{i=1}^{n} x_i \, d\overline{B}_i = \left(\frac{\partial B}{\partial T}\right)_{P,x} dT + \left(\frac{\partial B}{\partial P}\right)_{T,x} dP \tag{22.41}$$

where, in Eq. (22.41), x is a shorthand notation which indicates that $x_1, x_2, \ldots, x_n$ are all constant.

Equations (22.40) and (22.41) are known as the generalized Gibbs-Duhem relations. Clearly, the $(n + 2)$ intensive properties $\{T, P, \overline{B}_1, \overline{B}_2, \ldots, \overline{B}_n\}$ are related, as shown in Eqs. (22.40) and (22.41). We note that Eqs. (22.40) and (22.41) are particularly useful when T and P are constant. In that case, the right-hand sides in Eqs. (22.40) and (22.41) are both equal to zero, which yields:

$$\sum_{i=1}^{n} N_i \, d\overline{B}_i = 0, \text{ or } \sum_{i=1}^{n} x_i \, d\overline{B}_i = 0 \, (\text{At constant T and P}) \tag{22.42}$$

Equation (22.42) shows that, at constant T and P, given (n-1) $\overline{B}_i$s, we can calculate the nth one by integration.

Note that when $\underline{B} = \underline{G}$, Eq. (22.36) can be expressed as follows:

$$d\underline{B} = d\underline{G} = -\underline{S}dT + \underline{V}dP + \sum_{i=1}^{n} \overline{G}_i \, dN_i \tag{22.43}$$

Equation (22.43) shows that:

- $\overline{B}_i = \overline{G}_i = \left(\dfrac{\partial \underline{G}}{\partial N_i}\right)_{T,P,N_{j[i]}} = \mu_i$ $\hspace{2cm}$ (22.44)

- $\left(\dfrac{\partial \underline{B}}{\partial T}\right)_{P,N_i} = \left(\dfrac{\partial \underline{G}}{\partial T}\right)_{P,N_i} = -\underline{S}$ $\hspace{2cm}$ (22.45)

- $\left(\dfrac{\partial \underline{B}}{\partial P}\right)_{T,N_i} = \left(\dfrac{\partial \underline{G}}{\partial P}\right)_{T,N_i} = \underline{V}$ $\hspace{2cm}$ (22.46)

Using Eqs. (22.44), (22.45), and (22.46) in Eq. (22.40) yields:

$$\sum_{i=1}^{n} N_i \, d\mu_i = -\underline{S}dT + \underline{V}dP \tag{22.47}$$

Equation (22.47) is the celebrated Gibbs-Duhem equation for a mixture of n components.

Lecture 23

Mixture Equations of State, Mixture Departure Functions, Ideal Gas Mixtures, Ideal Solutions, and Sample Problem

23.1 Introduction

The material presented in this lecture is adapted from Chapter 9 in T&M. First, we will solve Sample Problem 23.1 to calculate the partial molar enthalpy of component 2, and the molar enthalpy of a binary liquid mixture of components 1 and 2, given the partial molar enthalpy of component 1 as a function of T, P, and the mole fraction of component 1. Second, we will discuss various equations of state (EOS) for gas mixtures, including the ideal gas mixture EOS, the van der Waals mixture EOS, the Peng-Robinson mixture EOS, and the virial mixture EOS. As needed, we will present several composition-dependent mixing rules to relate the parameters of the mixture EOS to those of the EOS corresponding to the various components comprising the mixture. We will see that, as expected, when the mixture reduces to a single component, the mixture EOS reduces to the EOS of the single component. Third, we will discuss how to calculate changes in the thermodynamic properties of gas mixtures by generalizing the attenuated state approach, first presented in Lecture 16, and the departure function approach, first presented in Lecture 19, from a single component (n = 1) to several components (n > 1). Specifically, the isothermal variations will be calculated using a mixture EOS, and the temperature variations will be calculated using mixture heat capacity data, obtained through a composition average of the heat capacities of the various components comprising the gas mixture. Finally, we will define an ideal gas, an ideal gas mixture, and an ideal solution in terms of the chemical potential of component i in each case.

© Springer Nature Switzerland AG 2020
D. Blankschtein, *Lectures in Classical Thermodynamics with an Introduction to Statistical Mechanics*, https://doi.org/10.1007/978-3-030-49198-7_23

23.2   Sample Problem 23.1

For a binary liquid mixture of components 1 and 2, if the partial molar enthalpy $\overline{H}_1$ is known as a function of T, P, and x_1, calculate $\overline{H}_2$ and H. The data $\overline{H}_1 = f(x_1)$ is available at constant T and P.

23.2.1   Solution

We begin by writing the generalized Gibbs-Duhem relation for $B = H$ at constant T and P, derived in Lecture 22, which for components 1 and 2, is given by:

$$x_1 d\overline{H}_1 + x_2 d\overline{H}_2 = 0 \qquad (23.1)$$

Because we are given $\overline{H}_1 = \overline{H}_1(T, P, x_1)$, we differentiate Eq. (23.1) with respect to x_1, at constant T and P. This yields:

$$x_1 \left(\frac{\partial \overline{H}_1}{\partial x_1} \right)_{T,P} + x_2 \left(\frac{\partial \overline{H}_2}{\partial x_1} \right)_{T,P} = 0 \qquad (23.2)$$

where $x_2 = 1 - x_1$. Rearranging Eq. (23.2) yields:

$$\left(\frac{\partial \overline{H}_2}{\partial x_1} \right)_{T,P} = - \frac{x_1}{1 - x_1} \left(\frac{\partial \overline{H}_1}{\partial x_1} \right)_{T,P} \qquad (23.3)$$

where the partial derivative of $\overline{H}_1$ with respect to x_1, at constant T and P, is known as a function of x_1.

Integrating Eq. (22.3) from $x_1 = 0$ (pure component 2) to x_1 yields:

$$\int_0^{x_1} \left(\frac{\partial \overline{H}_2}{\partial x_1} \right)_{T,P} dx_1 = \int_0^{x_1} (d\overline{H}_2)_{T,P} = \overline{H}_2(T, P, x_1) - \overline{H}_2(T, P, 0)$$

$$= - \int_0^{x_1} \left(\frac{x_1}{1 - x_1} \right) \left(\frac{\partial \overline{H}_1}{\partial x_1} \right)_{T,P} dx_1 \qquad (23.4)$$

where $\overline{H}_2(T, P, 0) = H_2(T, P)$ is the molar enthalpy of pure component 2. Rearranging Eq. (23.4) yields:

$$\overline{H}_2\left(T, P, x_1\right) = H_2(T, P) - \int_0^{x_1} \left(\frac{x_1}{1-x_1}\right)\left(\frac{\partial \overline{H}_1}{\partial x_1}\right)_{T,P} dx_1 \qquad (23.5)$$

Finally, given $\overline{H}_1(T, P, x_1)$, and having calculated $\overline{H}_2(T, P, x_1)$ using Eq. (23.5), we can "assemble" H as follows:

$$H = x_1 \overline{H}_1 + x_2 \overline{H}_2 \qquad (23.6)$$

Equation (23.6) shows that if we know $(n-1)$ $\overline{B}_i$s, we can compute the nth partial molar B and then assemble B. In this example, $B = H$, and $n = 2$.

23.3   Equations of State for Gas Mixtures

In this lecture, and in all the coming ones dealing with mixtures, the notation y will be used to denote gas-phase mole fractions, and the notation x will be used to denote condensed (liquid or solid)-phase mole fractions.

23.3.1   Ideal Gas (IG) Mixture EOS

$$P\underline{V} = NRT; \quad N = \sum_{i=1}^{n} N_i (\text{Extensive form}) \qquad (23.7)$$

$$PV = RT (\text{Intensive form}) \qquad (23.8)$$

All mixture EOS must approach the IG Mixture EOS in the limits:

$$P \to 0, \quad \underline{V} \to \infty, \quad \text{or } \rho \to 0 \qquad (23.9)$$

23.3.2   van der Waals (vdW) Mixture EOS

Extensive Form

$$P = \frac{NRT}{\underline{V} - Nb_m} - \frac{N^2 a_m}{\underline{V}^2} \qquad (23.10)$$

where b_m and a_m are mixture parameters. Specifically,

$$Nb_m = \sum_{i=1}^{n} N_i b_i; N^2 a_m = \left(\sum_{i=1}^{n} N_i a_i^{1/2} \right)^2 \qquad (23.11)$$

where b_i and a_i are the pure component parameters.

Intensive Form

$$P = \frac{RT}{V - b_m} - \frac{a_m}{V^2} \qquad (23.12)$$

where

$$b_m = \sum_{i=1}^{n} y_i b_i \, (\text{Arithmetic mean}) \qquad (23.13)$$

$$a_m^{1/2} = \sum_{i=1}^{n} y_i a_i^{1/2} \, (\text{``Geometric'' mean}) \qquad (23.14)$$

In Eqs. (23.13) and (23.14), $y_i = N_i/N$ is the mole fraction of component i, and b_i and a_i are pure component parameters.

23.3.3   Peng-Robinson (PR) Mixture EOS

$$P = \frac{NRT}{\underline{V} - Nb_m} - \frac{N^2 a_m}{V(\underline{V} + Nb_m) + Nb_m(\underline{V} - Nb_m)} \quad (\text{Extensive form})$$

$$(23.15)$$

where

$$b_m = \sum_{i=1}^{n} y_i b_i \quad (\text{Arithmetic mean}) \qquad (23.16)$$

$$a_m = \sum_{i=1}^{n} \sum_{j=1}^{n} y_i y_j a_{ij} \quad (\text{Composition-weighted average}) \qquad (23.17)$$

$$a_{ij} = \left(1 - \delta_{ij}\right)\left(a_{ii}\, a_{jj}\right)^{1/2}, \text{ for } i \neq j \tag{23.18}$$

In Eq. (23.18), δ_{ij} is a binary interaction parameter and is typically a small number (0.1). In addition,

$$a_{ii} = a_i = a_i\left(T_{ci}\right)\alpha_i\left(\omega_i, T_{ri}\right); T_{ri} = T/T_{ci} \tag{23.19}$$

where T_{ci} and T_{ri} are the critical temperature and the reduced temperature of pure component i, respectively. In Eq. (23.19),

$$\alpha_i = \left[1 - \kappa_i\left(1 - T_{ri}^{1/2}\right)\right]^2; \ a_i\left(T_{ci}\right) = 0.045724\,\frac{R^2 T_{ci}^{\,2}}{P_{ci}} \tag{23.20}$$

where

$$\kappa_i = f\left(\omega_i\right) = A + B\omega_i + C\omega_i^{\,2} \tag{23.21}$$

In Eq. (23.16),

$$b_i = 0.07780\,\frac{R T_{ci}}{P_{ci}} \tag{23.22}$$

The parameters a_{ii}, α_i, κ_i, and b_i are pure component parameters. In addition, when $\delta_{ij} = 0$, it follows that $(a_m)_{PR} = (a_m)_{vdW}$.

23.3.4   Virial Mixture EOS

$$P = \frac{NRT}{V} + \frac{B_m N^2 RT}{V^2} + \frac{C_m N^3 RT}{V^3} + \ldots \quad \text{(Extensive form)} \tag{23.23}$$

$$P = \frac{RT}{V} + \frac{B_m RT}{V^2} + \frac{C_m RT}{V^3} + \ldots \quad \text{(Intensive form)} \tag{23.24}$$

where

$$B_m = \sum_{i=1}^{n}\sum_{j=1}^{n} y_i y_j B_{ij} \quad \text{(Mixture second virial coefficient)} \tag{23.25}$$

and

For $i = j \Rightarrow B_{ii} = B_i$ (Second virial coefficient of pure component i)   (23.26)

$$\text{For } i \neq j \Rightarrow \text{(i) } B_{ij} = \left(B_{ii}B_{jj}\right)^{1/2} = \left(B_i B_j\right)^{1/2} \qquad (23.27)$$

$$\text{(ii) } B_{ij} = \left(1 - s_{ij}\right)^{1/2} \left(B_i B_j\right)^{1/2} \qquad (23.28)$$

In Eq. (23.28), s_{ij} is a binary interaction parameter which is different than δ_{ij}.

23.4   Calculation of Changes in the Thermodynamic Properties of Gas Mixtures

For a closed, multi-component gas mixture, or for an open, multi-component gas mixture at steady state, in the absence of chemical reactions, it follows that:

(i) The mole number, N_i, of component i is fixed for every $i = 1, 2, \ldots, n$ in the gas mixture.

(ii) Or, equivalently, the gas-phase mole fraction, y_i, is fixed for every $i = 1, 2, \ldots, n$ in the gas mixture.

In other words, in the absence of chemical reactions, the gas mixture composition, $\{y_1, y_2, \ldots, y_{n-1}\}$, is fixed as the mixture evolves from state 1 to state 2. According to the Corollary to Postulate I, we can characterize the equilibrium intensive state of the gas mixture by the set of $(n + 1)$ independent, intensive variables: $\{T, P, y_1, y_2, \ldots, y_{n-1}\}$.

To evaluate the change, $\Delta B_{1\rightarrow 2}$, we can proceed exactly as we did in the $(n = 1)$ case, albeit using a gas mixture EOS (at fixed composition) to evaluate the isothermal variations and C_{pm}^o (or C_{vm}^o) to evaluate the temperature variations. It is noteworthy that C_{pm}^o and C_{vm}^o are the mixture heat capacities at constant pressure and volume in the ideal gas (or attenuated) state, respectively.

The mixture heat capacities in the ideal gas (or attenuated) state are related to the pure component ideal gas heat capacities by a simple arithmetic-mean mixing rule. Specifically:

$$C_{pm}^o = \sum_{i=1}^{n} y_i C_{pi}^o \qquad (23.29)$$

$$C_{vm}^o = \sum_{i=1}^{n} y_i C_{vi}^o \qquad (23.30)$$

where C_{pi}^o and C_{vi}^o are the pure component i ideal gas heat capacities at constant pressure and volume, respectively.

For gas mixtures, we can use the mixture attenuated state approach, first introduced for a pure material in Lecture 16, as follows:

23.4.1 Mixture Attenuated State Approach

Imagine that we need to compute the change in the mixture molar property B when the mixture evolves from state 1, characterized by the $(n + 1)$ independent intensive variables $(T_1, P_1, y_1, y_2, \ldots, y_{n-1})$, to state 2, characterized by the $(n + 1)$ independent intensive variables $(T_2, P_2, y_1, y_2, \ldots, y_{n-1})$. To calculate $\Delta B_{1 \to 2}$, we can choose the (P-T) phase diagram shown on the left-hand side of Fig. 23.1 and then use the attenuated state approach that we discussed in Lecture 16 for a pure component fluid.

Similarly, imagine that we need to compute the change in the mixture molar property B when the mixture evolves from state 1, characterized by the $(n + 1)$ independent intensive variables $(T_1, V_1, y_1, y_2, \ldots, y_{n-1})$, to state 2, characterized by the $(n + 1)$ independent intensive variables $(T_2, V_2, y_1, y_2, \ldots, y_{n-1})$. To calculate $\Delta B_{1 \to 2}$, we can choose the (V-T) phase diagram shown on the right-hand side of Fig. 23.1 and then use the attenuated state approach that we discussed in Lecture 16 for a pure component fluid. The key observation is that in the two phase diagrams shown in Fig. 23.1, the mixture composition $(y_1, y_2, \ldots, y_{n-1})$ remains constant as the mixture evolves from state 1 to state 2. Consequently, the mixture heat capacities, C_{pm}^o and C_{vm}^o, as well as the parameters in the mixture EOS, can be evaluated at the fixed mixture composition, and do not change when the mixture evolves from state 1 to state 2.

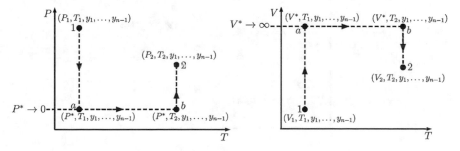

Fig. 23.1

It then follows (see the phase diagrams in Fig. 23.1), that:

$$\Delta B_{1\to 2} = \Delta B_{1\to a} + \Delta B_{a\to b} + \Delta B_{b\to 2} \tag{23.31}$$

where

(i) To compute $\Delta B_{1\to a}$ and $\Delta B_{b\to 2}$ (isothermal steps), we use a mixture EOS.

(ii) To compute $\Delta B_{a\to b}$ (T-variation), we use C^o_{pm} or C^o_{vm} data.

23.4.2   *Mixture Departure Function Approach*

If the mixture composition, $\{y_1, y_2, \ldots, y_{n-1}\}$, is fixed, then, the departure function of B for the mixture, $(DB)_m$, is defined as in the pure component case. Because most EOS are pressure explicit, it is convenient to define the mixture departure function using V, rather than P, as one of the independent intensive variables. Specifically,

$$(DB)_m = B(T, V, y_1, y_2, \ldots, y_{n-1}) - B^o(T, V^o, y_1, y_2, \ldots, y_{n-1}) \tag{23.32}$$

where

$$V^o = \frac{RT}{P} \tag{23.33}$$

In Eq. (23.32), the superscript o in B^o and V^o denotes an ideal gas mixture state. In addition, P corresponds to the actual mixture pressure.

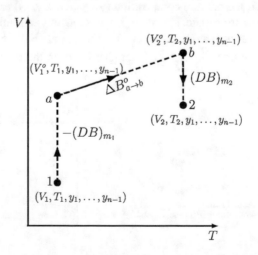

Fig. 23.2

To calculate $\Delta B_{1 \to 2}$ of the mixture using the departure function approach, we follow the three-step path shown in Fig. 23.2, where the mixture composition remains fixed as the mixture evolves from state 1 to state 2.

This yields:

$$\Delta B_{1 \to 2} = - (DB)_{m1} + \Delta B_{a \to b} + (DB)_{m2} \qquad (23.34)$$

where

(i) To compute $(DB)_{m1}$ and $(DB)_{m2}$, we use a mixture EOS.

(ii) To compute $\Delta B_{a \to b}$, we use C_{vm}^{o} and the ideal gas mixture EOS.

23.5   Ideal Gas Mixtures and Ideal Solutions

We saw that an ideal gas satisfies the following two requirements:

(1) It obeys the ideal gas EOS: $P\underline{V} = NRT$ or $PV = RT$

(2) $U = U(T)$ and $H = H(T)$

In an ideal gas mixture, requirement (2) is replaced by:

$$\overline{U}_i = \overline{U}_i(T) \text{ and } \overline{H}_i = \overline{H}_i(T) \qquad (23.35)$$

Next, we will define: (i) an ideal gas, (ii) an ideal gas mixture, and (iii) an ideal solution, by first defining the corresponding chemical potentials. We will then show that (i) and (ii) above satisfy requirements (1) and (2) above.

23.5.1   One Component (Pure, n = 1) Ideal Gas

$$\mu = \frac{G}{N} = G = \lambda(T) + RT\ln P \qquad (23.36)$$

where μ is the chemical potential, G is the molar Gibbs free energy, λ is only a function of temperature, and P is the gas pressure.

For a non-ideal, one component gas, we will see that P in Eq. (23.36) will be replaced by a new thermodynamic function – the fugacity of component i in the non-ideal gas, f(T, P).

23.5.2   Ideal Gas Mixture: For Component i

$$\mu_i = \overline{G}_i = \lambda_i(T) + RT\ln P_i \tag{23.37}$$

where μ_i is the chemical potential of component i, $\overline{G}_i$ is the partial molar Gibbs free energy of component i, λ_i is only a function of temperature and is specific to component i, and:

$$P_i = y_i P \tag{23.38}$$

is the partial pressure of component i, P is the gas mixture pressure, and y_i is the mole fraction of component i in the gas mixture.

For a non-ideal gas mixture, we will see that P_i in Eq. (23.37) will be replaced by a new thermodynamic function – the fugacity of component i in the non-ideal gas mixture, $\widehat{f}_i(T, P, y_1, \ldots, y_{n-1})$.

23.5.3   Ideal Solution: For Component i

$$\mu_i = \overline{G}_i = \Lambda_i(T, P) + RT\ln x_i \tag{23.39}$$

where Λ_i is a function of T and P and is specific to component i, and x_i is the mole fraction of component i in the condensed (liquid or solid) phase.

Note that when $x_i = 1$ (pure component i), Eq. (23.39) can be expressed as follows:

$$\mu_i = G_i = \Lambda_i(T, P) + 0 \Rightarrow \Lambda_i(T, P) = G_i(T, P) \tag{23.40}$$

Using the result in Eq. (23.40) in Eq. (23.39) yields:

$$\mu_i = \overline{G}_i = G_i(T, P) + RT\ln x_i \tag{23.41}$$

For a non-ideal solution, we will see that x_i in Eq. (23.41) will be replaced by a new thermodynamic function – the activity of component i in the non-ideal solution, $a_i(T, P, x_1, \ldots, x_{n-1})$.

Lecture 24

Mixing Functions, Excess Functions, and Sample Problems

24.1 Introduction

The material presented in this lecture is adapted from Chapter 9 in T&M. First, we will show that if the expressions for the chemical potentials presented in Lecture 23 apply, then, pure ideal gas behavior (see Sample Problem 24.1), ideal gas mixture behavior (see Sample Problem 24.2), and ideal solution behavior (see Sample Problem 24.3) are realized. Second, we will present expressions for G, S, H, U, and V of an ideal solution as a function of the solution composition. Third, we will show that it is often advantageous to calculate the deviation of a thermodynamic property from its value in a suitably chosen reference state, referred to as the mixing function, and then to compute the desired thermodynamic property by combining the mixing function with the value of the thermodynamic property in the reference state. Fourth, we will discuss reference states, with particular emphasis on the pure component reference state. Fifth, we will provide a physical interpretation of a mixing function when mixing three pure liquids to create a ternary liquid mixture. Sixth, we will discuss ideal mixing functions. Finally, we will show that it is often convenient to calculate the deviation of a thermodynamic property from its ideal solution value, at the same T, P, and composition, referred to as the excess function, and then to compute the desired thermodynamic property by combining the excess function with the value of the thermodynamic property in the ideal solution.

24.2 Sample Problem 24.1

Show that if $\mu = G(T, P) = \lambda(T) + RT\ln P$ for a one component gas, then, the gas is ideal because it satisfies the two requirements below:

(i) $PV = RT$

(ii) $U = U(T)$ and $H = H(T)$

© Springer Nature Switzerland AG 2020
D. Blankschtein, *Lectures in Classical Thermodynamics with an Introduction to Statistical Mechanics*, https://doi.org/10.1007/978-3-030-49198-7_24

24.2.1 Solution

Choosing T and P as the two independent intensive variables for a one component gas, it follows that $G = G(T,P)$. In addition, it follows that $dG = -SdT + VdP$, and therefore, that:

$$\left(\frac{\partial G}{\partial P}\right)_T = V \quad \text{(General result for n = 1)} \tag{24.1}$$

Using $\mu = G(T, P) = \lambda(T) + RT \ln P$ in Eq. (24.1), it follows that:

$$\left(\frac{\partial G}{\partial P}\right)_T = 0 + \frac{RT}{P} = \frac{RT}{P} \tag{24.2}$$

Equating Eqs. (24.1) and (24.2), including rearranging, yields:

$$PV = RT \quad \text{(Requirement (i) is satisfied)} \tag{24.3}$$

Next, we show that $H = H(T)$. Recall that:

$$G = H - TS \Rightarrow \frac{G}{T} = \frac{H}{T} - S \tag{24.4}$$

Taking the temperature partial derivative, at constant pressure, of the expression on the right-hand side of the arrow in Eq. (24.4) yields:

$$\frac{\partial}{\partial T}\left(\frac{G}{T}\right)_P = \frac{\partial}{\partial T}\left(\frac{H}{T}\right)_P - \left(\frac{\partial S}{\partial T}\right)_P \tag{24.5}$$

The first partial derivative on the right-hand side of Eq. (24.5) is given by:

$$\frac{\partial}{\partial T}\left(\frac{H}{T}\right)_P = \frac{1}{T}\left(\frac{\partial H}{\partial T}\right)_P - \left(\frac{H}{T^2}\right) \tag{24.6}$$

The second partial derivative on the right-hand side of Eq. (24.5) is given by:

$$\left(\frac{\partial S}{\partial T}\right)_P = \frac{C_P}{T} = \frac{1}{T}\left(\frac{\partial H}{\partial T}\right)_P \tag{24.7}$$

Using Eqs. (24.6) and (24.7) in Eq. (24.5), including cancelling the equal terms, yields:

$$\frac{\partial}{\partial T}\left(\frac{G}{T}\right)_P = -\frac{H}{T^2} \tag{24.8}$$

Equation (24.8) is known as the Gibbs-Helmholtz equation, and is a general relation between G and H for n = 1. If we know G(T,P), Eq.(24.8) allows us to compute H(T,P) by differentiation. Alternatively, if we know H(T,P), Eq. (24.8) allows us to compute G(T,P) by integration.

In extensive form, the Gibbs-Helmholtz equation is expressed as follows:

$$\frac{\partial}{\partial T}\left(\frac{G}{T}\right)_{P,N} = -\frac{H}{T^2} \tag{24.9}$$

According to the Problem Statement, for a one component (n = 1) Ideal Gas, $\mu = G(T,P) = \lambda(T) + RT\ln P$, and therefore:

$$\frac{G}{T} = \frac{\lambda(T)}{T} + R\ln P \tag{24.10}$$

Taking the partial derivative of Eq. (24.10) with respect to temperature, at constant pressure, yields:

$$\frac{\partial}{\partial T}\left(\frac{G}{T}\right)_P = \frac{\partial}{\partial T}\left(\frac{\lambda(T)}{T}\right) + 0 \tag{24.11}$$

Using Eq. (24.11) in Eq. (24.8), including rearranging, yields:

$$H = -T^2\frac{\partial}{\partial T}\left(\frac{G}{T}\right)_P = -T^2\frac{\partial}{\partial T}\left(\frac{\lambda(T)}{T}\right) \tag{24.12}$$

Equation (24.12) shows that H is only a function of temperature. In addition, because $U = H - PV = H(T) - RT$, it follows that $U = U(T)$. To derive the last result, we used the fact that, for an ideal gas, $PV = RT$.

24.3 Sample Problem 24.2

Show that if $\mu_i = \overline{G}_i = \lambda_i(T) + RT\ln(y_i P)$, the gas mixture is ideal, because it satisfies the following requirements:

(i) $\overline{V}_i = \dfrac{RT}{P}$ (Independent of y_i) $\tag{24.13}$

Equation (24.13) implies that:

$$V = \sum_{i=1}^{n} y_i \overline{V}_i = \frac{RT}{P} \sum_{i=1}^{n} y_i = \frac{RT}{P}(1) = RT/P \text{ (Ideal gas EOS)} \qquad (24.14)$$

(ii) $\overline{H}_i = H_i(T)$ and $\overline{U}_i = U_i(T)$ (Independent of y_i) $\qquad (24.15)$

24.3.1 Solution

In general,

$$\left(\frac{\partial \underline{G}}{\partial P}\right)_{T,N_i} = \underline{V} \Rightarrow \left(\frac{\partial \overline{G}_i}{\partial P}\right)_{T,y} = \overline{V}_i \qquad (24.16)$$

Recall that Eq. (24.16) is a general result for component i in the gas mixture.

According to the Problem Statement, in an ideal gas mixture, $\overline{G}_i = \lambda_i(T) + RT\ln(y_i P)$, where $y_i P$ is the partial pressure, P_i, of component i. Using this result in the expression on the right-hand side of the arrow in Eq. (24.16) yields:

$$\left(\frac{\partial \overline{G}_i}{\partial P}\right)_{T,y} = 0 + \frac{RT}{P} = \frac{RT}{P} \qquad (24.17)$$

Equating Eqs. (24.16) and (24.17) yields:

$$\overline{V}_i = \frac{RT}{P} \quad \text{(Independent of } y_i) \qquad (24.18)$$

It then follows that:

$$V = \sum_{i=1}^{n} y_i \overline{V}_i = \frac{RT}{P}\left(\sum_{i=1}^{n} y_i\right) = \frac{RT}{P} \Rightarrow PV$$

$$= RT \text{ (Ideal gas mixture EOS)} \qquad (24.19)$$

Next, starting with the Gibbs-Helmholtz equation in extensive form, repeated below for completeness (see Eq. (24.9)), it follows that:

$$\frac{\partial}{\partial T}\left(\frac{\underline{G}}{T}\right)_{P,N_i} = -\frac{\underline{H}}{T^2} \Rightarrow \frac{\partial}{\partial T}\left(\frac{\overline{G}_i}{T}\right)_{P,y} = -\frac{\overline{H}_i}{T^2} \qquad (24.20)$$

The relation on the right-hand side of the arrow in Eq. (24.20) applies to component i in the mixture. Recall that in Eq. (24.9), the subscript N_i is a shorthand notation indicating that every mole number N_i is kept constant, and in Eq. (24.20), the subscript y is a shorthand notation indicating that every mole fraction y_i is kept constant.

According to the Problem Statement, $\overline{G}_i$ in an ideal gas mixture is given by:

$$\overline{G}_i = \lambda_i(T) + RT\ln(y_iP) \tag{24.21}$$

Dividing Eq. (24.21) by T, and then taking the partial derivative of the resulting expression with respect to T, at constant P and mixture composition, yields:

$$\frac{\overline{G}_i}{T} = \frac{\lambda_i(T)}{T} + R\ln(y_iP) \Rightarrow \frac{\partial}{\partial T}\left(\frac{\overline{G}_i}{T}\right)_{P,y} = \frac{\partial}{\partial T}\left(\frac{\lambda_i(T)}{T}\right) + 0 \tag{24.22}$$

Using Eq. (24.22) in the expression on the right-hand side of the arrow in Eq. (24.20), including rearranging, yields:

$$\overline{H}_i = -T^2\frac{\partial}{\partial T}\left(\frac{\overline{G}_i}{T}\right)_{P,y} = -T^2\frac{\partial}{\partial T}\left(\frac{\lambda_i(T)}{T}\right) \tag{24.23}$$

Equation (24.23) shows that in an ideal gas mixture, $\overline{H}_i$ is only a function of T, and therefore, is equal to $H_i(T)$. Because $\overline{U}_i = \overline{H}_i - P\overline{V}_i$, and in an ideal gas mixture $P\overline{V}_i = RT$, and $\overline{H}_i = H_i(T)$, it follows that:

$$\overline{U}_i(T) - H_i(T) - RT \tag{24.24}$$

Equation (24.24) shows that, in an ideal gas mixture, $\overline{U}_i$ is only a function of T, and therefore, is equal to $U_i(T)$.

24.4 Sample Problem 24.3

As discussed in Lecture 23, in an ideal solution,

$$\mu_i = \overline{G}_i = G_i(T,P) + RT\ln x_i \tag{24.25}$$

Show that:

(i) $\overline{V}_i$ is not a function of composition, that is, $\overline{V}_i = V_i(T,P)$ (24.26)

(ii) $\overline{H}_i$ is not a function of composition, that is, $\overline{H}_i = H_i(T,P)$ (24.27)

(iii) $\overline{U}_i$ is not a function of composition, that is, $\overline{U}_i = U_i(T,P)$ (24.28)

In other words, show that the partial molar volume, the partial molar enthalpy, and the partial molar internal energy of component i in an ideal solution are equal to the molar values at the same T and P.

24.4.1   Solution

To prove (i), (ii), and (iii) above, we proceed as follows. In general, in a solution of n components, it follows that:

$$\left(\frac{\partial \overline{G}_i}{\partial P}\right)_{T,x} = \overline{V}_i \tag{24.29}$$

Recall that the subscript x in Eq. (24.29) indicates that all the mole fractions x_i are kept constant. Recall also that x_i denotes the mole fraction of component i in a condensed (liquid or solid) phase. On the other hand, y_i denotes the mole fraction of component i in the gas phase.

According to the Problem Statement, in an ideal solution, $\overline{G}_i = G_i(T,P) + RT\ln x_i$. Using this expression for $\overline{G}_i$ in Eq. (24.29) yields:

$$\left(\frac{\partial \overline{G}_i}{\partial P}\right)_{T,x} = \left(\frac{\partial G_i}{\partial P}\right)_T + 0 = V_i \Rightarrow \overline{V}_i = V_i(T,P) \tag{24.30}$$

In addition, using the expression for $\overline{G}_i$ given in the Problem Statement in the Gibbs-Helmholtz equation for a mixture yields:

$$\frac{\partial}{\partial T}\left(\frac{\overline{G}_i}{T}\right)_{P,x} = \frac{\partial}{\partial T}\left(\frac{G_i}{T}\right)_P = -\frac{H_i}{T^2} \Rightarrow \overline{H}_i = H_i(T,P) \tag{24.31}$$

Having shown that, in an ideal solution, $\overline{V}_i = V_i(T,P)$ and $\overline{H}_i = H_i(T,P)$, it follows that:

$$\overline{U}_i = \overline{H}_i - P\overline{V}_i = H_i - PV_i = U_i \Rightarrow \overline{U}_i = U_i(T,P) \tag{24.32}$$

24.5   Other Useful Relations for an Ideal Solution

$$\overline{G}_i = G_i + RT\ln x_i \tag{24.33}$$

and

$$\underline{G} = \underline{H} - T\underline{S} \Rightarrow \overline{G}_i = \overline{H}_i - T\overline{S}_i \quad \text{(Component i in the mixture)} \quad (24.34)$$

$$G_i = H_i - TS_i \quad \text{(Pure component i)} \quad (24.35)$$

Subtracting Eq. (24.35) from Eq. (24.34), including rearranging, yields:

$$\overline{G}_i - G_i = \left(\overline{H}_i - H_i\right) - T\left(\overline{S}_i - S_i\right) \Rightarrow \overline{S}_i = S_i(T, P) - R\ln x_i \quad (24.36)$$

where in Eq. (24.36), we have used the facts that, in an ideal solution, $\overline{G}_i - G_i = RT\ln x_i$, and $\overline{H}_i - H_i = 0$. The expression on the right-hand side of the arrow in Eq. (24.36) corresponds to the partial molar entropy of component i in an ideal solution.

24.6 Summary of Results for an Ideal Solution

$$G = \sum_{i=1}^{n} x_i \overline{G}_i = \sum_{i=1}^{n} x_i G_i(T, P) + RT \sum_{i=1}^{n} x_i \ln x_i \quad (24.37)$$

$$S = \sum_{i=1}^{n} x_i \overline{S}_i = \sum_{i=1}^{n} x_i S_i(T, P) - R \sum_{i=1}^{n} x_i \ln x_i \quad (24.38)$$

$$H = \sum_{i=1}^{n} x_i \overline{H}_i = \sum_{i=1}^{n} x_i H_i(T, P) \quad (24.39)$$

$$U = \sum_{i=1}^{n} x_i \overline{U}_i = \sum_{i=1}^{n} x_i U_i(T, P) \quad (24.40)$$

$$V = \sum_{i=1}^{n} x_i \overline{V}_i = \sum_{i=1}^{n} x_i V_i(T, P) \quad (24.41)$$

24.7 Mixing Functions

Sometimes, it is convenient to relate a mixture property, $\underline{B}$, to the value of that property in some reference state (RS) that can be real or hypothetical (for example, pure component RS, dilute mixture RS, etc.).

The difference between $\underline{B}$ and the value of $\underline{B}$ in the RS, denoted as $\underline{B}^\dagger$, is referred to as the Mixing $\underline{B}$, and denoted as $\Delta \underline{B}_{mix}$.

24.7.1 The Mixing B and Reference States

The defining equation is given by:

$$\Delta \underline{B}_{mix} = \underline{B}(T, P, N_1, \ldots, N_n) - \sum_{j=1}^{n} N_j \overline{B}_j^{\dagger}\left(T^{\dagger}, P^{\dagger}, x_j^{\dagger}, \ldots, x_{n-1}^{\dagger}\right) \qquad (24.42)$$

where N_j is the actual number of moles of component j in the mixture, and $\overline{B}_j^{\dagger}$ is the partial molar B of component j in the RS. Further, in Eq. (24.42), $\overline{B}_j^{\dagger}$ is not necessarily equal to $\overline{B}_j$.

In molar form, the mixing B can be expressed as follows:

$$\Delta B_{mix} = B(T, P, x_1, \ldots, x_{n-1}) - \sum_{j=1}^{n} x_j \overline{B}_j^{\dagger}\left(T^{\dagger}, P^{\dagger}, x_j^{\dagger}, \ldots, x_{n-1}^{\dagger}\right) \qquad (24.43)$$

where x_j is the actual mole fraction of component j in the mixture, and $\overline{B}_j^{\dagger}$ is the partial molar B of component j in the RS. Again, recall that $\overline{B}_j^{\dagger}$ is not necessarily equal to $\overline{B}_j$.

Moreover, $\Delta \underline{B}_{mix}$ (or ΔB_{mix}) is only specified when the RS of each component j has been specified, in addition to $T^{\dagger}$ and $P^{\dagger}$.

Clearly, for a mixing function to be useful, it should depend on the properties $(T, P, x_1, \ldots, x_{n-1})$ of the mixture that it is supposed to describe. In other words, we would like:

$$\Delta B_{mix} = f(T, P, x_1, \ldots, x_{n-1}) \qquad (24.44)$$

Equation (24.44) imposes restrictions on the RS variables. Three potential restrictions are discussed below:

1. $T^{\dagger} = T$, $P^{\dagger} = P$, $x_j^{\dagger} = x_j \Rightarrow \overline{B}_j^{\dagger} = \overline{B}_j^{\dagger}(T, P, x_1, \ldots, x_{n-1})$ and varies as the actual mixture conditions change.

2. $T^{\dagger} = T_o$, $P^{\dagger} = P_o$, $x_j^{\dagger} = x_{jo}$, where the subscript o denotes constant values in the RS. In this case, $\overline{B}_j^{\dagger} = \overline{B}_j^o(T_o, P_o, x_{10}, \ldots, x_{n-1o})$ is constant for every j, independent of variations in the actual mixture conditions.

3. In practice, the most common RS is the pure component RS at the same T, P, and state of aggregation (vapor, liquid, or solid) of the mixture, which we discuss next.

24.7.2 Pure Component Reference State for Component j

The pure component reference state is characterized by:

- Mixture state of aggregation
- $T^\dagger = T, \ P^\dagger = P$ (24.45)

- $x_j^\dagger = 1, x_i^\dagger = 0, \text{for } i \neq j$ (24.46)

- $\overline{B}_j^\dagger = B_j(T, P)$ (24.47)

24.7.3 Useful Relations for Mixing Functions

Because a mixing function is a thermodynamic property of the mixture, all the relations derived in previous lectures for $\underline{B}$ (or B) of a mixture also apply to $\Delta\underline{B}_{mix}$ (or ΔB_{mix}). For example,

$$\Delta\overline{B}_j = \left(\frac{\partial \Delta\underline{B}_{mix}}{\partial N_j}\right)_{T,P,N_{i[j]}} = \overline{B}_j - \overline{B}_j^\dagger \text{(Partial molar mixing B of component j)}$$

(24.48)

$$\Delta\underline{B}_{mix} = \sum_{j=1}^{n} N_j \, \Delta\overline{B}_j$$ (24.49)

$$\Delta B_{mix} = \sum_{j=1}^{n} x_j \, \Delta\overline{B}_j$$ (24.50)

Further, because

$$\Delta B_{mix} = \Delta B_{mix}(T, P, x_1, \ldots, x_{n-1})$$ (24.51)

it follows that:

$$d(\Delta B_{mix}) = \left(\frac{\partial(\Delta B_{mix})}{\partial T}\right)_{P,x} dT + \left(\frac{\partial(\Delta B_{mix})}{\partial P}\right)_{T,x} dP$$

$$+ \sum_{j=1}^{n-1} \left(\frac{\partial(\Delta B_{mix})}{\partial x_j}\right)_{T,P,x[j,n]} dx_j$$

(24.52)

In addition, the generalized Gibbs-Duhem relations apply as follows:

$$\sum_{i=1}^{n} x_i d\left(\Delta \overline{B}_i\right) = \left(\frac{\partial\left(\Delta B_{mix}\right)}{\partial T}\right)_{P,x} dT + \left(\frac{\partial\left(\Delta B_{mix}\right)}{\partial P}\right)_{T,x} dP \qquad (24.53)$$

At constant T and P, the generalized Gibbs-Duhem relations in Eq. (24.53) simplify to:

$$\sum_{i=1}^{n} x_i \left(\frac{\partial\left(\Delta \overline{B}_i\right)}{\partial x_j}\right)_{T,P,x[j,k]} = 0 \qquad (24.54)$$

As before, given (n-1) partial molar mixing Bs, one can calculate the remaining one, to within an arbitrary constant of integration, by integrating Eq. (24.54).

24.8   Mixing Functions: Mixing of Three Liquids at Constant T and P

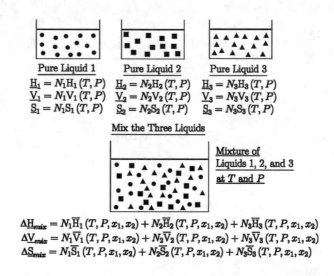

Pure Liquid 1

$\underline{H}_1 = N_1 H_1 (T, P)$
$\underline{V}_1 = N_1 V_1 (T, P)$
$\underline{S}_1 = N_1 S_1 (T, P)$

Pure Liquid 2

$\underline{H}_2 = N_2 H_2 (T, P)$
$\underline{V}_2 = N_2 V_2 (T, P)$
$\underline{S}_2 = N_2 S_2 (T, P)$

Pure Liquid 3

$\underline{H}_3 = N_3 H_3 (T, P)$
$\underline{V}_3 = N_3 V_3 (T, P)$
$\underline{S}_3 = N_3 S_3 (T, P)$

Mix the Three Liquids

Mixture of Liquids 1, 2, and 3 at T and P

$\Delta \underline{H}_{mix} = N_1 \overline{H}_1 (T, P, x_1, x_2) + N_2 \overline{H}_2 (T, P, x_1, x_2) + N_3 \overline{H}_3 (T, P, x_1, x_2)$
$\Delta \underline{V}_{mix} = N_1 \overline{V}_1 (T, P, x_1, x_2) + N_2 \overline{V}_2 (T, P, x_1, x_2) + N_3 \overline{V}_3 (T, P, x_1, x_2)$
$\Delta \underline{S}_{mix} = N_1 \overline{S}_1 (T, P, x_1, x_2) + N_2 \overline{S}_2 (T, P, x_1, x_2) + N_3 \overline{S}_3 (T, P, x_1, x_2)$

Fig. 24.1

The partial molar properties $\overline{H}_i$, $\overline{V}_i$, and $\overline{S}_i$ (i = 1, 2, 3) depend on the nature of the liquids i being mixed, as well as on the mixture composition (x_1, x_2), the temperature (T), and the pressure (P). Recall that $x_3 = 1 - x_1 - x_2$. The molar properties H_i, V_i, and S_i (i = 1, 2, 3) depend on the nature of liquid i, as well as on the temperature (T) and the pressure (P). In general:

$$\overline{H}_i (T, P, x_1, x_2) \neq H_i (T, P) \qquad (24.55)$$

$$\overline{V}_i(T, P, x_1, x_2) \neq V_i(T, P) \tag{24.56}$$

$$\overline{S}_i(T, P, x_1, x_2) \neq S_i(T, P) \tag{24.57}$$

where $i = 1, 2$, and 3.

The following three observations can be made (see Fig. 24.1):

1. The enthalpy of mixing, $\Delta \underline{H}_{mix}$, simply measures the difference between $\underline{H}_{mix}$ (in this example, the enthalpy of the ternary mixture) and the sum of the enthalpies of the three pure liquids (1, 2, and 3) which were mixed to create the ternary mixture, that is,

$$\Delta \underline{H}_{mix} = \underline{H}_{mix} - (\underline{H}_1 + \underline{H}_2 + \underline{H}_3) \tag{24.58}$$

where $\underline{H}_1 + \underline{H}_2 + \underline{H}_3$ is the pure component reference state enthalpy.

Expanding Eq. (24.58) in terms of the partial molar enthalpies and molar enthalpies of components 1, 2, and 3 yields:

$$\Delta \underline{H}_{mix} = N_1 \overline{H}_1 + N_2 \overline{H}_2 + N_3 \overline{H}_3 - (N_1 H_1 + N_2 H_2 + N_3 H_3) \tag{24.59}$$

Combining the N_1, N_2, and N_3 terms, we obtain:

$$\Delta \underline{H}_{mix} = N_1 (\overline{H}_1 - H_1) + N_2 (\overline{H}_2 - H_2) + N_3 (\overline{H}_3 - H_3) \tag{24.60}$$

or

$$\Delta \underline{H}_{mix} = N_1 \Delta \overline{H}_1 + N_2 \Delta \overline{H}_2 + N_3 \Delta \overline{H}_3 \tag{24.61}$$

where $\Delta \overline{H}_i$ is the partial molar enthalpy of mixing of component i (1, 2, and 3).

2. Similarly, the volume of mixing is given by:

$$\Delta \underline{V}_{mix} = \underline{V}_{mix} - (\underline{V}_1 + \underline{V}_2 + \underline{V}_3) \tag{24.62}$$

where $\underline{V}_1 + \underline{V}_2 + \underline{V}_3$ is the pure component reference state volume.

Similar to the enthalpy calculation presented above, Eq. (24.62) can be expressed as follows:

$$\Delta \underline{V}_{mix} = N_1 (\overline{V}_1 - V_1) + N_2 (\overline{V}_2 - V_2) + N_3 (\overline{V}_3 - V_3) \tag{24.63}$$

or

$$\Delta \underline{V}_{mix} = N_1 \, \Delta \overline{V}_1 + N_2 \, \Delta \overline{V}_2 + N_3 \, \Delta \overline{V}_3 \tag{24.64}$$

where $\Delta \overline{V}_i$ (i = 1, 2, 3) is the partial molar mixing volume of component i.

3. Finally, the entropy of mixing is given by:

$$\Delta \underline{S}_{mix} = \underline{S}_{mix} - (\underline{S}_1 + \underline{S}_2 + \underline{S}_3) \tag{24.65}$$

where $\underline{S}_1 + \underline{S}_2 + \underline{S}_3$ is the pure component reference state entropy.

Similar to the enthalpy and entropy of mixing calculations presented above, it follows that:

$$\Delta \underline{S}_{mix} = N_1 \left(\overline{S}_1 - S_1\right) + N_2 \left(\overline{S}_2 - S_2\right) + N_3 \left(\overline{S}_3 - S_3\right) \tag{24.66}$$

Equation (24.66) can also be expressed as follows:

$$\Delta \underline{S}_{mix} = N_1 \, \Delta \overline{S}_1 + N_2 \, \Delta \overline{S}_2 + N_3 \, \Delta \overline{S}_3 \tag{24.67}$$

where $\Delta \overline{S}_i$ (i = 1, 2, and 3) is the partial molar entropy of mixing of component i.

24.9   Ideal Solution Mixing Functions

For an ideal solution, we choose the pure component RS. Specifically,

$$\left\{ \begin{array}{l} T^\dagger = T, P^\dagger = P, x_j^\dagger = 1, x_i^\dagger = 0 \ (\text{for } i \neq j), \text{same} \\ \text{aggregation state as the mixture} \rightarrow \overline{B}_j^\dagger = B_j \left(T, P\right) \end{array} \right\} \tag{24.68}$$

Therefore, in an ideal solution, the Mixing B is given by:

$$\Delta B_{mix}^{ID} = B^{ID} - \sum_{j=1}^{n} x_j B_j \left(T, P\right) \tag{24.69}$$

Expressions for five ideal mixing functions follow:

$$\Delta G_{mix}^{ID} = RT \sum_{j=1}^{n} x_j \ln x_j; \quad \Delta \overline{G}_j^{ID} = RT \ln x_j \tag{24.70}$$

$$\Delta S_{mix}^{ID} = -RT \sum_{j=1}^{n} x_j \ln x_j; \quad \Delta \overline{S}_j^{ID} = -R\ln x_j \tag{24.71}$$

$$\Delta H_{mix}^{ID} = 0; \quad \Delta \overline{H}_j^{ID} = 0 \tag{24.72}$$

$$\Delta V_{mix}^{ID} = 0; \quad \Delta \overline{V}_j^{ID} = 0 \tag{24.73}$$

$$\Delta U_{mix}^{ID} = 0; \quad \Delta \overline{U}_j^{ID} = 0 \tag{24.74}$$

24.10 Excess Functions

The deviation of a mixture property, $\underline{B}$, from its ideal solution value at the same T, P, and composition $(x_1, \ldots, x_{n-1})$ as in the original mixture, $\underline{B}^{ID}$, is referred to as the Excess $\underline{B}$, and is given by:

$$\underline{B}^{EX} = \underline{B} - \underline{B}^{ID} \tag{24.75}$$

We know that:

$$\underline{B} = \sum_{j=1}^{n} N_j \overline{B}_j; \quad \underline{B}^{ID} = \sum_{j=1}^{n} N_j \overline{B}_j^{ID} \tag{24.76}$$

Using Eq. (24.76) in (24.75), we obtain:

$$\underline{B}^{EX} = \underline{B} - \underline{B}^{ID} = \sum_{j=1}^{n} N_j \left(\overline{B}_j - \overline{B}_j^{ID} \right) \tag{24.77}$$

Further, because

$$\underline{B}^{EX} = \sum_{j=1}^{n} N_j \overline{B}_j^{EX} \tag{24.78}$$

Equations (24.78) and (24.77) show that:

$$\overline{B}_j^{EX} = \overline{B}_j - \overline{B}_j^{ID} \text{(Partial molar excess B of component j)} \tag{24.79}$$

Clearly, every relation derived earlier for $\underline{B}$ (or B) applies to $\underline{B}^{EX}$ (or B^{EX}). In particular, we can define excess mixing functions, $\Delta\underline{B}_{mix}^{EX}$, as follows:

$$\Delta\underline{B}_{mix}^{EX} = \Delta\underline{B}_{mix} - \Delta\underline{B}_{mix}^{ID} \tag{24.80}$$

Expanding Eq. (24.80) in terms of partial molar properties yields:

$$\sum_{j=1}^{n} N_j \, \Delta\overline{B}_j^{EX} = \sum_{j=1}^{n} N_j \, \Delta\overline{B}_j - \sum_{j=1}^{n} N_j \, \Delta\overline{B}_j^{ID} \tag{24.81}$$

Equation (24.81) shows that:

$$\Delta\overline{B}_j^{EX} = \Delta\overline{B}_j - \Delta\overline{B}_j^{ID} \text{(Partial molar excess mixing B of component j)} \tag{24.82}$$

One can show that unlike $\Delta\underline{B}_{mix}$ or $\Delta\underline{B}_{mix}^{ID}$, an excess function is independent of the RS, provided that the same RS is used for $\Delta\underline{B}_{mix}$ and $\Delta\underline{B}_{mix}^{ID}$. Because for $\Delta\underline{B}_{mix}^{ID}$ we always choose the pure component RS, the same RS will also be used for $\Delta\underline{B}_{mix}$, unless specified otherwise, for the calculation of $\underline{B}^{EX}$.

The proof of this last statement is presented below. Choosing the pure component reference state yields:

$$\Delta\overline{B}_j = \overline{B}_j - B_j \tag{24.83}$$

and

$$\Delta\overline{B}_j^{ID} = \overline{B}_j^{ID} - B_j \tag{24.84}$$

Subtracting Eq. (24.84) from (Eq. 24.83), including cancelling the two equal terms, yields:

$$\Delta\overline{B}_j - \Delta\overline{B}_j^{ID} = \Delta\overline{B}_j^{EX} = \overline{B}_j - \overline{B}_j^{ID} = \overline{B}_j^{EX} \tag{24.85}$$

or

$$\Delta\overline{B}_j^{EX} = \overline{B}_j^{EX} \tag{24.86}$$

Equation (24.86) shows that the partial molar excess mixing B of component j is equal to the partial molar excess B of component j.

The equality in Eq. (24.86) ensures that:

$$\Delta \underline{B}^{EX}_{mix} = \underline{B}^{EX} \quad (\text{The excess mixing B is equal to the excess B}) \quad (24.87)$$

Lecture 25

Ideal Solution, Regular Solution, and Athermal Solution Behaviors, and Fugacity and Fugacity Coefficient

25.1 Introduction

The material presented in this lecture is adapted from Chapter 9 in T&M. First, we will discuss ideal solution behavior, including presenting a molecular interpretation of ideality, followed by discussing regular solution and athermal solution behaviors. Second, we will introduce the concept of fugacity of component i in a non-ideal gas mixture, including showing that when the pressure approaches zero, the fugacity of component i approaches the partial pressure of component i. We will also define the ratio of the mixture fugacity of component i and the partial pressure of component i as the fugacity coefficient of component i. Clearly, in an ideal gas mixture, the fugacity of component i is equal to the partial pressure of component i, and the fugacity coefficient of component i is equal to unity. For a one component gas, the fugacity is equal to the pressure, and their ratio is equal to the fugacity coefficient of the one component gas, which is equal to unity when the one component gas is ideal. Second, we will derive expressions for the variations of the fugacity with pressure and temperature, both for a pure component gas and for a gas mixture. These expressions will be used when we discuss the differential approach to phase equilibria. Third, we will discuss how to calculate fugacities using an EOS approach, both for a pure component gas and for a gas mixture. Fourth, we will derive the generalized Gibbs-Duhem relation for the fugacities, which is particularly useful when T and P are constant, and allows us to calculate an unknown fugacity if we know the remaining (n-1) independent fugacities. Finally, we will derive the Lewis and Randal Rule, which relates the fugacity of component i in an ideal solution to the product of the fugacity of pure component i, at the same T and P as those in the ideal solution, and the mole fraction of component i in the ideal solution.

© Springer Nature Switzerland AG 2020
D. Blankschtein, *Lectures in Classical Thermodynamics with an Introduction to Statistical Mechanics*, https://doi.org/10.1007/978-3-030-49198-7_25

25.2   Ideal Solution Behavior

For an ideal solution, where the pure component reference state is used for every component, all the excess functions are zero, that is, $\underline{B}_{EX}^{ID} = \underline{B}_{ID} - \underline{B}_{ID} = 0$.

In Fig. 25.1, we present a simple molecular model of an ideal binary solution where both the excess entropy and the excess enthalpy are zero. Indeed, when the "colored blind" black and gray billiard balls of equal radius are mixed, "they cannot differentiate if they are in a mixture or by themselves," which corresponds to ideal mixing.

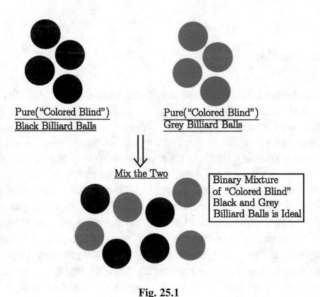

Pure("Colored Blind")   Pure("Colored Blind")
Black Billiard Balls   Grey Billiard Balls

Mix the Two

Binary Mixture
of "Colored Blind"
Black and Grey
Billiard Balls is Ideal

Fig. 25.1

25.3   Regular Solution Behavior

(a) $\overline{S}_j^{EX} = 0$, for every component j in the solution. Accordingly,

$$\underline{S}^{EX} = \sum_{j=1}^{n} N_j \overline{S}_j^{EX} = 0 \qquad (25.1)$$

Equation (25.1) shows that the excess entropy of a regular solution is zero. In other words, from the entropy point of view, the solution behaves ideally. This typically occurs when the various components in the solution are similar in size.

(b) $\overline{H}_j^{EX} \neq 0$, for every component j in the solution. Therefore, it follows that:

$$\underline{H}^{EX} \neq 0 \tag{25.2}$$

The result in Eq. (25.2) shows that the excess enthalpy of a regular solution is non-zero. In other words, from the enthalpy (interactions) point of view, an athermal solution is not ideal.

Using Eqs. (25.1) and (25.2), it follows that:

(c) $\underline{G}^{EX} = \underline{H}^{EX} - T\underline{S}^{EX} = \underline{H}^{EX} = \Delta\underline{H}^{EX}_{mix}$ \hfill (25.3)

25.4 Athermal Solution Behavior

(a) $\overline{H}_j^{EX} = 0$, for every component j in the solution. Therefore, it follows that:

$$\underline{H}^{EX} = 0 \tag{25.4}$$

Equation (25.4) shows that from the enthalpy (interactions) point of view, an athermal solution behaves ideally.

(b) $\overline{S}_j^{EX} \neq 0$, for every component j in the solution. Therefore, it follows that:

$$\underline{S}^{EX} \neq 0 \tag{25.5}$$

Equation (25.5) shows that the excess entropy of an athermal solution is non-zero. In other words, from the entropy point of view, an athermal solution is not ideal.

Using Eqs. (25.4) and (25.5), it follows that:

(c) $\underline{G}^{EX} = \underline{H}^{EX} - T\underline{S}^{EX} = - T\underline{S}^{EX} = -T\Delta\underline{S}^{EX}$ \hfill (25.6)

25.5 Fugacity and Fugacity Coefficient

We know that for a one component (n = 1) ideal gas (IG) i:

$$\mu_i^{IG}(T, P) = G_i(T, P) = \lambda_i(T) + RT\ln P \tag{25.7}$$

In Eq. (25.7), the units are implicitly specified by the ideal gas (IG) state condition contained in the $\lambda_i(T)$ term. For example:

$$\lambda_i(T) = \mu_i^{IG}\,(T, P = 1 \text{ bar or } 1 \text{ atm}) \tag{25.9}$$

If gas i is not ideal, we introduce a new thermodynamic function which replaces P, and is called fugacity, $f_i(T,P)$. The fugacity of pure component i is an intensive property and, according to the Corollary to Postulate I, depends on the $(n + 1) = (1 + 1) = 2$ independent intensive variables T and P. For a non-ideal gas i, it follows that:

$$\mu_i(T, P) = G_i(T, P) = \lambda_i(T) + RT\ln f_i(T, P) \tag{25.10}$$

Clearly, in the limit when $P \to P^* \approx 0$, the gas must behave ideally, and therefore:

$$\lim_{P \to 0} \left(\frac{f_i(T, P)}{P}\right) = 1 \quad \text{(For pure gas i)} \tag{25.11}$$

We also saw that for gas i in an ideal gas mixture (IGM), it follows that:

$$\mu_i^{IGM}\,(T, P, y_i) = \overline{G}_i(T, P, y_i) = \lambda_i(T) + RT\ln P_i \tag{25.12}$$

where $P_i = y_i P$ (Partial pressure of i).

In Eq. (25.12), μ_i^{IGM} depends only on y_i of gas i. In principle, it should depend on $\{y_1, y_2, \ldots, y_{n-1}\}$, not only on y_i. If the gas mixture is not ideal, we replace P_i by the fugacity, $\widehat{f}_i\,(T, P, y_1, \ldots, y_{n-1})$, of gas i in the mixture. It then follows that:

$$\mu_i\,(T, P, y_1, \ldots, y_{n-1}) = \overline{G}_i(T, P, y_1, \ldots, y_{n-1})$$
$$= \lambda_i(T) + RT\ln\widehat{f}_i\,(T, P, y_1, \ldots, y_{n-1}) \tag{25.13}$$

In the limit when $P \to P^* \approx 0$, the gas mixture must behave ideally, and therefore:

$$\lim_{P \to 0} \left(\frac{\widehat{f}_i}{P_i}\right) = 1 \quad \text{(For gas i in the mixture)} \tag{25.14}$$

Next, we will derive useful relations involving the variations of $\widehat{f}_i$ and f_i with pressure and temperature.

25.5.1 *Variations of $\widehat{f}_i$ and f_i with Pressure*

We know that, in general,

$$\underline{G} = \underline{G}\,(T, P, N_1, \ldots, N_n) \Rightarrow \left(\frac{\partial \underline{G}}{\partial P}\right)_{T, N_i} = \underline{V} \Rightarrow \left(\frac{\partial \overline{G}_i}{\partial P}\right)_{T, y} = \overline{V}_i \qquad (25.15)$$

We also know that:

$$\overline{G}_i = \lambda_i(T) + RT\ln\widehat{f}_i \qquad (25.16)$$

Using Eq. (25.16) in Eq. (25.15) yields:

$$\left(\frac{\partial \overline{G}_i}{\partial P}\right)_{T, y} = 0 + RT\left(\frac{\partial \ln \widehat{f}_i}{\partial P}\right)_{T, y} = \overline{V}_i \qquad (25.17)$$

Rearranging Eq. (25.17) yields:

$$\left(\frac{\partial \ln \widehat{f}_i}{\partial P}\right)_{T, y} = \frac{\overline{V}_i}{RT} \quad \text{(For component i in the mixture)} \qquad (25.18)$$

Recall that both $\widehat{f}_i$ and $\overline{V}_i$ are intensive variables which depend on $\{T, P, y_1, \ldots, y_{n-1}\}$.

In a similar manner, we can show that for pure gas i:

$$\left(\frac{\partial \ln f_i}{\partial P}\right)_T = \frac{V_i}{RT} \quad \text{(For pure component i)} \qquad (25.19)$$

Recall that both f_i and V_i are intensive variables which depend on T and P.

25.5.2 *Variations of $\widehat{f}_i$ and f_i with Temperature*

We saw that, in general, component i in a mixture must satisfy the Gibbs-Helmholtz equation, given by:

$$\left(\frac{\partial \left(\overline{G}_i / T\right)}{\partial T}\right)_{P, y} = -\frac{\overline{H}_i}{T^2} \qquad (25.20)$$

We know that:

$$\overline{G}_i = \lambda_i(T) + RT\ln\widehat{f}_i \tag{25.21}$$

In order to use Eq. (25.21) in Eq. (25.20), we need to compute:

$$\left(\frac{\partial\lambda_i(T)}{\partial T}\right)_{P,y} \tag{25.22}$$

where keeping P and y constant in the partial derivative is redundant, because λ_i depends only on temperature.

Because we have no information about $\lambda_i(T)$, it would be convenient to eliminate $\lambda_i(T)$ altogether. The "trick" is to calculate $\overline{G}_i$ for an ideal gas mixture, denoted by $\overline{G}_i{}^\circ$, and then to subtract $\overline{G}_i{}^\circ$ from $\overline{G}_i$. Specifically,

$$\overline{G}_i{}^\circ = \lambda_i(T) + RT\ln P_i; \quad P_i = y_iP \tag{25.23}$$

Subtracting Eq. (25.23) from Eq. (25.21), including cancelling the equal terms, we obtain:

$$\overline{G}_i - \overline{G}_i{}^\circ = RT\ln\left(\frac{\widehat{f}_i}{P_i}\right) \quad \text{(For component i in the mixture)} \tag{25.24}$$

Dividing Eq. (25.24) by RT, and then taking the partial derivative with respect to T, at constant P and y, yields:

$$\frac{\partial}{\partial T}\left[(\overline{G}_i - \overline{G}_i{}^\circ)/RT\right]_{P,y} = -\frac{(\overline{H}_i - \overline{H}_i{}^\circ)}{RT^2} = \left(\frac{\partial\ln\left(\widehat{f}_i/y_iP\right)}{\partial T}\right)_{P,y}$$

$$= \left(\frac{\partial\ln\widehat{f}_i}{\partial T}\right)_{P,y} \tag{25.25}$$

Equating the second and the last terms in Eq. (25.25), including rearranging, yields:

$$\left(\frac{\partial\ln\widehat{f}_i}{\partial T}\right)_{P,y} = -\left(\frac{(\overline{H}_i - \overline{H}_i{}^\circ)}{RT^2}\right)_{P,y} \tag{25.26}$$

Recall that in an ideal gas mixture, $\overline{H}_i{}^\circ = H_i{}^\circ(T)$, independent of y and P. Accordingly, Eq. (25.26) can be expressed as follows:

$$\left(\frac{\partial \ln \widehat{f}_i}{\partial T}\right)_{P,y} = -\frac{\left(\overline{H}_i - H_i^{\,\circ}(T)\right)}{RT^2} \qquad \text{(For component i in the mixture)} \qquad (25.27)$$

Recall that both $\widehat{f}_i$ and $\overline{H}_i$ in Eq. (25.27) are intensive variables which depend on $\{T, P, y_1, \ldots, y_{n-1}\}$.

In a similar manner, we can show that for pure gas i:

$$\left(\frac{\partial \ln f_i}{\partial T}\right)_{P} = -\frac{\left(H_i - H_i^{\,\circ}(T)\right)}{RT^2} \qquad \text{(For pure component i)} \qquad (25.28)$$

Recall that both f_i and H_i in Eq. (25.28) are intensive variables which depend on T and P.

25.6   Other Relations Involving Fugacities

The set of $(n + 2)$ intensive variables, $\left\{T, P, \widehat{f}_1, \ldots, \widehat{f}_n\right\}$, is not independent. Indeed, according to the Corollary to Postulate I, only $(n + 1)$ intensive variables are independent. Therefore, these $(n + 2)$ intensive variables are related by a generalized Gibbs-Duhem relation. Specifically,

$$\sum_{i=1}^{n} x_i d\left(\ln \widehat{f}_i\right) = -\left[\sum_{i=1}^{n} x_i \frac{\left(\overline{H}_i - H_i^{\,\circ}\right)}{RT^2}\right] dT + \left(\sum_{i=1}^{n} \frac{x_i \overline{V}_i}{RT}\right) dP \qquad (25.29)$$

At constant T and P, Eq. (25.29) yields:

$$\sum_{i=1}^{n} x_i \left(\frac{\partial \ln \widehat{f}_i}{\partial x_j}\right)_{T,P,x[j,k]} = 0 \qquad (25.30)$$

Equation (25.30) shows that, given $(n-1)$ of the $\widehat{f}_i$s, the nth one can be calculated by integration, to within an arbitrary constant of integration.

25.7   Calculation of Fugacity

The calculation of $\widehat{f}_i$ or f_i makes use of an EOS, where this is not restricted to gases, if the EOS used can accurately describe the volumetric behavior of liquids. Multiplying both sides of Eq. (25.18) by dP yields:

$$\left(\frac{\partial \ln \widehat{f_i}}{\partial P}\right)_{T,y} dP = d\left(\ln \widehat{f_i}\right)\Big|_{T,y} = \left(\frac{\overline{V}_i}{RT}\right) dP \tag{25.31}$$

Integrating the second and the third terms in Eq. (25.31) with respect to P from $P^* \to 0$ to P yields:

$$\int_{P^* \to 0}^{P} d\left(\ln \widehat{f_i}\right)\Big|_{T,y} = \ln\left(\frac{\widehat{f_i}}{y_i P^*}\right) = \int_{P^* \to 0}^{P} \left(\frac{\overline{V}_i}{RT}\right) dP \tag{25.32}$$

In the limit $P^* \to 0$, the mixture behaves ideally, and:

$$\overline{V}_i \to \frac{RT}{P} \tag{25.33}$$

In that case, the pressure integral in Eq. (25.32) diverges logarithmically when $P^* \to 0$. To take care of this divergence, as we have done in the $(n = 1)$ case, we subtract:

$$\int_{P^* \to 0}^{P} \frac{dP}{P} = \ln\left(\frac{P}{P^*}\right) \tag{25.34}$$

from the second and the third terms in Eq. (25.32). This yields:

$$\ln\left(\frac{\widehat{f_i}}{y_i P^*}\right) - \ln\left(\frac{P}{P^*}\right) = \ln\left(\frac{\widehat{f_i}}{y_i P}\right) = \ln\left(\frac{\widehat{f_i}}{P_i}\right) = \int_{0}^{P} \left(\frac{\overline{V}_i}{RT} - \frac{1}{P}\right) dP \tag{25.35}$$

where the integral in Eq. (25.35) is well-behaved when $P = 0$.

The ratio, $\widehat{f_i}/P_i$, in Eq. (25.35) measures deviations from the ideal mixture behavior and is known as the fugacity coefficient of component i in the mixture. Specifically,

$$\widehat{\phi}_i = \frac{\widehat{f_i}}{P_i} = \frac{\widehat{f_i}}{y_i P} \tag{25.36}$$

where $\widehat{\phi}_i = \widehat{\phi}_i(T, P, y_1, \ldots, y_{n-1})$, and $\widehat{\phi}_i^{IGM} = 1$. Using Eq. (25.36) in the third term in Eq. (25.35), including equating it with the last term in Eq. (25.35) and then multiplying both terms by RT, yields:

$$RT\ln \widehat{\phi}_i = RT\ln\left(\frac{\widehat{f}_i}{y_iP}\right) = \int_0^P \left(\overline{V}_i - \frac{RT}{P}\right)dP \qquad (25.37)$$

We can use Eq. (25.37) if we have access to a volume-explicit EOS, that is, if we are given:

$$V = \underline{V}(T, P, N_1, \ldots, N_n) \qquad (25.38)$$

from which we can compute:

$$\overline{V}_i = (\partial \underline{V}/\partial N_i)_{T,P,Nj[i]} \qquad (25.39)$$

Unfortunately, most EOS are pressure explicit, that is, they provide: $P = P(T, \underline{V}, N_1, \ldots, N_n)$. In that case, we can use a different equation, which relies on a pressure-explicit EOS, to compute $\widehat{\phi}_i$. Alternatively, we can use the triple-product rule to calculate $\overline{V}_i$ in terms of the pressure-explicit EOS. Specifically,

$$\overline{V}_i = \left(\frac{\partial \underline{V}}{\partial N_i}\right)_{T,P,Nj[i]} = -\frac{\left(\frac{\partial P}{\partial N_i}\right)_{T,\underline{V},Nj[i]}}{\left(\frac{\partial P}{\partial \underline{V}}\right)_{T,N_i}} \qquad (25.40)$$

It is important to recognize that, in Eq. (25.40), the parameters in the mixture EOS depend on $\{N_1, \ldots, N_n\}$ when calculating the partial derivative $(\partial P/\partial N_i)_{T,\underline{V},Nj[i]}$.

If we have access to a pressure-explicit EOS, that is, to $P = P(T, \underline{V}, N_1, \ldots, N_n)$, we can derive a different equation to compute $\widehat{\phi}_i$. Specifically,

$$RT\ln\widehat{\phi}_i = RT\ln\left(\frac{\widehat{f}_i}{y_iP}\right) = -RT\ln Z - \int_\infty^{\underline{V}} \left[\left(\frac{\partial P}{\partial N_i}\right)_{T,\underline{V},Nj[i]} - \frac{RT}{\underline{V}}\right]d\underline{V} \qquad (25.41)$$

In Eq. (25.41), Z is the compressibility factor, and is given by $P\underline{V}/NRT$. Again, we recognize that the mixture EOS parameters depend on $\{N_1, N_2, \ldots, N_n\}$, when calculating:

$$\left(\frac{\partial P}{\partial N_i}\right)_{T,P,Nj[i]} \qquad (25.42)$$

in Eq. (25.41). In a similar manner, we can show that for pure gas i:

$$RT\ln\phi_i = RT\ln\left(\frac{f_i}{P}\right) = \int\limits_0^P \left(V_i - \frac{RT}{P}\right)dP \qquad (25.43)$$

In Eq. (25.43), ϕ_i is the fugacity coefficient of pure gas i. Recall that for a pure ideal gas, $\phi_i^{IG} = 1$. Further, Eq. (25.43) is particularly useful when we have access to $\underline{V} = \underline{V}(T, P, N)$, that is, to a volume-explicit EOS. Finally, recall that both ϕ_i and V_i are intensive variables which depend on T and P.

If we have access to $P = P(T,\underline{V},N)$, that is, to a pressure-explicit EOS, we can show that:

$$RT\ln\phi_i = -RT\ln Z - \int\limits_\infty^{\underline{V}} \left[\left(\frac{\partial P}{\partial N}\right)_{T,\underline{V}} - \frac{RT}{\underline{V}}\right]d\underline{V} \qquad (25.44)$$

In Eq. (25.44), Z is the compressibility factor, given by $Z = P\underline{V}/NRT$.

25.8 The Lewis and Randall Rule

As stated earlier, in principle, the fugacity can also be used to describe a liquid mixture. In particular, using Eq. (25.21) for an ideal solution yields:

$$\overline{G}_i^{ID} = \lambda_i(T) + RT\ln\widehat{f}_i^{ID} \qquad (25.45)$$

For pure component i, it follows that (see Eq. (25.10)):

$$G_i(T,P) = \lambda_i(T) + RT\ln f_i(T,P) \qquad (25.46)$$

Subtracting Eq. (25.46) from Eq. (25.45), including cancelling the equal terms and rearranging, yields:

$$\therefore \overline{G}_i^{ID} - G_i(T,P) = RT\ln\left(\frac{\widehat{f}_i^{ID}}{f_i(T,P)}\right) \qquad (25.49)$$

For an ideal solution, it follows that:

$$\overline{G}_i^{ID} - G_i(T,P) = RT\ln x_i \qquad (25.50)$$

A comparison of Eqs. (25.49) and (25.50) shows that:

$$\widehat{f}_i^{ID} = f_i(T, P)x_i \qquad (25.51)$$

The relation in Eq. (25.51) is known as the Lewis and Randall Rule. It simply states that the fugacity of component i in an ideal solution can be obtained by multiplying the pure component i fugacity at the same T and P as in the ideal solution, by the mole fraction of component i in the ideal solution. Interestingly, $\widehat{f}_i^{ID}$ in Eq. (25.51) depends only on x_i, and not on the mole fractions of the other components present in the ideal solution!

Lecture 26

Activity, Activity Coefficient, and Sample Problems

26.1 Introduction

The material presented in this lecture is adapted from Chapter 9 in T&M. First, we will introduce a new thermodynamic function known as the activity of component i, which will describe deviations from the ideal solution behavior. Second, we will define the product of RT and the natural logarithm of the activity of component i as the difference between the partial molar Gibbs free energy of component i in the actual non-ideal solution and the partial molar Gibbs free energy of component i in a reference state having the same temperature as that of the non-ideal solution. Third, we will revisit the pure component reference state, originally introduced in Lecture 24, and use it in this lecture and beyond. Fourth, we will show that if we choose the pure component reference state, when the solution approaches ideality, the activity of component i is equal to the mole fraction of component i. We will also define the ratio of the activity of component i and the mole fraction of component i as the activity coefficient, which will approach unity when the solution approaches ideality. Fifth, we will discuss how to calculate the activity of component i using a mathematical model for the excess Gibbs free energy of mixing. Sixth, we will solve Sample Problem 26.1 to calculate the activity coefficients of components 1 and 2 in a non-ideal binary solution for which a mathematical model for the excess Gibbs free energy of mixing is known, including discussing the model properties when the solution composition varies from 0 (pure component 1) to 0.5 (equimolar composition) to 1 (pure component 2). We will also discuss, mathematically and graphically, deviations from ideality as reflected in the activity coefficients of components 1 and 2 as a function of the solution composition. Specifically, we will show that the activity coefficient is equal to 1 in an ideal solution, exhibits positive deviations from ideality when it is larger than 1, and exhibits negative deviations from ideality when it is smaller than 1. Finally, we will solve Sample Problem 26.2 to calculate the activity coefficient of component 2 in a non-ideal binary solution as a function of T,

© Springer Nature Switzerland AG 2020
D. Blankschtein, *Lectures in Classical Thermodynamics with an Introduction to Statistical Mechanics*, https://doi.org/10.1007/978-3-030-49198-7_26

P, and the solution composition, given the activity coefficient of component 1 as a function of T, P, and the solution composition.

26.2 Activity and Activity Coefficient

Recall that to calculate the fugacity of component i in the mixture, $\widehat{f}_i$, one can use the equations derived earlier which make use of a pressure-explicit or a volume-explicit mixture EOS. Unfortunately, EOS are best suited to describe the volumetric behavior of gas mixtures. To deal with liquid (or even solid) mixtures, it is convenient to develop a new approach, which is based on describing the deviations from the ideal solution behavior using a new thermodynamic function known as the activity. The role played by the activity in a non-ideal solution is analogous to that played by the fugacity in a non-ideal gas mixture.

We introduce the activity of component i in the solution, a_i, as the difference in the partial molar Gibbs free energy of component i between the real state and a reference state. The reference state (RS) is denoted by †, and as will be explained below, we choose $T^\dagger = T$ of the actual solution, while $x_i^\dagger \neq x_i$ (for every i), and $P^\dagger \neq P$. Specifically,

$$RT\ln a_i = \overline{G}_i(T, P, x_1, \ldots, x_{n-1}) - \overline{G}_i^\dagger\left(T, P^\dagger, x_1^\dagger, \ldots, x_{n-1}^\dagger\right) \tag{26.1}$$

Recall that we can also express $\overline{G}_i$ and $\overline{G}_i^\dagger$ in terms of fugacities as follows:

$$\overline{G}_i = \lambda_i(T) + RT\ln\widehat{f}_i \tag{26.2}$$

$$\overline{G}_i^\dagger = \lambda_i(T) + RT\ln\widehat{f}_i^\dagger \tag{26.3}$$

Subtracting Eq. (26.3) from Eq. (26.2), cancelling the equal terms and rearranging, yields:

$$\overline{G}_i - \overline{G}_i^\dagger = RT\ln\left(\frac{\widehat{f}_i}{\widehat{f}_i^\dagger}\right) \tag{26.4}$$

Using Eq. (26.4) in Eq. (26.1) yields:

$$RTln\left(\frac{\widehat{f_i}}{\widehat{f_i^\dagger}}\right) = RTlna_i \Rightarrow a_i = \frac{\widehat{f_i}}{\widehat{f_i^\dagger}} \tag{26.5}$$

The last term in Eq. (26.5) shows that the activity of component i in the solution is equal to the ratio of the fugacities of component i in the real state and in the RS, denoted by †. Clearly, the choice of RS determines the activity a_i. A number of RSs are available. We will deal mainly with the pure component RS, originally introduced in Lecture 24, and discussed again for completeness next.

26.3 Pure Component Reference State

Recall that the pure component reference state has the following attributes:

- Same state of aggregation as the solution
- $P^\dagger = P$ (Recall that $T^\dagger = T$)
- $x_i^\dagger = 1, x_j^\dagger = 0$, for $j \neq i$
- $\widehat{f_i^\dagger} = f_i(T, P)$ and $\overline{G}_i^\dagger = G_i(T, P)$

Therefore, it follows that for this RS:

$$a_i = \frac{\widehat{f_i}(T, P, x_1, \ldots, x_{n-1})}{f_i(T, P)} \tag{26.6}$$

or, rearranging Eq. (26.6), that:

$$\widehat{f_i} = f_i a_i \tag{26.7}$$

Equation (26.7) shows that if we know a_i, we can calculate $\widehat{f_i}$. In addition, combining Eqs. (26.4), for the pure component RS, with Eq. (26.6) yields:

$$RTlna_i = \overline{G}_i - G_i$$
$$= \Delta\overline{G}_i \quad \text{(Partial molar mixing Gibbs free energy of i)} \tag{26.8}$$

where $a_i = a_i(T, P, x_1, \ldots, x_{n-1})$.

Equation (26.6) shows that the pure component RS is symmetric for every component i, in the sense that:

$$\lim_{x_i \to 1} a_i = \frac{f_i(T,P)}{f_i(T,P)} = 1 \quad \text{(For every i)} \tag{26.9}$$

There are also asymmetric RSs, like the infinite-dilution RS, for which this is different (see below). For an Ideal (ID) Solution, we saw that:

$$\Delta \overline{G}_i^{ID} = \overline{G}_i^{ID} - G_i = RT\ln x_i = RT\ln a_i^{ID} \tag{26.10}$$

Comparison of the last two terms in Eq. (26.10) shows that the activity of component i in an ideal solution is equal to x_i.

When the solution is not ideal, the ratio, a_i/x_i, is not unity. Therefore, the deviation of this ratio from unity serves as a quantitative measure of the solution non-ideality. The ratio is known as the activity coefficient of component i, and is given by:

$$\gamma_i = \frac{a_i}{x_i}; \; a_i = \frac{\widehat{f}_i}{f_i(T,P)} \Rightarrow \gamma_i = \frac{\widehat{f}_i}{x_i f_i(T,P)} \tag{26.11}$$

The expressions in Eq. (26.11) are all based on the pure component RS. The last term in Eq. (26.11) shows that, for pure component i, $x_i = 1$, $\widehat{f}_i = f_i(T,P)$, and γ_i (pure i) = 1. In other words, for the pure component RS:

$$\lim_{x_i \to 1} \gamma_i = 1 \tag{26.12}$$

For an Ideal (ID) Solution, the last term in Eq. (26.11) is given by:

$$\gamma_i^{ID} = \frac{\widehat{f}_i^{ID}}{x_i f_i(T,P)} \tag{26.13}$$

According to the Lewis and Randall Rule introduced in Lecture 25, it follows that:

$$\widehat{f}_i^{ID} = x_i f_i(T,P) \tag{26.14}$$

Using Eq. (26.14) in Eq. (26.13) shows that for an ideal solution:

$$\gamma_i^{ID} = 1 \tag{26.15}$$

26.4   Calculation of Activity

When the pure component RS is used, the activity coefficient can be related to the excess Gibbs free energy of mixing. To show this, we begin with (see the first term in Eq. (26.11)):

$$a_i = \gamma_i x_i \tag{26.16}$$

Taking the natural logarithm of Eq. (26.16), including multiplying the resulting expression by RT, and using Eq. (26.8), yields:

$$RT\ln a_i = RT\ln(\gamma_i x_i) = \overline{G}_i - G_i = \Delta\overline{G}_i \tag{26.17}$$

For an Ideal (ID) Solution, Eq. (26.17) can be expressed as follows:

$$RT\ln a_i^{ID} = RT\ln\left(\gamma_i^{ID} x_i\right) = \overline{G}_i^{ID} - G_i = \Delta\overline{G}_i^{ID} \tag{26.18}$$

Subtracting Eq. (26.18) from Eq. (26.17), including rearranging, yields:

$$RT\ln\left(\frac{\gamma_i}{\gamma_i^{ID}}\right) = \Delta\overline{G}_i - \Delta\overline{G}_i^{ID} = \Delta\overline{G}_i^{EX} = \overline{G}_i - \overline{G}_i^{ID} - \overline{G}_i^{EX} \tag{26.19}$$

In the derivation of Eq. (26.19), we have used relations presented in Lecture 25. Recalling that $\gamma_i^{ID} = 1$, Eq. (26.19) can be written as follows:

$$RT\ln\gamma_i = \Delta\overline{G}_i^{EX} = \overline{G}_i^{EX} \tag{26.20}$$

Equation (26.20) shows that RT times the natural logarithm of the activity coefficient of component i in the solution is equal to the partial molar excess Gibbs free energy of mixing of component i, which is equal to the partial molar excess Gibbs free energy of component i.

Equation (26.20) is very useful, because it permits calculation of γ_i (and, therefore, of $a_i = \gamma_i x_i$) for a non-ideal solution if either $\underline{G}^{EX}$ or $\Delta\underline{G}^{EX}$ are known.

Specifically, given:

(i) $\Delta\underline{G}^{EX} = \Delta\underline{G}^{EX}(T, P, N_1, \dots, N_n) \Rightarrow RT\ln\gamma_i = \Delta\overline{G}_i^{EX}$

$$= \left(\frac{\partial \Delta\underline{G}^{EX}}{\partial N_i}\right)_{T,P,N_{j[i]}} \tag{26.21}$$

(ii) $\underline{G}^{EX} = \underline{G}^{EX}(T, P, N_1, \ldots, N_n) \Rightarrow RT\ln\gamma_i = \overline{G}_i^{EX}$

$$= \left(\frac{\partial \underline{G}^{EX}}{\partial N_i}\right)_{T,P,N_{j[i]}} \tag{26.22}$$

Because $\Delta\overline{G}_i^{EX} = RT\ln\gamma_i$ and $\Delta\underline{G}^{EX} = \sum\limits_{i=1}^{n} N_i \Delta\overline{G}_i^{EX}$, it follows that:

$$\Delta\underline{G}^{EX} = \underline{G}^{EX} = RT\sum\limits_{i=1}^{n} N_i\ln\gamma_i \tag{26.23}$$

The set of $(n + 2)$ intensive variables, $\{T, P, \gamma_1, \ldots, \gamma_n\}$, satisfies the following generalized Gibbs-Duhem relation (gGDr) at constant T and P:

$$\sum\limits_{i=1}^{n} x_i d\ln\gamma_i = 0 \quad \text{(Constant T and P)} \tag{26.24}$$

Equation (26.24) shows that, given $(n-1)$ γ_is at constant T and P, the gGDr can be used to compute the nth γ_i by integration, to within an arbitrary constant of integration.

One can also show that:

$$\left(\frac{\partial \ln\gamma_i}{\partial T}\right)_{P,x} = -\frac{\Delta\overline{H}_i}{RT^2} \tag{26.25}$$

$$\left(\frac{\partial \ln\gamma_i}{\partial P}\right)_{T,x} = \frac{\Delta\overline{V}_i}{RT} \tag{26.26}$$

Typically, for liquids and solids, $\Delta\overline{H}_i$ and $\Delta\overline{V}_i$ are small, and therefore, γ_i is a weak function of T and P.

26.5 Sample Problem 26.1

For a binary solution of components 1 and 2, it is known that:

$$\frac{\Delta\underline{G}^{EX}}{RT} = NCx_1x_2 \tag{26.27}$$

where C is a parameter which does not depend on composition (it may depend on T and P), and $N = N_1 + N_2$.

(i) Show that $\Delta \underline{G}^{EX}$ is symmetric around $x_1 = x_2 = 0.5$.

(ii) Calculate $\gamma_1(T, P, x_2)$ and $\gamma_2(T, P, x_2)$, and discuss the type of nonideality.

26.5.1   Solution

(i) Equation (26.27) shows that $\Delta \underline{G}^{EX}$ vanishes when $x_1 = 0$ (pure 2) and when $x_2 = 0$ (pure 1). This is a manifestation of us choosing the pure component RS to calculate $\Delta \underline{G}^{EX}$.

Further, $\Delta \underline{G}^{EX}$ attains its maximum value when $x_1 = x_2 = 0.5$, and is symmetric in x_1 (or x_2) around 0.5 (see Fig. 26.1). In Fig. 26.1, we replaced x by X for better visualization.

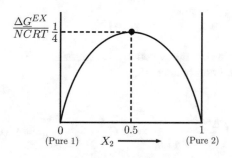

Fig. 26.1

In fact, any mixing function whose RS is the pure components must vanish in the limit $x_i = 1$ for each component i.

(ii) To calculate γ_1 and γ_2, we use the expression derived above which we repeat below for completeness:

$$RT\ln\gamma_i = \Delta \overline{G}_i^{EX} = \left(\frac{\partial \Delta \underline{G}^{EX}}{\partial N_i} \right)_{T, P, N_{j[i]}} \tag{26.28}$$

It is useful to first rewrite $\Delta\underline{G}^{EX}$ in Eq. (26.28) as an explicit function of N_1 and N_2:

$$\frac{\Delta\underline{G}^{EX}}{RT} = CNx_1x_2 = CN\left(\frac{N_1}{N}\right)\left(\frac{N_2}{N}\right) = \frac{CN_1N_2}{N} \tag{26.29}$$

or

$$\frac{\Delta\underline{G}^{EX}}{RT} = \frac{CN_1N_2}{(N_1 + N_2)} \tag{26.30}$$

Using Eq. (26.30) in the last term in Eq. (26.28) for $i = 1$, we obtain:

$$RT\ln\gamma_1 = \left(\frac{\Delta\underline{G}^{EX}}{\partial N_1}\right)_{T,P,N_2} = RTC\left(\frac{N_2}{N_1 + N_2}\right)^2 = RTCx_2^2 \tag{26.31}$$

or

$$\gamma_1 = e^{Cx_2^2} \tag{26.32}$$

Equation (26.32) shows that when $x_2 = 0$ (pure 1), γ_1 $(x_1 = 1) = 1$, as required when we use the pure component RS.

Similarly,

$$RT\ln\gamma_2 = \left(\frac{\Delta\underline{G}^{EX}}{\partial N_2}\right)_{T,P,N_1} = RTC\left(\frac{N_1}{N_1 + N_2}\right)^2 = RTCx_1^2 \tag{26.33}$$

or

$$\gamma_2 = e^{C(1-x_2)^2} \tag{26.34}$$

Equation (26.34) shows that when $x_2 = 1$ (pure 2), γ_2 $(x_2 = 1) = 1$, as required when we use the pure component RS

Because x_2 varies between 0 and 1 (as does $x_1 = 1 - x_2$), Eqs. (26.32) and (26.34) show that, at fixed T and P, γ_1 and γ_2 are both >1 over the entire composition range. We refer to this behavior as a positive deviation from ideality, because:

$$\gamma_i^{ID} = 1 \quad \text{(For } i = 1 \text{ and 2)} \tag{26.35}$$

If $\gamma_i < 1$, the solution is said to exhibit a negative deviation from ideality. We can plot the variations of γ_1 and γ_2 with x_2, at constant T and P. To this end, let us assume that $C = 1$. In that case, Eqs. (26.32) and (26.34) yield:

$$\gamma_1 = e^{x_2^2}; \ \gamma_2 = e^{(1-x_2)^2} \tag{26.36}$$

Figure 26.2 shows how γ_1 and γ_2 in Eq. (26.36) vary as a function of x_2, at constant T and P. Again, for better visualization, in Fig. 26.2, we replaced x by X.

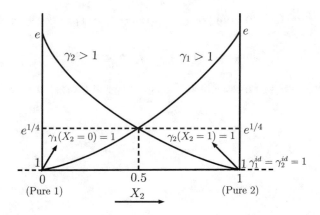

Fig. 26.2

26.6   Sample Problem 26.2

Given $\gamma_1 = e^{Cx_2^2}$, calculate $\gamma_2(T, P, x_2)$.

26.6.1   Solution

To solve this problem, we simply use the generalized Gibbs-Duhem relation for the activity coefficients of a binary mixture at constant T and P. Specifically,

$$x_1 d\ln\gamma_1 + x_2 d\ln\gamma_2 = 0 (\text{Const.T and P}) \Rightarrow d\ln\gamma_2 = -\frac{x_1}{x_2} d\ln\gamma_1$$

$$= -\frac{(1-x_2)}{x_2} d\ln\gamma_1 \tag{26.37}$$

Based on the Problem Statement,

$$d\ln\gamma_1 = 2Cx_2dx_2 \tag{26.38}$$

Using the expression for $d\ln\gamma_1$ in Eq. (26.38) in the expression for $d\ln\gamma_2$ in Eq. (26.37) yields:

$$d\ln\gamma_2 = d\left[C(1 - x_2)^2\right] \tag{26.39}$$

Integrating Eq. (26.39) from $[x_2 = 1,\ \gamma_2 (x_2 = 1) = 1]$ to $[x_2, \gamma_2 (x_2)]$ yields:

$$\ln\gamma_2 = C(1\text{-}x_2)^2 \text{ or } \gamma_2 = e^{C(1-x_2)^2} \tag{26.40}$$

As expected, the γ_2 expressions in Eqs. (26.40) and (26.34) are identical.

Lecture 27

Criteria of Phase Equilibria, and the Gibbs Phase Rule

27.1 Introduction

The material presented in this lecture is adapted from Chapter 15 in T&M. First, we will discuss the infinite-dilution reference state typically used to describe solutions where the solvent dominates (that is, the solvent mole fraction approaches unity, and the solutes are present at mole fractions which approach zero). In this context, we will discuss colligative properties, including freezing-point depression, boiling-point elevation, vapor-pressure lowering, and osmotic pressure. Second, we will discuss the thermodynamic criteria of phase equilibria for a system consisting of π phases, each containing n components, and separated by boundaries that are open, diathermal, and movable. Accordingly, the entire composite system is simple. Third, we will show that, at thermodynamic equilibrium, the system is in thermal, mechanical, and diffusional equilibrium for every component present in each phase. Fourth, we will present a derivation of the celebrated Gibbs Phase Rule, which specifies the number of independent intensive variables that we need to select in order to unambiguously characterize the intensive equilibrium thermodynamic state of the system. Note that the Gibbs Phase Rule does not specify which intensive variables we need to select, but only their number. Finally, we will show how the Gibbs Phase Rule needs to be generalized when the multi-phase, multi-component system is not simple, including discussing the number of independent extensive variables that we need to select to fully characterize the extensive equilibrium thermodynamic state of the system.

27.2 Use of Other Reference States

How do we model the behavior of a solution, where one of the components (say #1) is present at a very high mole fraction ($x_1 \rightarrow 1$, referred to as the solvent), while all the other components (referred to as the solutes) are present at very low mole fractions

© Springer Nature Switzerland AG 2020
D. Blankschtein, *Lectures in Classical Thermodynamics with an Introduction to Statistical Mechanics*, https://doi.org/10.1007/978-3-030-49198-7_27

($x_i \to 0$, for $i \neq 1$)? This is, of course, what occurs in dilute solutions. As an illustration, consider a dilute binary solution of sugar (the solute, denoted as component 2) and water (the solvent, denoted as component 1) at a fixed temperature T and pressure P. Experimentally, we know that adding small concentrations of sugar ($x_2 \to 0$) to water results in:

- Vapor − Pressure (vp) Lowering : $P_{vp}^o(T, x_2) < P_{vp}^o(T)$ (27.1)

- Freezing (f) − Point Depression : $T_f^o(P, x_2) < T_f^o(P)$ (27.2)

- Boiling (b) − Point Elevation : $T_b^o(P, x_2) > T_b^o(P)$ (27.3)

where in Eqs. (27.1), (27.2), and (27.3), the variables on the right-hand side of the inequality signs denote the vapor pressure, the freezing temperature, and the boiling temperature of the solvent (water), respectively.

- Osmotic Pressure : To be discussed below   (27.4)

The four solution properties in Eqs. (27.1), (27.2), (27.3), and (27.4) are known as colligative properties, namely, properties which in dilute solution are proportional to the solute concentration.

Because in the dilute sugar-water solution there is a clear asymmetry between the solvent ($x_1 \to 1$), and the solute ($x_2 \to 0$), it makes sense to use an asymmetric reference state (RS), rather than to use the pure component RS for both the solute and the solvent (as we did in previous lectures). Indeed, we will continue to use the pure component RS for the solvent, but will use the infinite-dilution RS for the solute. Therefore, the resulting RS is asymmetric, as shown in Fig. 27.1.

In the last entry of the left column of Fig. 27.1, the dilute solution behaves ideally (solvent wise), that is, $\gamma_1 = 1$ when $x_1 \to 1$. On the other hand, in the last entry of the right column of Fig. 27.1, the dilute solution behaves ideally (solute wise), that is, $\gamma_2 = 1$ when $x_2 \to 0$.

27.3   Phase Equilibria: Introduction

Next, we will discuss scenarios in which two or more phases are in equilibrium. Phase equilibria are important in many industrial applications of relevance to chemical engineers, including distillation, absorption, and extraction, where two

Solvent (#1) RS	Solute (#2) RS
• Liquid State of Aggregation	• Liquid State of Aggregation
• $T^\dagger = T$	• $T^\dagger = T$
• $P^\dagger = P$	• $P^\dagger = P$
• $x_1^\dagger = 1$	• $x_2^\dagger = 0$
• $\mu_1 = \overline{G}_1 = G_1(T,P) + RT \ln a_1$	• $\mu_2 = \overline{G}_2 = G_2(T,P) + RT \ln a_2$
• $\lim_{x_1 \to 1} a_1 = x_1$ The Dilute Solution behaves Ideally (Solvent Wise) when $x_1 \to 1$ • $\lim_{x_1 \to 1} \gamma_1 = 1$	• $\lim_{x_2 \to 0} a_2 = x_2$ The Dilute Solution behaves Ideally (Solute Wise) when $x_2 \to 0$ • $\lim_{x_2 \to 0} \gamma_2 = 1$

Fig. 27.1

phases are in contact. In some cases, three phases are in contact, for example, at the solid/liquid/vapor equilibrium observed at the triple point. The coexisting phases can be: Liquid/Vapor (L/V), Liquid/Liquid (L/L), Solid/Vapor (S/V), Solid/Liquid (S/L), and Solid/Solid (S/S).

In the remainder of this lecture, and in Lectures 28, 29, and 30, we will discuss:

1. The general thermodynamic criteria of phase equilibria
2. The Gibbs Phase Rule (GPR)
3. The differential approach to phase equilibria
4. The integral approach to phase equilibria
5. Sample Problems which illustrate the implementation of the new material in 1 to 4 above

27.4   Criteria of Phase Equilibria

Consider π phases in equilibrium, where each phase contains n components. All the internal boundaries are open, diathermal, and movable. The entire composite system is therefore simple. Further, the entire composite system is surrounded by a closed, adiabatic, and rigid boundary, and therefore, is isolated.

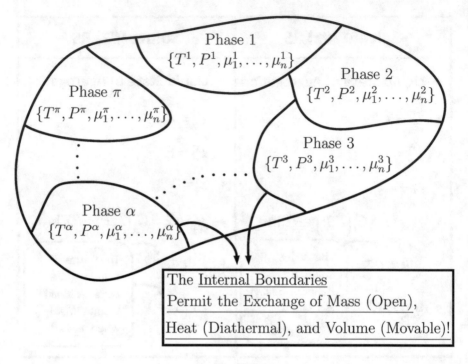

Fig. 27.2

With each phase α, we can associate the following set of $(n + 2)$ intensive variables:

$$\{T^\alpha, P^\alpha, \mu_1^\alpha, \mu_2^\alpha, \ldots, \mu_n^\alpha\} \tag{27.5}$$

which, as we know, are not independent, because they are related by the Gibbs-Duhem equation in phase α. Figure 27.2 illustrates the phases and variables characterizing the system under consideration.

For the simple, composite, multi-phase, multi-component system depicted in Fig. 27.2, the following equilibrium conditions apply.

27.4.1   Thermal Equilibrium

$$T^1 = T^2 = \ldots = T^\alpha = \ldots = T^\pi \tag{27.6}$$

Equation (27.6) indicates that thermal equilibrium implies no temperature gradients, and therefore, no heat transfer between the coexisting phases.

27.4.2   Mechanical Equilibrium

$$P^1 = P^2 = \ldots = P^\alpha = \ldots = P^\pi \qquad (27.7)$$

Equation (27.7) indicates that mechanical equilibrium implies no pressure gradients, and therefore, no volume transfer between the coexisting phases.

27.4.3   Diffusional Equilibrium

$$\mu_i^{\,1} = \mu_i^{\,2} = \ldots = \mu_i^{\,\alpha} = \ldots = \mu_i^{\,\pi} \ (i = 1, \ldots, n) \qquad (27.8)$$

Equation (27.8) indicates that diffusional equilibrium implies no chemical potential gradients for every component i, and therefore, no mass transfer of every component i between the coexisting phases.

What happens to the conditions of thermodynamic equilibrium in Eqs. (27.6), (27.7), and (27.8) if the multi-phase system is not simple? As an illustration, consider a binary water (•) + sugar (x) solution coexisting with pure water (•) across a membrane which only allows free passage of water. The resulting osmometry cell is shown in Fig. 27.3.

In Fig. 27.3, the resulting two-phase system is not simple because the membrane is rigid (no pressure equality is possible), and is impermeable to the sugar molecules (no sugar chemical potential equality is possible). Accordingly, only two conditions of thermodynamic equilibrium apply. Specifically,

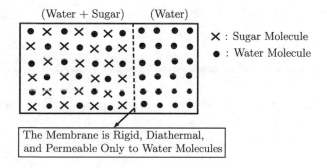

Fig. 27.3

$$T_{\text{Pure Water}} = T_{\text{Sugar Solution}} \tag{27.9}$$

$$\mu_w^{\text{Pure Water}} = \mu_w^{\text{Sugar Solution}} \tag{27.10}$$

Because the membrane is rigid, no pressure equality can be established at equilibrium. In fact, we can show that the pressure in the sugar solution compartment exceeds the pressure in the pure water compartment by an amount known as the osmotic pressure, π, which is the fourth colligative property discussed earlier. Specifically,

$$P_{\text{Sugar Solution}} > P_{\text{Pure Water}} \Rightarrow P_{\text{Sugar Solution}} - P_{\text{Pure Water}} = \pi \tag{27.11}$$

27.5   The Gibbs Phase Rule

We would like to determine how many independent intensive variables need to be specified to fully describe the intensive thermodynamic equilibrium state of a simple composite system consisting of π phases, each carrying n components, in thermodynamic equilibrium. The answer is embodied in the celebrated Gibbs Phase Rule.

The following derivation of the Gibbs Phase Rule involves four steps:

1. Characterize phase alpha (α) using the following set of $(n + 2)$ intensive variables:

$$\left\{ T^\alpha, P^\alpha, \mu_1^\alpha, \mu_2^\alpha, \ldots, \mu_n^\alpha \right\} \tag{27.12}$$

2. Impose the Gibbs-Duhem equation in phase α:

$$\underline{S}^\alpha dT^\alpha - \underline{V}^\alpha dP^\alpha + \sum_{i=1}^{n} N_i^\alpha d\mu_i^\alpha = 0 \quad (\alpha = 1, 2, \ldots, \pi) \tag{27.13}$$

3. Impose the conditions of thermal, mechanical, and diffusional equilibrium:

$$T^1 = T^2 = \ldots T^\alpha = \ldots = T^\pi \equiv T \quad \text{(Thermal equilibrium)} \tag{27.14}$$

$$P^1 = P^2 = \ldots = P^\alpha = \ldots = P^\pi \equiv P \quad \text{(Mechanical equilibrium)} \quad (27.15)$$

$$\mu_i^1 = \mu_i^2 = \ldots = \mu_i^\alpha = \ldots = \mu_i^\pi \equiv \mu_i \; (i = 1, 2, \ldots, n)$$
(Diffusional equilibrium)
$$\quad (27.16)$$

4. Relate the set of $(n + 2)$ intensive variables

$$\{T, P, \mu_1, \ldots, \mu_n\} \quad (27.17)$$

by π Gibbs-Duhem equations, one for each phase α. Specifically,

$$\underline{S}^\alpha dT - \underline{V}^\alpha dP + \sum_{i=1}^n N_i^\alpha d\mu_i = 0 \quad (\alpha = 1, 2, \ldots, \pi) \quad (27.18)$$

In other words, out of the $(n + 2)$ intensive variables, only $(n + 2 - \pi)$ are independent. This number is referred to as the variance, and is denoted by L, where,

$$L = n + 2 - \pi \quad (27.19)$$

Equation (27.19) is the Gibbs Phase Rule (GPR) and was originally derived by J.W. Gibbs in 1875. Note that if there are additional constraints (e.g., chemical reactions), L is decreased further by the number of additional constraints, r, that is, $L = n + 2 - \pi - r$.

A simpler and illuminating derivation of the Gibbs Phase Rule, which is particularly useful if the systems considered are not simple, is presented next. However, to better understand the derivation for a non-simple system, we will first deal with a simple system, in which all the internal boundaries are open, diathermal, and movable. Using the Corollary to Postulate I, we select $(n + 1)$ independent intensive variables to characterize each of the π coexisting phases of the simple system. As discussed in Part II, these $\pi(n + 1)$ intensive variables are not independent. Indeed, the $\pi(n + 1)$ intensive variables are related by the conditions of thermal equilibrium (TE), mechanical equilibrium (ME), and diffusional equilibrium (DE) which apply. If the multi-phase system is simple, all the conditions of TE, ME, and DE apply. It then follows that:

(i)   TE : $T^1 = T^2 = \ldots = T^\pi$, Imposes $(\pi - 1)$ constraints   (27.20)

(ii)   ME : $P^1 = P^2 = \ldots = P^\pi$, Imposes $(\pi - 1)$ constraints   (27.21)

(iii) $\mu_i^1 = \mu_i^2 = \ldots = \mu_i^\alpha = \ldots = \mu_i^\pi$, Imposes $(\pi - 1)n$ constraints   (27.22)

The total number of constraints relating the $\pi(n + 1)$ intensive variables is therefore given by the sum of all the constraints in Eqs. (27.20), (27.21), and (27.22), that is, by:

$$\text{Total number of TE, ME, and DE constraints}$$
$$= (\pi - 1) + (\pi - 1) + (\pi - 1)n = (\pi - 1)(n + 2) \qquad (27.23)$$

It then follows that the number of independent intensive variables, L, is given by:

$$L = \pi(n + 1) - (\pi - 1)(n + 2) \qquad (27.24)$$

Rearranging Eq. (27.24), including cancelling the equal terms, yields:

$$L = n + 2 - \pi \quad \text{(Gibbs Phase Rule for a simple system)} \qquad (27.25)$$

As expected, Eq. (27.25) is identical to Eq. (27.19).

If the multi-phase system is not simple, then, not all the (TE + ME + DE) constraints apply, because some of the internal boundaries are closed, adiabatic, or rigid. Accordingly, L will be determined by:

$$L = \pi(n + 1) - (\text{Number of constraints which apply}) \qquad (27.26)$$

Equation (27.26) shows that, because there is no universal expression for the number of constraints which apply, there is no universal expression for L in the case of a multi-phase, non-simple system. Each case needs to be carefully examined to determine the number of constraints which apply.

Returning to a simple system, Postulate I requires that we specify $(n + 2)$ independent properties. If we choose L of these $(n + 2)$ to be intensive, then, the number of extensive ones, E, is given by:

$$E = (n + 2) - L = (n + 2) - (n + 2 - \pi - r) = n + 2 - n - 2 + \pi + r \quad (27.27)$$

or

$$E = \pi + r \quad \text{(For a simple system)} \qquad (27.28)$$

For example, for:

$$\left.\begin{cases} n = 1 \\ r = 0 \\ \pi \end{cases}\right\} \Rightarrow E = \pi + 0 = \pi \qquad (27.29)$$

If $\pi = 1$, then $E = 1$, and a natural choice of extensive variable is the number of moles, N. In addition, $L = 1 + 2 - 1 = 2$, and a natural choice of the two independent intensive variables is T and P.

Lecture 28

Application of the Gibbs Phase Rule, Azeotrope, and Sample Problem

28.1 Introduction

The material presented in this lecture is adapted from Chapter 15 in T&M. First, we will utilize the Gibbs Phase Rule derived in Lecture 27 to elucidate the (P-T) phase diagram of a one component ($n = 1$) substance, in the absence of chemical reactions ($r = 0$), which can exist in Solid (S), Liquid (L), and Vapor (V) phases. In each of the three phases, T and P can vary independently, and as a result, the system is referred to as divariant. Second, Solid and Vapor, Solid and Liquid, and Liquid and Vapor can coexist along the S/V, S/L, and L/V equilibrium lines, respectively. Along each of these lines, $P = f(T)$, and the system is referred to as monovariant. The S/V, S/L, and L/V monovariant lines meet at a special point, known as the triple point, which is invariant. The L/V monovariant line begins at the triple point and ends at another special point, known as the critical point, which is invariant. Finally, we will qualitatively solve Sample Problem 28.1 to examine the boiling of a binary liquid mixture coexisting with its binary vapor mixture. Specifically, we will utilize the Gibbs Phase Rule to understand qualitatively how the mixture vapor pressure changes as a function of the solution composition at constant temperature, including discussing the pressure-composition phase diagram, and the azeotropic point.

28.2 The Gibbs Phase Rule for a Pure Substance

As discussed in Lecture 27, for a pure substance ($n = 1$) existing in one phase ($\pi = 1$), for example, Solid (S), Liquid (L), or Vapor (V), in the absence of chemical reactions ($r = 0$), the variance is given by:

$$L\,(n = 1, \pi = 1, r = 0) = n + 2 - \pi - r = 2 \tag{28.1}$$

© Springer Nature Switzerland AG 2020
D. Blankschtein, *Lectures in Classical Thermodynamics with an Introduction to Statistical Mechanics*, https://doi.org/10.1007/978-3-030-49198-7_28

In other words, $L = 2$, and therefore, we can vary at most two independent intensive variables, for example, T and P, in the (P-T) phase diagram. The system is said to be divariant.

For a pure substance ($n = 1$) coexisting in two phases ($\pi = 2$), for example, Liquid/Vapor (L/V), Solid/Liquid (S/L), or Solid/Vapor (S/V), in the absence of chemical reactions ($r = 0$), the variance is given by:

$$L\ (n = 1, \pi = 2, r = 0) = n + 2 - \pi - r = 1 \qquad (28.2)$$

In other words, $L = 1$, and therefore, we can only vary one intensive variable along the L/V, S/L, or S/V coexistence lines, that is, $P = P(T)$, in a (P-T) phase diagram, and the system is said to be monovariant.

For a pure substance ($n = 1$) coexisting in three phases ($\pi = 3$), for example, Solid/Liquid/Vapor (S/L/V), in the absence of chemical reactions ($r = 0$), the variance is given by:

$$L\ (n = 1, \pi = 3, r = 0) = n + 2 - \pi - r = 0 \qquad (28.3)$$

In other words, $L = 0$, which corresponds to the invariant triple point (TP) in the (P-T) phase diagram.

The L/V coexistence line begins at the triple point (TP) and terminates at the critical point (CP). Therefore, the Gibbs Phase Rule helps us understand qualitatively the typical (P-T) phase diagram of a pure substance (see Fig. 28.1).

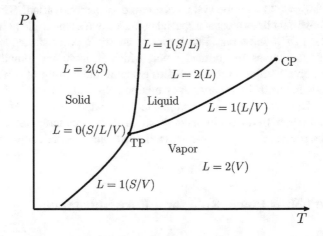

Fig. 28.1

Interestingly, as discussed in Part I, at the CP of a pure substance, two constraints $s = 2$, where s denotes the number of additional constraints which need to be enforced. Specifically,

$$\left(\frac{\partial P}{\partial V}\right)_T = 0 \text{ and } \left(\frac{\partial^2 P}{\partial V^2}\right)_T = 0 \tag{28.4}$$

Accordingly, at the critical point of a pure substance, it follows that L (n = 1, $\pi = 1$, s = 2) = n + 2 − π − s = 0, indicating that the critical point is invariant.

28.3 Sample Problem 28.1

Boiling of a binary liquid mixture coexisting with its binary vapor mixture. Examine qualitatively how the mixture vapor pressure varies with the mixture composition at constant temperature.

28.3.1 Solution

For the liquid-vapor equilibrium of a binary mixture, it follows that: n = 2, $\pi = 2$, r = 0 $\Rightarrow$ L = n + 2 − π − r = 2. Accordingly, we need to specify two independent intensive variables, and any other intensive variable should be uniquely determined. In this Sample Problem, we are asked to examine how the vapor pressure of the binary (liquid) mixture varies with T and x_1 (the liquid mole fraction of component 1) or with T and y_1 (the vapor mole fraction of component 1).

To be specific, we will assume that, at temperature T, the vapor pressure of component 1 [$P^o_{vp1}(T)$] is larger than the vapor pressure of component 2 [$P^o_{vp2}(T)$]. Without yet solving the liquid-vapor equilibrium problem quantitatively (we will learn how to do that in Lectures 29 and 30), let us examine how P_{vp} varies with x_1 or y_1 at constant T. Fig. 28.2 depicts the P versus x_1 or P versus y_1 phase diagram at a given temperature (T).

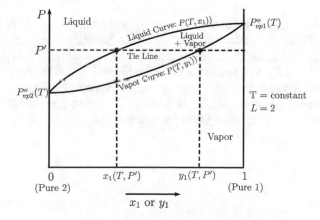

Fig. 28.2

Figure 28.2 shows that for $P^o_{vp2}(T) < P' < P^o_{vp1}(T)$, the system is unstable, and separates into a liquid phase at $x_1(T, P')$ coexisting with a vapor phase at $y_1(T, P')$. At each T and P', there is a unique value of $x_1(T, P')$ and $y_1(T, P')$, as required by the Gibbs Phase Rule, because $L = 2$. Clearly, as shown in Fig. 28.2, the Gibbs Phase Rule helps us understand qualitatively the pressure-composition phase diagram of this binary mixture.

There is a special temperature, known as the azeotropic temperature, at which $T = T_{az}$ and $P = P_{az}$, where the liquid and vapor mole fractions are equal, that is:

$$\text{At } T_{az}, P_{az} \quad \Rightarrow \quad \begin{matrix} x^{az}_1 = y^{az}_1 \\ x^{az}_2 = y^{az}_2 \end{matrix} \tag{28.5}$$

At the azeotrope, the functions $P(T, x_1)$ or $P(T, y_1)$ are no longer monotonic. The corresponding phase diagram is shown in Fig. 28.3.

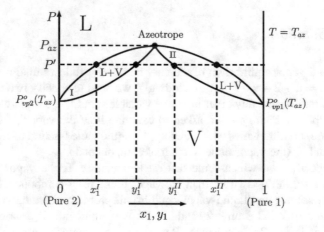

Fig. 28.3

If we choose $(T_{az}$ and $P')$ as the two independent intensive variables (as required by the Gibbs Phase Rule, where $L = 2$), we do not know if we are located on Lobe I or on Lobe II (see Fig. 28.3). Therefore, a better choice in this case is (T_{az}, x^I_1) or (T_{az}, y^I_1), or (T_{az}, x^{II}_1) or (T_{az}, y^{II}_1), which locates us on Lobes I or II. This then enables us to unambiguously characterize the intensive thermodynamic equilibrium state of the system. This example involving the azeotrope shows that great care is sometimes required when one applies the Gibbs Phase Rule.

Lecture 29

Differential Approach to Phase Equilibria, Pressure-Temperature-Composition Relations, Clausius-Clapeyron Equation, and Sample Problem

29.1 Introduction

The material presented in this lecture is adapted from Chapter 15 in T&M. First, we will solve Sample Problem 29.1 to discuss how to model the phase equilibria of a binary liquid mixture coexisting with its binary vapor mixture. Recall that, in Lecture 28, we discussed various aspects of this problem qualitatively in the context of the Gibbs Phase Rule. To solve this problem quantitatively, we can pursue two modeling approaches: (i) the Differential Approach to Phase Equilibria, which equates the differentials of the natural logarithms of the fugacities of every component present in the binary liquid mixture and in the binary vapor mixture, and (ii) the Integral Approach to Phase Equilibria, which equates the fugacities of every component present in the binary liquid mixture and in the binary vapor mixture. In this lecture, we will discuss approach (i) in great detail, and reserve our discussion of approach (ii) to Lecture 30. Specifically, to implement approach (i), we will expand the differentials of the natural logarithms of the fugacities as a function of temperature, pressure, and mixture composition, using the expressions derived in Lecture 25. Doing so, we will derive two equations relating the four unknowns, dT, dP, the differential of the liquid mole fraction of component 1, and the differential of the vapor mole fraction of component 1. According to the Gibbs Phase Rule, for this two-phase, two-component system in the absence of chemical reactions, the number of independent intensive variables is L = 2. This implies that by solving the two equations simultaneously, given two intensive variables, for example, T and P, we will be able to uniquely determine the liquid and the vapor mole fractions of component 1. Third, we will discuss eight simplifications of the two equations which will lead to simple solutions. Finally, we will solve the two equations simultaneously, including discussing the types of predictions that we can make.

© Springer Nature Switzerland AG 2020
D. Blankschtein, *Lectures in Classical Thermodynamics with an Introduction to Statistical Mechanics*, https://doi.org/10.1007/978-3-030-49198-7_29

29.2 Sample Problem 29.1

Model the phase equilibria of a binary liquid mixture coexisting with its binary vapor mixture.

29.2.1 Solution

To solve this phase equilibria problem, we will undertake the following steps:

1. Draw the system and specify the independent intensive variables in each phase using the Gibbs Phase Rule, as shown in Fig. 29.1.

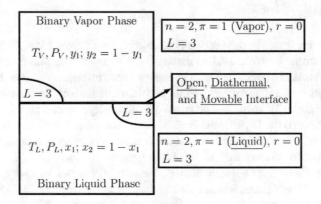

Fig. 29.1

2. Determine the variance of the two-phase system:

$$L = n + 2 - \pi - r \tag{29.1}$$

$$n = 2, \pi = 2 \ (\text{Liquid} + \text{Vapor}), r = 0 \Rightarrow L = 2 \tag{29.2}$$

According to the Gibbs Phase Rule, $L = 2$ (see Eq. (29.2)), and therefore, if we choose two independent intensive variables, all the other intensive variables must be uniquely determined. Clearly, out of the six intensive variables, $\{T_V, P_V, y_1, T_L, P_L, x_1\}$, in Fig. 29.1, only two are independent.

3. Impose the conditions of phase equilibria:

(a) Thermal Equilibrium

$$T_L = T_V = T \tag{29.3}$$

(b) Mechanical Equilibrium

$$P_L = P_V = P \tag{29.4}$$

(c) Diffusional Equilibrium

$$\mu_1^L = \mu_1^V \Rightarrow \lambda_1(T) + RT\ln\widehat{f}_1^L = \lambda_1(T) + RT\ln\widehat{f}_1^V \Rightarrow \widehat{f}_1^L = \widehat{f}_1^V \tag{29.5}$$

$$\mu_2^L = \mu_2^V \Rightarrow \lambda_2(T) + RT\ln\widehat{f}_2^L = \lambda_2(T) + RT\ln\widehat{f}_2^V \Rightarrow \widehat{f}_2^L = \widehat{f}_2^V \tag{29.6}$$

Recall that:

$$\widehat{f}_1^L = \widehat{f}_1^L(T, P, x_1); \widehat{f}_2^L = \widehat{f}_2^L(T, P, x_2) \tag{29.7}$$

and

$$\widehat{f}_1^V = \widehat{f}_1^V(T, P, y_1); \widehat{f}_2^V = \widehat{f}_2^V(T, P, y_2) \tag{29.8}$$

where in Eqs. (29.7) and (29.8), $x_2 = 1 - x_1$, and $y_2 = 1 - y_1$.

4. Solve the fugacity equations:

(a) Use the Integral Approach

In the Integral Approach, we equate the fugacity of component i (1 and 2) in each coexisting phase (L and V). Specifically,

$$\widehat{f}_1^L(T, P, x_1) = \widehat{f}_1^V(T, P, y_1) \tag{29.9}$$

$$\widehat{f}_2^L(T, P, x_2) = \widehat{f}_2^V(T, P, y_2) \tag{29.10}$$

where $x_2 = 1 - x_1$, and $y_2 = 1 - y_1$.

As we discussed in Part II, to model the liquid fugacities, $\widehat{f}_1^L$ and $\widehat{f}_2^L$, we use an activity coefficient approach based on a model for ΔG^{EX}. In addition, to model the vapor fugacities, $\widehat{f}_1^V$ and $\widehat{f}_2^V$, we use a fugacity coefficient approach based on an Equation of State.

In this interesting Sample Problem, there are four unknowns: T, P, x_1, and y_1, which are related by the two fugacity equalities in Eqs. (29.9) and (29.10). This indicates that in order to obtain a unique solution, we must specify two out of the four unknown intensive variables, as required by the Gibbs Phase Rule, which indicated that L = 2. For example, if we choose T and P, we can calculate the liquid composition x_1 and the vapor composition y_1.

(b) Use the Differential Approach

In the differential approach, we equate $d\left(\ln \widehat{f}_i^{\,L}\right)$ with $d\left(\ln \widehat{f}_i^{\,V}\right)$ for i = 1 and 2, that is:

$$d\left(\ln \widehat{f}_1^{\,L}\right) = d\left(\ln \widehat{f}_1^{\,V}\right) \tag{29.11}$$

and

$$d\left(\ln \widehat{f}_2^{\,L}\right) = d\left(\ln \widehat{f}_2^{\,V}\right) \tag{29.12}$$

and we then expand each $d\left(\ln \widehat{f}_i^{\,\alpha}\right)$ term (for i = 1 and 2, and α = L and V) in terms of T, P, and composition. As we discussed in Part II, because $\widehat{f}_i^{\,V} = \widehat{f}_i^{\,V}(T, P, y_i)$ for i = 1 and 2, it follows that:

$$d\left(\ln \widehat{f}_i^{\,V}\right) = \left(\frac{\partial \ln \widehat{f}_i^{\,V}}{\partial T}\right)_{P, y_i} dT + \left(\frac{\partial \ln \widehat{f}_i^{\,V}}{\partial P}\right)_{T, y_i} dP + \left(\frac{\partial \ln \widehat{f}_i^{\,V}}{\partial y_i}\right)_{T, P} dy_i \tag{29.13}$$

We should recognize that if n > 2, additional dy_i terms will appear in Eq. (29.13). Further, because $\widehat{f}_i^{\,L} = \widehat{f}_i^{\,L}(T, P, x_i)$ for i = 1 and 2, it follows that:

$$d\left(\ln \widehat{f}_i^{\,L}\right) = \left(\frac{\partial \ln \widehat{f}_i^{\,L}}{\partial T}\right)_{P, x_i} dT + \left(\frac{\partial \ln \widehat{f}_i^{\,L}}{\partial P}\right)_{T, x_i} dP + \left(\frac{\partial \ln \widehat{f}_i^{\,L}}{\partial x_i}\right)_{T, P} dx_i \tag{29.14}$$

Again, if n > 2, additional dx_i terms will appear in Eq. (29.14).
As shown in Part II,

$$\left(\frac{\partial \ln \widehat{f}_i^{\,V}}{\partial T}\right)_{P, y_i} = -\left(\frac{\overline{H}_i^V - H_i^\circ(T)}{RT^2}\right) \tag{29.15}$$

$$\left(\frac{\partial \ln \widehat{f}_i^L}{\partial T}\right)_{P,x_i} = -\left(\frac{\overline{H}_i^L - H_i^o(T)}{RT^2}\right) \tag{29.16}$$

$$\left(\frac{\partial \ln \widehat{f}_i^V}{\partial P}\right)_{T,y_i} = \left(\frac{\overline{V}_i^V}{RT}\right) \tag{29.17}$$

$$\left(\frac{\partial \ln \widehat{f}_i^L}{\partial P}\right)_{T,x_i} = \left(\frac{\overline{V}_i^L}{RT}\right) \tag{29.18}$$

Using Eqs. (29.15)–(29.18) in $d\left(\ln \widehat{f}_i^V\right) = d\left(\ln \widehat{f}_i^L\right)$, for $i = 1$ and 2 (see Eqs. (29.13) and (29.14)) yields the following equation:

$$-\left(\frac{\overline{H}_i^V - H_i^o(T)}{RT^2}\right)dT + \left(\frac{\overline{V}_i^V}{RT}\right)dP + \left(\frac{\partial \ln \widehat{f}_i^V}{\partial y_i}\right)_{T,P} dy_i =$$

$$-\left(\frac{\overline{H}_i^L - H_i^o(T)}{RT^2}\right)dT + \left(\frac{\overline{V}_i^L}{RT}\right)dP + \left(\frac{\partial \ln \widehat{f}_i^L}{\partial x_i}\right)_{T,P} dx_i \tag{29.19}$$

Combining the dT and dP terms in Eq. (29.19) yields:

$$-\left(\frac{\overline{H}_i^V - \overline{H}_i^L}{RT^2}\right)dT + \left(\frac{\overline{V}_i^V - \overline{V}_i^L}{RT}\right)dP + \left(\frac{\partial \ln \widehat{f}_i^V}{\partial y_i}\right)_{T,P} dy_i$$

$$-\left(\frac{\partial \ln \widehat{f}_i^L}{\partial x_i}\right)_{T,P} dx_i$$

$$= 0 \tag{29.20}$$

where $i = 1$ and 2, $dy_2 = -dy_1$ and $dx_2 = -dx_1$.

Equation (29.20), for both $i = 1$ and $i = 2$, relates the four unknown intensive variables T, P, x_1, and y_1 Therefore, to obtain a unique solution, we must specify two of these four unknowns in order to determine the remaining two unknowns. This is fully consistent with the Gibbs Phase Rule which indicated that $L = 2$.

Next, we write Eq. (29.20) separately for $i = 1$ and $i = 2$ in terms of the four unknowns: T, P, x_1, and y_1. This yields:

$$-\left(\frac{\overline{H}_1^V - \overline{H}_1^L}{RT^2}\right)dT + \left(\frac{\overline{V}_1^V - \overline{V}_1^L}{RT}\right)dP + \left(\frac{\partial \ln \hat{f}_1^V}{\partial y_1}\right)_{T,P} dy_1$$

$$-\left(\frac{\partial \ln \hat{f}_1^L}{\partial x_1}\right)_{T,P} dx_1$$

$$= 0 \tag{29.21}$$

and

$$-\left(\frac{\overline{H}_2^V - \overline{H}_2^L}{RT^2}\right)dT + \left(\frac{\overline{V}_2^V - \overline{V}_2^L}{RT}\right)dP + \left(\frac{\partial \ln \hat{f}_2^V}{\partial y_1}\right)_{T,P} dy_1$$

$$-\left(\frac{\partial \ln \hat{f}_2^L}{\partial x_1}\right)_{T,P} dx_1$$

$$= 0 \tag{29.22}$$

where in Eq. (29.22), we replaced dy_2 by $-dy_1$, ∂y_2 by $-\partial y_1$, dx_2 by $-dx_1$, and ∂x_2 by $-\partial x_1$. Before we discuss the general solution of Eqs. (29.21) and (29.22), we first consider several simplifications.

29.3   Simplifications of Eqs. (29.21) and (29.22)

1. Only component 1 is volatile (that is, evaporates). In that case, the vapor phase is pure component 1, so that only Eq. (29.21) applies, where:

$$y_1 = 1, dy_1 = 0, \overline{H}_1^V = H_1^V(T, P), \text{ and } \overline{V}_1^V = V_1^V(T, P) \tag{29.23}$$

2. Only component 2 is volatile (that is, evaporates). In that case, the vapor phase is pure component 2, so that only Eq. (29.22) applies, where:

$$y_2 = 1, dy_2 = 0, \overline{H}_2^V = H_2^V(T, P), \text{ and } \overline{V}_2^V = V_2^V(T, P) \tag{29.24}$$

3. Components 1 and 2 are volatile (that is, both evaporate), but the vapor mixture composition is constant. In that case, both Eqs. (29.21) and (29.22) apply, where:

$$y_1 = \text{const.}, y_2 = 1 - y_1 = \text{const.} \Rightarrow dy_1 = 0, dy_2 = 0$$

4. Components 1 and 2 are volatile (that is, both evaporate), but the liquid mixture composition is constant. Once again, both Eqs. (29.1) and (29.2) apply, where:

$$x_1 = \text{const.}, x_2 = 1 - x_1 = \text{const.} \Rightarrow dx_1 = 0, dx_2 = 0 \qquad (29.25)$$

5. Pressure is kept constant, so that $dP = 0$ in Eqs. (29.21) and (29.22).
6. Temperature is kept constant, so that $dT = 0$ in Eqs. (29.21) and (29.22).
7. The vapor mixture is ideal. In that case:

$$\left(\hat{f}_i^V\right)^{ID} = P_i = y_i P, \text{ so that } \left(\frac{\partial \ln \hat{f}_i^V}{\partial y_i}\right)_{T,P}^{ID} = \frac{1}{y_i}, \text{ for } i = 1 \text{ and } 2 \qquad (29.26)$$

If the vapor mixture is not ideal, the dependence of the partial derivative in Eq. (29.26) on composition is more complex, and as discussed in Part II, can be calculated using an EOS for the vapor mixture.

8. The liquid mixture is ideal. In that case, the Lewis and Randall Rule introduced in Part II applies and states that:

$$\left(\hat{f}_i^L\right)^{ID} = x_i f_i^L(T, P), \text{ so that } \left(\frac{\partial \ln \hat{f}_i^L}{\partial x_i}\right)_{T,P}^{ID} = \frac{1}{x_i}, \text{ for } i = 1 \text{ and } 2 \qquad (29.27)$$

If the liquid mixture is not ideal, the dependence of the partial derivative in Eq. (29.27) on composition is more complex, and as discussed in Part II, can be calculated using an excess Gibbs free energy of mixing approach to evaluate the liquid mixture fugacities. Recall that, in general:

$$\hat{f}_i^L = \gamma_i^L x_i f_i(T, P); RT \ln \gamma_i^L = \left(\frac{\partial \Delta G^{EX}}{\partial N_i^L}\right)_{T,P,N_{j[i]}^L} \qquad (29.28)$$

where γ_i^L depends on T, P, and $x_1, ..., x_{n-1}$.

In the various problems that we will encounter, some (and if we are fortunate, all) of the simplifications (1) to (8) above may apply.

Next, let us return to Eqs. (29.21) and (29.22), and combine them to eliminate one of the mole fractions. For example, imagine that we want to calculate how the vapor mixture composition, y_1, varies with T and P (recall that, according to the Gibbs Phase Rule, $L = 2$). To obtain an equation relating dT, dP, and dy_1, we proceed as follows:

- Multiply Eq. (29.21) by x_1
- Multiply Eq. (29.22) by x_2
- Add the resulting equations

This yields:

$$-\left\{\frac{x_1\left(\overline{H}_1^V - \overline{H}_1^L\right) + x_2\left(\overline{H}_2^V - \overline{H}_2^L\right)}{RT^2}\right\}dT + \left\{\frac{x_1\left(\overline{V}_1^V - \overline{V}_1^L\right) + x_2\left(\overline{V}_2^V - \overline{V}_2^L\right)}{RT}\right\}dP+$$

$$\left\{x_1\left(\frac{\partial \ln \widehat{f}_1^V}{\partial y_1}\right)_{T,P} + x_2\left(\frac{\partial \ln \widehat{f}_2^V}{\partial y_1}\right)_{T,P}\right\}dy_1$$

$$-\left\{x_1\left(\frac{\partial \ln \widehat{f}_1^L}{\partial x_1}\right)_{T,P} + x_2\left(\frac{\partial \ln \widehat{f}_2^V}{\partial x_1}\right)_{T,P}\right\}dx_1 = 0$$

$$(29.29)$$

In Eq. (29.29), the last term in the curly brackets multiplying dx_1 is equal to zero. Indeed, as discussed in Part II, this term corresponds to the gGDr for $\widehat{f}_1^L$ and $\widehat{f}_2^L$ at constant T and P.

The coefficient of the dT term in Eq. (29.29) has the following simple physical interpretation. Consider the vaporization of N_1 moles of component 1 and N_2 moles of component 2 from a binary liquid mixture to a binary vapor mixture, where $N_1 + N_2 = N$ is fixed. In that case:

$$\underline{H}_L = N_1 \overline{H}_1^L + N_2 \overline{H}_2^L \text{ and } \underline{H}_V = N_1 \overline{H}_1^V + N_2 \overline{H}_2^V \qquad (29.30)$$

and

$$\frac{\underline{H}_V - \underline{H}_L}{(N_1 + N_2)} \equiv \Delta H_{vap}^{mix} = x_1\left(\overline{H}_1^V - \overline{H}_1^L\right) + x_2\left(\overline{H}_2^V - \overline{H}_2^L\right) \qquad (29.31)$$

where ΔH_{vap}^{mix} is the Mixture Molar Enthalpy of Vaporization.

In a similar manner, the coefficient of the dP term in Eq. (29.29) is related to the Mixture Molar Volume of Vaporization, ΔV_{vap}^{mix}, that is,

$$\frac{\underline{V}_V - \underline{V}_L}{(N_1 + N_2)} \equiv \Delta V_{vap}^{mix} = x_1\left(\overline{V}_1^V - \overline{V}_1^L\right) + x_2\left(\overline{V}_2^V - \overline{V}_2^L\right) \qquad (29.32)$$

Using Eqs. (29.31) and (29.32) in Eq. (29.29), including setting the last term to zero, we obtain:

$$-\left(\frac{\Delta H_{vap}^{mix}}{RT^2}\right)dT + \left(\frac{\Delta V_{vap}^{mix}}{RT}\right)dP$$

$$+\left\{x_1\left(\frac{\partial \ln \widehat{f}_1^V}{\partial y_1}\right)_{T,P} + x_2\left(\frac{\partial \ln \widehat{f}_2^V}{\partial y_1}\right)_{T,P}\right\}dy_1 = 0 \qquad (29.33)$$

Integrating Eq. (29.33), we can calculate: (i) $y_1 = y_1(T,P)$, (ii) $T = T(P, y_1)$, or (iii) $P = P(T, y_1)$, consistent with the Gibbs Phase Rule which requires that $L = 2$.

Lecture 30

Pure Liquid in Equilibrium with Its Pure Vapor, Integral Approach to Phase Equilibria, Composition Models, and Sample Problems

30.1 Introduction

The material presented in this lecture is adapted from Chapter 15 in T&M. First, we will complete our discussion of the differential approach to phase equilibria that we began in Lecture 29. In particular, we will derive the Clapeyron equation, including simplifying it to obtain the Clausius-Clapeyron equation which models the variation of pressure with temperature for a pure ($n = 1$) material along the solid-vapor, solid-liquid, and liquid-vapor equilibrium lines. Second, we will solve Sample Problem 30.1, where in the context of the integral approach to phase equilibria, we will discuss all aspects of the phase equilibria of a multi-component ($n > 1$) liquid mixture coexisting with its multi-component ($n > 1$) vapor mixture. To this end, we will utilize the equalities of the fugacities of each of the n components present in each phase to model the conditions of diffusional equilibrium. Specifically, gas-phase fugacities will be modeled using a fugacity-coefficient approach based on an EOS. Further, liquid-phase fugacities will be modeled using an activity-coefficient approach based on a model for the excess Gibbs free energy of mixing. Third, we will show how to relate the pure component fugacity at temperature T and pressure P to the pure component fugacity at temperature T and the vapor pressure of the same component at temperature T. The resulting relation will involve introducing the Poynting correction. Fourth, using the results above, we will derive n equations modeling the diffusional equilibrium of the n components, where these equations relate the 2n unknowns, T, P, the (n-1) liquid mixture mole fractions, and the (n-1) vapor mixture mole fractions. In addition, using the Gibbs Phase Rule, we will show that $L = n$. This implies that if we specify n of the 2n unknowns, we should be able to uniquely calculate the n remaining ones by simultaneously solving the n equations discussed above. Fifth, we will discuss four simplifications of the n equations which will lead to simpler solutions, including the celebrated Raoult's law. Finally, we will

© Springer Nature Switzerland AG 2020
D. Blankschtein, *Lectures in Classical Thermodynamics with an Introduction to Statistical Mechanics*, https://doi.org/10.1007/978-3-030-49198-7_30

present several mathematical models, known as composition models, that can be used to evaluate the excess Gibbs free energy of mixing.

30.2   From Lecture 29

For completeness, below, we present again Eqs. (29.21) and (29.22), derived in the context of the differential approach to phase equilibria in Lecture 29, which relate the four unknowns, dT, dP, dx_1, and dy_1. Specifically,

$$-\left(\frac{\overline{H}_1^V - \overline{H}_1^L}{RT^2}\right)dT + \left(\frac{\overline{V}_1^V - \overline{V}_1^L}{RT}\right)dP + \left(\frac{\partial \ln \widehat{f}_1^V}{\partial y_1}\right)_{T,P} dy_1$$

$$-\left(\frac{\partial \ln \widehat{f}_1^L}{\partial x_1}\right)_{T,P} dx_1$$

$$= 0 \tag{29.21}$$

and

$$-\left(\frac{\overline{H}_2^V - \overline{H}_2^L}{RT^2}\right)dT + \left(\frac{\overline{V}_2^V - \overline{V}_2^L}{RT}\right)dP + \left(\frac{\partial \ln \widehat{f}_2^V}{\partial y_1}\right)_{T,P} dy_1$$

$$-\left(\frac{\partial \ln \widehat{f}_2^L}{\partial x_1}\right)_{T,P} dx_1$$

$$= 0 \tag{29.22}$$

where in Eq. (29.22), $dy_2 = - dy_1$, $dx_2 = - dx_1$, $\partial y_2 = - \partial y_1$, and $\partial x_2 = - \partial x_1$.

30.3   Pure Liquid in Equilibrium with Its Pure Vapor

First, let us consider a practically relevant application of Eqs. (29.21) and (29.22), that is, a pure liquid in equilibrium with its pure vapor. Because we are dealing with a pure ($n = 1$) system, we can select to work with either Eq. (29.21) or Eq. (29.22). If we select Eq. (29.21), then, $y_1 = 1$, $x_1 = 1$, $\overline{V}_1 = V_1$ (molar volume of pure component 1), and $\overline{H}_1 = H_1$ (molar enthalpy of pure component 1). In that case, Eq. (29.21) yields:

$$-\left(\frac{H_1^V\text{-}H_1^L}{RT^2}\right)dT + \left(\frac{V_1^V\text{-}V_1^L}{RT}\right)dP + 0 = 0 \qquad (30.1)$$

where the numerator of the term multiplying dT is $-\Delta H_{vap}$, minus the Molar Enthalpy of Vaporization of pure component 1, and the numerator of the term multiplying dP is ΔV_{vap}, the Molar Volume of Vaporization of pure component 1. Accordingly, Eq. (30.1) can be expressed as follows:

$$-\left(\frac{\Delta H_{vap}}{RT^2}\right)dT + \left(\frac{\Delta V_{vap}}{RT}\right)dP = 0 \qquad (30.2)$$

Rearranging Eq. (30.2), we obtain an equation which describes how P varies with T along the L/V equilibrium monovariant (L = 1) line in a Pressure (P)-Temperature (T) phase diagram (see Fig. 30.1). This equation is known as the Clapeyron Equation.

The Clapeyron Equation also describes how P varies with T along the Solid/Liquid [S/L] and the Solid/Vapor [S/V] equilibrium monovariant lines. Specifically:

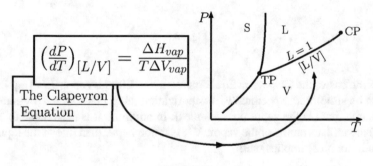

Fig. 30.1

$$\left(\frac{dP}{dT}\right)_{[S/L]} = \frac{\Delta H_{fus}}{T\Delta V_{fus}} \qquad (30.3)$$

where ΔH_{fus} is the Molar Enthalpy of Fusion, and ΔV_{fus} is the Molar Volume of Fusion. In addition,

$$\left(\frac{dP}{dT}\right)_{[S/V]} = \frac{\Delta H_{sub}}{T\Delta V_{sub}} \qquad (30.4)$$

where ΔH_{sub} is the Molar Enthalpy of Sublimation, and ΔV_{sub} is the Molar Volume of Sublimation. The (P-T) phase diagram depicted in Fig. 30.2 shows the various phases (L = 2), equilibrium lines (L = 1), the triple point (TP, L = 0)), and the critical point (CP, L = 0).

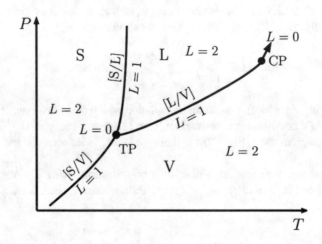

Fig. 30.2

In some cases, the Clapeyron Equation along the Liquid/Vapor [L/V] equilibrium line can be simplified. Specifically, if vaporization takes place at low pressures, one can assume that (1) the vapor phase is ideal. In addition, it is reasonable to assume that (2) the molar volume of the vapor, V^V, is much larger than that of the liquid, V^L. Assumption (1) implies that:

$$V^V = \frac{RT}{P} \text{ (Ideal gas behavior of the vapor)} \tag{30.5}$$

Assumption (2) implies that:

$$\Delta V_{vap} = V^V - V^L \approx V^V = \frac{RT}{P} \tag{30.6}$$

where we have used Eq. (30.5). Utilizing assumptions (1) and (2) above in the [L/V] Clapeyron Equation (see Fig. 30.1) yields:

$$\left(\frac{dP}{dT}\right)_{[L/V]} = \frac{\Delta H_{vap}}{T\Delta V_{vap}} = \frac{\Delta H_{vap}}{T\left(\frac{RT}{P}\right)} \qquad (30.7)$$

Rearranging Eq. (30.7), we obtain:

$$\Rightarrow \left(\frac{dP/P}{dT/T^2}\right)_{[L/V]} = \frac{\Delta H_{vap}}{R} \Rightarrow \left[\frac{d(\ln P)}{d(1/T)}\right]_{[L/V]} = -\frac{\Delta H_{vap}}{R} \qquad (30.8)$$

or

$$\left[\frac{d(\ln P)}{d(1/T)}\right]_{[L/V]} = -\frac{\Delta H_{vap}}{R} \qquad (30.9)$$

It is noteworthy that the vapor pressure is often denoted as $P_{L/V}$, P_{vp}, or P_{sat}. Equation (30.9) is known as the Clausius-Clapeyron equation, and can be plotted as shown in Fig. 30.3.

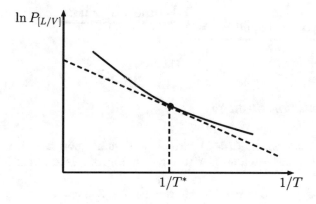

Fig. 30.3

Figure 30.3 and Eq. (30.9) show that the tangent to the line at any value of 1/T, say, at $1/T^*$ (the dashed line), corresponds to the value of $-\Delta H_{vap}/R$ at T^*. We should recognize that while the Clapeyron equation is general, the Clausius-Clapeyron equation is an approximation.

30.4   Integral Approach to Phase Equilibria

30.4.1   Sample Problem 30.1

Discuss all aspects of the phase equilibria of a multi-component (n > 1) liquid mixture coexisting with its multi-component (n > 1) vapor mixture in the absence of chemical reactions (r = 0) (see Fig. 30.4).

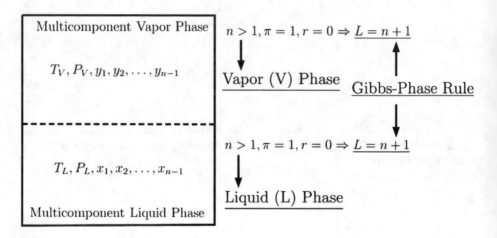

Fig. 30.4

30.4.2   Solution Strategy

For the multi-component (n > 1), two-phase (liquid/vapor; $\pi = 2$) system in the absence of chemical reactions (r = 0) shown in Fig. 30.4, the variance is given by

$$L = n + 2 - \pi - r = n + 2 - 2 - 0 = n \qquad (30.10)$$

In other words, out of the 2(n + 1) intensive variables, $\{T_V, P_V, y_1, \ldots, y_{n-1}; T_L, P_L, x_1, \ldots, x_{n-1}\}$, only n are independent.

Next, we impose the conditions of phase equilibria:

(a) Thermal Equilibrium

$$T_L = T_V \equiv T \qquad (30.11)$$

(b) Mechanical Equilibrium

$$P_L = P_V \equiv P \qquad (30.12)$$

(c) Diffusional Equilibrium

$$\mu_i^L = \mu_i^V, \ i = 1, 2, \ldots, n \tag{30.13}$$

As discussed in Part II, Eq. (30.13) can be expressed using fugacities instead of chemical potentials. This yields:

$$\lambda_i(T) + RT\ln\widehat{f}_i^L = \lambda_i(T) + RT\ln\widehat{f}_i^V, i = 1, 2, \ldots, n \tag{30.14}$$

Cancelling the λ_i (T)s in Eq. (30.14) yields:

$$\widehat{f}_i^L = \widehat{f}_i^V, i = 1, 2, \ldots, n \tag{30.15}$$

where $\widehat{f}_i^L = \widehat{f}_i^L(T, P, x_1, \ldots, x_{n-1})$ and $\widehat{f}_i^V = \widehat{f}_i^V(T, P, y_1, \ldots, y_{n-1})$. Given $L = n$ of the 2n intensive variables, $\{T, P, x_1, \ldots, x_{n-1}; y_1, \ldots, y_{n-1}\}$, the n fugacity equations in Eq. (30.15) can be solved simultaneously to calculate the remaining n intensive variables.

30.4.3   Calculation of Vapor Mixture Fugacities

In the integral approach to phase equilibria, we calculate the vapor mixture fugacities, $\widehat{f}_i^V$, using the fugacity-coefficient approach based on a reliable mixture EOS (see Lecture 25). Specifically:

$$\widehat{\phi}_i^V = \frac{\widehat{f}_i^V}{P_i} \Rightarrow \widehat{f}_i^V = Py_i\widehat{\phi}_i^V \tag{30.16}$$

or, more explicitly:

$$\widehat{f}_i^V = y_i\widehat{\phi}_i^V (T, P, y_1, \ldots, y_{n-1}) P \tag{30.17}$$

Recall that $\widehat{\phi}_i^V$ in Eqs. (30.16) and (30.17) can be calculated using a Volume-Explicit (VE) or a Pressure-Explicit (PE) mixture EOS.

30.4.4   Calculation of Liquid Mixture Fugacities

In the integral approach to phase equilibria, we calculate the liquid mixture fugacities, $\widehat{f}_i^L$, using the activity-coefficient approach based on a model for the excess Gibbs free energy of mixing, $\Delta \underline{G}^{EX}$, of the liquid mixture (see Lecture 26). Specifically,

$$\gamma_1^L = \frac{a_i^L}{x_i} = \frac{\widehat{f}_i^L / f_1^L(T, P)}{x_i} \tag{30.18}$$

where $f_i^L(T, P)$ is the pure component i liquid fugacity. Below, we will show how to calculate these fugacities using a creative strategy. Equation (30.18) can be expressed as follows:

$$\widehat{f}_i^L = x_i \gamma_i^L(T, P, x_1, \ldots, x_{n-1}) f_i^L(T, P) \tag{30.19}$$

Recall that γ_1^L can be calculated using the relation:

$$RT\ln\gamma_i^L = \left(\frac{\partial \Delta \underline{G}^{EX}}{\partial N_i^L}\right)_{T, P, N_{j[i]}^L} \tag{30.20}$$

Using the expressions for $\widehat{f}_i^V$ and $\widehat{f}_i^L$ in Eqs. (30.17) and (30.19), respectively, in Eq. (30.15) yields the desired n equations of the integral approach. Specifically,

$$y_i \widehat{\phi}_i^V(T, P, y_1, \ldots, y_{n-1})P = x_i \gamma_i^L(T, P, x_1, \ldots, x_{n-1}) f_i^L(T, P), (i = 1, 2, \ldots, n) \tag{30.21}$$

where the 2n unknowns, $\{T, P, y_1, \ldots, y_{n-1}; x_1, \ldots, x_{n-1}\}$, are related by the n equations given in Eq. (30.21). The Gibbs Phase Rule indicates that $L = n$. Therefore, if we specify $L = n$ of these unknowns, we can uniquely determine the remaining n by simultaneously solving the n equations in Eq. (30.21). In fact, the selection of the n variables that we specify determines the types of phase equilibria calculations that we can carry out. For example, if we specify $\{T, y_1, \ldots, y_{n-1}\}$, we can calculate $\{P, x_1, \ldots, x_{n-1}\}$.

30.4.5 Calculation of Pure Component i Liquid Fugacity

Typically, EOS for liquids are not reliable. Consequently, it is not recommended to calculate $f_i^L(T, P)$ using a pure component fugacity coefficient approach based on an EOS. Instead, we recommend to replace $f_i^L(T, P)$ by a vapor fugacity, which can be readily calculated if we have access to a reliable EOS for the vapor. We can do that at conditions where pure liquid i is at equilibrium with pure vapor i. This occurs at temperature, T, and a pressure which is equal to the vapor pressure, $P_{vpi}(T)$, of pure liquid i (see Fig. 30.5, including the boxed equation).

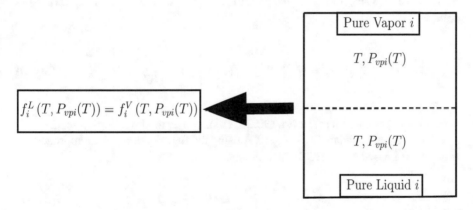

Fig. 30.5

However, we need to calculate f_i^L at $P \neq P_{vpi}(T)$, which requires relating:

$$f_i^L(T, P) \text{ to } f_i^L(T, P_{vpi}(T)) = f_i^V(T, P_{vpi}(T)) \qquad (30.22)$$

We can readily do so by recalling that:

$$\left(\frac{\partial \ln f_i^L}{\partial P}\right)_T = \frac{V_i^L}{RT} \Rightarrow d\left(\ln f_i^L\right)\Big|_T = \left(\frac{V_i^L}{RT}\right) dP \qquad (30.23)$$

where V_i^L is the molar volume of pure liquid i.

Integrating the differential relation to the right of the arrow in Eq. (30.23), from $P_{vpi}(T)$ to P, yields:

$$\ln\left[\frac{f_i^L(T,P)}{f_i^L(T,P_{vpi}(T))}\right] = \int\limits_{P_{vpi}(T)}^{P}\left(\frac{V_i^L}{RT}\right)dP \qquad (30.24)$$

or

$$f_i^L(T,P) = f_i^L(T,P_{vpi}(T))\exp\left[\int\limits_{P_{vpi}(T)}^{P}\left(\frac{V_i^L}{RT}\right)dP\right] \qquad (30.25)$$

Because $f_i^L(T,P_{vpi}(T))$ is equal to $f_i^V(T,P_{vpi}(T))$, we can express Eq. (30.25) as follows:

$$f_i^L(T,P) = f_i^V(T,P_{vpi}(T))\exp\left[\int\limits_{P_{vpi}(T)}^{P}\left(\frac{V_i^L}{RT}\right)dP\right] \qquad (30.26)$$

where the exponential term in Eq. (30.26) is known as the Poynting correction.

Next, we can calculate $f_i^V(T,P_{vpi}(T))$ using the fugacity coefficient approach for pure vapor i based on an EOS. Specifically,

$$\phi_i^V = \frac{f_i^V(T,P)}{P}, \text{ for } P = P_{vpi}(T) \qquad (30.27)$$

Rearranging Eq. (30.27), we obtain:

$$f_i^V(T,P_{vpi}(T)) = \phi_i^V(T,P_{vpi}(T))P_{vpi}(T) \qquad (30.28)$$

Using Eq. (30.28) in Eq. (30.26), and then using the resulting equation in Eq. (30.21), we obtain the set of n equations relating the 2n intensive variables, $\{T,P,y_1,\ldots,y_{n-1}; x_1,\ldots,x_{n-1}\}$, in the context of the integral approach to phase equilibria. Specifically,

$$y_i\widehat{\phi}_i^V(T,P,y_1,\ldots,y_{n-1})P = x_i\gamma_i^L(T,P,x_1,\ldots,x_{n-1})\phi_i^V(T,P_{vpi}(T))P_{vpi}(T)C_i \qquad (30.29)$$

where

$$C_i = \exp \left[\int_{P_{vpi}(T)}^{P} \left(\frac{V_i^L}{RT} \right) dP \right] \quad \text{(Poynting correction for } i = 1, 2, \ldots, n) \quad (30.30)$$

Typically, $C_i \approx 1$, but we must always check that this is indeed the case. For example, we can first assume that $C_i = 1$, calculate P, and then carry out the integration in Eq. (30.30) to verify that it is indeed ≈ 1.

30.4.6   Simplifications of Eq. (30.29)

(1) The vapor mixture is ideal (ID): $\left(\phi_i^V \right)^{ID} = 1$, for every i

(2) The liquid mixture is ideal: $\left(\gamma_i^L \right)^{ID} = 1$, for every i

(3) Pure gas i is ideal: $\left(\phi_i^V \right)^{ID} = 1$, for every i

(4) The Poynting correction is unity: $C_i = 1$, for every i

If simplifications (1)–(4) all apply, Eq. (30.29) reduces to:

$$y_i P = P_i = x_i P_{vpi}(T), \quad i = 1, 2, \ldots, n \quad \text{(Raoult's law)} \quad (30.31)$$

If only simplifications (1), (3), and (4) apply, Eq. (30.29) reduces to:

$$y_i P = x_i \gamma_i^L(T, P, x_1, \ldots, x_{n-1}) P_{vpi}(T), \quad i = 1, 2, \ldots, n \quad \text{(Modified Raoult's law)} \tag{30.32}$$

To calculate γ_i^L in Eqs. (30.29) and (30.32), we require mathematical models for $\Delta \underline{G}^{EX}$, that can be used as follows:

$$RT \ln \gamma_i^L = \left(\frac{\partial \Delta \underline{G}^{EX}}{\partial N_i^L} \right)_{T, P, N_{j[i]}^L} \tag{30.33}$$

Several mathematical models for $\Delta \underline{G}^{EX}$, for both binary mixtures (n = 2) and multi-component mixtures (n > 2), are presented in the following tables (all adapted from T&M).

30.4.7 Models for the Excess Gibbs Free Energy of Mixing of n = 2 Mixtures

Type and name	Operating equations
First-order polynomial	
Two-suffix[a] Margules	$\Delta G^{EX} = Ax_1x_2$
	Binary parameters = A
	$RT\ln\gamma_1 = Ax_2^2$
	$RT\ln\gamma_2 = Ax_1^2$
Three-suffix[a] Margules	$\Delta G^{EX} = x_1x_2\,[A + B(x_1 - x_2)]$
	Binary parameters = A, B
	$RT\ln\gamma_1 = (A + 3B)x_2^2 - 4Bx_2^3$
	$RT\ln\gamma_2 = (A - 3B)x_1^2 - 4Bx_1^3$
van Laar	$\Delta G^{EX} = \dfrac{Ax_1x_2}{x_1(A/B)+x_2}$
	Binary parameters = A, B
	$RT\ln\gamma_1 = A\left(1 + \dfrac{Ax_1}{Bx_2}\right)^{-2}$
	$RT\ln\gamma_2 = B\left(1 + \dfrac{Bx_2}{Ax_1}\right)^{-2}$
Four suffix[a] Margules	$\Delta G^{EX} = x_1x_2\,[A + B(x_1 - x_2) + C(x_1 - x_2)^2]$
	Binary parameters = A, B, C
	$RT\ln\gamma_1 = (A + 3B + 5C)x_2^2 - 4(B + 4C)x_2^3 + 12Cx_2^4$
	$RT\ln\gamma_2 = (A - 3B + 5C)x_1^2 - 4(B - 4C)x_1^3 + 12Cx_1^4$
General Redlich-Kister	$\dfrac{\Delta G^{EX}}{RT} = x_1x_2\left[B + C(x_1 - x_2) + D(x_1 - x_2)^2 + \ldots\right]$
	Binary parameters = B, C, D, ...
Volume-fraction based	
Hildebrand-Scatchard	$\Delta G^{EX} = \dfrac{(\delta_1 - \delta_2)^2}{\left(\dfrac{1}{v_1x_1} + \dfrac{1}{v_2x_2}\right)}$
	V_1, V_2 = molar volumes of pure 1 and 2
	Pure component parameters = δ_1, δ_2
	$\quad RT\ln\gamma_1 = V_1\Phi_2^{*2}(\delta_1 - \delta_2)^2$
	$\quad RT\ln\gamma_2 = V_2\Phi_1^{*2}(\delta_1 - \delta_2)^2$
	$\Phi_i^* = \dfrac{V_ix_i}{V_1x_1 + V_2x_2}\, i = 1, 2$
Flory-Huggins	$\dfrac{\Delta G^{EX}}{RT} = x_1\ln\dfrac{\Phi_1^*}{x_1} + x_2\ln\dfrac{\Phi_2^*}{x_2} + \dfrac{\chi\,\Phi_1^*\Phi_2^*(N_1+rN_2)}{(N_1+N_2)}$
	$\Phi_1^* = \dfrac{N_1}{N_1+rN_2}\;\;\Phi_2^* = \dfrac{rN_2}{N_1+rN_2}$
	r = chain length of polymer = number of monomer units
	Binary parameters = χ; 2(1 − solvent/monomer; 2 − polymer)

(continued)

Type and name	Operating equations
General Wohl expansion	$\dfrac{\Delta G^{EX}}{RT} = (x_1 q_1 + x_2 q_2)\big[2a_{12}z_1 z_2 + 3a_{112}z_1^2 z_2 + 3a_{122}z_1 z_2^2 +$ $4a_{1112}z_1^3 z_2 + 4a_{1222}z_1 z_2^3 + 6a_{1122}z_1^2 z_2^2 + \ldots\big]$ $\left.\begin{array}{l} z_1 \equiv \dfrac{x_1 q_1}{x_1 q_1 + x_2 q_2} \\[2mm] z_2 \equiv \dfrac{x_2 q_2}{x_1 q_1 + x_2 q_2} \end{array}\right\}$ effective volume fraction q_i = volume fraction size of pure i (i = 1, 2) Binary parameters = $a_{12}, a_{112}, a_{122}, \ldots$
Local-composition based Wilson	$\dfrac{\Delta G^{EX}}{RT} = -x_1 \ln(x_1 + \Lambda_{12}x_2) - x_2 \ln(x_2 + \Lambda_{21}x_1)$ Binary parameters = $\Lambda_{12}, \Lambda_{21}$ $\ln \gamma_1 = -\ln(x_1 + \Lambda_{12}x_2) + \beta x_2$ $\ln \gamma_2 = -\ln(x_2 + \Lambda_{21}x_1) + \beta x_1$ $\beta = \left(\dfrac{\Lambda_{12}}{x_1 + \Lambda_{12}x_2} - \dfrac{\Lambda_{21}}{\Lambda_{21}x_1 + x_2}\right)$
TK-Wilson[f]	$\dfrac{\Delta G^{EX}}{RT} = x_1 \ln \dfrac{(x_1 + V_2 x_2/V_1)}{(x_1 + \Lambda_{12}x_2)} + x_2 \ln \dfrac{(V_1 x_1/V_2 + x_2)}{(\Lambda_{21}x_1 + x_2)}$ V_1, V_2 = molar volumes of pure 1 and 2 β same as for Wilson model Binary parameters = $\Lambda_{12}, \Lambda_{21}$ $\ln \gamma_1 = \ln \dfrac{(x_1 + V_2 x_2/V_1)}{(x_1 + \Lambda_{12}x_2)} + (\beta - \beta^*)x_2$ $\beta^* = \dfrac{V_2/V_1}{(x_1 + V_2 x_2/x_1)} - \dfrac{V_1/V_2}{(V_1 x_1/V_2 + x_2)}$
NRTL[b]	$\dfrac{\Delta G^{EX}}{RT} = x_1 x_2 \left(\dfrac{\tau_{21}G_{21}}{x_1 + x_2 G_{21}} + \dfrac{\tau_{12}G_{12}}{x_2 + x_1 G_{12}}\right)$ where $\tau_{12} = \dfrac{\Delta g_{12}}{RT}$, $\tau_{21} = \dfrac{\Delta g_{21}}{RT}$ $\ln G_{12} = -\alpha_{12}\tau_{12}$, $\ln G_{21} = -\alpha_{12}\tau_{21}$ Binary parameters = $\Delta g_{12}, \Delta g_{21}, \alpha_{12}$[c] $\ln \gamma_1 = x_2^2 \left[\tau_{21}\left(\dfrac{G_{21}}{x_1 + x_2 G_{21}}\right)^2 + \dfrac{\tau_{12}G_{12}}{(x_2 + x_1 G_{12})^2}\right]$ $\ln \gamma_2 = x_1^2 \left[\tau_{12}\left(\dfrac{G_{12}}{x_2 + x_1 G_{12}}\right)^2 + \dfrac{\tau_{21}G_{21}}{(x_1 + x_2 G_{21})^2}\right]$
UNIQUAC[d]	$\Delta G^{EX} = \Delta G^{EX}(\text{combinatorial}) + \Delta G^{EX}(\text{residual})$ $\dfrac{\Delta G^{EX}(\text{combinatorial})}{RT} = x_1 \ln \dfrac{\Phi_1^*}{x_1} + x_2 \ln \dfrac{\Phi_2^*}{x_2} + \dfrac{z}{2}\left(q_1 x_1 \ln \dfrac{\theta_1}{\Phi_1^*} + q_2 x_2 \ln \dfrac{\theta_1}{\Phi_2^*}\right)$ $\dfrac{\Delta G^{EX}(\text{residual})}{RT} = -q_1' x_1 \ln\left[\theta_1' + \theta_2'\tau_{21}\right] - q_2' x_2 \ln\left[\theta_2' + \theta_1'\tau_{12}\right]$ $\Phi_1^* \equiv \dfrac{x_1 r_1}{x_1 r_1 + x_2 r_2}$, $\theta_1 \equiv \dfrac{x_1 q_1}{x_1 q_1 + x_2 q_2}$ $\theta_1' \equiv \dfrac{x_1 q_1'}{x_1 q_1' + x_2 q_2'}$ $\ln \tau_{21} \equiv -\dfrac{\Delta u_{21}}{RT}$, $\ln \tau_{12} \equiv -\dfrac{\Delta u_{12}}{RT}$

(continued)

Type and name	Operating equations
	r, q, and q' are pure component parameters and coordination number z = 10
	Binary parameters $= \Delta u_{12}$ and Δu_{21} [e]
	$\ln\gamma_i = \ln \dfrac{\Phi_i^*}{x_i} + \dfrac{z}{2}q_i \ln \dfrac{\theta_i}{\Phi_i^*} + \Phi_j^*\left(l_i - \dfrac{r_i}{r_j}l_j\right) - q_i' \ln\left(\theta_i' + \theta_j'\tau_{ji}\right)$
	$+\theta_j'q_i'\left(\dfrac{\tau_{ji}}{\theta_i' + \theta_j'\tau_{ji}} - \dfrac{\tau_{ij}}{\theta_i' + \theta_i'\tau_{ij}}\right)$
	where $i = 1, j = 2$ or $i = 2, j = 1$
	$l_i \dfrac{z}{2}(r_i - q_i) - (r_i - 1); l_j \dfrac{z}{2}\left(r_j - q_j\right) - (r_j - 1)$

[a] Two-suffix signifies that the expansion for ΔG^{EX} is quadratic in mole fraction. Three-suffix signifies a third-order, and four-suffix signifies a fourth-order equation.
[b] NRTL = non-random two-liquid model.
[c] $\Delta g_{12} = g_{12} - g_{22}; \Delta g_{21} = g_{21} - g_{11}$
[d] UNIQUAC = universal quasi-chemical activity coefficient model.
[e] $\Delta u_{12} = u_{12} - u_{22}; \Delta u_{21} = u_{21} - u_{11}$.
[f] works for liquid-liquid systems.
Sources: Prausnitz *et al.* (1986), 2nd ed, Chapter 6 and Walas (1985), Chapter 4 where the Margules, Redlich-Kister, van Laar, Wilson, TK-Wilson, Wohl, Hildebrand-Scatchard, NRTL, and UNIQUAC equations are discussed.

30.4.8   Models for the Excess Gibbs Free Energy of Mixing of n > 2 Mixtures

Model	$\ln \gamma_i$	Parameters required
Two-Suffix Margules	$\displaystyle\sum_{k=1}^n \sum_{j=1}^n \left[A_{ki} - \frac{1}{2}A_{kj}\right]x_k x_j$ (binary interactions only)	$A_{jj} = A_{kk} = 0$ $A_{kj} = A_{jk}$
Wilson	$\displaystyle 1 - \ln\left[\sum_{j=1}^n \Lambda_{ij}x_j\right] - \sum_{k=1}^n \left[\frac{[\Lambda_{ki}x_k]}{\sum_{j=1}^n \Lambda_{kj}x_j}\right]$	$\Lambda_{ii} = \Lambda_{jj} = 1$ $\Lambda_{ij} = \frac{V_j}{V_i}\exp\left(\frac{-\lambda_{ij}}{RT}\right)$
TK-Wilson	$\displaystyle -\ln\left[\sum_{j=1}^n \Lambda_{ij}x_j\right] - \sum_{k=1}^n \left[\frac{[\Lambda_{ki}x_k]}{\sum_{j=1}^n \Lambda_{kj}x_j}\right]$ $\displaystyle + \ln\left[\sum_{j=1}^n V_j x_j / V_i\right] + \sum_{k=1}^n \left[\frac{(V_i/V_k)x_k}{\sum_{j=1}^n (V_j/V_k)x_j}\right]$	Λ_{ij} and Λ_{ji} are defined as in Wilson model
NRTL	$\displaystyle \frac{\sum_{j=1}^n \tau_{ji}G_{ji}x_j}{\sum_{k=1}^n G_{ki}x_k} + \sum_{j=1}^n \left[\frac{G_{ij}x_j}{\sum_{k=1}^n G_{kj}x_k}\right]\left[\tau_{ij} - \frac{\sum_{m=1}^n \tau_{mj}G_{mj}x_m}{\sum_{k=1}^n G_{kj}x_k}\right]$	$\tau_{ji} = \frac{g_{ji}-g_{ii}}{RT}$ $\tau_{ii} = \tau_{jj} = 0$ $G_{ji} = \exp(-\alpha_{ji}\,\tau_{ji})$ $G_{ii} = G_{jj} = 1.0$

(continued)

Model	$\ln \gamma_i$	Parameters required
UNIQUAC	$\ln\gamma_i = \ln\gamma_i^C + \ln\gamma_i^R$	$\tau_{ji} = \exp\left[-\dfrac{(u_{ji}-u_{ii})}{RT}\right]$
	$\ln\gamma_i^C = \ln\dfrac{\Phi_i^*}{x_i} + \dfrac{z}{2}q_i \ln\dfrac{\theta_i}{\Phi_i^*} + l_i - \dfrac{\Phi_i^*}{x_i}\sum\limits_{j=1}^{n} x_j l_j$	$\tau_{ii} = \tau_{jj} = 1.0$
	$\ln\gamma_i^R = q_i\left[1 - \ln\left[\sum\limits_{j=1}^{n}\theta_j\tau_{ji}\right] - \sum\limits_{j=1}^{n}\left[\dfrac{\theta_j\tau_{ij}}{\sum\limits_{k=1}^{n}\theta_k\tau_{kj}}\right]\right]$	$l_i = \dfrac{z}{2}(r_i - q_i) - (r_i - 1)$
	$\Phi_i^* = \dfrac{r_i x_i}{\sum\limits_{k=1}^{n} r_k x_k}$ and $\theta_i = \dfrac{q_i x_i}{\sum\limits_{k=1}^{n} q_k x_k}$	$z = 10$ (usually)

Lecture 31

Chemical Reaction Equilibria: Stoichiometric Formulation and Sample Problem

31.1 Introduction

The material presented in this lecture is adapted from Chapter 16 in T&M. So far in Part II, we have discussed multi-component ($n > 1$) and multi-phase ($\pi > 1$) systems under the assumption that the various chemical species comprising the system do not participate in any chemical reactions. In Lectures 31, 32, 33, 34, 35, and 36, we will allow the various chemical species to participate in chemical reactions, including generalizing the thermodynamic formulation accordingly. In this lecture, we will first contrast the calculation of changes in the thermodynamic properties of a binary mixture consisting of two inert components with that of a binary mixture where the two components participate in a dissociation reaction. Second, we will discuss the stoichiometric formulation of chemical reaction equilibria. Specifically, we will introduce stoichiometric numbers, stoichiometric coefficients, and the extent of reaction. Finally, we will solve Sample Problem 31.1, where 2 moles of methane and 3 moles of water are introduced into a closed chemical reactor, where they undergo two chemical reactions simultaneously to produce carbon monoxide, carbon dioxide, and hydrogen. Given the initial mole numbers of the reactants (methane and water), we will calculate the two extents of reaction, the final mole numbers of methane, water, carbon monoxide, carbon dioxide, and hydrogen, and their corresponding mole fractions.

31.2 Contrasting the Calculation of Changes in Thermodynamic Properties With and Without Chemical Reactions

When chemical reactions are involved, it is desirable to calculate equilibrium conversions, including the effect of temperature, pressure, and reactant composition on these conversions. We should stress that the calculation of reaction rates is

© Springer Nature Switzerland AG 2020
D. Blankschtein, *Lectures in Classical Thermodynamics with an Introduction to Statistical Mechanics*, https://doi.org/10.1007/978-3-030-49198-7_31

outside the realm of thermodynamics. It is also practically important to determine the effects of temperature, pressure, and ratio of reactants on the equilibrium conversion of chemical reactions in order to optimize them. Finally there is an important need to calculate changes in thermodynamic properties when chemical reactions are involved in the thermodynamic process. A comparison of Case I and Case II below will illustrate this important need.

31.2.1   Case I: Closed Binary System of Inert Components 1 and 2

In Case I, in the set of variables, T, P, N_1^o, N_2^o, the initial mole numbers, N_1^o and N_2^o, remain fixed when the binary mixture evolves from the initial state i to the final state f. As a result, the change in the extensive property $\underline{B}$ as the binary mixture evolves from the initial state i to the final state f is given by:

$$\Delta \underline{B}_{i \to f} = \underline{B}_f(T_f, P_f, N_1^o, N_2^o) - \underline{B}_i(T_i, P_i, N_1^o, N_2^o) \tag{31.1}$$

where

$$\{N_1^o, N_2^o\}_i = \{N_1^o, N_2^o\}_f \tag{31.2}$$

31.2.2   Case II: Closed Binary System of Components 1 and 2 Undergoing a Dissociation Reaction

In this case, in the set of variables, T, P, N_1, and N_2, the initial mole numbers, N_1^o and N_2^o, change as the binary mixture evolves from the initial state i to the final state f. For example, suppose that a dissociation reaction takes place as follows: $I_2 \rightleftarrows 2I$, where $I_2 = 1$, and $I = 2$.

As we vary T and P, N_1^o and N_2^o will change to N_1 and N_2, where as we will show:

$$N_1 = N_1(T, P, N_1^o, N_2^o) \text{ and } N_2 = N_2(T, P, N_1^o, N_2^o) \tag{31.3}$$

As a result, the change in the extensive property $\underline{B}$ as the binary mixture evolves from the initial state i to the final state f is given by:

$$\Delta \underline{B}_{i \to f} = \underline{B}_f(T_f, P_f, N_1, N_2) - \underline{B}_i(T_i, P_i, N_1^o, N_2^o) \qquad (31.4)$$

where

$$N_1 \equiv N_1^f \neq N_1^o, \text{ and } N_2 \equiv N_2^f \neq N_2^o \qquad (31.5)$$

31.3   Stoichiometric Formulation

As an illustration, we begin with the following simple, stoichiometrically-balanced chemical reaction:

$$CH_4 + H_2O \rightleftarrows CO + 3H_2 \qquad (31.6)$$

where the coefficients multiplying the various species i, including a sign convention, are referred to as stoichiometric numbers, and are denoted by υ_i. The convention for the sign of υ_i is as follows:

(1) $\upsilon_i > 0$, For products
(2) $\upsilon_i < 0$, For reactants $\qquad (31.7)$
(3) $\upsilon_i = 0$, For inert species

Specifically, for the chemical reaction in Eq. (31.6), it follows that:

$$\upsilon_{CH_4} = -1, \upsilon_{H_2O} = -1, \text{For the two reactants} \qquad (31.8)$$

$$\upsilon_{CO} = +1, \upsilon_{H_2} = +3, \text{For the two products} \qquad (31.9)$$

For the chemical reaction in Eq. (31.6), the changes in the numbers of moles of the species present are in direct proportion to the stoichiometric numbers. For example, if 0.5 moles of CH_4 disappear by reaction, then, 0.5 moles of H_2O must also disappear. In addition, simultaneously, 0.5 moles of CO and 1.5 moles of H_2 are formed by reaction, where:

$$\frac{\Delta N_{CH_4}}{\upsilon_{CH_4}} = \frac{\Delta N_{H_2O}}{\upsilon_{H_2O}} = \frac{\Delta N_{CO}}{\upsilon_{CO}} = \frac{\Delta N_{H_2}}{\upsilon_{H_2}} \qquad (31.10)$$

or

$$\frac{-0.5}{-1} = \frac{-0.5}{-1} = \frac{+0.5}{+1} = \frac{+1.5}{+3} \tag{31.11}$$

For a general chemical reaction:

$$|\upsilon_1|A_1 + |\upsilon_2|A_2 + \ldots \rightleftarrows |\upsilon_3|A_3 + |\upsilon_4|A_4 + \ldots$$

where the $|\upsilon_i|$s are the stoichiometric coefficients, and the A_is stand for the chemical formulas of the species. As shown in Eq. (31.7), the υ_is themselves (with the sign included) are referred to as stoichiometric numbers.

For a stoichiometrically-balanced chemical reaction, if we consider a differential amount of reaction, we can write:

$$\frac{dN_1}{\upsilon_1} = \frac{dN_2}{\upsilon_2} = \frac{dN_3}{\upsilon_3} = \ldots = \frac{dN_n}{\upsilon_n} \tag{31.12}$$

where each term is related to an amount, or an extent, of reaction, as represented by a change in the number of moles of a chemical species. Because all the n ratios in Eq. (31.12) are equal, they can be identified collectively with a single quantity, $d\xi$, where the variable ξ is referred to as the extent of reaction, or the reaction coordinate. In other words,

$$dN_j = \upsilon_j \, d\xi (j = 1, 2, \ldots, n) \tag{31.13}$$

In Eq. (31.13), n is the number of reactive species, and includes inert species, for which $\upsilon_j = 0$.

If several chemical reactions take place simultaneously, then, for reaction r, there is an extent of reaction ξ_r. In that case, component j participating in chemical reaction r will satisfy:

$$dN_{jr} = \upsilon_{jr} \, d\xi_r (r = 1, 2, \ldots, m) \tag{31.14}$$

where m is the number of independent chemical reactions.

Clearly, if we add up the changes in the dN_{jr}s over all the chemical reactions in which component j participates, we obtain the total change in the number of moles of component j. Specifically,

$$dN_j = \sum_{r=1}^{m} dN_{jr} = \sum_{r=1}^{m} \upsilon_{jr} \, d\xi_r \tag{31.15}$$

where Eq. (31.14) was used. In Eq. (31.15), the sum is over the m independent chemical reactions, and (j = 1, 2, ..., n), including inert species for which $\upsilon_{jr} = 0$.

Integrating Eq. (31.15) from ($\xi_r = 0$, $N_j = N_{jo}$) to some (ξ_r, N_j), we obtain:

$$N_j = N_j(\text{Initial}) + \sum_{r=1}^{m} \upsilon_{jr}\xi_r \tag{31.16}$$

where N_j (Initial) $= N_{jo}$, and $j = 1, 2, \ldots, n$. In Eq. (31.16), N_j represents the number of moles of component j in the equilibrium mixture, and depends on the number of moles of component j initially charged, N_{jo}, as modified by all the m chemical reactions in which component j participates.

Summing over all the species in Eq. (31.16), including the inert ones for which $\upsilon_{jr} = 0$, we obtain:

$$N = \sum_{j=1}^{n} N_j = \sum_{j=1}^{n} \left(N_{jo} + \sum_{r=1}^{m} \upsilon_{jr}\xi_r \right) \tag{31.17}$$

or

$$N = \sum_{j=1}^{n} N_{jo} + \sum_{j=1}^{n} \sum_{r=1}^{m} \upsilon_{jr}\xi_r \tag{31.18}$$

Let us first deal with the double summation in Eq. (31.18), that is, with:

$$\sum_{j=1}^{n} \sum_{r=1}^{m} \upsilon_{jr}\xi_r = \sum_{r=1}^{m} \left(\sum_{j=1}^{n} \upsilon_{jr} \right) \xi_r \tag{31.19}$$

Denoting $\sum_{j=1}^{n} \upsilon_{jr} = \upsilon_r$, it follows that:

$$\sum_{r=1}^{n} \left(\sum_{j=1}^{m} \upsilon_{jr} \right) \xi_r = \sum_{r=1}^{n} \upsilon_r \xi_r \tag{31.20}$$

Using Eq. (31.20) in Eq. (31.18) yields:

$$N = N_o + \sum_{r=1}^{m} \upsilon_r \xi_r \tag{31.21}$$

where

$$N = \sum_{j=1}^{n} N_j, \, N_o = \sum_{j=1}^{n} N_{jo}, \, \upsilon_r = \sum_{j=1}^{n} \upsilon_{jr} \tag{31.22}$$

31.4   Important Remark

Equation (31.22) shows that although the system is closed, N changes with respect to N_o. This follows because we are considering a balance on the components, instead of a balance on the individual atoms. If we did consider a balance on the individual atoms, then, for a closed system,

$$N_{Atom,i}^{Initial} = Constant \tag{31.23}$$

irrespective of any chemical reactions.

The mole fraction of component j in the mixture is obtained by taking the ratio of N_j in Eq. (31.16) and N in Eq. (31.21). Specifically,

$$y_j = \frac{N_j}{N} = \left(\frac{N_{jo} + \sum_{r=1}^{m} \upsilon_{jr}\xi_r}{N_o + \sum_{r=1}^{m} \upsilon_r\xi_r} \right) \tag{31.24}$$

31.5   Sample Problem 31.1

Consider a reactor in which we initially charge 2 moles of methane (CH_4) and 3 moles of water (H_2O). It is known that the following two chemical reactions take place simultaneously:

$$CH_4 + H_2O \rightleftarrows CO + 3H_2, r = 1 \tag{31.25}$$

$$CH_4 + 2H_2O \rightleftarrows CO_2 + 4H_2, r = 2 \tag{31.26}$$

If initially

$$N_{CH_4}^o = 2\,moles, N_{H_2O}^o = 3\,moles, and\, N_{CO}^o = N_{CO_2}^o = N_{H_2}^o = 0 \tag{31.27}$$

Determine the dependence of the five N_js and y_js on ξ_1 and ξ_2 as the two chemical reactions proceed.

31.5.1   Solution

It is useful to construct a stoichiometric table to keep track of the various v_{jr}'s corresponding to the two simultaneous chemical reactions. Specifically,

j \ r	Reaction 1, $r = 1$, v_{j1}	Reaction 2, $r = 2$, v_{j2}
$CH_4 \equiv 1$	$v_{11} = -1$	$v_{12} = -1$
$H_2O \equiv 2$	$v_{21} = -1$	$v_{22} = -2$
$CO \equiv 3$	$v_{31} = +1$	$v_{32} = 0$
$CO_2 \equiv 4$	$v_{41} = 0$	$v_{42} = +1$
$H_2 \equiv 5$	$v_{51} = +3$	$v_{52} = +4$
$v_r = \displaystyle\sum_{j=1}^{5} v_{jr}$	$v_1 = +2$	$v_2 = +2$

We also know that:

$$N^o_{CH_4} = 2\,\text{moles} \tag{31.28}$$

$$N^o_{H_2O} = 3\,\text{moles} \tag{31.29}$$

$$N^o_{CO} = 0 \tag{31.30}$$

$$N^o_{CO_2} = 0 \tag{31.31}$$

$$N^o_{H_2} = 0 \tag{31.32}$$

Using the stoichiometric table above, and the definitions of N_j and y_j above, we obtain the following result:

$$N = \sum_{j=1}^{5}\left(N_j^o\right) + \sum_{r=1}^{2} \upsilon_r \xi_r$$

$N_j = N_j^o + \sum_{r=1}^{2} \upsilon_{jr}\xi_r$	$y_j = \dfrac{N_j}{N} = \dfrac{N_j^o + \sum_{r=1}^{2}\upsilon_{jr}\xi_r}{N}$
$N_{CH_4} \equiv N_1 = 2 - \xi_1 - \xi_2$	$y_{CH_4} \equiv y_1 = \dfrac{2 - \xi_1 - \xi_2}{5 + 2\xi_1 + 2\xi_2}$
$N_{H_2O} \equiv N_2 = 3 - \xi_1 - 2\xi_2$	$y_{H_2O} \equiv y_2 = \dfrac{3 - \xi_1 - 2\xi_2}{5 + 2\xi_1 + 2\xi_2}$
$N_{CO} \equiv N_3 = 0 + \xi_1 + 0\xi_2 = \xi_1$	$y_{CO} \equiv y_3 = \dfrac{\xi_1}{5 + 2\xi_1 + 2\xi_2}$
$N_{CO_2} \equiv N_4 = 0 + 0\xi_1 + \xi_2 = \xi_2$	$y_{CO_2} \equiv y_4 = \dfrac{\xi_2}{5 + 2\xi_1 + 2\xi_2}$
$N_{H_2} \equiv N_5 = 0 + 3\xi_1 + 4\xi_2$	$y_{H_2} \equiv y_5 = \dfrac{3\xi_1 + 4\xi_2}{5 + 2\xi_1 + 2\xi_2}$
$N = \sum_{j=1}^{5} N_j = 5 + 2\xi_1 + 2\xi_2$	$\sum_{j=1}^{5} y_j = \dfrac{5 + 2\xi_1 + 2\xi_2}{5 + 2\xi_1 + 2\xi_2} = 1$

It is noteworthy that both N_j and y_j depend on ξ_1 and ξ_2. Therefore, as the two chemical reactions proceed forward, both N_j and y_j change for each j.

Lecture 32

Criterion of Chemical Reaction Equilibria, Standard States, and Equilibrium Constants for Gas-Phase Chemical Reactions

32.1 Introduction

The material presented in this lecture is adapted from Chapter 16 in T&M. First, we will derive the criterion of chemical reaction equilibria, which establishes a relation between the chemical potentials of every component i (1, 2, . . ., n) participating in chemical reaction r (1, 2, . . ., m). Second, expressing the chemical potentials in terms of fugacities, and using these expressions in the criterion of chemical reaction equilibria, we will derive an expression for the equilibrium constant associated with chemical reaction r, expressed in terms of the product of the ratios of the fugacities of component i in the actual mixture and in a suitably chosen reference state, both at the same temperature, raised to the power of the stoichiometric number associated with component i in chemical reaction r. Third, we will define the standard molar Gibbs free energy of reaction r and show that the equilibrium constant associated with chemical reaction r can be expressed as the exponential of minus the standard molar Gibbs free energy of reaction r divided by RT. Fourth, we will focus on a single chemical reaction (m = 1), and discuss the selection of reference states to model gas-phase, liquid-phase, and solid-phase chemical reactions, where the last two are collectively referred to as condensed-phase chemical reactions. In particular, we will discuss the selection of the reference state pressure, including its effect on the equilibrium constant. Fifth, for a gas-phase chemical reaction, we will decompose the fugacity of component i into a product of the fugacity coefficient of component i in the gas mixture, the mixture pressure, and the mole fraction of component i in the gas mixture. Using this fugacity decomposition, we will show that the equilibrium constant can be expressed as a product of contributions from the fugacity coefficients, the gas mixture mole fractions, and the pressure. Finally, we will consider the limit of an ideal gas mixture, including deriving an expression for the equilibrium constant.

© Springer Nature Switzerland AG 2020

D. Blankschtein, *Lectures in Classical Thermodynamics with an Introduction to Statistical Mechanics*, https://doi.org/10.1007/978-3-030-49198-7_32

32.2 Derivation of the Criterion of Chemical Reaction Equilibria

Consider a multi-component ($n > 1$), single-phase ($\pi = 1$) system consisting of $\{N_1, N_2, \ldots, N_n\}$ moles at T and P. The Gibbs free energy, $\underline{G}$, is a function of the ($n + 2$) variables $\{T, P, N_1, N_2, \ldots, N_n\}$, i.e., $\underline{G} = \underline{G}(T, P, N_1, N_2, \ldots, N_n)$, and its differential is given by:

$$d\underline{G} = -\underline{S}dT + \underline{V}dP + \sum_{j=1}^{n} \mu_j dN_j \qquad (32.1)$$

In Eq. (32.1), in the absence of chemical reactions, the various mole numbers, $\{N_1, N_2, \ldots, N_n\}$, are free to vary in any manner that we choose, except for the possible constraint that:

$$\sum_{j=1}^{n} dN_j = 0 \qquad (32.2)$$

if the overall system is closed. However, in the presence of m independent chemical reactions, this is no longer the case, and the various dN_js are all related to each other through the extents of reaction, ξ_r. Specifically, as shown in Lecture 31,

$$dN_j = \sum_{r=1}^{m} \upsilon_{jr} \, d\xi_r \qquad (32.3)$$

Using Eq. (32.3) in Eq. (32.1) yields:

$$d\underline{G} = -\underline{S}dT + \underline{V}dP + \sum_{r=1}^{m} \left(\sum_{j=1}^{n} \upsilon_{jr} \mu_j \right) d\xi_r \qquad (32.4)$$

Equation (32.4) shows that $\underline{G} = \underline{G}(T, P, \xi_1, \ldots, \xi_m)$. As a result, the variations of the various ξ_rs in $d\underline{G}$ are independent.

At equilibrium, at constant T and P, the Gibbs free energy must attain its minimum value. This requires that the first derivative of $\underline{G}$ with respect to ξ_s, keeping all the other ξ_rs $\neq \xi_s$, T, and P constant, be zero, that is:

$$\left(\frac{\partial \underline{G}}{\partial \xi_s} \right)_{T,P,\xi_{r \neq s}} = 0 \qquad (32.5)$$

Figure 32.1 helps us understand Eq. (32.5):

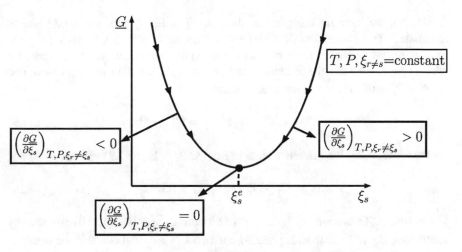

Fig. 32.1

Using Eq. (32.4) in Eq. (32.5), it follows that:

$$\left(\frac{\partial \underline{G}}{\partial \xi_s}\right)_{T,P,\xi_{r\neq s}} = \sum_{j=1}^{n} \upsilon_{js}\, \mu_j \qquad (32.6)$$

Equations (32.5) and (32.6) show that, at equilibrium, we have:

$$\sum_{j=1}^{n} \upsilon_{jr}\, \mu_j = 0 \qquad (32.7)$$

where we replaced the dummy index s by r. Equation (32.7) is the condition of chemical reaction equilibria, a central result which relates the chemical potentials of the n components participating in chemical reaction r.

32.3   Derivation of the Equilibrium Constant for Chemical Reaction r

Next, we can express μ_j in Eq. (32.7) using its relation to the fugacity of component j in the mixture, $\hat{f}_j$. Specifically, as we showed in Part II,

$$\mu_j = \overline{G}_j = \lambda_j(T) + RT\ln\widehat{f}_j \qquad (32.8)$$

Because we have no information about $\lambda_j(T)$ in Eq. (32.8), we would like to eliminate it. To this end, we select a reference state, denoted by (o), for which $T^o = T$. In addition, we select the reference state to be a pure component reference state, for which $x_j^o = 1$. Note that the reference state pressure, P^o, is left arbitrary for now, that is, $P^o \neq P$. In this reference state (RS), we have:

$$x_j^o = 1, T^o = T, P^o, f_j^o = f_j^o(T, P^o) \qquad (32.9)$$

Therefore, using the reference state in Eq. (32.9) in Eq. (32.8) yields:

$$\mu_j^o = G_j^o(T, P^o) = \lambda_j(T) + RT\ln f_j^o \qquad (32.10)$$

Because $\lambda_j(T)$ appears in both Eq. (32.8) for μ_j and Eq. (32.10) for μ_j^o, by subtracting Eq. (32.10) from Eq. (32.8), the two $\lambda_j(T)$s cancel out, and we obtain:

$$\mu_j = G_j^o(T, P^o) + RT\ln\left\{\frac{\widehat{f}_j(T, P, \text{Composition})}{f_j^o(T, P^o)}\right\} \qquad (32.11)$$

Next, we can use Eq. (32.11) for μ_j in the condition of chemical reaction equilibria, Eq. (32.7). This yields:

$$\sum_{j=1}^{n} \upsilon_{jr}\left(G_j^o + RT\ln\left(\frac{\widehat{f}_j}{f_j^o}\right)\right) = 0 \qquad (32.12)$$

Rearranging Eq. (32.12) yields:

$$-\left(\sum_{j=1}^{n} \upsilon_{jr}G_j^o\right) = RT\sum_{j=1}^{n} \upsilon_{jr}\ln\left(\frac{\widehat{f}_j}{f_j^o}\right) \qquad (32.13)$$

First, let us work on the right-hand side of Eq. (32.13). Specifically,

$$RT\sum_{j=1}^{n} \upsilon_{jr}\ln\left(\frac{\widehat{f}_j}{f_j^o}\right) = RT\sum_{j=1}^{n}\ln\left[\left(\frac{\widehat{f}_j}{f_j^o}\right)^{\upsilon_{jr}}\right] = RT\ln\left\{\prod_{j=1}^{n}\left(\frac{\widehat{f}_j}{f_j^o}\right)^{\upsilon_{jr}}\right\} \qquad (32.14)$$

where we used the equality:

$$\sum_{j=1}^{n} (\ln) = \ln \left(\prod_{j=1}^{n} \right) \tag{32.15}$$

In Eq. (32.14), the product of the fugacity ratios for chemical reaction r is referred to as the Equilibrium Constant for Chemical Reaction r, K_r, that is:

$$K_r = \prod_{j=1}^{n} \left(\frac{\hat{f}_j(T, P, \text{Composition})}{f_j^o(T, P)} \right)^{\upsilon_{jr}} \tag{32.16}$$

where Composition denotes the mole fractions of the various components in the mixture, as related through the extents of reaction.

Second, let us work on the left-hand side of Eq. (32.13). The summation over j from 1 to n represents the molar Gibbs free energy difference between the products and the reactants in the RS, or the Standard State (o), and is referred to as the Standard Molar Gibbs Free Energy of Reaction, given by:

$$\Delta G_r^o = \sum_{j=1}^{n} \upsilon_{jr} G_j^o(T, P_o) \tag{32.17}$$

Using Eqs. (32.17) and (32.16) in Eq. (32.13), we obtain:

$$-\Delta G_r^o = RT\ln K_r \tag{32.18}$$

In summary:

$$K_r = \exp\left(\frac{-\Delta G_r^o}{RT} \right) \tag{32.19}$$

$$\Delta G_r^o = \sum_{j=1}^{n} \upsilon_{jr} G_j^o(T, P^o) \tag{32.20}$$

$$K_r = \prod_{j=1}^{n} \left(\frac{\hat{f}_j(T, P, \text{Composition})}{f_j^o(T, P^o)} \right)^{\upsilon_{jr}} \tag{32.21}$$

where r = 1, 2, ..., m (Number of independent chemical reactions).

In Eqs. (32.19)–(32.21), G_j^o, f_j^o, and $\widehat{f}_j$ are constant throughout the mixture, irrespective of the chemical reactions in which component j participates. As a result, these three quantities do not depend on r.

32.4 Derivation of the Equilibrium Constant for a Single Chemical Reaction

For the special case of a single chemical reaction, we can eliminate the index r, and express Eqs. (32.19)–(32.21) as follows:

$$K = \exp\left(\frac{-\Delta G^o}{RT}\right) \tag{32.22}$$

$$\Delta G = \sum_{j=1}^{n} \upsilon_j G_j^o\,(T, P^o) \tag{32.23}$$

$$K = \prod_{j=1}^{n}\left(\frac{\widehat{f}_j(T, P, \text{Composition})}{f_j^o\,(T, P^o)}\right)^{\upsilon_j} \tag{32.24}$$

32.5 Discussion of Standard States for Gas-Phase, Liquid-Phase, and Solid-Phase Chemical Reactions

In Eqs. (32.22)–(32.24), ΔG^o and K depend on the choice of standard state, particularly on the choice of P^o. Next, we will also assume that all the stoichiometric numbers are known. In order to calculate the standard molar Gibbs free energy of reaction,

$$\Delta G^o = \sum_{j=1}^{n} \upsilon_j G_j^o \tag{32.25}$$

or the chemical potential of component j,

$$\mu_j = G_j^o + RT\ln\left(\frac{\hat{f}_i}{f_j^o}\right) \tag{32.26}$$

we need to know the values of G_j^o for all the non-inert components, for which $\upsilon_j \neq 0$. Therefore, the standard state (o) needs to be specified precisely. As we saw earlier, the standard-state temperature, T^o, was chosen to be equal to the system temperature, T, in order to eliminate the unknown function $\lambda_j(T)$. In addition, the standard-state pressure, P^o, the standard-state composition (the various x_j^os in the standard state), and the state of aggregation of the standard state (gas, liquid, or solid) may be chosen for convenience.

It is common to denote the state of aggregation of the standard state as follows:

$$|\upsilon_j|A_j(g) + |\upsilon_k|A_k(l) + |\upsilon_q|A_q(s) = 0 \tag{32.27}$$

where g (gas), l (liquid), and s (solid) designate the state of aggregation of the standard state. Below, we discuss each standard state separately:

1. (g): Indicates the following standard state (o)

Pure Ideal Gas j, $x_j^o = 1$
$T^o = T$ (System temperature)
$P^o = 1$ bar $\Rightarrow f_j^o = 1$ bar (Unit fugacity standard state)

2. (l): Indicates the following standard state (o)

Pure Liquid k, $x_k^o = 1$
$T^o = T$ (System temperature)
$P^o = P$ (System pressure), or P_{vpk} (T), or 1 bar

where vp denotes vapor pressure.

3. (s): Indicates the following standard state (o)

Pure Solid q, $x_q^o = 1$, at the most stable solid state
$T^o = T$ (System temperature)
$P^o = P$ (System pressure), or P_{vpq} (T), or 1 bar

Typically, we will choose P^o to be either 1 bar or P_{vpj} (T) for component j, so that the standard-state properties, $\Delta G^o(T, P^o)$ and $f_j^o(T, P^o)$, depend only on temperature, and not on pressure. However, in some cases, we can choose $P^o = P$, such that, when the system pressure, P, changes, P^o will also change.

When $P^o = 1$ bar or $P_{vpj}(T)$, it follows that $K = \exp(-\Delta G^o(T, P^o) / RT)$ depends solely on the temperature, T. On the other hand, when $P^o = P$, it follows that $K = \exp(-\Delta G^o(T, P) / RT)$ depends on both the temperature, T, and the pressure, P.

32.6   Comments on the Standard-State Pressure

As we have just shown, the standard-state pressure, P^o, is intimately connected with the pressure dependence of the equilibrium constant, K. Indeed,

$$K(T^o, P^o) = \exp\left(\frac{-\Delta G^o(T^o, P^o)}{RT}\right) \tag{32.28}$$

is strictly a property of the standard state (o). As discussed, we typically choose $T^o = T$ and $P^o = 1$ bar. In that case, K depends only on T. However, sometimes, P^o is chosen to be $P_{vpj}(T)$ in the case of liquids or solids. In that case, one needs to use $G_j^o(T, P_{vpj}(T))$ to calculate the molar Gibbs free energy of component j in the standard state. If only $G_j^o(T, 1 \text{ bar})$ is known, then, we need to calculate how $G_j^o(T, P_{vpj}(T))$ is related to $G_j^o(T, 1 \text{ bar})$ using a relation derived in Part I. Specifically,

$$\left(\frac{\partial G_j^o(T, P)}{\partial P}\right)_T = V_j^o(T, P) \;\Rightarrow\; dG_j^o = V_j^o(T, P) dP \tag{32.29}$$

Integrating the differential relation to the right of the arrow in Eq. (32.29), from $P = 1$ bar to $P = P_{vpj}(T)$, yields:

$$G_j^o(T, P_{vpj}(T)) = G_j^o(T, 1 \text{ bar}) + \int_{1 \text{ bar}}^{P_{vpj}(T)} V_j^o(T, P) \, dP \tag{32.30}$$

Recall that in Eqs. (32.29) and (32.30), $V_j^o(T, P)$ is the molar volume of component j in the standard state (o) at T and P. If for some of the n components we choose $P^o = P$, then, we need to compute $G_j^o(T, P)$ in the standard state in terms of $G_j^o(T, 1 \text{ bar})$ using Eq. (32.30), where $P_{vpj}(T)$ is replaced by P, that is,

$$G_j^o(T, P) = G_j^o(T, 1 \text{ bar}) + \int_{1 \text{bar}}^{P} V_j^o(T, P) dP \tag{32.31}$$

If ΔG^o is known at 1 bar, the simplest approach is to choose $P^o = 1$ bar for every component j in the standard state.

32.7 Decomposition of the Equilibrium Constant into Contributions from the Fugacity Coefficients, the Gas Mixture Mole Fractions, and the Pressure

In the case of a single chemical reaction, we showed that (see Eq. (32.24), repeated below for clarity):

$$K = \prod_{j=1}^{n} \left(\frac{\widehat{f}_j(T, P, \text{Composition})}{f_j^o(T, P^o)} \right)^{\upsilon_j} \tag{32.32}$$

It is convenient to express K in Eq. (32.32) in terms of the mole fractions of the various components participating in the chemical reaction. To this end, it is useful to express the mixture fugacity of component j, $\widehat{f}_j$, as a function of concentrations using the fugacity-coefficient approach based on an equation of state which we discussed in Part II. This approach is particularly useful when dealing with gas-phase chemical reactions. On the other hand, when dealing with liquid-phase or solid-phase chemical reactions, it is convenient to express $\widehat{f}_j$ as a function of concentrations using the activity-coefficient approach based on a model for the excess Gibbs free energy of mixing, ΔG^{EX}, which we also discussed in Part II. We will carry out this calculation in Lecture 33. Recall that if component j is present in two or more phases, at thermodynamic equilibrium, we can use the fugacity of component j in any of the coexisting phases, because they are all equal.

Focusing on a single gas (vapor)-phase chemical reaction, we can express the fugacity of component j in the gas mixture as follows:

$$\widehat{f}_j = \widehat{f}_j(T, P, y_1, y_2, \ldots, y_{n-1}) = \widehat{\phi}_j(T, P, y_1, y_2, \ldots, y_{n-1}) P y_j \tag{32.33}$$

In Eq. (32.33), the (n-1) gas mole fractions, $\{y_1, y_2, \ldots, y_{n-1}\}$, are dependent, because they are determined by the extent of reaction, ξ, and $\widehat{\phi}_j$ is the fugacity coefficient of component j.

To simplify the notation, we can rewrite Eq. (32.33) as follows:

$$\widehat{f}_j = \widehat{\phi}_j P y_j \tag{32.34}$$

We also know that:

$$\widehat{f}_j^o = 1 \text{ bar} \tag{32.35}$$

Dividing Eq. (32.34) by Eq. (32.35) we obtain:

$$\frac{\widehat{f}_j}{f_j^o} = \frac{\widehat{\phi}_j P y_j}{1 \text{ bar}}$$

(32.36)

where in Eqs. (32.34) and (32.36), P is in units of bar. From the definition of the equilibrium constant, K, in Eq. (32.32), it follows that:

$$K = \prod_{j=1}^{n}\left(\frac{\widehat{f}_j}{f_j^o}\right)^{\upsilon_j} = \prod_{j=1}^{n}\left(\widehat{f}_j P y_j\right)^{\upsilon_j}$$

(32.37)

or

$$K = K_\phi K_y K_P$$

(32.38)

where

$$K_\phi = \prod_{j=1}^{n}\widehat{\phi}_j^{\upsilon_j}$$

(32.39)

where K_ϕ reflects the non-idealities of mixing in the gas mixture. Recall that if the gas mixture is ideal, then, $\widehat{\phi}_j^{\text{ID}} = 1$ and $K_\phi^{\text{ID}} = 1$.

Further, in Eq. (32.38):

$$K_y = \prod_{j=1}^{n} y_j^{\upsilon_j}$$

(32.40)

Recall that the y_j's are all related to a single extent of reaction, ξ, and are therefore dependent.

Finally, in Eq. (32.38):

$$K_P = P^{\upsilon}, \text{ where } \upsilon = \sum_{j=1}^{n}\upsilon_j$$

(32.41)

where P is given in units of bar, because we have used the fact that $\widehat{f}_j^o = 1$ bar.

In addition, $K_y K_P = K_y P^{\upsilon}$, which we can express as follows:

$$K_y P^{\upsilon} = \prod_{j=1}^{n}\left(y_j P\right)^{\upsilon_j} = \prod_{j=1}^{n} P_j^{\upsilon_j}$$

(32.42)

where P_j is the partial pressure of gas j in the gas mixture. Accordingly,

$$K = K_\phi \prod_{j=1}^{n} P_j^{\nu_j} \qquad (32.43)$$

If the gas mixture is ideal, it follows that:

$$\hat{\phi}_j^o = 1 \Rightarrow K_\phi^{ID} = 1 \qquad (32.44)$$

Using the result to the right of the arrow in Eq. (32.44) in Eq. (32.43), we obtain:

$$K^{ID} = \prod_{j=1}^{n} P_j^{\nu_j} \qquad (32.45)$$

Lecture 33

Equilibrium Constants for Condensed-Phase Chemical Reactions, Response of Chemical Reactions to Temperature, and Le Chatelier's Principle

33.1 Introduction

The material presented in this lecture is adapted from Chapter 16 in T&M. In addition, Tables 33.1 and 33.2 are adapted from *Introduction to Chemical Engineering Thermodynamics*, Fourth Edition, by J.M. Smith and H.C. Van Ness, McGraw-Hill Book Company, NY (1987). First, we will complete our discussion of equilibrium constants that we began in Lecture 32. Specifically, we will discuss the equilibrium constant of a condensed (liquid or solid)-phase chemical reaction. Second, we will discuss how to determine the standard molar Gibbs free energy of reaction, including decomposing it into enthalpic and entropic contributions which can be calculated using results presented in Part I. Knowledge of the standard molar Gibbs free energy of reaction will then allow us to determine the equilibrium constant of the chemical reaction. Third, we will discuss how a chemical reaction responds to temperature, including classifying it as exothermic or endothermic. Finally, we will discuss Le Chatelier's Principle.

33.2 Derivation of the Equilibrium Constant for a Condensed-Phase Chemical Reaction

As discussed in Part II, the fugacity of component j in a condensed (liquid or solid) mixture is given by:

$$\widehat{f}_j(T, P, x_1, x_2, \ldots, x_{n-1}) = \gamma_j(T, P, x_1, x_2, \ldots, x_{n-1}) x_j f_j(T, P) \tag{33.1}$$

where the mole fractions, $\{x_1, x_2, \ldots, x_{n-1}\}$, are all related to a single extent of reaction. Further, in the standard state (o), we have:

© Springer Nature Switzerland AG 2020   341
D. Blankschtein, *Lectures in Classical Thermodynamics with an Introduction to Statistical Mechanics*, https://doi.org/10.1007/978-3-030-49198-7_33

$$f_j^o = f_j^o(T, P^o), \text{ where } P^o = P, P_{vpj}(T), \text{ or } 1 \text{ bar} \tag{33.2}$$

Equations (33.1) and (33.2) indicate that:

$$\frac{\widehat{f_j}}{f_j^o} = \frac{\gamma_j(T, P, x_1, x_2, x_{n-1}) \, x_j \, f_j(T, P)}{f_j^o(T, P^o)} \tag{33.3}$$

In Eq. (33.3), we can relate $f_j(T, P)$ to $f_j^o(T, P^o)$ using results presented in Part II. Specifically,

$$\left(\frac{\partial \ln f_j}{\partial P}\right)_T = \frac{V_j}{RT} \Rightarrow f_j(T, P) = f_j^o(T, P^o) \exp\left[\int_{P^o}^{P}\left(\frac{V_j}{RT}\right)dP\right] \tag{33.4}$$

In Eq. (33.4), $V_j(T, P)$ is the molar volume of component j, and the exponential term is referred to as the Poynting correction.

Using the definition of the equilibrium constant, K, introduced in Lecture 32, and Eq. (33.3), we obtain:

$$K = \prod_{j=1}^{n}\left(\frac{\widehat{f_i}}{f_j^o}\right)^{\upsilon_j} = K_\gamma K_x K_P \tag{33.5}$$

where

$$K_\gamma = \prod_{j=1}^{n}\gamma_j^{\upsilon_j} \tag{33.6}$$

where K_γ reflects the nonidealities of mixing in a liquid or solid (condensed) phase. If the mixture is ideal (ID), then, $\gamma_j^{ID} = 1$, and $K_\gamma = 1$.

It also follows that K_x in Eq. (33.5) is given by:

$$K_x = \prod_{j=1}^{n}x_j^{\upsilon_j} \tag{33.7}$$

where all the x_js in Eq. (33.7) are related to a single extent of reaction. In addition, the x_js affect both K_x and K_γ.

Finally, K_P in Eq. (33.5) is given by:

$$K_P = \prod_{j=1}^{n} \exp\left[\upsilon_j \int_{P^o}^{P}\left(\frac{V_j}{RT}\right)dP\right] \tag{33.8}$$

Recall that K is intimately related to the standard state (o), because:

$$K = \exp[-\Delta G^\circ(T, P^\circ)/RT] \tag{33.9}$$

Equation (33.9) indicates that in order to determine the equilibrium constant K, we need to determine the standard molar Gibbs free energy of reaction. Below, we will discuss how to do that.

33.3 Determination of the Standard Molar Gibbs Free Energy of Reaction

Because direct experimental data for K is only available for very few simple chemical reactions, it is more typical to determine K using Eq. (33.9) from available data on:

$$\Delta G^\circ(T, P^\circ) = \sum_{j=1}^{n} \upsilon_j G_j^\circ(T, P^\circ) \tag{33.10}$$

Alternatively, $\Delta G^\circ(T, P^\circ)$ can be obtained if enthalpy and entropy data in the standard state (o) is available. Specifically, as discussed in Part I,

$$G_j^\circ = H_j^\circ - TS_j^\circ \Rightarrow \upsilon_j G_j^\circ = \upsilon_j H_j^\circ - T\upsilon_j S_j^\circ \Rightarrow \sum_{j=1}^{n} \upsilon_j G_j^\circ \equiv \Delta G^\circ \tag{33.11}$$

It then follows that:

$$\Delta G^\circ = \sum_{j=1}^{n} \upsilon_j H_j^\circ - T \sum_{j=1}^{n} \upsilon_j S_j^\circ \Rightarrow \Delta G^\circ(T, P^\circ)$$
$$= \Delta H^\circ(T, P^\circ) - T\Delta S^\circ(T, P^\circ) \tag{33.12}$$

In Eq. (33.12),

$\Delta G^\circ(T, P^\circ)$:   Standard molar Gibbs free energy of reaction
$\Delta H^\circ(T, P^\circ)$:   Standard molar enthalpy of reaction, also known as the standard heat of reaction, Q_r
ΔS°:   Standard molar entropy of reaction

Because it is not practical to measure ΔG° (or ΔH° and ΔS°) for every chemical reaction, tables are available for a large number of compounds reporting the standard molar Gibbs free energy of formation and the standard molar enthalpy of formation, or the standard heat of formation, of the species from the elements. In these tables, the function ΔG° becomes, for each species j, ΔG_{fj}°, and similarly, ΔH° for each species j, becomes ΔH_{fj}°, where f denotes formation. To obtain ΔG° and ΔH° from the given ΔG_{fj}° and ΔH_{fj}°, respectively, we simply utilize:

$$\Delta G^{\circ} = \sum_{j=1}^{n} \upsilon_j \Delta G_{fj}^{\circ} \quad \text{and} \quad \Delta H^{\circ} = \sum_{j=1}^{n} \upsilon_j \Delta H_{fj}^{\circ} \tag{33.13}$$

Usually, tables for ΔG_{fj}° and ΔH_{fj}° are available only at 298 K. If ΔG° or ΔH° are needed at other temperatures, we can compute these using temperature integrations (see below).

For pure elements, by convention, we choose:

$$\Delta G_{fj}^{\circ} = 0, \Delta H_{fj}^{\circ} = 0, \text{at all Ts} \tag{33.14}$$

For pure elements that are solids at the temperature of interest, the crystal form must be specified. For example, the standard state of carbon is based on graphite, and only for graphite are $\Delta G_f^{\circ} = 0$ and $\Delta H_f^{\circ} = 0$. Should other forms of carbon be present in the system, then, ΔG_f° and ΔH_f° for these forms of carbon are not zero.

Tabulations are available that list ΔG_{fi}° and ΔH_{fi}° for several components (usually at 25 °C or 298 K), see Tables 33.1 and 33.2 below.

If ΔH° is known at one temperature, it can be found at any other temperature by integration using heat capacity data. Specifically, as discussed in Part I,

$$dH_j^{\circ} = C_{pj}^{\circ} dT \tag{33.15}$$

where C_{pj}° is the heat capacity at constant pressure of pure component j in the standard state (o) at T and P°. It then follows that:

$$\upsilon_j dH_j^{\circ} = d\left(\upsilon_j H_j^{\circ}\right) = \upsilon_j C_{pj}^{\circ} dT \Rightarrow \sum_{j=1}^{n} d\left(\upsilon_j H_j^{\circ}\right) = d\left(\sum_{j=1}^{n} \upsilon_j H_j^{\circ}\right)$$

$$\Rightarrow \left(\sum_{j=1}^{n} \upsilon_j C_{pj}^{\circ}\right) dT \tag{33.16}$$

where the first summation over j in the large brackets is the standard molar enthalpy of reaction (ΔH°), and the second summation over j in the large brackets is the standard molar heat capacity of reaction at constant pressure $\left(\Delta C_p^{\circ}\right)$. Equating the last two terms in Eq. (33.16) yields:

$$d(\Delta H^{\circ}) = \Delta C_p^{\circ} dT \tag{33.17}$$

Table 33.1 Standard Gibbs Free Energy of reaction at 298 K (Joules per mole of the substance formed)

Chemical species		State (Note 2)	ΔG^{ϕ}_{f20g}
Paraffins:			
Methane	CH_4	g	−50,460
Ethane	C_2H_6	g	−31,855
Propane	C_3H_8	g	−24,290
n-Butane	C_4H_{10}	g	−16,570
n-Pentane	C_5H_{12}	g	−8,650
n-Hexane	C_6H_{14}	g	150
n-Heptane	C_7H_{16}	g	8,260
n-Octane	C_8H_{18}	g	16,260
1-Alkenes:			
Ethylene	C_2H_4	g	68,460
Propylene	C_3H_6	g	62,205
1-Butene	C_4H_8	g	70,340
1-Pentene	C_5H_{10}	g	78,410
1-Hexene	C_6H_{12}	g	86,830
Miscellaneous organics:			
Acetaldehyde	C_2H_4O	g	−128,860
Acetic acid	$C_2H_4O_2$	l	−389,900
Acetylene	C_2H_2	g	209,970
Benzene	C_6H_6	g	129,665
Benzene	C_6H_6	l	124,520
1,3-Butadiene	C_4H_6	g	149,795
Cyclohexane	C_6H_{12}	g	31,920
Cyclohexane	C_6H_{12}	l	26,850
1,2-Ethanediol	$C_2H_6O_2$	l	−323,080
Ethanol	C_2H_6O	g	−168,490
Ethanol	C_2H_6O	l	−174,780
Ethylbenzene	C_8H_{10}	g	130,890
Ethylene oxide	C_2H_4O	g	−13,010
Formaldehyde	CH_2O	g	−102,530
Methanol	CH_4O	g	−161,960
Methanol	CH_4O	l	−166,270
Methylcyclohexane	C_7H_{14}	g	27,480
Methylcyclohexane	C_7H_{14}	l	20,560
Styrene	C_8H^8	g	213,900
Toluene	C_7H_8	g	122,050
Toluene	C_7H_8	l	113,630

Table 33.2 Standard heats of formation at 25 °C (Joules per mole of the substance formed)

Chemical species		State	ΔH°_{298}
Paraffins:			
Methane	CH_4	g	−74,520
Ethane	C_2H_6	g	−83,820
Propane	C_3H_8	g	−104,680
n-Butane	C_4H_{10}	g	−125,790
n-Pentane	C_5H_{12}	g	−146,760
n-Hexane	C_6H_{14}	g	−166,920
n-Heptane	C_7H_{16}	g	−187,780
n-Octane	C_8H_{18}	g	−208,750
1-Alkenes:			
Ethylene	C_2H_4	g	52,510
Propylene	C_3H_6	g	19,710
1-Butene	C_4H_8	g	−540
1-Pentene	C_5H_{10}	g	−21,280
1-Hexene	C_6H_{12}	g	−41,950
1-Heptene	C_7H_{14}	g	−62,760
Miscellaneous organics:			
Acetaldehyde	C_2H_4O	g	−166,190
Acetic acid	$C_2H_4O_2$	l	−484,500
Acetylene	C_2H_2	g	227,480
Benzene	C_6H_6	g	82,930
Benzene	C_6H_6	l	49,080
1,3-Butadiene	C_4H_6	g	109,240
Cyclohexane	C_6H_{12}	g	−123,140
Cyclohexane	C_6H_{12}	l	−156,230
1,2-Ethanediol	$C_2H_6O_2$	l	−454,800
Ethanol	C_2H_6O	g	−235,100
Ethanol	C_2H_6O	l	−277,690
Ethylbenzene	C_8H_{10}	g	29,920
Ethylene oxide	C_2H_4O	g	−52,630
Formaldehyde	CH_2O	g	−108,570
Methanol	CH_4O	g	−200,660
Methanol	CH_4O	l	−238,660
Methylcyclohexane	C_7H_{14}	g	−154,770
Methylcyclohexane	C_7H_{14}	l	−190,160
Styrene	C_8H_8	g	147,360
Toluene	C_7H_8	g	50,170
Toluene	C_7H_8	l	12,180

Integrating Eq. (33.17) from T_1 to T yields:

$$\Delta H^\circ(T, P^\circ) = \Delta H^\circ(T_1, P^\circ) + \int_{T_1}^{T} \Delta C_p^\circ(T, P^\circ) dT \tag{33.18}$$

Typically, T_1 in Eq. (33.18) is 298 K.

If ΔH° is known as a function of temperature, we can obtain ΔG° as a function of temperature using the Gibbs-Helmholtz equation discussed in Part I. Recall that:

$$\frac{d}{dT}\left(\frac{G}{T}\right)_{P^\circ} = -\frac{H}{T^2} \tag{33.19}$$

and therefore,

$$\frac{d}{dT}\left(\frac{\Delta G^\circ}{T}\right)_{P^\circ} = -\frac{\Delta H^\circ}{T^2} \tag{33.20}$$

We can first calculate how ΔH° varies with T using Eq. (33.18). Subsequently, we can use $\Delta H^\circ(T, P^\circ)$ so deduced in Eq. (33.20) to compute how ΔG° varies with T. Specifically, integrating Eq. (33.20) between T_1 and T, we obtain:

$$\int_{T_1}^{T} \frac{d}{dT}\left(\frac{\Delta G^\circ}{T}\right)_{P^\circ} dT = \int_{T_1}^{T} d\left(\frac{\Delta G^\circ}{T}\right)_{P^\circ} = \frac{\Delta G^\circ(T, P^\circ)}{T} - \frac{\Delta G^\circ(T_1, P^\circ)}{T_1}$$

$$= \int_{T_1}^{T} \left(\frac{-\Delta H^\circ}{T^2}\right) dT \tag{33.21}$$

or

$$\frac{\Delta G^\circ(T, P^\circ)}{T} = \frac{\Delta G^\circ(T_1, P^\circ)}{T_1} - \int_{T_1}^{T} \left(\frac{\Delta H^\circ}{T^2}\right) dT \tag{33.22}$$

Using Eq. (33.22) in Eq. (33.9) we can then calculate:

$$K(T, P^\circ) = \exp\left(\frac{-\Delta G^\circ(T, P^\circ)}{RT}\right) \tag{33.23}$$

33.4   Response of Chemical Reactions to Changes in Temperature and Pressure

We would like to know how a change in temperature, or in pressure, affects the equilibrium conversion of reactants into products. In other words, does an increase in temperature (T) or in pressure (P) shift the chemical reaction in the direction of the products (the desired outcome), or does it shift the chemical reaction in the direction of the reactants (the undesired outcome)? This information is, of course, essential from a practical viewpoint, because we need to know if we should operate at a higher T or P in order to convert more reactants into products. We will first discuss the effect of T on the equilibrium conversion. The effect of P on the equilibrium conversion will be discussed in Lecture 34.

33.5   How Does a Chemical Reaction Respond to Temperature?

We know that in the standard state (o), we have:

$$\Delta G^o = \Delta G^o(T, P^o); \; K = \exp\left[\frac{-\Delta G^o}{RT}\right]; \; \text{and} \; \ln K = -\frac{\Delta G^o}{RT} \qquad (33.24)$$

It then follows that:

$$\left(\frac{\partial \ln K}{\partial T}\right)_{P^o} = -\frac{1}{R}\left[\frac{\partial}{\partial T}\left(\frac{\Delta G^o}{T}\right)\right]_{P^o} = \frac{1}{R}\left(\frac{\Delta H^o}{T^2}\right) \qquad (33.25)$$

where Eq. (33.20) was used. Accordingly,

$$\left(\frac{\partial \ln K}{\partial T}\right)_{P^o} = \frac{\Delta H^o}{RT^2} \qquad (33.26)$$

or

$$\left(\frac{\partial \ln K}{\partial(1/T)}\right)_{P^o} = -\frac{\Delta H^o}{R} \qquad (33.27)$$

Equation (33.27) shows that if we plot $\ln K$ versus $1/T$, at constant P^o, the slope of the resulting curve at any T is equal to $(-\Delta H^o/R)$ at that T.

Chemical reactions are classified as follows:

(a)   Exothermic, if $\Delta H^\circ < 0$
(b)   Endothermic, if $\Delta H^\circ > 0$

In Part I, we showed that the First Law of Thermodynamics for a closed system is given by:

$$d\underline{U} = \delta Q + \delta W \tag{33.28}$$

where $d\underline{U}$ is the differential change in the internal energy of the system, δQ is the differential heat absorbed by the system, and δW is the differential work done on the system. Assuming $Pd\underline{V}$-type work, it follows that:

$$\delta W = -Pd\underline{V} \tag{33.29}$$

Using Eq. (33.29) in Eq. (33.28) yields:

$$d\underline{U} = \delta Q - Pd\underline{V} \tag{33.30}$$

or

$$d\underline{U} + Pd\underline{V} = \delta Q \tag{33.31}$$

If we assume a constant pressure (isobaric) process, Eq. (33.31) can be written as follows:

$$d(\underline{U} + P\underline{V}) = \delta Q|_P \tag{33.32}$$

or

$$\delta Q|_P = d\underline{H} \tag{33.33}$$

Equation (33.33) helps us understand that, at constant pressure, the standard molar enthalpy of reaction, ΔH°, is equal to the standard heat of reaction, Q_r°. This result shows that:

(i)   If $\Delta H^\circ < 0$, this implies that $Q_r^\circ < 0$. According to our definition of heat flow in Part I, this indicates that the system rejects heat, Q_r°, to the environment. This explains why the chemical reaction in this case is referred to as exothermic.
(ii) If, on the other hand, $\Delta H^\circ > 0$, this implies that $Q_r^\circ > 0$. According to our definition of heat flow in Part I, this indicates that the system absorbs heat, Q_r°, from the environment. This explains why the chemical reaction in this case is referred to as endothermic.

To summarize, for an exothermic reaction, $\Delta H^\circ < 0$, and therefore, the slope of $\ln K$ versus $1/T$ is positive. On the other hand, for an endothermic reaction, $\Delta H^\circ > 0$,

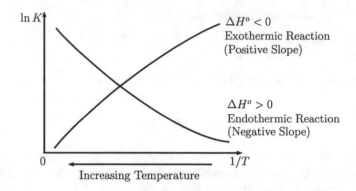

Fig. 33.1

and therefore, the slope of lnK versus 1/T is negative. Figure 33.1 depicts these two lnK versus 1/T behaviors.

Interestingly, only if ΔH° is independent of temperature, the lnK versus 1/T curves are straight lines.

Equation (33.27), as well as Fig. 33.1, indicate that when $\Delta H^\circ < 0$ (exothermic reaction), an increase in temperature, that is, a decrease in 1/T, brings about a decrease in lnK, and hence, in K. Therefore, for an exothermic reaction, an increase in temperature shifts the reaction in the direction of the reactants. The opposite occurs for an endothermic reaction ($\Delta H^\circ > 0$), for which an increase in temperature, or a decrease in 1/T, brings about an increase in lnK, and hence, in K. This shifts the reaction in the direction of the products.

33.6 Le Chatelier's Principle

This principle states that: "A system at equilibrium, when subjected to a perturbation, responds in a manner that tends to eliminate the effect of the perturbation."

Therefore, for an endothermic reaction, when we perturb the system by increasing the temperature, the system responds by shifting the reaction toward the products whose enthalpy is higher than that of the reactants ($\Delta H^\circ > 0$). As a result, some of the heat input goes to the reaction, thereby cooling down the system and lowering its temperature.

On the other hand, for an exothermic reaction, when we perturb the system by increasing the temperature, the system responds by shifting the reaction towards the reactants, whose enthalpy is higher than that of the products ($\Delta H^\circ < 0$). As a result, some of the heat provided to increase the temperature goes into the reaction, thereby cooling down the system and lowering its temperature.

Returning to Eq. (33.27), we can obtain another useful result. Specifically,

$$\left(\frac{\partial \ln K}{\partial (1/T)}\right)_{P^o} = -\frac{\Delta H^o}{R} \Rightarrow \int_{T_1}^{T_2} \partial \ln K \Bigg|_{P^o} = \int_{T_1}^{T_2} \left(\frac{-\Delta H^o(T, P^o)}{R}\right) d(1/T)$$

or

$$K(T_2, P^o) = K(T_1, P^o) \exp\left[-\int_{T_1}^{T_2}\left(\frac{\Delta H^o(T, P^o)}{R}\right) d(1/T)\right] \qquad (33.34)$$

where in Eq. (33.34), T_1 is typically equal to 298 K. In addition, Eq. (33.34) enables a direct evaluation of $K(T_2, P^o)$ if $\Delta H^o(T, P^o)$ is known.

If ΔH^o is a constant independent of T, then, Eq. (33.34) can be readily integrated to yield:

$$K(T_2, P^o) = K(T_1, P^o) \exp\left[\frac{-\Delta H^o}{R}\left(\frac{1}{T_2} - \frac{1}{T_1}\right)\right] \qquad (33.35)$$

Lecture 34

Response of Chemical Reactions to Pressure, and Sample Problems

34.1 Introduction

The material presented in this lecture is adapted from Chapter 16 in T&M. First, we will discuss how a chemical reaction responds to pressure. Second, we will solve Sample Problem 34.1 to calculate the effect of pressure on the equilibrium gas mixture composition. Finally, we will solve Sample Problem 34.2 to calculate the equilibrium conversion for a gas mixture undergoing a dissociation reaction.

34.2 How Does a Chemical Reaction Respond to Pressure?

In Lecture 33, we saw that the equilibrium constant, K, depends on the standard molar Gibbs free energy of reaction, ΔG°, which is a function of T° and P°, where the reference-state temperature, T°, was chosen to be equal to the system temperature, T. If the reference-state pressure, P°, is chosen to be 1 bar or P_{vpj} (T), the vapor pressure of component j at temperature T, then, ΔG° is not a function of pressure. In that case:

$$\ln K = -\frac{\Delta G^\circ(T, P^\circ)}{RT} \Rightarrow \left(\frac{\partial \ln K}{\partial P}\right)_T = -\frac{1}{RT}\left(\frac{\partial \Delta G^\circ}{\partial P}\right)_T = 0 \qquad (34.1)$$

However, if P° is chosen to be equal to the system pressure, P, then, $\Delta G^\circ = \Delta G^\circ$ (T, P) depends explicitly on P, and therefore, K will also depend on P. Specifically:

$$\ln K = -\frac{\Delta G^\circ(T, P)}{RT} \Rightarrow \left(\frac{\partial \ln K}{\partial P}\right)_T = -\frac{1}{RT}\left(\frac{\partial \Delta G^\circ}{\partial P}\right)_T \qquad (34.2)$$

© Springer Nature Switzerland AG 2020
D. Blankschtein, *Lectures in Classical Thermodynamics with an Introduction to Statistical Mechanics*, https://doi.org/10.1007/978-3-030-49198-7_34

Recall that:

$$\Delta G^{\circ}(T,P) = \sum_{j=1}^{n} \upsilon_j G_j^{\circ}(T,P) \qquad (34.3)$$

and, therefore, that:

$$\left(\frac{\partial \Delta G^{\circ}(T,P)}{\partial P}\right)_T = \sum_{j=1}^{n} \upsilon_j \left(\frac{\partial G_j^{\circ}(T,P)}{\partial P}\right)_T \qquad (34.4)$$

In Eq. (34.4), $\left(\partial G_j^{\circ}(T,P)/\partial P\right)_T = V_j^{\circ}(T,P)$, the molar volume of component j in the standard state (o) at T and P. Using Eq. (34.4) in Eq. (34.2) yields:

$$\left(\frac{\partial \ln K}{\partial P}\right)_T = -\frac{1}{RT} \sum_{j=1}^{n} \upsilon_j V_j^{\circ}(T,P) \qquad (34.5)$$

where the sum over j is equal to $\Delta V^{\circ}(T,P)$, the standard molar volume of reaction. We can therefore express Eq. (34.5) as follows:

$$\left(\frac{\partial \ln K}{\partial P}\right)_T = -\frac{\Delta V^{\circ}(T,P)}{RT} \qquad (34.6)$$

Equation (34.6) shows that:

(i) If $\Delta V^{\circ} < 0$, then, if P increases, K also increases, and the products are favored.
(ii) If $\Delta V^{\circ} > 0$, then, if P increases, K decreases, and the reactants are favored.

Figure 34.1 illustrates behaviors (i) and (ii) above.
Recall that in Fig. 34.1, T is constant.

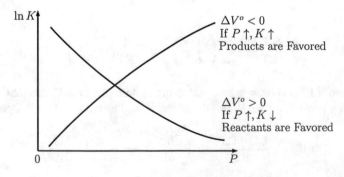

Fig. 34.1

34.3 Sample Problem 34.1

When the standard-state pressure P° is 1 bar or P_{vpj} (T), the vapor pressure of component j at temperature T, we showed that K is independent of P. Nevertheless, there can still be a pressure dependence of the equilibrium mixture composition for the various components participating in the chemical reaction. With this in mind, consider the following gas-phase chemical reaction involving the four gasses A, B, C, and D:

$$|\upsilon_A|A(g) + |\upsilon_B|B(g) \leftrightarrows |\upsilon_C|C(g) + |\upsilon_D|D(g) \qquad (34.7)$$

for which the mixture of gases is ideal. Calculate the variation of the equilibrium constant K_y with pressure P, including plotting your result.

34.3.1 Solution

First, we can use the fugacity-coefficient formulation discussed earlier to obtain:

$$\widehat{f}_j = \widehat{\phi}_j y_j P, \text{ where } \widehat{\phi}_j = \widehat{\phi}_j^{ID} = 1 \Rightarrow \widehat{f}_j = \widehat{f}_j^{ID} = y_j P, \text{ for } j = A, B, C, \text{ and } D \quad (34.8)$$

For this gas-phase chemical reaction:

$$T^\circ = T, P^\circ = 1\,\text{bar} \Rightarrow f_j^\circ = 1\,\text{bar}, \quad \text{for } j = A, B, C, \text{ and } D \qquad (34.9)$$

By combining Eqs. (34.8) and (34.9), we obtain:

$$\left(\frac{\widehat{f}_j}{f_j^\circ}\right) = y_j\left(\frac{P}{1\,\text{bar}}\right), \quad \text{for } j = A, B, C, \text{ and } D \qquad (34.10)$$

The equilibrium constant for the gas-phase chemical reaction considered here is given by:

$$K = \frac{\left(\widehat{f}_C/f_C^\circ\right)^{|\upsilon_C|}\left(\widehat{f}_D/f_D^\circ\right)^{|\upsilon_D|}}{\left(\widehat{f}_A/f_A^\circ\right)^{|\upsilon_A|}\left(\widehat{f}_B/f_B^\circ\right)^{|\upsilon_B|}} \qquad (34.11)$$

Using Eq. (34.10) for A, B, C, and D in Eq. (34.11) yields:

$$K = \frac{[y_C(P/1\,bar)]^{|v_C|}[y_D(P/1\,bar)]^{|v_D|}}{[y_A(P/1\,bar)]^{|v_A|}[y_B(P/1\,bar)]^{|v_B|}} \tag{34.12}$$

Equation (34.12) can also be expressed as follows:

$$K = K_y K_P \tag{34.13}$$

where, because the mixture of gases A, B, C, and D is ideal, $K_\phi = 1$, and does not appear on the right-hand side of Eq. (34.13). In addition,

$$K = \exp\frac{[-\Delta G^\circ(T, 1\,bar)]}{RT} \tag{34.14}$$

$$K_y = \frac{[y_C^{|v_C|} y_D^{|v_D|}]}{[y_A^{|v_A|} y_B^{|v_B|}]} \tag{34.15}$$

and

$$K_P = (P/1\,bar)^{\Delta v}, \text{ where } \Delta v = |v_C| + |v_D| - |v_A| - |v_B| \tag{34.16}$$

Because $P^\circ = 1$ bar, we know that ΔG°, and therefore, K, are independent of P (see Eq. (34.14)). Accordingly:

$$\left(\frac{\partial K}{\partial P}\right)_T = 0 \tag{34.17}$$

However, for Eq. (34.17) to be valid, Eq. (34.13) indicates that K_y must depend on P to counterbalance the dependence of K_P on P (see Eq. (34.16)).

Differentiating Eq. (34.13) with respect to P, at constant T, yields:

$$\left(\frac{\partial K}{\partial P}\right)_T = \left(\frac{\partial K_y}{\partial P}\right)_T K_P + \left(\frac{\partial K_P}{\partial P}\right)_T K_y \tag{34.18}$$

Differentiating Eq. (34.16) with respect to P yields:

$$\left(\frac{\partial K_P}{\partial P}\right)_T = \Delta v \left(\frac{P}{1\,bar}\right)^{\Delta v - 1} \tag{34.19}$$

where holding T constant is redundant.

Combining Eqs. (34.17), (34.18), and (34.19) yields:

$$0 = \left(\frac{\partial K_y}{\partial P}\right)_T \left(\frac{P}{1\,\text{bar}}\right)^{\Delta\upsilon} + K_y\left[\Delta\upsilon\left(\frac{P}{1\,\text{bar}}\right)^{\Delta\upsilon-1}\right] \qquad (34.20)$$

or

$$0 = \left(\frac{P}{1\,\text{bar}}\right)^{\Delta\upsilon-1}\left\{\left(\frac{\partial K_y}{\partial P}\right)_T \left(\frac{P}{1\,\text{bar}}\right) + \Delta\upsilon K_y\right\} \qquad (34.21)$$

Equation (34.21) shows that because P is not zero, the two terms inside the curly brackets must add up to zero, that is:

$$\left(\frac{\partial K_y}{\partial P}\right)_T \left(\frac{P}{1\,\text{bar}}\right) + \Delta\upsilon K_y = 0 \qquad (34.22)$$

Rearranging Eq. (34.22) yields:

$$\left(\frac{\partial K_y}{K_y}\right)_T + \left(\frac{\partial P}{P}\right)_T \Delta\upsilon = 0 \qquad (34.23)$$

or

$$(\partial \ln K_y)_T + (\partial \ln P)_T \Delta\upsilon = 0 \qquad (34.24)$$

or

$$\left(\frac{\partial \ln K_y}{\partial \ln P}\right)_T = -\Delta\upsilon \qquad (34.25)$$

where $\Delta\upsilon$ is a constant number, and therefore, the slope of the $\ln K_y$ versus $\ln P$ curve is constant, giving rise to a straight line.

Figure 34.2 below helps visualize the predicted behavior.

Figure 34.2 shows that when P increases, the equilibrium constant K_y increases if $\Delta\upsilon < 0$, thus favoring the products. On the other hand, if $\Delta\upsilon > 0$, when P increases, the equilibrium constant K_y decreases, thus favoring the reactants. Note that if $\Delta\upsilon = 0$, K_y is independent of P.

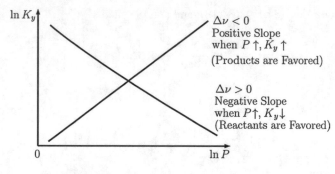

Fig. 34.2

34.4   Sample Problem 34.2

Consider a closed reactor which is charged initially with 1 mole of pure $I_2(g)$, and which is maintained at 800 °C and a very low pressure, P. It is known that the following dissociation reaction occurs:

$$I_2(g) \rightleftarrows 2I(g) \tag{34.26}$$

where

- $N_{I_2}^o \equiv N_1^o = 1$ mole
- $N_I^o \equiv N_2^o = 0$ mole
- $T = 800°C = 1073\,K$

- $P =$ is known, and is very low

- $\left\{ \begin{array}{l} \Delta H^o = 156.6\,kJ\,mole \\ \Delta S^o = 108.4\,J/mole\,K \end{array} \right\}$
- We can therefore calculate ΔG^o from : $(\Delta H^o - T\Delta S^o)$.

Compute the mole fractions of I_2 and I at equilibrium, as well as the equilibrium extent of reaction, ξ, and the mole numbers, N_{I_2} and N_I, at equilibrium.

34.4.1 Solution

Because the dissociation process involves a gas-phase chemical reaction, we choose the following standard state (o):

* $T^o = T$

* $P^o = 1$ bar
* Pure gases [$j = 1$ (I_2) and $j = 2$ (I)] in the ideal gas state, for which, $f_j^o = 1$ bar (for $j = 1$ and 2)

Next, we invoke the Gibbs Phase Rule given by:

$$L = n + 2 - \pi - r \qquad (34.27)$$

where

$$n = 2, \pi = 1, r = 1 \Rightarrow L = 2 + 2 - 1 - 1 = 2 \qquad (34.28)$$

The Gibbs Phase Rule indicates that if we fix two independent intensive variables ($L = 2$), for example, T and P (both known in Sample Problem 34.2), we should be able to calculate any other intensive variables, such as, y_1 and y_2. In other words, we should be able to calculate:

$$y_1 = y_1(T, P) \text{ and } y_2 = y_2(T, P) \qquad (34.29)$$

Regarding the extent of reaction, ξ, it is an extensive property as we defined it. Overall, for this simple dissociation reaction, according to Postulate 1, we need to specify a total of ($n + 2$) independent variables. If we specify L of these to be intensive, then, the remaining ($n + 2 - L$) variables should be extensive. In this case, $n = 2$, $L = 2$, and therefore, we need to specify $2 + 2 - 2 = 2$ extensive variables, to fully characterize both the intensive and the extensive equilibrium thermodynamic state of the system. In particular, to calculate the extensive properties ξ, N_1, and N_2, in addition to T and P, two intensive properties, we also specify the two extensive properties, N_1^o and N_2^o. Accordingly, the set of ($n + 2$) = 4 independent variables includes $\{T, P, N_1^o, N_2^o\}$, which will allow us to calculate any other intensive as well as extensive property of interest. In Sample Problem 34.2, we are asked to calculate:

* $y_1 = y_1(T, P)$
* $y_2 = y_2(T, P)$
* $\xi = \xi(T, P, N_1^o, N_2^o)$
* $N_1 = N_1(T, P, N_1^o, N_2^o)$
* $N_2 = N_2(T, P, N_1^o, N_2^o)$

In order to calculate $y_1(T, P)$ and $y_2(T, P)$, we use the expression which relates K to K_ϕ, K_y, and K_P. Specifically,

$$K(T) = \exp\left[\frac{-\Delta G^o(T)}{RT}\right] = K_\phi K_y K_P \qquad (34.30)$$

In Eq. (34.30), $\Delta G°(T) = \Delta H°(T) - T\Delta S°(T)$, and we have assumed that $P° = 1$ bar, and therefore, K, $\Delta H°$, $\Delta S°$, and $\Delta G°$ do not depend on P. Because we know $\Delta H°(T)$ and $\Delta S°(T)$, we also know $\Delta G°(T)$, and therefore, we also know K(T).

For the given dissociation reaction:

$$I_2(g) \leftrightarrows 2I(g) \tag{34.31}$$

we have:

$$\upsilon_1 = -1, \upsilon_2 = +2, \ \Delta\upsilon = \upsilon_1 + \upsilon_2 = -1 + 2 = 1 \tag{34.32}$$

Using $\Delta\upsilon = 1$ (see Eq. (34.32)) in the defining equation for K_P (see Eq. (34.16)) yields:

$$K_P = P^{\Delta\upsilon} = P \tag{34.33}$$

where P is in units of bars.

Because we are told that the pressure, P, is very low, we can model the binary gas mixture as being ideal. Therefore, as we assumed earlier:

$$\left. \begin{array}{l} \hat{\phi}_1 = \hat{\phi}_1^{ID} = 1 \\ \hat{\phi}_2 = \hat{\phi}_2^{ID} = 1 \end{array} \right\} \Rightarrow K_\phi = K_\phi^{ID} = 1 \tag{34.34}$$

Further, for the dissociation reaction in Eq. (34.31), we have:

$$K_y = \frac{y_2^2}{y_1} \tag{34.35}$$

Using Eqs. (34.34), (34.35), and (34.33) in Eq. (34.30) yields:

$$K(T) = (1)\left(\frac{y_2^2}{y_1}\right)(P) \tag{34.36}$$

Rearranging Eq. (34.36) yields:

$$\left[K(T)P^{-1}\right] = \frac{y_2^2}{y_1} = \frac{y_2^2}{(1 - y_2)} \tag{34.37}$$

Equation (34.37) is a quadratic equation for y_2. Because we know K as a function of T, and we know P, the solution to Eq. (34.37) will yield $y_2(T, P)$ and $y_1(T, P) = 1 - y_2(T, P)$. This is, of course, consistent with the Gibbs Phase Rule, which indicates that given $L = 2$ intensive variables (T and P in the present case), any other intensive variables like y_1 and y_2 should also be uniquely determined.

Rearranging Eq. (34.37), we obtain:

$$y_2^2 + \left(KP^{-1}\right)y_2 - \left(KP^{-1}\right) = 0 \qquad (34.38)$$

The solution of the quadratic equation (34.38) is given by:

$$y_2 = \frac{-\left(KP^{-1}\right) \pm \left[\left(KP^{-1}\right)^2 + 4\left(KP^{-1}\right)\right]^{1/2}}{2} \qquad (34.39)$$

where we need to choose the positive root, because choosing the negative root yields $y_2 < 0$, which is unphysical, because a mole fraction cannot be negative.

Choosing the positive root, and factoring out $\left(KP^{-1}\right)$ in Eq. (34.39), yields:

$$y_2(T, P) = \left(\frac{K(T)}{2P}\right)\left\{\left(1 + \frac{4P}{K(T)}\right)^{1/2} - 1\right\}, y_1(T, P) = 1 - y_2(T, P) \qquad (34.40)$$

Equation (34.40) shows that, consistent with the Gibbs Phase Rule, the two intensive variables, y_1 and y_2, depend on the two independent intensive variables, T and P, chosen here to describe the intensive equilibrium thermodynamic state of the binary gas mixture considered.

Next, to predict the equilibrium extent of reaction, ξ, we can proceed in two ways:

1. Express y_2 (or y_1) in terms of N_1^o (known), N_2^o (known), and ξ, and then use the expression for y_2 (or y_1) given in Eq. (34.40).
2. Express, from the outset, y_1 or y_2 in terms of N_1^o, N_2^o, and ξ, and then, solve directly for ξ using Eq. (34.37).

Of course, approaches (1) and (2) above will lead to identical predictions for ξ. However, as we will see, in some cases, approach (2) may be more convenient. To implement approaches (1) or (2) above requires first relating y_1 and y_2 to N_1^o, N_2^o, and ξ. Recall that for the dissociation reaction considered here, $\upsilon_1 = -1$ and $\upsilon_2 = +2$. It then follows that:

$$N_1 = N_1^o + \upsilon_1\xi = N_1^o - \xi \qquad (34.41)$$

$$N_2 = N_2^o + \upsilon_2\xi = N_2^o + 2\xi \qquad (34.42)$$

$$N_1 + N_2 = \left(N_1^o + N_2^o\right) + \xi \tag{34.43}$$

Using Eqs. (34.41), (34.42), and (34.43) in the definitions of y_1 and y_2 yields:

$$y_1 = \frac{N_1}{N_1 + N_2} = \frac{\left(N_1^o - \xi\right)}{\left(N_1^o + N_2^o\right) + \xi} \tag{34.44}$$

$$y_2 = \frac{N_2}{N_1 + N_2} = \frac{\left(N_2^o + 2\xi\right)}{\left(N_1^o + N_2^o\right) + \xi} \tag{34.45}$$

Having derived expressions for y_1 and y_2 in terms of N_1^o, N_2^o, and ξ, let us first pursue approach (1) above. Specifically, using Eq. (34.45) for y_2 in Eq. (34.40) yields:

$$y_2 = \frac{\left(N_2^o + 2\xi\right)}{\left(N_1^o + N_2^o\right) + \xi} = \left(\frac{K(T)}{2P}\right)\left\{\left(1 + \frac{4P}{K(T)}\right)^{1/2} - 1\right\} \tag{34.46}$$

where T, P, and K(T) are all known.

Solving for ξ in Eq. (34.46) yields $\xi = \xi\left(T, P, N_1^o, N_2^o\right)$, where T, P, N_1^o, N_2^o are $(n + 2) = (2 + 2) = 4$ independent variables, as required by Postulate I.

We know that $N_1^o = 1$ mole and $N_2^o = 0$ mole. Using these values in Eq. (34.46) yields:

$$y_2 = \frac{(0 + 2\xi)}{1 + \xi} \Rightarrow y_2 = \frac{2\xi}{(1 + \xi)} \Rightarrow \xi = \frac{y_2}{(2 - y_2)} \tag{34.47}$$

Next, we use y_2 in Eq. (34.40) in Eq. (34.47), which yields:

$$\xi = \frac{(K/2P)\left\{(1 + 4P/K)^{1/2} - 1\right\}}{2 - (K/2P)\left\{(1 + 4P/K)^{1/2} - 1\right\}} \tag{34.48}$$

The denominator in Eq. (34.48) can be simplified as follows:

$$2 - (K/2P)\left\{(1 + 4P/K)^{1/2} - 1\right\} \Rightarrow$$

$$2\left\{1 - (K/4P)(1 + 4P/K)^{1/2} + K/4P\right\} \Rightarrow$$

$$2\left\{(1 + K/4P) - (K/4P)(1 + 4P/K)^{1/2}\right\} \Rightarrow$$

$$2(K/4P)\left\{(1 + 4P/K) - (1 + 4P/K)^{1/2}\right\} \Rightarrow$$

$$2(K/4P)(1 + 4P/K)^{1/2} \left\{ (1 + 4P/K)^{1/2} - 1 \right\} \qquad (34.49)$$

Replacing the denominator in Eq. (34.48) by Eq. (34.49), and cancelling the appropriate terms, yields:

$$\xi = (1 + 4P/K)^{1/2} \Rightarrow \xi = \left(\frac{K(T)}{K(T) + 4P} \right)^{1/2} \qquad (34.50)$$

Next, let us make use of approach (2) above to calculate ξ. We know that:

$$y_1 = \frac{(1 - \xi)}{(1 + \xi)} \text{ and } y_2 = \frac{2\xi}{(1 + \xi)} \qquad (34.51)$$

Using the two mole fractions in Eq. (34.51) directly in Eq. (34.37), we obtain:

$$\left[K(T) P^{-1} \right] = \frac{y_2}{y_1} = \frac{\left(\frac{2\xi}{(1+\xi)} \right)^2}{\frac{(1-\xi)}{1+\xi}} \Rightarrow K(T)P^{-1} = \frac{4\xi^2}{(1 + \xi)(1 - \xi)}$$

$$= \frac{4\xi^2}{(1 - \xi^2)} \qquad (34.52)$$

Solving Eq. (34.52) for ξ yields:

$$4\xi^2 = \left[K(T) P^{-1} \right] - \left[K(T)P^{-1} \right]\xi^2 \Rightarrow \xi^2 = \frac{\left[K(T)P^{-1} \right]}{4 + \left[K(T)P^{-1} \right]} \Rightarrow$$

$$\xi^2 = \frac{K(T)}{K(T) + 4P} \Rightarrow \xi = \left(\frac{K(T)}{K(T) + 4P} \right)^{1/2} \qquad (34.53)$$

As expected, approaches (1) and (2) above yield the same result for ξ (see Eqs. (34.50) and (34.53)).

Having calculated ξ, we can readily calculate N_1 and N_2 using the known values of N_1^o and N_2^o. Specifically,

$$N_1 = N_1^o - \xi = 1 - \xi; N_2 - N_2^o + 2\xi = 2\xi \qquad (34.54)$$

At the given temperature of T = 800 °C = 1073 K, we can calculate, as well as plot, how the extent of reaction, ξ, varies with P from P = 1 bar. We already showed that $\xi = [K(T)/(K(T) + 4P)]^{1/2}$, where T = 1073 K. Accordingly, at P = 1 bar, Eq. (34.53) shows that:

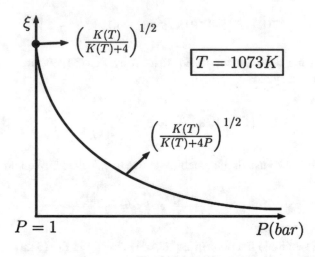

Fig. 34.3

$$\xi(1\,\text{bar}) = \left(\frac{K(T)}{K(T) + 4}\right)^{1/2} \tag{34.55}$$

Further, as $P \gg 1$, Eq. (34.53) shows that $\xi \ll 1$, and approaches zero when $P \to \infty$. It then follows that when $P \to \infty$, $\xi \to 0$, $y_2 = 2\xi/(1 + \xi) \to 0$, and $y_1 = (1 + \xi)/(1 + \xi) \to 1$. In other words, when $P \to \infty$, the dissociation reaction shifts 100% backwards in the direction of component 1 (I_2). Recall that K does not depend on pressure. Figure 34.3 illustrates this behavior:

It is interesting that the behavior of ξ with P, at constant T, is expected based on the pressure dependence of K_y. Recall that:

$$\left(\frac{\partial \ln K_y}{\partial \ln P}\right)_T = -\Delta v = -1 \tag{34.56}$$

As a result, as P increases, K_y decreases, so that the reactant (I_2 in this dissociation reaction) is preferred. In other words, as $P \to \infty$, $\xi \to 0$, $y_1 \to 1$, and $y_2 \to 0$, and there is no dissociation of 1 (I_2) into 2 (2I).

Finally, we can calculate a few properties based on the data given in Sample Problem 34.2. Specifically,

$$\left\{ \begin{array}{l} \text{At T} = 800^{\circ}\text{C} = 1073\text{K,} \\ \Delta\text{H}^{\circ} = 156.6\,\text{KJ/mole,} \\ \Delta\text{S}^{\circ} \ = 0.1084\,\text{KJ/moleK,} \end{array} \right\} \begin{array}{l} \therefore\Delta\text{G}^{\circ} = \Delta\text{H}^{\circ} - \text{T}\Delta\text{S}^{\circ} = 40.29\,\text{KJ/mole} \\ (\Delta\text{G}^{\circ}/\text{RT}) = 4.51 \\ \text{K(T} = 1073\text{K}, \text{P}^{\circ} = 1\,\text{bar}) = \text{e}^{-4.51} \end{array}$$

$$(34.57)$$

Lecture 35

The Gibbs Phase Rule for Chemically-Reacting Systems and Sample Problem

35.1 Introduction

In this lecture, we will solve Sample Problem 35.1, an illuminating problem that deals with the partial decomposition of calcium carbonate solid into calcium oxide solid and carbon dioxide gas in a closed chemical reactor. We are asked to calculate the equilibrium pressure, the extent of reaction, and the equilibrium mole numbers of the three species present in the reactor. For this purpose, we will follow the chemical reaction equilibria approach, including formulating the Gibbs Phase Rule for chemically reacting systems.

35.2 Sample Problem 35.1

A closed chemical reactor (see Fig. 35.1), which is initially evacuated, is loaded with $CaCO_3(s)$. It is known that $CaCO_3(s)$ partially decomposes into $CaO(s)$ and CO_2 (g) according to the following chemical reaction:

$$CaCO_3(s) \leftrightarrows CaO(s) + CO_2(g) \qquad (35.1)$$

You are asked to calculate:

1. The equilibrium gas pressure, P.
2. The extent of reaction, ξ.
3. The equilibrium mole numbers of the three species, $N^s{}_{CaCO_3}$, $N^s{}_{CaO}$, and $N^g{}_{CO_2}$.

We are given:

(i) $\Delta G^\circ(T)$ – Standard Molar Gibbs Free Energy of Reaction

(ii) $V^s_{CaCO_3}$ – Molar volume of $CaCO_3$ solid

© Springer Nature Switzerland AG 2020
D. Blankschtein, *Lectures in Classical Thermodynamics with an Introduction to Statistical Mechanics*, https://doi.org/10.1007/978-3-030-49198-7_35

(iii) V^s_{CaO} – Molar volume of CaO solid

(iv) An EOS for the gas

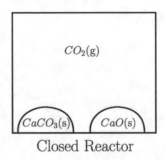

Closed Reactor

Fig. 35.1

For the decomposition reaction in Eq. (35.1), it follows that:

$$
\left\{
\begin{array}{l}
\upsilon_{CaCO_3} = -1 \\
\upsilon_{CaO} = +1 \\
\upsilon_{CO_2} = +1
\end{array}
\right\}
\Delta\upsilon = -1 + 1 + 1 = +1 \tag{35.2}
$$

35.2.1   Solution Strategy

To solve this Sample Problem, we will use the Chemical Reaction Equilibria Approach, where the overall system is simple, Postulate I applies, and the conventional Gibbs Phase Rule can be used. In this case, we treat the system as consisting of $n = 3$ components, $\pi = 3$ phases (two solid and one gas), where a single decomposition reaction takes place ($r = 1$). The conventional Gibbs Phase Rule indicates that:

$$
n = 3, \pi = 3, r = 1 \Rightarrow L = n + 2 - \pi - r = 3 + 2 - 3 - 1 \Rightarrow L = 1 \tag{35.3}
$$

Consistent with Eq. (35.3), it is convenient to specify T as the single intensive variable. It should then be possible to compute the equilibrium pressure, P, which is another intensive variable. Treating this system as simple, Postulate I applies and indicates that we need to specify $(n + 2) = (3 + 2) = 5$ independent variables to fully characterize both the intensive and the extensive equilibrium thermodynamic state of the system. Because $L = 1$, we need to specify $(n + 2) - L = (3 + 2) - 1 = 5 - 1 = 4$ independent extensive variables. A convenient choice includes the initial moles of $CaCO_3$ solid, the initial moles of CaO solid, the initial moles of CO_2 gas, and the total volume of the reactor, that is,

$$N_{CaCO_3}^{so}, N_{CaO}^{so}, N_{CO_2}^{go}, \text{ and } \underline{V} \qquad (35.4)$$

The set of five independent variables

$$\left\{ T, N_{CaCO_3}^{so}, N_{CaO}^{so}, N_{CO_2}^{go}, \underline{V} \right\} \qquad (35.5)$$

will then allow us to compute what the Sample Problem requests: P (Intensive), ξ (Extensive), $N_{CaCO_3}^{s}$ (Extensive), N_{CaO}^{s} (Extensive), and $N_{CO_2}^{g}$ (Extensive).

35.2.2 Selection of Standard States

(i) For $CO_2(g)$

- Pure $CO_2(g)$ in an ideal gas state
- $T^{\circ} = T$
- $P^{\circ} = 1$ bar, $f_{CO_2}^{go} = 1$ bar

(ii) For $CaCO_3(s)$

- Pure $CaCO_3(s)$
- $T^{\circ} = T$
- $P^{\circ} = 1$ bar, $f_{CaCO_3}^{so} = f_{CaCO_3}^{so}(T, 1 \text{ bar})$

(iii) For $CaO(s)$

- Pure $CaO(s)$
- $T^{\circ} = T$
- $P^{\circ} = 1$ bar, $f_{CaO}^{so} = f_{CaO}^{so}(T, 1 \text{ bar})$

35.2.3   Remarks

We know that $\Delta G^\circ = \Delta G^\circ(T^\circ, P^\circ)$. If we choose $P^\circ = 1$ bar, then, $\Delta G^\circ(T^\circ, 1\ \text{bar})$ is usually available in tabular form for $T^\circ = 298$ K. If $P^\circ = P$, then, we need to calculate $\Delta G^\circ(T^\circ, P)$, where if $\Delta G^\circ(T^\circ, 1\ \text{bar})$ is known, we need to implement a pressure correction to relate $\Delta G^\circ(T^\circ, 1\ \text{bar})$ to $\Delta G^\circ(T^\circ, P)$, as discussed in Part I. Later in this lecture, we will discuss in more detail what happens if we choose $P^\circ = P$ for the three components under consideration: $CaCO_3(s)$, $CaO(s)$, and $CO_2(g)$.

35.2.4   Evaluation of Fugacities

In this Sample Problem, three pure $(n = 1)$ phases are involved: $CaCO_3(s)$, $CaO(s)$, and $CO_2(g)$. The fugacity of each phase will therefore depend on $(n + 1) = (1 + 1) = 2$ independent, intensive variables, consistent with the Corollary to Postulate I. These two variables are T and P. We can therefore write:

$$f^s_{CaCO_3} = f^s_{CaCO_3}(T, P), f^s_{CaO} = f^s_{CaO}(T, P), f^g_{CO_2} = f^g_{CO_2}(T, P) \quad (35.6)$$

35.2.5   Calculation of the Equilibrium Constant

Having determined the three fugacities at T and P, as well as in the reference state at T and 1 bar, we can write down the expression for the Equilibrium Constant, K, as follows:

$$K = K(T) = \prod_{j=1,2,3} \left(f_j/f^\circ\right)^{\upsilon_j} \quad (35.7)$$

where

$$1 = CaO(s), 2 = CO_2(g), 3 = CaCO_3(s) \quad (35.8)$$

$$\upsilon_1 = 1, \upsilon_2 = 1, \upsilon_3 = -1 \quad (35.9)$$

Using the expressions for f_j and f°_j (j = 1, 2, and 3) in the expression for K in Eq. (35.7) yields:

$$K(T) = \frac{\left(\frac{f^s_{CaO}(T,P)}{f^{so}_{CaO}(T,1\,bar)}\right)^1 \left(\frac{f^g_{CO_2}(T,P)}{1\,bar}\right)^1}{\left(\frac{f^s_{CaCO_3}(T,P)}{f^{so}_{CaCO_3}(T,1\,bar)}\right)^1} \tag{35.10}$$

where if all the components involved in the chemical reaction are pure, then, Eq. (35.10) shows that ξ does not appear in the expression for K, because it is independent of composition. In that case, the K expression provides a unique relation between T and P.

The fugacity ratio of each pure solid can be evaluated from the Poynting correction, where we need to know the molar volume of each solid as a function of T and P. Specifically,

$$\frac{f^s_{CaO}(T,P)}{f^{so}_{CaO}(T,1\,bar)} = \exp\left\{\int_{1\,bar}^{P} \left(V^s_{CaO}(T,P)/RT\right)dP\right\} \tag{35.11}$$

and

$$\frac{f^s_{CaCO_3}(T,P)}{f^{so}_{CaCO_3}(T,1\,bar)} = \exp\left\{\int_{1\,bar}^{P} \left(V^s_{CaCO_3}(T,P)/RT\right)dP\right\} \tag{35.12}$$

The fugacity of $CO_2(g)$ is evaluated using the fugacity coefficient approach that we discussed in Part II. Specifically,

$$f^g_{CO_2}(T,P) = \phi^g_{CO_2}(T,P)P \tag{35.13}$$

where we can compute the fugacity coefficient, $\phi^g_{CO_2}(T,P)$, given a suitable volumetric EOS for $CO_2(g)$.

Using Eqs. (35.11), (35.12), and (35.13) in Eq. (35.10), we obtain:

$$K(T) = \left\{ \exp \int_{1\,bar}^{P} \left(\frac{V^s_{CaO}(T,P) - V^s_{CaCO_3}(T,P)}{RT} \right) dP \right\} f^g_{CO_2}(T,P) \left(\frac{P}{1\,bar} \right)$$

$$(35.14)$$

Given T, in order to calculate P using Eq. (35.14), we need access to:

- $\Delta G^\circ(T, 1\,bar) \Rightarrow K(T)$
- $V^s_{CaO}(T,P)$ and $V^s_{CaCO_3}(T,P)$
- An EOS for $CO_2(g) \Rightarrow \phi^g_{CO_2}(T,P)$

As an illustration of how to use Eq. (35.14) to calculate P, we make the following approximations:

(i) If P is low, the gas (CO_2) behaves ideally, and hence, $\phi^{g,ID}_{CO_2}(T,P) = 1$

(ii) If the Poynting corrections can be neglected, then, the exponential term is unity

Using (i) and (ii) above, Eq. (35.14) reduces to:

$$P = K(T)\,bar \qquad (35.15)$$

To calculate the Extent of Reaction, ξ (an extensive property), we proceed as follows. Because the gas pressure is low, we can assume that $CO_2(g)$ behaves ideally. Otherwise, we require a more robust EOS for $CO_2(g)$. It then follows that:

$$P\underline{V}^g_{CO_2} = N^g_{CO_2} RT \qquad (35.16)$$

We also know that the volumes of the three phases add up to $\underline{V}$, that is:

$$\underline{V} = \underline{V}^g_{CO_2} + \underline{V}^s_{CaCO_3} + \underline{V}^s_{CaO} \qquad (35.17)$$

Further, we know that:

$$N^g_{CO_2} = N^{go}_{CO_2} + \xi \qquad (35.18)$$

Using Eq. (35.18) in Eq. (35.16) and then solving for $\underline{V}\,^g_{CO_2}$ yields:

$$\underline{V}\,^g_{CO_2} = \left(\frac{RT}{P}\right)\left(N\,^{go}_{CO_2} + \xi\right) \tag{35.19}$$

In addition,

$$\underline{V}\,^s_{CaO} = V\,^s_{CaO}N\,^s_{CaO} = V\,^s_{CaO}\left(N\,^s_{CaO} + \xi\right) \tag{35.20}$$

$$\underline{V}\,^s_{CaCO_3} = V\,^s_{CaCO_3}N\,^s_{CaCO_3} = V\,^s_{CaCO_3}\left(N\,^s_{CaCO_3} - \xi\right) \tag{35.21}$$

Using Eqs. (35.19), (35.20), and (35.21) in Eq. (35.17) yields:

$$\underline{V} = \left(\frac{RT}{P}\right)\left[N\,^{go}_{CO_2} + \xi\right] + V\,^s_{CaO}\left[N\,^{so}_{CaO} + \xi\right] + V\,^s_{CaCO_3}\left[N\,^{so}_{CaCO_3} - \xi\right] \tag{35.22}$$

Solving Eq. (35.22) for ξ yields:

$$\xi = \underline{V} - \left\{\frac{(RT/P)N\,^{go}_{CO_2} + \underline{V}\,^s_{CaO}N\,^{so}_{CaO} + V\,^s_{CaCO_3}N\,^{so}_{CaCO_3}}{(RT/P) + \left[V\,^s_{CaO} - V\,^{so}_{CaO_3}\right]}\right\} \tag{35.23}$$

where

$$T,\ \underline{V},\ N\,^{go}_{CO_2}, N\,^{so}_{CaO},\ N\,^{so}_{CaCO_3}, V\,^s_{CaO}, \text{and } V\,^s_{CaCO_3} \tag{35.24}$$

are all known. Knowing ξ, we can readily calculate:

$$N\,^s_{CaO} = N\,^{so}_{CaO} + \xi,\ N\,^s_{CaCO_3} = N\,^{so}_{CaCO_3} - \xi, \text{and } N\,^g_{CO_2} = N\,^{go}_{CO_2} + \xi \tag{35.25}$$

35.2.6   Comment on the Standard-State Pressure

As we have seen, the standard-state pressure, P^o, is intimately connected with the pressure dependence of the equilibrium constant, $K(T^o, P^o)$, that is,

$$K(T^o, P^o) = \exp\left[\frac{-\Delta G^o(T^o, P^o)}{RT}\right] \tag{35.26}$$

As stressed earlier, Eq. (35.26) clearly shows that $K(T^o, P^o)$ is indeed a property of the standard state (o). As we have seen, we typically choose $T^o = T$ (the system temperature), and $P^o = 1$ bar. As a result, $\Delta G^o(T, 1\,\text{bar})$, and $K = K(T)$ as shown in Eq. (35.26) above. However, sometimes P^o is chosen to be $P_{vpj}(T)$ for component j. In that case, we need to compute $G_j^o(T, P_{vpj}(T))$ in order to calculate:

$$\Delta G^o(T) = \sum_{j=1}^{n} \upsilon_j G_j^o\left(T, P_{vpj}\right) \tag{35.27}$$

If we know $G_j^o(T, 1\,\text{bar})$, we can use the relation derived in Part I relating the variation of $G_j^o(T, P)$ with P, at constant T, to the molar volume of component j, that is,

$$\left(\frac{\partial G_j^o(T, P)}{\partial P}\right)_T = V_j^o(T, P) \tag{35.28}$$

Integration of Eq. (35.28) with respect to P, at constant T, then yields:

$$G_j^o\left(T, P_{vpj}(T)\right) = G_j^o(T, 1\,\text{bar}) + \int_{1\,\text{bar}}^{P_{vpj}(T)} V_j^o(T, P')dP' \tag{35.29}$$

If $P^o = P$ for some components j (say, solids), then, Eq. (35.29) is modified as follows:

$$G_j^o(T, P) = G_j^o(T, 1\,\text{bar}) + \int_{1\,\text{bar}}^{P} V_j^o(T, P')dP' \tag{35.30}$$

Lecture 36

Effect of Chemical Reaction Equilibria on Changes in Thermodynamic Properties and Sample Problem

36.1 Introduction

The material presented in this lecture is adapted from Chapter 16 in T&M.

- An important goal in Parts I and II of the book is to learn how to calculate changes in thermodynamic properties of pure component (n = 1) and multi-component (n > 1) systems as these evolve from some initial state (i) to some final state (f). That is, to learn how to calculate: $\Delta \underline{U}_{i \to f}, \Delta \underline{H}_{i \to f}, \Delta \underline{S}_{i \to f}, \Delta \underline{G}_{i \to f}$, etc. So far, we have accomplished this goal in the absence of chemical reactions.

- Here, we will consider cases where some of the components in the system undergo chemical reactions. As a result, the system equilibrium composition is no longer constant and has to be determined as a function of the system initial composition. In addition, the thermodynamic properties of the system have to be calculated as they respond to the composition changes. To illustrate how to carry out such calculations, in this lecture, we will solve an interesting Sample Problem which integrates concepts and equations presented in Parts I and II of the book.

36.2 Sample Problem 36.1: Production of Sulfuric Acid by the Contact Process

In the manufacture of sulfuric acid by the contact process, elemental sulfur is burned with air (assumed to be a mixture of O_2 and N_2) to form SO_2, which is then further oxidized to form SO_3. The SO_3 then reacts with water to produce sulfuric acid (H_2SO_4).

Assume that the product gas stream from the sulfur burner contains 9 mole % of SO_2, 80 mole % of nitrogen, and 11 mole % of oxygen. The product gas stream is

© Springer Nature Switzerland AG 2020
D. Blankschtein, *Lectures in Classical Thermodynamics with an Introduction to Statistical Mechanics*, https://doi.org/10.1007/978-3-030-49198-7_36

subsequently cooled to 723 K and passed over a catalyst bed to convert SO_2 into SO_3 according to the following chemical reaction:

$$SO_2(g) + 0.5O_2(g) \rightleftarrows SO_3(g) \qquad (36.1)$$

(a) Choose 1 mole of the SO_2-O_2-N_2 product gas stream from the sulfur burner as a basis, and derive a relation between the equilibrium constant and the extent of reaction. Assume that the pressure is maintained constant at one bar, and that the gas mixture is ideal.

(b) Using the data given in Table 36.1, derive a relation between the equilibrium constant and the temperature.

Table 36.1

	ΔH_f° (298K) (J/mol)	ΔG_f° (298K) (J/mol)	C_p° (J/molK)
$SO_3(g)$	-3.954×10^5	-3.705×10^5	60.19
$SO_2(g)$	-2.970×10^5	-3.005×10^5	45.21
$O_2(g)$	0	0	29.96
$N_2(g)$	0	0	29.96

Recall that C_p° is the heat capacity at constant pressure and can be assumed to be constant.

(c) If the SO_2-O_2-N_2 mixture is fed to the SO_3 reactor at 723 K and if this reactor is adiabatic and operates at steady state, derive a relation between the outlet temperature and the extent of reaction. Indicate how you would calculate the outlet temperature, the extent of reaction, and the mole % of SO_2 left. Assume ideal gas behavior of the inlet and the outlet gas mixtures.

36.3   Solution Strategy

- When we solve problems of this type, the first question that we should ask is: Do we have sufficient information to solve the problem? Specifically, we are searching for two unknowns, T_{out} and ξ, and in order to find them, we require two equations that we can solve simultaneously. The first equation is the *equilibrium constant relation*, and the second equation is *the First Law of Thermodynamics for an Open System*. We therefore know, a priori, that we will be able to solve this problem!

In this Sample Problem, there are two basic steps, as depicted in Figure 36.1

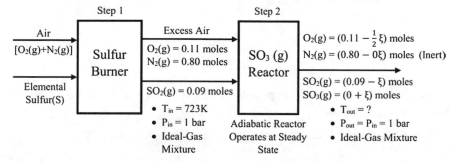

Inlet Streams

Outlet Stream

$$SO_2(g) + \frac{1}{2}O_2(g) \rightleftharpoons SO_3(g)$$

$$\nu_{SO_2} = -1, \ \nu_{O_2} = -1/2, \ \nu_{SO_3} = +1, \ \nu_{N_2} = 0 \ (\text{Inert})$$

Fig. 36.1

- Because we are dealing with a *gas-phase chemical reaction*, we choose the following standard state (o):

  - $T^o = T$
  - $P^o = 1$ bar
  - Pure ideal gas for each gas (j = O_2, N_2 (inert), SO_2, SO_3) and $f_j^o = 1$ bar for each j

It then follows that $\Delta G^o = \Delta G^o(T)$ and $K = K(T)$, where, for simplicity, we omitted the 1 bar notation in ΔG^o and K.

- For the gas-phase chemical reaction in Eq. (36.1), the *equilibrium constant* is given by:

$$K(T) = K_\phi K_y K_P \qquad (36.2)$$

- Because the gas mixture is ideal (ID), it follows that:

$$\hat{\phi}_{SO_2}^{ID} = 1, \hat{\phi}_{O_2}^{ID} = 1, \hat{\phi}_{SO_3}^{ID} = 1 \Rightarrow K_\phi = K_\phi^{ID} = 1 \qquad (36.3)$$

- Given the gas-phase chemical reaction in Eq. (36.1), it follows that:

$$K_y = \frac{\left(y_{SO_3}\right)^1}{\left(y_{SO_2}\right)^1 \left(y_{O_2}\right)^1} \tag{36.4}$$

Recall that N_2 is an inert component, that is, $\nu_{N_2} = 0$, and therefore, it does not appear in Eq. (36.4) for K_y.

- Given the gas-phase chemical reaction in Eq. (36.1), we know that:

$$y_{SO_3} = \frac{N_{SO_3}}{N} = \frac{\xi}{\left(1 - \frac{1}{2}\xi\right)} \tag{36.5}$$

$$y_{SO_2} = \frac{N_{SO_2}}{N} = \frac{(0.09 - \xi)}{\left(1 - \frac{1}{2}\xi\right)} \tag{36.6}$$

$$y_{O_2} = \frac{N_{O_2}}{N} = \frac{\left(0.11 - \frac{1}{2}\xi\right)}{\left(1 - \frac{1}{2}\xi\right)} \tag{36.7}$$

$$y_{N_2} = \frac{N_{N_2}}{N} = \frac{0.80}{\left(1 - \frac{1}{2}\xi\right)} \tag{36.8}$$

- We also know that:

$$K_P = \left(\frac{P}{1\,\text{bar}}\right)^{\Delta\nu} \text{with } \Delta\nu = -(1/2) \tag{36.9}$$

However, because $P = 1$ bar, Eq. (36.9) yields:

$$K_P = 1 \tag{36.10}$$

- Using Eqs. (36.3) and (36.10)) in Eq. (36.2), we obtain:

$$K(T) = K_y \tag{36.11}$$

- Using Eq. (36.4) for K_y, along with Eqs. (36.5), (36.6), and (36.7) for y_{SO3}, y_{SO2}, and y_{O2}, respectively, in Eq. (36.11) yields:

$$K(T) = K_y = \frac{\left(\xi/\left(1 - \frac{1}{2}\xi\right)\right)^1}{\left[(0.09 - \xi)/\left(1 - \frac{1}{2}\xi\right)\right]^1 \left[0.11 - \frac{1}{2}\xi)/\left(1 - \frac{1}{2}\xi\right)\right]^{1/2}}$$

or

$$K(T) = \frac{(\xi)(1 - 0.5\xi)^{1/2}}{(0.09 - \xi)(0.11 - 0.5\xi)^{1/2}} \qquad (36.12)$$

Equation (36.12) provides the first needed relation between the *two unknowns*, T and ξ. However, to utilize Eq. (36.12), we first need to evaluate K(T). Fortunately, we can do that using some of the data provided in the Problem Statement, including Table 36.1.

36.4   Evaluation of K(T)

We begin from the expression for K(T) given in terms of $\Delta G^\circ(T)$, where:

$$K(T) = \exp\left[-\frac{\Delta G^\circ(T)}{RT}\right] \qquad (36.13)$$

In Table 36.1, we are given data at T = 298 K. Using this data, we can calculate $\Delta G^\circ(298\,K)$, and then, using Eq. (36.13), we can calculate K(298 K). However, we need to calculate K(T) at T $\neq$ 298 K. To this end, we can utilize the *standard-state molar enthalpy of reaction*, $\Delta H^\circ(T)$, which we can calculate using the data given in Table 36.1 (see below).

- First, we calculate $\Delta G^\circ(298\,K)$ as follows:

$$\Delta G^\circ(298\,K) = \sum_{\substack{i=SO_3,\\SO_2,\\O_2,\\N_2}} \nu_i \Delta G^\circ_{fi}(298\,K) \qquad (36.14)$$

Using the data in Table 36.1 in Eq. (36.14), we obtain:

$\Delta G^\circ(298\ K) =$

$$\left\{ \underbrace{[(+1)(-3.705^*10^5)]}_{SO_3} + \underbrace{[(-1)(-3.005^*10^5)]}_{SO_2} + \underbrace{\left[\left(-\frac{1}{2}\right)(0)\right]}_{O_2} + \underbrace{[(0)(0)]}_{N_2\ (Inert)} \right\} \frac{J}{mol}$$

$$(36.15)$$

or

$$\Delta G^\circ(298 \ K) = -7 \times 10^4 J/mol \qquad (36.16)$$

Using Eq. (36.16) and T = 298 K in Eq. (36.13) yields:

$$K(298\,K) = \exp\left[-\frac{\Delta G^\circ(298\,K)}{R(298\,K)}\right]$$

or

$$K(298\,K) = \exp(28.253) \qquad (36.17)$$

To find K(T) given K(298 K), we use the following relation derived in Lecture 33:

$$\left(\frac{\partial \ln K}{\partial T}\right)_P = \frac{\Delta H^\circ}{RT^2} \qquad (36.18)$$

Multiplying both sides of Eq. (36.18) by dT′, integrating from 298 K to T, and exponentiating yields:

$$K(T) = K(298\,K) \exp\left[\int_{298\,K}^{T} \left(\frac{\Delta H^\circ(T')}{RT'^2}\right) dT'\right] \qquad (36.19)$$

Note that in Eq. (36.17), holding P constant in $(\partial \ln K/\partial T)_P$ is redundant, because $K = K(T, P^\circ)$ and does not depend on P.

Equation (36.19) shows that if we know $\Delta H^\circ(T)$, we can calculate K(T). To this end, we can first compute $\Delta H^\circ(298\ K)$ using the data given in Table 36.1 and, then, calculate $\Delta H^\circ(T)$ using the temperature derivative of $\Delta H^\circ(T)$ given by:

$$\left(\frac{\partial \Delta H^\circ}{\partial T}\right)_P = \Delta C_P^\circ \qquad (36.20)$$

where, again, holding P constant in Eq. (36.20) is redundant, because $\Delta H^\circ = \Delta H^\circ(T, P^\circ)$, and does not depend on P. Multiplying both sides of Eq. (36.20) by dT′ and then integrating from 298 K to T yields (recall that according to Table 36.1, C_P° and, therefore, ΔC_P°, do not depend on T):

$$\Delta H^\circ(T) = \Delta H^\circ(298\,K) + \int_{298\,K}^{T} \Delta C_P^\circ \, dT' \qquad (36.21)$$

In the Problem Statement, we are told that all the standard-state heat capacities at constant pressure (C_{Pi}°, for i = SO$_3$, SO$_2$, O$_2$, and N$_2$) are *independent of temperature* and given by their values at 298 K (see Table 36.1).

As a result, in Eqs. (36.20) and (36.21), the standard-state molar heat capacity of reaction at constant pressure, ΔC_P°, *is also independent of temperature* and can be calculated as follows:

$$\Delta C_P^\circ = \Delta C_P^\circ(298\,\text{K}) = \sum_{\substack{i=SO_3,\\SO_2,\\O_2,\\N_2}} \nu_i C_{P_i}^\circ(298\,\text{K}) \qquad (36.22)$$

Using the values of $C_{P_i}^\circ(298\,\text{K})$ and ν_i given in the Problem Statement, including Table 36.1, in Eq. (36.22) yields:

$$\Delta C_P^\circ = \left\{ \overbrace{(+1)(60.19)}^{SO_3} + \overbrace{(-1)(45.21)}^{SO_2} + \overbrace{\left(-\frac{1}{2}\right)(29.96)}^{O_2} + \underbrace{(0)(29.96)}_{N_2\text{-Inert}} \right\} \text{J/mol}\,\text{K}$$

$$(36.23)$$

or

$$\Delta C_P^\circ = 0! \qquad (36.24)$$

Equations (36.21) and (36.24) show that ΔH° *is independent of temperature* and *is given by its value at* 298 K, which we calculate below using the data given in the Problem Statement, including Table 36.1:

$$\Delta H^\circ(298\,\text{K}) = \sum_{\substack{i=SO_3,\\SO_2,\\O_2,N_2}} \nu_i H_{if}^\circ(298\,\text{K}) \qquad (36.25)$$

$\Delta H^\circ(298\,\text{K}) =$

$$\left\{ \underbrace{(+1)(-39.54 \times 10^4)}_{SO_3} + \underbrace{(-1)(-29.7 \times 10^4)}_{SO_2} + \underbrace{\left(-\frac{1}{2}\right)(0)}_{O_2} + \underbrace{(0)(0)}_{N_2\text{-Inert}} \right\} \frac{\text{J}}{\text{mol}}$$

$$(36.26)$$

or

$$\Delta H^\circ(298\,\text{K}) = -9.84 \times 10^4 \frac{\text{J}}{\text{mol}} \qquad (36.27)$$

Note that Eq. (36.27) corresponds to an *exothermic reaction*. Using Eq. (36.17) in Eq. (36.19), along with Eq. (36.27), and rearranging yields:

$$K(T) = \exp\left\{-11.463 + \frac{1.1835 \times 10^4 K}{T(\text{in K})}\right\} \qquad (36.28)$$

Using Eq. (36.28) for K(T) in Eq. (36.12) yields:

$$\exp\left\{-11.463 + \frac{1.1835 \times 10^4 K}{T(\text{in K})}\right\} = \left[(\xi)(1 - 0.5\xi)^{1/2}\right] / \left[(0.09 - \xi)\left(0.11 - \frac{1}{2}\xi\right)^{1/2}\right]$$
$$(36.29)$$

Equation (36.29) provides an equation relating the two desired unknowns, T and ξ, where all the inputs are known. Clearly, we need a second independent equation involving the two unknowns, T and ξ, that we can solve along with Eq. (36.29) to uniquely determine T and ξ at the outlet.

36.5   Derivation of the Second Equation Relating T and ξ

According to Postulate I, to fully characterize the system under consideration which is *simple*, we need to *specify* (n + 2) *independent variables*, which in this problem, where n = 4 (SO_3, SO_2, O_2, and N_2), is equal to (4 + 2) = 6. Based on the description in the Sample Problem, it is convenient to choose the following *six independent variables*:

$$\left\{T, P, N^o_{SO_3}, N^o_{SO_2}, N^o_{O_2}, N^o_{N_2}\right\} \qquad (36.30)$$

The *inlet (in) stream* is then fully characterized in terms of:

$$\left\{T_{in}, P_{in}, N^{in}_{SO_2} = N^o_{SO_2}, N^{in}_{SO_3} = N^o_{SO_3} = 0, N^{in}_{O_2} = N^o_{O_2}, N^{in}_{N_2} = N^o_{N_2}\right\} \qquad (36.31)$$

Note that all the inputs in Eq. (36.31) are known!
The *outlet (out) stream* is fully characterized in terms of:

$$\left\{T_{out} = ?, P_{out} = P_{in}, N^{out}_{SO_2} = N^o_{SO_2} - \xi_{out}, N^{out}_{SO_3} = \xi_{out}, N^{out}_{O_2} = N^o_{O_2} - \frac{1}{2}\xi_{out}, N^{out}_{N_2} = N^o_{N_2}\right\} \qquad (36.32)$$

Note that in Eq. (36.32), the only unknown inputs are T_{out} and ξ_{out}.
As stressed above, in addition to Eq. (36.29) which relates the two unknowns $T = T_{out}$ and $\xi = \xi_{out}$, we need to write down a second independent equation which

also relates T_{out} and ξ_{out}. This equation is provided by the First Law of Thermodynamics describing the operation of the *steady-state, open, adiabatic* reactor.

First Law of Thermodynamics Analysis of the Gas Mixture in the Reactor

- System: Gas Mixture in the SO_3 Reactor, <u>Well-Mixed</u> at all times.
- <u>Boundaries</u>:
  - <u>Open</u> (Inlet and Outlet Streams)
  - <u>Adiabatic</u> ($\delta Q = 0$)
  - <u>Rigid</u> ($\delta W = 0$)

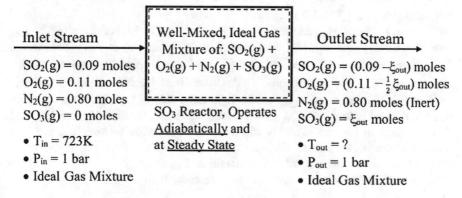

Inlet Stream

$SO_2(g) = 0.09$ moles
$O_2(g) = 0.11$ moles
$N_2(g) = 0.80$ moles
$SO_3(g) = 0$ moles

- $T_{in} = 723K$
- $P_{in} = 1$ bar
- Ideal Gas Mixture

Well-Mixed, Ideal Gas Mixture of: $SO_2(g) + O_2(g) + N_2(g) + SO_3(g)$

SO_3 Reactor, Operates Adiabatically and at <u>Steady State</u>

Outlet Stream

$SO_2(g) = (0.09 - \xi_{out})$ moles
$O_2(g) = (0.11 - \frac{1}{2}\xi_{out})$ moles
$N_2(g) = 0.80$ moles (Inert)
$SO_3(g) = \xi_{out}$ moles

- $T_{out} = ?$
- $P_{out} = 1$ bar
- Ideal Gas Mixture

Fig. 36.2

In addition, as per the Problem Statement, the reactor *operates isobarically*, that is, $P_{out} = P_{in}$. Next, we carry out a First Law of Thermodynamics analysis of the gas mixture in the reactor (see Fig. 36.2).

The gas mixture in the SO_3 reactor is well-mixed at all times, and therefore, it occupies the entire volume of the reactor. As a result, no $Pd\underline{V}$-type work is incurred ($\delta W = 0$). In addition, the reactor operates adiabatically, and therefore, the gas mixture has no heat interactions ($\delta Q = 0$). The gas mixture is a simple, open system, and therefore, the differential form of *the First Law of Thermodynamics for an Open System* applies. Specifically:

$$\underbrace{d\underline{U}}_{0\left(\substack{\text{Steady}\\\text{State}}\right)} = \underbrace{\delta Q}_{0\left(\text{Adiabatic}\right)} + \underbrace{\delta W}_{0\left(\substack{\text{Well}\\\text{Mixed}}\right)} + d\underline{H}_{in} - d\underline{H}_{out} \qquad (36.33)$$

Accordingly, after rearranging, Eq. (36.33) becomes:

$$\underline{dH}_{in} = \underline{dH}_{out} \tag{36.34}$$

In integral form, Eq. (36.34) is given by:

$$\underline{H}_{in}\left(T_{in}, P_{in}, N_{SO_3}^{in}, N_{SO_2}^{in}, N_{O_2}^{in}, N_{N_2}^{in}\right) = \underline{H}_{out}\left(T_{out}, P_{out}, N_{SO_3}^{out}, N_{SO_2}^{out}, N_{O_2}^{out}, N_{N_2}^{out}\right) \tag{36.35}$$

Recall that in Eq. (36.35), the $(n + 2) = (4 + 2) = 6$ variables $\left\{T_{in}, P_{in}, N_{SO_3}^{in}, N_{SO_2}^{in}, N_{O_2}^{in}, N_{N_2}^{in}\right\}$ determining $\underline{H}_{in}$ *are all known*. Accordingly, the left-hand side of Eq. (36.35) is known!

On the other hand, $\underline{H}_{out}$ on the right-hand side of Eq. (36.35) is determined by T_{out} (*not known*), P_{out} (*known*), $N_{SO_3}^{out}$ (determined by $N_{SO_3}^{in}$ (*known*) and ξ_{out} (*unknown*)), $N_{SO_2}^{out}$ (determined by $N_{SO_2}^{in}$ (*known*) and ξ_{out} (*unknown*)), $N_{O_2}^{out}$ (determined by $N_{O_2}^{in}$ (*known*) and ξ_{out} (*unknown*)), and $N_{N_2}^{out}$ (determined by $N_{N_2}^{in}$ (*known*) and ξ_{out} (*unknown*)). In other words, $\underline{H}_{out}$ on the right-hand side of Eq. (36.35) depends on the two desired unknowns, T_{out} and ξ_{out}! We will next calculate $\underline{H}_{in}$ and $\underline{H}_{out}$ and then use them in Eq. (36.35). This will result in an equation relating T_{out} and ξ_{out}. This equation, when solved simultaneously with Eq. (36.29), will allow us to uniquely determine the two unknown quantities, T_{out} and ξ_{out}.

In general, for an n component mixture, we know that:

$$\underline{H} = \sum_{i=1}^{n} N_i \overline{H}_i \tag{36.36}$$

Because we are told that the inlet and outlet gas mixtures are ideal, it follows that the partial molar enthalpy of component i is equal to the molar enthalpy of component i and depends solely on the temperature. Specifically,

$$\overline{H}_i = H_i(T) \tag{36.37}$$

Using Eq. (36.37) for $i = SO_3$, SO_2, O_2, and N_2 in Eq. (36.36) for the inlet gas mixture, we obtain:

$$\underline{H}_{in} = \sum_{\substack{i=SO_2, \\ SO_2, \\ O_2, \\ N_2}} N_i^{in} H_i(T_{in}) \tag{36.38}$$

where $N_{SO_2}^{in} = 0.09\,moles$, $N_{SO_3}^{in} = 0\,moles$, $N_{O_2}^{in} = 0.11\,moles$, and $N_{N_2}^{in} = 0.80\,moles$. Using these four mole numbers in Eq. (36.38), we obtain:

$$\underline{H}_{in}(T_{in}) = 0.09 H_{SO_2}(T_{in}) + 0.11 H_{O_2}(T_{in}) + 0.80 H_{N_2}(T_{in}) \tag{36.39}$$

Carrying out a similar analysis for $\underline{H}_{out}$ yields:

$$\underline{H}_{out} = \sum_{\substack{i=SO_2, \\ SO_2, \\ O_2, N_2}} N_i^{out} H_i(T_{out}) \tag{36.40}$$

where $N_{SO_2}^{out} = (0.99 - \xi_{out})$ moles, $N_{SO_3}^{out} = \xi_{out}$ moles, $N_{O_2}^{out} = \left(0.11 - \frac{1}{2}\xi_{out}\right)$ moles, and $N_{N_2}^{out} = 0.80$ moles. Using these four mole numbers in Eq. (36.40) yields:

$$\underline{H}_{out}(T_{out}) = (0.09 - \xi_{out}) H_{SO_2}(T_{out}) + \xi_{out} H_{SO_3}(T_{out})$$
$$+ \left(0.11 - \frac{1}{2}\xi_{out}\right) H_{O_2}(T_{out}) + 0.80 H_{N_2}(T_{out}) \tag{36.41}$$

In Eq. (36.41), combining all the terms which depend on ξ_{out} and all the terms which do not depend on ξ_{out}, we obtain:

$$\underline{H}_{out}(T_{out}) = [0.09 H_{SO_2} + 0.11 H_{O_2} + 0.80 H_{N_2}]$$
$$+ \xi_{out} \left[H_{SO_3} - H_{SO_2} - \frac{1}{2}H_{O_2}\right] \tag{36.42}$$

Note that in Eq. (36.42), $H_{SO_3} = H_{SO_3}(T_{out})$, $H_{SO_2} = H_{SO_2}(T_{out})$, $H_{O_2} = H_{O_2}(T_{out})$, and $H_{N_2} = H_{N_2}(T_{out})$. In addition, note that in Eq. (36.42),

$$H_{SO_3}(T_{out}) - H_{SO_2}(T_{out}) - \frac{1}{2}H_{O_2}(T_{out})$$
$$= \Delta H^\circ(T_{out}) = \Delta H^\circ(298\,K) \tag{36.43}$$

where $\Delta H^\circ(298\ K)$ is the *standard molar enthalpy of reaction*, shown above to be independent of temperature and evaluated at 298 K using the given data.

In addition, a comparison of the first term on the right-hand side of Eq. (36.42) with Eq. (36.39) shows that:

$$[0.09 H_{SO_2}(T_{out}) + 0.11 H_{O_2}(T_{out}) + 0.80 H_{N_2}(T_{out})] = \underline{H}_{in}(T_{out}) \tag{36.44}$$

Using Eqs. (36.44) and (36.43) in Eq. (36.42) yields:

$$\underline{H}_{out}(T_{out}) = \underline{H}_{in}(T_{out}) + \xi_{out}\Delta H^\circ(298\,K) \tag{36.45}$$

An examination of Eq. (36.45) shows that:

(i) If there is no chemical reaction, that is, if $\xi_{out} = 0$, then,

$$\underline{H}_{out}(T_{out}) = \underline{H}_{in}(T_{out}) \tag{36.46}$$

We also know that (see Eq. (36.35)):

$$\underline{H}_{out}(T_{out}) = \underline{H}_{in}(T_{in}) \tag{36.47}$$

A comparison of Eqs. (36.46) and (36.47) shows that in the absence of the chemical reaction, $T_{out} = T_{in}$!

(ii) On the other hand, in the presence of the chemical reaction, that is, if $\xi_{out} \neq 0$, and because we have shown that $\Delta H^{\circ}(298\,K) < 0$ (*exothermic reaction*), a comparison of Eqs. (36.45) and (36.47) shows that:

$$\underline{H}_{out}(T_{out}) = \underline{H}_{in}(T_{out}) + \xi_{out}\underbrace{\Delta H^{\circ}(298\,K)}_{<0} = \underline{H}_{in}(T_{in}) \tag{36.48}$$

Equation (36.48) clearly shows that $T_{out} > T_{in}$! This result is expected, because the *exothermic chemical reaction releases heat*.

Next, we rewrite Eq. (36.48) as follows:

$$\underline{H}_{in}(T_{in}) - \underline{H}_{in}(T_{out}) = \xi_{out}\Delta H^{\circ}(298\,K) \tag{36.49}$$

where $\underline{H}_{in}(T_{in}) - \underline{H}_{in}(T_{out})$ in Eq. (36.49) is given by:

$$\begin{aligned} \underline{H}_{in}(T_{in}) - \underline{H}_{in}(T_{out}) &= 0.09[H_{SO_2}(T_{in}) - H_{SO_2}(T_{out})] \\ &+ 0.11[H_{O_2}(T_{in}) - H_{O_2}(T_{out})] + 0.80[H_{N_2}(T_{in}) - H_{N_2}(T_{out})] \end{aligned} \tag{36.50}$$

For each pure gas i (SO_2, O_2, and N_2) in Eq. (36.50), we can compute the quantity, $[H_i(T_{in}) - H_i(T_{out})]$, using the constant C_{pi}° values given in Table 36.1. Specifically:

$$H_i(T_{in}) - H_i(T_{out}) = \int_{T_{out}}^{T_{in}} C_{pi}^{\circ}\, dT = C_{pi}^{\circ}[T_{in} - T_{out}] \tag{36.51}$$

Using Eq. (36.51) for i = SO_3, O_2, and N_2 in Eq. (36.50) yields:

$$\underline{H}_{in}(T_{in}) - \underline{H}_{in}(T_{out}) = (T_{in} - T_{out})$$

$$\bullet \left(0.09 C_{PSO_2}^{\circ} + 0.11 C_{PO_2}^{\circ} + 0.80 C_{PN_2}^{\circ} \right) \tag{36.52}$$

Using Eq. (36.52) in Eq. (36.49) and then solving for T_{out} yields:

$$T_{out} = T_{in} - \left(\frac{\Delta H^\circ (298\,K)}{0.09 C^\circ_{PSO_2} + 0.11 C^\circ_{PO_2} + 0.80 C^\circ_{PN_2}} \right) \xi_{out} \tag{36.53}$$

Using $T_{in} = 723$ K, ΔH° (298 K) given in Eq. (36.27) and the C°_{pi} values given in Table 36.1 in Eq. (36.53) yields:

$$T_{out} = (723 + 3140.5\,\xi_{out})K \tag{36.54}$$

As expected for an *exothermic chemical reaction*, Eq. (36.54) shows that $T_{out} > T_{in} = 723$ K.

Equation (36.54) can now be solved simultaneously with Eq. (36.29) to obtain the following equation for ξ_{out}:

$$\frac{(\xi_{out})(1-0.5\xi_{out})^{1/2}}{(0.09-\xi_{out})(0.11-0.5\xi_{out})^{1/2}} = \exp\left[-11.463 + \left(1.1835 \times 10^4\right)/(723+3140.5\xi_{out})\right] \tag{36.55}$$

Solving Eq. (36.55), for example, by iteration, yields:

$$\xi_{out} = 0.0555 \tag{36.56}$$

Using Eq. (36.56) in Eq. (36.54) yields:

$$T_{out} = \left[723 + \underbrace{(3140.5)(0.0555)}_{174.30} \right] K \tag{36.57}$$

or

$$T_{out} = 897.3\,K \tag{36.58}$$

Again, as expected, $T_{out} > T_{in} = 723$K.
Finally, the % of SO_2 left is given by:

$$\left[\frac{(0.09 - 0.0555)}{0.09} \right] \cdot 100\% \simeq 38.3\% \tag{36.59}$$

Lecture 37

Review of Part II and Sample Problem

37.1 Introduction

In this lecture, we will first review the topics covered in Part II and then solve Sample Problem 37.1, an interesting problem which will help crystallize many of the concepts and methodologies presented in Part II.

37.2 Partial Molar Properties

$$d\underline{B} = \left(\frac{\partial \underline{B}}{\partial T}\right)_{P,N_i} dT + \left(\frac{\partial \underline{B}}{\partial P}\right)_{T,N_i} dP + \sum_i \left(\frac{\partial \underline{B}}{\partial N_i}\right)_{T,P,N_{j[i]}} dN_i \qquad (37.1)$$

$$\overline{B}_i \equiv \left(\frac{\partial \underline{B}}{\partial N_i}\right)_{T,P,N_{j[i]}} \qquad (37.2)$$

$$\underline{B} = \sum_i N_i \overline{B}_i \qquad (37.3)$$

$$B = \sum_i x_i \overline{B}_i \qquad (37.4)$$

In addition, equations of state (EOS) for mixtures, including mixing rules for the EOS parameters, mixture heat capacities, and the attenuated state and departure function approaches for mixtures.

© Springer Nature Switzerland AG 2020
D. Blankschtein, *Lectures in Classical Thermodynamics with an Introduction to Statistical Mechanics*, https://doi.org/10.1007/978-3-030-49198-7_37

37.3   Generalized Gibbs-Duhem Relations for Mixtures

$$\sum_i N_i d\overline{B}_i = \left(\frac{\partial \underline{\underline{B}}}{\partial T}\right)_{P,N_i} dT + \left(\frac{\partial \underline{\underline{B}}}{\partial P}\right)_{T,N_i} dP \tag{37.5}$$

37.4   Gibbs-Helmholtz Relation

$$\frac{\partial}{\partial T}\left(\frac{G}{T}\right)_P = -\frac{H}{T^2} \tag{37.6}$$

$$\frac{\partial}{\partial T}\left(\frac{\underline{G}}{T}\right)_{P,N} = -\frac{\underline{H}}{T^2} \tag{37.7}$$

37.5   Mixing Functions

$$\Delta \underline{B}_{mix} = \underline{B} - \sum_i N_i \overline{B}_i^{\dagger} = \sum_i N_i \Delta \overline{B}_i \tag{37.8}$$

$$\Delta B_{mix} = B - \sum_i x_i \overline{B}_i^{\dagger} = \sum_i x_i \Delta \overline{B}_i \tag{37.9}$$

$$\Delta \overline{B}_i = \left(\frac{\partial(\Delta \underline{B}_{mix})}{\partial N_i}\right)_{T,P,N_{j[i]}} = \overline{B}_i - \overline{B}_i^{\dagger} \tag{37.10}$$

Reference States
1. Pure Component: $\overline{B}_i^{\dagger} = B_i(T, P)$
2. Infinite Dilution: $\overline{B}_i^{\dagger} = \overline{B}_i(T, P, x_i \to 0)$

37.6 Ideal Gas Mixtures

$$\overline{G}_i = \lambda_i(T) + RT\ln(y_i P) \tag{37.11}$$

$$\overline{V}_i = \frac{RT}{P} \tag{37.12}$$

$$\overline{U}_i = U_i(T) \tag{37.13}$$

$$\overline{H}_i = H_i(T) \tag{37.14}$$

$$\Delta H_{mix} = 0 \quad \text{(Pure component reference state)} \tag{37.15}$$

$$\Delta V_{mix} = 0 \quad \text{(Pure component reference state)} \tag{37.16}$$

$$\Delta S_{mix} = -R\sum_i x_i \ln x_i \quad \text{(Pure component reference state)} \tag{37.17}$$

$$\Delta \overline{S}_i = -R\ln x_i \quad \text{(Pure component reference state)} \tag{37.18}$$

$$\Delta G_{mix} = RT\sum_i x_i \ln x_i \quad \text{(Pure component reference state)} \tag{37.19}$$

$$\Delta \overline{G}_i = RT\ln x_i \quad \text{(Pure component reference state)} \tag{37.20}$$

37.7 Ideal Solutions

$$\overline{G}_i = \Lambda_i(T, P) + RT\ln x_i = G_i(T, P) + RT\ln x_i \tag{37.21}$$

$$\overline{V}_i = V_i(T, P) \tag{37.22}$$

$$\overline{U}_i = U_i(T, P) \tag{37.23}$$

$$\overline{H}_i = H_i(T, P) \tag{37.24}$$

$$\Delta H_{mix} = \Delta H_{mix}^{ID} = 0 \quad \text{(Pure component reference state)} \tag{37.25}$$

$$\Delta V_{mix} = \Delta V_{mix}^{ID} = 0 \quad \text{(Pure component reference state)} \tag{37.26}$$

$$\Delta S_{mix} = \Delta S_{mix}^{ID} = -R\sum_i x_i \ln x_i \quad \text{(Pure component reference state)} \tag{37.27}$$

$$\Delta \overline{S}_i = \Delta \overline{S}_i^{ID} = -R\ln x_i \quad \text{(Pure component reference state)} \tag{37.28}$$

$$\Delta G_{mix} = \Delta G_{mix}^{ID} = RT\sum_i x_i \ln x_i \quad \text{(Pure component reference state)} \tag{37.29}$$

$$\Delta \overline{G}_i = \Delta \overline{G}_i^{ID} = RT\ln x_i \quad \text{(Pure component reference state)} \tag{37.30}$$

37.8   Excess Functions

$$\underline{B}^{EX} = \underline{B} - \underline{B}^{ID} \tag{37.31}$$

$$B^{EX} = B - B^{ID} \tag{37.32}$$

$$\overline{B}_i^{EX} = \overline{B}_i - \overline{B}_i^{ID} \tag{37.33}$$

$$\Delta \underline{B}^{EX}_{mix} = \Delta \underline{B}_{mix} - \Delta \underline{B}^{ID}_{mix} \tag{37.34}$$

$$\Delta \overline{B}^{EX}_i = \Delta \overline{B}_i - \Delta \overline{B}^{ID}_i \tag{37.35}$$

$$\Delta \underline{B}^{EX}_{mix} = \underline{B}^{EX} \tag{37.36}$$

$$\Delta \overline{B}^{EX}_i = \overline{B}^{EX}_i \tag{37.37}$$

Regular Solution

$$\Delta S^{EX}_{mix} = 0 \tag{37.38}$$

Athermal Solution

$$\Delta H^{EX}_{mix} = 0 \tag{37.39}$$

37.9   Fugacity

$$G_i = \lambda_i(T) + RT\ln f_i \quad \text{(For pure component i)} \tag{37.40}$$

$$\overline{G}_i = \lambda_i(T) + RT\ln \widehat{f}_i \quad \text{(For component i in a mixture)} \tag{37.41}$$

Limits of Ideality

$$\lim_{P \to 0} \left(\frac{f_i}{P} \right) = 1 \quad \text{(For pure component i)} \tag{37.42}$$

$$\lim_{P \to 0} \left(\frac{\widehat{f}_i}{y_i P} \right) = 1 \quad \text{(For component i in a mixture)} \tag{37.43}$$

37.10   Variation of Fugacity with Temperature and Pressure

$$\left(\frac{\partial \ln f_i}{\partial P}\right)_{T,N_i} = \left(\frac{\partial (G_i/RT)}{\partial P}\right)_{T,N_i} = \frac{V_i}{RT} \tag{37.44}$$

$$\left(\frac{\partial \ln \widehat{f_i}}{\partial P}\right)_{T,y_i} = \left(\frac{\partial (\overline{G}_i/RT)}{\partial P}\right)_{T,y_i} = \frac{\overline{V}_i}{RT} \tag{37.45}$$

$$\left(\frac{\partial \ln f_i}{\partial T}\right)_{P,N_i} = \left(\frac{\partial ((G_i - G_i^0)/RT)}{\partial T}\right)_{P,N_i} = -\frac{H_i - H_i^0}{RT^2} \tag{37.46}$$

$$\left(\frac{\partial \ln \widehat{f_i}}{\partial T}\right)_{P,y_i} = \left(\frac{\partial \left(\left(\overline{G}_i - \overline{G}_i^0\right)/RT\right)}{\partial T}\right)_{P,y_i} = -\frac{\overline{H}_i - H_i^0}{RT^2} \tag{37.47}$$

37.11   Generalized Gibbs-Duhem Relation for Fugacities

$$\sum_i x_i d\ln\widehat{f_i} = -\sum_i x_i\left(\frac{\overline{H}_i - H_i^0}{RT^2}\right)dT + \sum_i x_i\left(\frac{\overline{V}_i}{RT}\right)dP \tag{37.48}$$

37.12   Fugacity Coefficient

$$\phi_i \equiv \frac{f_i}{P} \tag{37.49}$$

$$\widehat{\phi}_i \equiv \frac{\widehat{f_i}}{y_i P} \tag{37.50}$$

$$RT\ln\phi_i = \int\limits_0^P \left(V_i - \frac{RT}{P}\right)dP \qquad (37.51)$$

$$RT\ln\widehat{\phi}_i = \int\limits_0^P \left(\overline{V}_i - \frac{RT}{P}\right)dP \qquad (37.52)$$

$$RT\ln\phi_i = -RT\ln Z - \int\limits_\infty^{\underline{V}} \left(\left(\frac{\partial P}{\partial N}\right)_{T,\underline{V}} - \frac{RT}{\underline{V}}\right)d\underline{V} \qquad (37.53)$$

$$RT\ln\widehat{\phi}_i = -RT\ln Z - \int\limits_\infty^{\underline{V}} \left(\left(\frac{\partial P}{\partial N_i}\right)_{T,\underline{V},N_{j[i]}} - \frac{RT}{\underline{V}}\right)d\underline{V} \qquad (37.54)$$

37.13 Lewis and Randall Rule

$$\widehat{f}_i^{ID} = f_i(T,P)x_i \qquad (37.55)$$

37.14 Activity

$$RT\ln a_i = \overline{G}_i - \overline{G}_i^\dagger = \Delta\overline{G}_i \qquad (37.56)$$

$$a_i \equiv \frac{\widehat{f}_i}{\widehat{f}_i^\dagger} \qquad (37.57)$$

For an Ideal Solution (Pure Component Reference State)

$$a_i = x_i \qquad (37.58)$$

37.15 Activity Coefficient

$$\gamma_i \equiv \frac{a_i}{x_i} = \frac{\widehat{f_i}}{x_i f_i^\dagger} \tag{37.59}$$

$$RT\ln\gamma_i = \Delta\overline{G}_i^{EX} = \overline{G}_i^{EX} \tag{37.60}$$

37.16 Variation of Activity Coefficient with Temperature and Pressure

$$\left(\frac{\partial\ln\gamma_i}{\partial P}\right)_{T,X} = \left(\frac{\partial\left(\Delta\overline{G}_i^{EX}/RT\right)}{\partial P}\right)_{T,X} = \frac{\Delta\overline{V}_i}{RT} \tag{37.61}$$

$$\left(\frac{\partial\ln\gamma_i}{\partial T}\right)_{P,X} = \left(\frac{\partial\left(\Delta\overline{G}_i^{EX}/RT\right)}{\partial T}\right)_{X,P} = -\frac{\Delta\overline{H}_i}{RT^2} \tag{37.62}$$

37.17 Generalized Gibbs-Duhem Relation for Activity Coefficients

$$\sum_i x_i d\ln\gamma_i = -\left(\frac{\Delta H_{mix}}{RT^2}\right)dT + \left(\frac{\Delta V_{mix}}{RT}\right)dP \tag{37.63}$$

37.18 Conditions for Thermodynamic Phase Equilibria

$$\textbf{Thermal Equilibrium}: T^\alpha = T^\beta = T^\gamma = \ldots = T^\pi \tag{37.64}$$

$$\text{Mechanical Equilibrium}: P^\alpha = P^\beta = P^\gamma = \ldots = P^\pi \qquad (37.65)$$

$$\text{Diffusional Equilibrium}: \mu_i^\alpha = \mu_i^\beta = \mu_i^\gamma = \ldots = \mu_i^\pi (i = 1, 2, \ldots, n) \qquad (37.66)$$

37.19 Gibbs Phase Rule

$$L = n + 2 - \pi - r - s \qquad (37.67)$$

where n is the number of components, π is the number of phases, r is the number of independent chemical reactions, and s is the number of additional constraints.

In addition, understanding phase diagrams.

37.20 Differential Approach to Phase Equilibria

$$\widehat{f}_i^\alpha = \widehat{f}_i^\beta = \ldots = \widehat{f}_i^\pi \Rightarrow d \ln \widehat{f}_i^\alpha = d \ln \widehat{f}_i^\beta = \ldots = d \ln \widehat{f}_i^\pi \qquad (37.68)$$

37.21 Dependence of Fugacitities on Temperature, Pressure, and Mixture Composition

$$
d\ln\widehat{f}_i^\alpha = \left(\frac{\partial \ln\widehat{f}_i^\alpha}{\partial T}\right)_{P,x_i^\alpha} dT + \left(\frac{\partial \ln\widehat{f}_i^\alpha}{\partial P}\right)_{T,x_i^\alpha} dP + \sum_{i=1}^{n-1} \left(\frac{\partial \ln\widehat{f}_i^\alpha}{\partial x_i^\alpha}\right)_{T,P,x_{j[i,n]}^\alpha} dx_i^\alpha
$$

$$
\Rightarrow d\ln\widehat{f}_i^\alpha = -\left(\frac{\overline{H}_i^\alpha - H_i^0}{RT^2}\right) dT + \left(\frac{\overline{V}_i^\alpha}{RT}\right) dP + \sum_{i-1}^{n-1} \left(\frac{\partial \ln\widehat{f}_i^\alpha}{\partial x_i^\alpha}\right)_{T,P,x_{j[i,n]}^\alpha} dx_i^\alpha
$$

$$(37.69)$$

37.22   Integral Approach to Phase Equilibria

$$\widehat{f}_i^{\alpha} = \widehat{f}_i^{\beta} = \ldots = \widehat{f}_i^{\pi} \tag{37.70}$$

When only two phases ($\pi = 2$) are in thermodynamic equilibrium, say, α = Vapor (V) and β = Liquid (L), each containing n components, we can compute the vapor fugacity of component i using an EOS approach, and the liquid fugacity of component i using an Excess Gibbs Free Energy model. In that case, Eq. (37.70) with α = V and β = L is given by:

$$y_i \widehat{\phi}_i^V (T, P, y_1, \ldots, y_{n-1}) P = x_i \gamma_i^L (T, P, x_1, \ldots, x_{n-1}) \phi_i^V (T, P_{vpi}(T)) P_{vpi}(T) C_i \tag{37.71}$$

where the Poynting correction C_i is given by:

$$C_i = \exp \left[\int_{P_{vpi}(T)}^{P} \left(\frac{V_i^L}{RT} \right) dP \right] \text{(Poynting correction for i} = 1, 2, \ldots, n) \tag{37.72}$$

In addition, knowledge of how to simplify Eq. (37.72) under various equilibrium conditions.

37.23   Pressure-Temperature Relations

Clapeyron Equation

$$\left(\frac{dP}{dT} \right)_{[L/V]} = \frac{\Delta H_{vap}}{T \Delta V_{vap}} \tag{37.73}$$

Clausius-Clapeyron Equation

$$\left[\frac{d(\ln P)}{d(1/T)} \right]_{[L/V]} = -\frac{\Delta H_{vap}}{R} \tag{37.74}$$

Recall that the Clausius-Clapeyron Equation assumes that: (1) the vapor phase is ideal, and (2) the molar volume of the vapor is much larger than that of the liquid.

37.24 Stoichiometric Formulation for Chemical Reactions

$$\frac{dn_{i,r}}{\nu_{i,r}} = \frac{dn_{j,r}}{\nu_{j,r}} = \ldots = d\xi_r \text{ for all species in all independent chemical reactions, r}$$

$$(37.75)$$

37.25 Equilibrium Constant

Criteria of Chemical Reaction Equilibria

$$\sum_j \nu_{j,r}\mu_j = 0 \quad (r = 1, 2, \ldots, m) \tag{37.76}$$

$$K(T, P^\circ) = \prod_i \left(\frac{\hat{f}_i(T, P, y_1, y_2, \ldots, y_{n-1})}{f_i^\circ(T, P^\circ)}\right)^{\nu_i} = \exp\left(-\frac{\Delta G^\circ(T, P^\circ)}{RT}\right) \tag{37.77}$$

Note that all the terms in Eq. (37.77) are evaluated at the system temperature. Furthermore, the reference-state fugacity and standard molar Gibbs free energy of reaction are evaluated at the same reference pressure.

37.26 Typical Reference States for Gas, Liquid, and Solid

Gas: Pure ideal gas at the system temperature and at 1 bar pressure.
Liquid: Pure liquid at the system temperature and at 1 bar pressure, or at its vapor pressure, or at the system pressure.
Solid: Pure solid in its most stable crystal state at the system temperature and 1 bar pressure, or at its vapor pressure, or at the system pressure.

37.27 Equilibrium Constant for Gases Undergoing a Single Chemical Reaction

$$K(T, 1 \text{ bar}) = \left(\prod_i \hat{\phi}_i^{\nu_i}\right)\left(\prod_i y_i^{\nu_i}\right)\prod_i \left(\frac{P}{1 \text{ bar}}\right)^{\nu_i} = K_\phi K_y K_P \tag{37.78}$$

37.28   Equilibrium Constants for Liquids and Solids

If the reference-state pressure Po is chosen to be 1bar or $P_{vpi}(T)$, then:

$$K(T, P_o) = \left(\prod_i x_i^{\nu_i}\right)\left(\prod_i \gamma_i^{\nu_i}\right)\prod_i \exp\left(\nu_i \int_{P_0}^{P} \frac{V_i}{RT} dP\right)$$

$$= K_x K_\gamma \exp\left(\sum_i \nu_i \int_{P_0}^{P} \frac{V_i}{RT} dP\right) \tag{37.79}$$

If the reference-state pressure Po is chosen to be the system pressure P, then:

$$K(T, P) = \left(\prod_i x_i^{\nu_i}\right)\left(\prod_i \gamma_i^{\nu_i}\right) = K_x K_\gamma \tag{37.80}$$

37.29   Calculation of the Standard Molar Gibbs Free Energy of Reaction

(i) Using the standard Gibbs free energy of formation:

$$\Delta G^\circ(T, P^\circ) = \sum_i \nu_i \Delta G_{f,i}^\circ(T, P^\circ) \tag{37.81}$$

(ii) Using: $\Delta G^\circ(T^*, P^\circ)$, $\Delta H_{f,i}^\circ(T^*, P^\circ)$, and $C_{P,i}^\circ(T, P^\circ)$

$$\frac{\Delta G^\circ(T, P^\circ)}{T} = \frac{\Delta G^\circ(T^*, P^\circ)}{T^*} - \int_{T^*}^{T}\left(\frac{\Delta H^\circ(T, P^\circ)}{T^2}\right)dT \tag{37.82}$$

$$\Delta H^\circ(T, P^\circ) = \sum_i \nu_i \Delta H_{f,i}^\circ(T, P^\circ) \tag{37.83}$$

$$\Delta H_{f,i}^\circ(T, P^\circ) = \Delta H_{f,i}^\circ(T^*, P^\circ) + \int_{T^*}^{T} C_{P,i}^\circ(T, P^\circ)dT \tag{37.84}$$

37.30   Variation of the Equilibrium Constant with Temperature and Pressure

$$\left(\frac{\partial \ln K}{\partial T}\right)_P = \frac{\Delta H^\circ}{RT^2} \tag{37.85}$$

$$\left(\frac{\partial \ln K}{\partial P}\right)_T = 0 \quad \text{if} \quad P_0 = 1 \text{ bar or } P_{vpi} \tag{37.86}$$

$$\left(\frac{\partial \ln K}{\partial P}\right)_T = -\frac{\Delta V}{RT} \quad \text{if} \quad P_0 = P \tag{37.87}$$

37.31   Sample Problem 37.1

For the solution of this problem, you can assume that:

1. Liquid and vapor mixtures are ideal.
2. Liquid hydrocarbons are completely miscible.
3. Liquid hydrocarbon and water are completely immiscible.

(a) Calculate at what pressure will a liquid mixture consisting of droplets of benzene and toluene dispersed in water begin to boil at 40 °C (see Fig. 37.1). The following information is provided:

   (i) The liquid mixture consists of 2 moles of water, 0.5 moles of benzene, and 1 mole of toluene

   (ii) The pure component vapor pressures at 40 °C are:

   - Water, 55.3 mmHg
   - Benzene, 181.1 mmHg
   - Toluene, 59.1 mmHg

   (iii) The pure component molar volumes at 40 °C are:

   - Water, 18 cm^3/mol
   - Benzene, 89.4 cm^3/mol
   - Toluene, 106.5 cm^3/mol

37.31.1 Solution

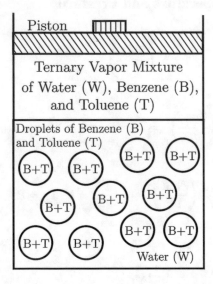

Fig. 37.1

1. We are dealing with a three-phase system, which is not simple, because the oil mixture (B + T) in the oil droplets and the continuous water (w) phase are fully immiscible (see Fig. 37.1)! We can represent and model the three-phase system in the following useful way (see Fig. 37.2):

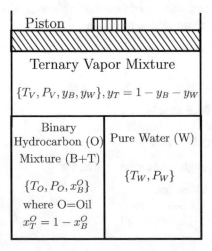

Fig. 37.2

2. Use the Gibbs Phase Rule in each simple phase to determine the number of independent intensive variables in each phase:

Ternary Vapor Mixture

$$n = 3, \pi = 1, r = 0 \Rightarrow L = n + 2 - \pi - r = 3 + 2 - 1 - 0 = 4 \qquad (37.88)$$

We choose these four intensive variables to be: $\{T_v, P_v, y_B, y_w\}$.
where, $v = $ vapor (see Fig. 37.2).

Pure Water Phase

$$n = 1, \pi = 1, r = 0 \Rightarrow L = n + 2 - \pi - r = 1 + 2 - 1 - 0 = 2 \qquad (37.89)$$

We choose these two variables to be: $\{T_w, P_w\}$ (see Fig. 37.2).

Binary Hydrocarbon Mixture

$$n = 2, \pi = 1, r = 0 \Rightarrow L = n + 2 - \pi - r = 2 + 2 - 1 - 0 = 3 \qquad (37.90)$$

We choose these three variables to be: $\{T_o, P_o, X_B^o\}$ (see Fig. 37.2).

3. Determine the variance (L) of the phase equilibrium system using the generalized Gibbs approach for composite systems. The system is not simple because there is an impermeable barrier between the water (w) and oil (o) phases. Therefore, we need to use the generalized Gibbs approach to determine L. Specifically,

(a) Total number of intensive variables in the three phacses:

$$\underbrace{T_v, P_v, y_B, y_T}_{\text{Vapor phase, L=4}}; \quad \underbrace{T_o, P_o, X_B^o}_{\text{Oil phase, L=3}}; \quad \underbrace{T_w, P_w}_{\text{Water phase, L=2}} \qquad (37.91)$$

(b) Total conditions of thermodynamic equilibrium which apply:

- T.E.: $T_v = T_o = T_w$ (2 conditions)
- M.E.: $P_v = P_o = P_w$ (2 conditions)
- D.E.: $\hat{f}_w^w = \hat{f}_w^v; \hat{f}_B^o = \hat{f}_B^v; \hat{f}_T^o = \hat{f}_T^v$ (3 conditions)

It then follows that:

$$L = (a) - (b) = \underbrace{(4 + 3 + 2)}_{9} - \underbrace{(2 + 2 + 3)}_{7} = 2 \qquad (37.92)$$

or

$$L = 2 \tag{37.93}$$

4. We next solve the three fugacity equations:

$$f_w^w = \widehat{f}_w^v \tag{37.94}$$

$$\widehat{f}_B^o = \widehat{f}_B^v \tag{37.95}$$

$$\widehat{f}_T^o = \widehat{f}_T^v \tag{37.96}$$

Because of the information provided in the Problem Statement, it is convenient to use the Integral Approach to Phase Equilibria. Specifically,

$$f_w^w = f_w^w\big[T, P_{vpw}(T)\big]C_w = \widehat{f}_w^v = y_w\widehat{\phi}_w^v P \tag{37.97}$$

where the Poynting correction for water is given by:

$$C_w = \exp\left[\int_{P_{vpw}(T)}^{P} \left(\frac{V_w}{RT}\right)dP\right] \tag{37.98}$$

Because at equilibrium,

$$f_w^w\big[T, P_{vpw}(T)\big] = f_w^v\big[T, P_{vpw}(T)\big] \tag{37.99}$$

Equation (37.97) can be rewritten as follows:

$$f_w^v\big[T, P_{vpw}(T)\big]C_w = y_w\widehat{\phi}_w^v P \tag{37.100}$$

We also know that:

$$f_w^v\big[T, P_{vpw}(T)\big] = \widehat{\phi}_w^v P_{vpw}(T) \tag{37.101}$$

Using Eqs. (37.101) and (37.100) yields:

$$\widehat{\phi}_w^v P_{vpw}(T)C_w = y_w\widehat{\phi}_w^v P \tag{37.102}$$

Because the vapor mixture is ideal, it follows that:

$$\hat{\phi}_w^v = \left(\hat{\phi}_w^v\right)^{ID} = 1 \tag{37.103}$$

Because $\hat{\phi}_w^v$ is the fugacity coefficient of pure water vapor at $T = 40\,°C = 313.15\,K$ and P_{vpw} (313.15 K) = 55.3 mmHg, and this vapor pressure is low, we can assume that the pure water vapor phase also behaves ideally, that is, that:

$$\hat{\phi}_w^v = 1 \tag{37.104}$$

Using Eqs. (37.103) and (37.104) in Eq. (37.102) then yields:

$$P_{vpw}(T)C_w = y_w P \tag{37.105}$$

We will also assume, and check later, that $C_w = 1$. In that case, Eq. (37.105) yields:

$$P_{vpw}(T) = y_w P \quad \text{(Raoult's law for water)} \tag{37.106}$$

We next deal with the phase equilibria of benzene and toluene in the binary oil (o) phase and the ternary vapor phase. For this purpose, we use an activity coefficient approach for benzene and toluene in the binary oil (o) phase, and a fugacity coefficient approach for benzene and toluene in the ternary vapor (v) phase. For benzene we obtain:

$$\hat{f}_B^o = X_B^o \gamma_B^o f_B^o(T, P)$$

$$\hat{f}_B^v = y_B \hat{\phi}_B^v \tag{37.107}$$

$$\therefore \hat{f}_B^o = \hat{f}_B^v \Rightarrow X_B^o \gamma_B^o f_B^o(T, P) = y_B \hat{\phi}_B^v P$$

Similarly, for toluene, we obtain:

$$\hat{f}_T^o = X_T^o \gamma_T^o f_T^o(T, P) = f_T^v = y_T \phi_T^v P \tag{37.108}$$

Like in the case for water, we can use the Poynting corrections for benzene and toluene, that is:

$$f_i^o(T, P) = f_i^o\left(T, P_{vpi}(T)\right)C_i, i = B \text{ and } T \tag{37.109}$$

Using Eq. (37.109), for i = B, in Eq. (37.107) yields:

$$X_B^o \gamma_B^o f_B^o\left(T, P_{vpB}(T)\right)C_B = y_B \hat{\phi}_B^v P \tag{37.110}$$

Using Eq. (37.109), for i = T, in Eq. (37.108) yields:

$$X_T^o \gamma_T^o f_T^o\left(T, P_{vpT}(T)\right) C_T = y_T \widehat{\phi}_T^v P \qquad (37.111)$$

Like in the case of water,

$$f_i^o\left(T, P_{vpi}(T)\right) = f_i^v\left(T, P_{vpi}(T)\right) = \phi_i^v P_{vpi}(T), i = B \text{ and } T \qquad (37.112)$$

Using Eq. (37.112), for i = B, in Eq. (37.110) yields:

$$X_B^o \gamma_B^o \phi_B^v P_{vpB}(T) C_B = y_B \widehat{\phi}_B^v P \qquad (37.113)$$

Using Eq. (37.112), for i = T, in Eq. (37.111) yields:

$$X_T^o \gamma_T^o \phi_T^v P_{vpT}(T) C_T = y_T \widehat{\phi}_T^v P \qquad (37.114)$$

Like in the case for water, we can assume that:

$$\phi_B^v = 1, \ \phi_T^v = 1, \ \widehat{\phi}_B^v = 1, \widehat{\phi}_T^v = 1, C_B = 1, \text{and } C_T = 1 \qquad (37.115)$$

In addition, the (B + T) binary mixture is ideal, and therefore, $\gamma_B^o = 1$ and $\gamma_T^o = 1$. Using all these simplifications in Eqs. (37.113) and (37.114) yields:

$$X_B^o P_{vpB}(T) = y_B P \qquad (37.116)$$

$$X_T^o P_{vpT}(T) = y_T P \qquad (37.117)$$

Adding up Eqs. (37.106), (37.116), and (37.117) we obtain:

$$P_{vpw}(T) + X_B^o P_{vpB}(T) + X_T^o P_{vpT}(T) = P(y_w + y_B + y_T) \qquad (37.118)$$

where $X_B^o + X_T^o = 1$. Equation (37.118) shows that given T and X_B^o or T and X_T^o, two independent intensive variables, we can compute the dependent intensive variable, P, consistent with the generalized Gibbs Phase Rule Approach which indicated that L = 2!

At the conditions given (see Eq. (37.119)), and using the pure component vapor pressure values provided in the Problem Statement (see Eq. (37.120)):

For $T = 40°C = 313.15$ K, $X_B^o = \left(\dfrac{0.5}{0.5 + 1}\right)$, $X_T^o = \left(\dfrac{1}{0.5 + 1}\right)$,   (37.119)

$P_{vpw} = 55.3$ mmHg, $P_{vpB} = 181.1$ mmHg, and $P_{vpT} = 50.1$ mmHg,   (37.120)

(a) We predict that the mixture will begin to boil at a pressure P given by:

$$P = \left[55.3 + \left(\dfrac{0.5}{0.5 + 1}\right)181.1 + \left(\dfrac{1}{0.5 + 1}\right)50.1\right] \text{mmHg, or P}$$

$$= 155.1 \text{ mmHg} \tag{37.121}$$

We can next check that, as assumed, the Poynting corrections C_W, C_B, and C_T are close to unity. Indeed, the interested reader is encouraged to show that:

$C_W = 1.00007$
$C_B = 0.99988$
$C_T = 1.00050$

Food for Thought Calculate the composition of the first vapor bubble that forms.

Part III
Introduction to Statistical Mechanics

Lecture 38

Statistical Mechanics, Canonical Ensemble, Probability and the Boltzmann Factor, and Canonical Partition Function

38.1 Introduction

The material presented in this lecture is adapted from Chapter 3 in M&S. First, we will discuss the Canonical ensemble and the Boltzmann factor. Second, we will calculate the probability that a system in the Canonical ensemble is in quantum state j with energy $\underline{E}_j(N,\underline{V})$. Finally, we will provide a physical interpretation of the Canonical partition function.

Statistical mechanics studies macroscopic systems from a microscopic, or molecular, viewpoint. The goal of statistical mechanics is both to understand and to predict macroscopic behavior given the properties of the individual molecules comprising the system, including their interactions.

As we showed in Parts I and II, thermodynamics provides mathematical relations between experimental properties of macroscopic systems at equilibrium. However, it provides no information about the magnitude of any of these properties. Furthermore, thermodynamics does not seek to connect the relations that it describes to molecular models. As such, thermodynamics is limited by its inability to calculate, at the molecular level, physical properties of the type discussed in Parts I and II, including $\underline{U}$, $\underline{S}$, $\underline{H}$, $\underline{A}$, $\underline{G}$, C_v, μ, etc., or to provide molecular interpretations of its governing equations.

When our goal is to formulate a molecular theory that can accomplish the above, we enter the realm of statistical mechanics, which assumes the existence of molecules, to both calculate and interpret thermodynamic behavior from a molecular perspective. As the title of the book indicates, in Part III, we will present an introductory, albeit rigorous, exposure to the fundamentals of statistical mechanics. We hope that this introductory exposure will sufficiently spark the interest of the readers to pursue a broader immersion into this increasingly relevant subject for chemical and mechanical engineers and for chemists, physicists, and materials scientists.

© Springer Nature Switzerland AG 2020
D. Blankschtein, *Lectures in Classical Thermodynamics with an Introduction to Statistical Mechanics*, https://doi.org/10.1007/978-3-030-49198-7_38

In Part III, when possible, connections will be established with many of the concepts and methodologies that we discussed in Parts I and II. Specifically, we will present (i) a statistical mechanical interpretation of the First, Second, and Third Laws of Thermodynamics; (ii) a statistical mechanical derivation of the ideal gas EOS, the virial EOS, and the van der Waals EOS; (iii) a statistical mechanical description of ideal binary liquid mixtures using a lattice theory approach; and (iv) a statistical mechanical formulation of chemical reaction equilibria.

38.2 Canonical Ensemble and the Boltzmann Factor

Consider a macroscopic system, for example, a liter of oxygen, a gallon of water, or a kilogram of nickel. From a mechanical viewpoint, such macroscopic systems may be described by specifying the number of molecules, N; the volume, $\underline{V}$; and the interactions operating between the molecules. Although N is of the order of Avogadro's number, from a quantum mechanical perspective, each system may be described in terms of its Hamiltonian operator and associated wave functions, which depend on the coordinates of all the molecules comprising the system.

In Part III, we will not derive the fundamental equations of quantum mechanics. Instead, we refer the interested reader to an introductory textbook on this fascinating subject. However, we will provide the most important quantum mechanical tools needed to obtain the statistical mechanical results presented in Part III. We begin with the celebrated Schrödinger equation for an N-body (molecule) system, given by:

$$H_N \psi_j = \underline{E}_j \psi_j, j = 1, 2, 3, \ldots \tag{38.1}$$

where H_N is the Hamiltonian operator of the N-body (molecule) system, ψ_j is the quantum mechanical wave function associated with quantum state j, and $\underline{E}_j$ is the quantum mechanical energy (eigenvalue) associated with quantum state j. It is noteworthy that the discrete energy, $\underline{E}_j$, of quantum state j (1, 2, 3,...) depends on N and $\underline{V}$, that is:

$$\underline{E}_j = \underline{E}_j(N, \underline{V}) \tag{38.2}$$

For the special case of a non-interacting system (e.g., an ideal gas), because the molecules are independent, the total energy, $\underline{E}_j(N, \underline{V})$, can be expressed as a sum of the individual energies of each of the N molecules comprising the system, that is,

$$\underline{E}_j(N, \underline{V}) = \varepsilon_1 + \varepsilon_2 + \ldots + \varepsilon_N \tag{38.3}$$

where ε_i is the energy of molecule i (i = 1, 2, ..., N).

Next, we would like to determine the probability, p_j, that a system will be in quantum state j having energy, $\underline{E}_j(N, \underline{V})$. As we will show in Lecture 39, knowledge of this probability is essential because it will allow us to calculate average thermodynamic properties of a macroscopic system, including those discussed in Parts I and II. To calculate the desired probability, we consider a very large collection of identical systems in thermal contact with each other, as well as with an infinite heat reservoir maintained at a constant temperature T. We should recognize that each system in the ensemble has the same values of N, $\underline{V}$, and T but is likely to be in a

The Canonical Ensemble

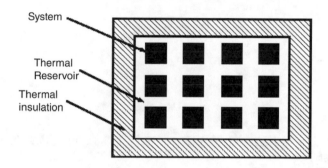

System

Thermal Reservoir

Thermal insulation

Fig. 38.1

diffcrent quantum state, consistent with the values of N and $\underline{V}$. Such a collection of systems is referred to as an ensemble and when N, $\underline{V}$, and T are specified is referred to as the Canonical ensemble, illustrated in Fig. 38.1.

In Fig. 38.1, the number of systems (s) in quantum state j with energy $\underline{E}_j(N,\underline{V})$ is denoted as s_j, and the total number of systems in the ensemble is denoted as S. Because the ensemble is a conceptual construction, S can be chosen to be as large as desired. This will become important when we define statistical probabilities (see below). Clearly, the following relation applies:

$$\sum_j s_j = S \qquad (38.4)$$

Next, we would like to calculate the relative number of systems in the Canonical ensemble in each quantum state j. As an illustration, we consider quantum states 1 and 2, having energies $\underline{E}_1(N,\underline{V})$ and $\underline{E}_2(N,\underline{V})$, respectively. Clearly, the relative number of systems in quantum states 1 and 2 must depend on $\underline{E}_1$ and $\underline{E}_2$, so that:

$$\frac{s_2}{s_1} = f(\underline{E}_1, \underline{E}_2) \qquad (38.5)$$

where the function f will be determined below.

Because energy is always defined relative to a zero of energy, the dependence of the function, f, on $\underline{E}_1$ and $\underline{E}_2$ is expected to have the following form:

$$f(\underline{E}_1, \underline{E}_2) = f(\underline{E}_1 - \underline{E}_2) \tag{38.6}$$

Using Eq. (38.6) in Eq. (38.5), we obtain:

$$\frac{s_2}{s_1} = f(\underline{E}_1 - \underline{E}_2) \tag{38.7}$$

Because Eq. (38.7) also applies to quantum states 3 and 2 and 3 and 1, it follows that:

$$\frac{s_3}{s_2} = f(\underline{E}_2 - \underline{E}_3) \tag{38.8}$$

and

$$\frac{s_3}{s_1} = f(\underline{E}_1 - \underline{E}_3) \tag{38.9}$$

Because $s_3/s_1 = (s_2/s_1) \cdot (s_3/s_2)$, Eqs. (38.7), (38.8), and (38.9) indicate that f should satisfy:

$$f(\underline{E}_1 - \underline{E}_3) = f(\underline{E}_1 - \underline{E}_2) f(\underline{E}_2 - \underline{E}_3) \tag{38.10}$$

Recalling that the exponential function satisfies $e^{x + y} = e^x \cdot e^y$, Eq. (38.10) suggests that:

$$f(\underline{E}) = e^{\beta \underline{E}} \tag{38.11}$$

where β is an arbitrary constant to be determined later.

For any quantum states n and m, using Eq. (38.11), with $\underline{E} \rightarrow \underline{E}_m - \underline{E}_n$, in Eq. (38.7), with $2 \rightarrow n$ and $1 \rightarrow m$, yields:

$$\frac{s_n}{s_m} = e^{\beta(\underline{E}_m - \underline{E}_n)} \tag{38.12}$$

The form of Eq. (38.12) implies that:

$$s_j = Ce^{-\beta \underline{E}_j} \tag{38.13}$$

where j is either quantum state n or m and C is a constant to be determined below.

38.3   Probability That a System in the Canonical Ensemble Is in Quantum State j with Energy $\underline{E}_j(N, \underline{V})$

Equation (38.13) has two unknown quantities, C and β, that we need to determine. Determining C is quite simple. Indeed, summing both sides of Eq. (38.13) with respect to j and recalling that $\sum_j s_j$ is equal to the total number of systems in the

Canonical ensemble, S, we obtain:

$$\sum_j s_j = S = C \sum_j e^{-\beta E_j} \tag{38.14}$$

or

$$C = \frac{S}{\sum_j e^{-\beta E_j}} \tag{38.15}$$

Using Eq. (38.15) in Eq. (38.13), including rearranging, yields:

$$\frac{s_j}{S} = \frac{e^{-\beta E_j}}{\sum_j e^{-\beta E_j}} \tag{38.16}$$

In Eq. (38.16), the ratio s_j/S is the fraction of systems in the Canonical ensemble that are in quantum state j with energy $\underline{E}_j$. In the limit of large S, which we can certainly take because our ensemble may chosen to be as large as we would like it to be, the ratio s_j/S becomes the statistical probability that a randomly chosen system in the Canonical ensemble will be in quantum state j with energy $\underline{E}_j(N, \underline{V})$. Denoting this probability as p_j, it follows that:

$$p_j = \frac{e^{-\beta E_j}}{\sum_j e^{-\beta E_j}} \tag{38.17}$$

Equation (38.17) is a central result of statistical mechanics because, as we will show in the following lectures, it will allow us to calculate average thermodynamic properties of macroscopic systems.

It is customary to denote the denominator in Eq. (38.17) as Q, where:

$$Q(N, \underline{V}, \beta) = \sum_j e^{-\beta E_j(N, \underline{V})} \tag{38.18}$$

In a coming lecture, we will show that:

$$\beta = \frac{1}{k_B T} \tag{38.19}$$

where k_B is the Boltzmann constant $(1.381 \times 10^{-23} \text{J/K})$ and T is the Kelvin temperature.

Using Eqs. (38.19) and (38.18) in Eq. (38.17) yields:

$$p_j = \frac{e^{-E_j(N,\underline{V})/k_B T}}{Q(N, \underline{V}, T)} \tag{38.20}$$

Equation (38.20) can also be expressed in terms of β, as follows:

$$p_j = \frac{e^{-\beta E_j(N,\underline{V})}}{Q(N, \underline{V}, \beta)} \tag{38.21}$$

The function $Q(N, \underline{V}, \beta)$, or $Q(N, \underline{V}, T)$, is known as the partition function of the system in the Canonical ensemble representation or, in short, as the Canonical partition function. In the coming lectures, we will show how to calculate all the thermodynamic properties of a macroscopic system if we know $Q(N, \underline{V}, \beta)$ or $Q(N, \underline{V}, T)$. Specifically, we will learn how to calculate the Canonical partition function for a number of interesting systems.

38.4   Physical Interpretation of the Canonical Partition Function

According to Eq. (38.18), Q is a sum of Boltzmann factors, $e^{-E_j/k_B T}$, that determine how molecules are partitioned throughout the accessible quantum states of the system.

For simplicity, if we assume that the energy of the ground state (1) is zero, that is, if $\underline{E}_1 = 0$, then:

$$Q = \sum_{j=1}^{t} e^{-E_j/k_B T} = 1 + e^{-E_2/k_B T} + e^{-E_3/k_B T} + \cdots + e^{-E_t/k_B T} \tag{38.22}$$

where t is the number of accessible quantum states, which is determined by the underlying physics of the system.

The Canonical partition function, Q, accounts for the number of quantum states that are effectively accessible to the system, out of the t accessible quantum states. To better understand what this means, it is instructive to consider two limiting cases:

(i) The energies are small, or equivalently, the temperatures are high, that is, $\underline{E}_j/k_BT \to 0$, for all js > 1. Accordingly,

$$Q = \overbrace{1 + 1 + 1 + \cdots + 1}^{t \text{ Times}} = t \tag{38.23}$$

and

$$p_j = \frac{e^{-E_j/k_BT}}{Q} = \frac{1}{t} \tag{38.24}$$

Equation (38.24) indicates that, in this limit, all the t states become accessible with equal probability, 1/t.

(ii) The energies are large, or equivalently, the temperatures are low, that is, $\underline{E}_j/k_BT \to \infty$, for all js > 1. Accordingly,

$$Q = 1 + \overbrace{0 + 0 + 0 + \cdots + 0}^{(t-1) \text{ Times}} = 1 \tag{38.25}$$

and

$$\begin{aligned} p_1 &= \frac{1}{1} = 1 \\ p_{j>1} &= \frac{0}{1} = 0 \end{aligned} \tag{38.26}$$

Equation (38.26) indicates that, in this limit, only the ground state is accessible.

In conclusion, the magnitude of $\underline{E}_j$ relative to k_BT, namely, $\underline{E}_j/k_BT$, determines whether or not quantum state j is effectively accessible. Indeed, quantum states that possess energies that are higher than k_BT are relatively inaccessible and unpopulated at temperature T. On the other hand, quantum states that possess energies that are lower than k_BT are accessible and well populated at temperature, T. This criterion will be utilized extensively in the coming lectures.

Lecture 39

Calculation of Average Thermodynamic Properties Using the Canonical Partition Function and Treatment of Distinguishable and Indistinguishable Molecules

39.1 Introduction

The material presented in this lecture is adapted from Chapter 3 in M&S. First, we will utilize material presented in Lecture 38 to calculate various Canonical ensemble-averaged thermodynamic properties of the system, including the average energy, the average heat capacity at constant volume, and the average pressure. Second, we will derive an expression for the Canonical partition function of a system of independent and distinguishable molecules. Third, we will derive an expression for the Canonical partition function of a system of independent and indistinguishable molecules, including stressing the role of the Pauli exclusion principle when carrying out the summation over the indistinguishable quantum states. Fourth, we will decompose the molecular Canonical partition function into contributions from the translational, vibrational, rotational, and electronic degrees of freedom. Finally, we will contrast energy states and energy levels, including discussing the degeneracy of an energy level.

39.2 Calculation of the Average Energy of a Macroscopic System

In Lecture 38, we saw that in the Canonical ensemble, the probability that a system is in quantum state j is given by:

$$p_j(N, \underline{V}, \beta) = \frac{\exp(-\beta \underline{E}_j(N, \underline{V}))}{Q(N, \underline{V}, \beta)} \tag{39.1}$$

where Q is the Canonical partition function, given by:

© Springer Nature Switzerland AG 2020
D. Blankschtein, *Lectures in Classical Thermodynamics with an Introduction to Statistical Mechanics*, https://doi.org/10.1007/978-3-030-49198-7_39

$$Q(N, \underline{V}, \beta) = \sum_j \exp\left(-\beta \underline{E}_j(N, \underline{V})\right) \tag{39.2}$$

Using the well-known statistical definition of an average property, the Canonical ensemble-averaged energy, $\langle E \rangle$, which is equal to the experimentally observed energy, $\underline{U}$, is given by:

$$\underline{U} = \langle E \rangle = \sum_j p_j \underline{E}_j = \sum_j \frac{\underline{E}_j(N, \underline{V}) e^{-\beta \underline{E}_j(N, \underline{V})}}{Q(N, \underline{V}, \beta)} \tag{39.3}$$

where Eq. (39.1) was used. Although the quantum mechanical energies, $\underline{E}_j$, do not depend on temperature, the Canonical ensemble-averaged energy, $\langle E \rangle = \underline{U}$, depends on T or equivalently on β, as well as on N and $\underline{V}$.

It is possible to express $\langle E \rangle = \underline{U}$ in Eq. (39.3) entirely in terms of $Q(N, \underline{V}, \beta)$. Specifically, differentiating $\ln Q(N, \underline{V}, \beta)$ with respect to β, at constant N and $\underline{V}$, yields:

$$\left(\frac{\partial \ln Q(N, \underline{V}, \beta)}{\partial \beta}\right)_{\underline{V}, N} = -\sum_j \frac{\underline{E}_j(N, \underline{V}) e^{-\beta \underline{E}_j(N, \underline{V})}}{Q(N, \underline{V}, \beta)} \tag{39.4}$$

A comparison of Eqs. (39.3) and (39.4) shows that:

$$\underline{U} = \langle E \rangle = -\left(\frac{\partial \ln Q(N, \underline{V}, \beta)}{\partial \beta}\right)_{\underline{V}, N} \tag{39.5}$$

We can also express $\langle E \rangle = \underline{U}$ as a temperature derivative, rather than as a β derivative. Using the fact that:

$$\frac{\partial}{\partial \beta} = -k_B T^2 \frac{\partial}{\partial T} \tag{39.6}$$

and then using Eq. (39.6) in Eq. (39.5), we obtain:

$$\underline{U} = \langle E \rangle = k_B T^2 \left(\frac{\partial \ln Q(N, \underline{V}, T)}{\partial T}\right)_{\underline{V}, N} \tag{39.7}$$

39.3 Calculation of the Average Heat Capacity at Constant Volume of a Macroscopic System

Recall that the heat capacity at constant volume, C_V, was introduced in Part I as follows:

$$C_V = \left(\frac{\partial U}{\partial T}\right)_V = \frac{1}{N}\left(\frac{\partial U}{\partial T}\right)_{V,N} \tag{39.8}$$

Using the fact that the internal energy of a macroscopic system, $\underline{U} = \langle E \rangle$, it follows that:

$$C_V = \left(\frac{\partial \langle E \rangle}{\partial T}\right)_V = \frac{1}{N}\left(\frac{\partial \langle E \rangle}{\partial T}\right)_{\underline{V},N} \tag{39.9}$$

We can use Eq. (39.7), along with Eq. (39.2), to calculate C_V using Eq. (39.9).

39.4 Calculation of the Average Pressure of a Macroscopic System

In one of the coming lectures, we will show that the pressure of a macroscopic system in quantum state j is given by:

$$P_j(N, \underline{V}) = -\left(\frac{\partial E_j}{\partial \underline{V}}\right)_N \tag{39.10}$$

Equation (39.10) is analogous to the definition of the pressure presented in Part I, where:

$$P = -(\partial \underline{U}/\partial \underline{V})_{\underline{S},N} \tag{39.11}$$

Accepting Eq. (39.10) for now, the Canonical ensemble-averaged pressure, $\langle P \rangle$, which is equal to the experimentally observed pressure, P, is given by:

$$P = \langle P \rangle = \sum_j p_j(N, \underline{V}, \beta) P_j(N, \underline{V}) \tag{39.12}$$

Using Eq. (39.1) for p_j and Eq. (39.10) for P_j, in Eq. (39.12), yields:

$$P = \langle P \rangle = \sum_j -\left(\frac{\partial E_j}{\partial \underline{V}}\right)_N \frac{e^{-\beta E_j(N,\underline{V})}}{Q(N,\underline{V},\beta)} \tag{39.13}$$

We can express $P = \langle P \rangle$ solely in terms of Q as follows. Starting from the definition of Q in Eq. (39.2), and differentiating Q with respect to $\underline{V}$, keeping N and β constant, yields:

$$\left(\frac{\partial Q}{\partial \underline{V}}\right)_{N,\beta} = -\beta \sum_j \left(\frac{\partial E_j}{\partial \underline{V}}\right)_N e^{-\beta E_j(N,\underline{V})} \tag{39.14}$$

A comparison of Eqs. (39.13) and (39.14) shows that:

$$P = \langle P \rangle = \frac{k_B T}{Q(N,\underline{V},\beta)}\left(\frac{\partial Q}{\partial \underline{V}}\right)_{\beta,N} \tag{39.15}$$

Equation (39.15) can also be expressed as follows:

$$P = \langle P \rangle = k_B T\left(\frac{\partial \ln Q(N,\underline{V},\beta)}{\partial \underline{V}}\right)_{\beta,N} \tag{39.16}$$

39.5 Canonical Partition Function of a System of Independent and Distinguishable Molecules

The general results that we have presented so far in Part III are valid for arbitrary systems. In order to actually use these results, we need to calculate the Canonical partition function, Q. For this purpose, we need to know the set of energy eigenvalues, $\{E_j(N,\underline{V})\}$, for the N-body (molecule) Schrödinger equation. In general, because of the complexity of this interacting multi-molecule system, this turns out to be an intractable problem. However, for many important systems, the total energy of the N-molecule system can be expressed as a sum of individual energies associated with each molecule comprising the system. This approximation, when applicable, leads to a great simplification in the evaluation of the Canonical partition function, Q.

For a system of independent, distinguishable molecules, we will denote the energies of the individual molecules as $\{\varepsilon_j^a\}$, where a denotes the molecule in question (they are distinguishable) and j denotes the quantum state of the molecule in question.

For such a system, the total energy, $\underline{E}_{ijk...}(N, \underline{V})$, can be written as follows:

$$\underline{E}_{ijk...}(N, \underline{V}) = \underbrace{\varepsilon_i^a(\underline{V}) + \varepsilon_j^b(\underline{V}) + \varepsilon_k^c(\underline{V}) + \cdots}_{N\ \text{Terms}} \tag{39.17}$$

In principle, because the ε_is in Eq. (39.17) depend on $\underline{V}$, they should carry an underbar. However, to simplify the notation, we have not included the underbars. In addition, as stressed above, the temperature does not appear at the quantum mechanical level.

Using Eq. (39.17), the Canonical partition function of the system can be written as follows:

$$Q(N, \underline{V}, T) = \sum_{ijk...} e^{-\beta \underline{E}_{i,j,k,...}} = \sum_{i,j,k,...} e^{-\beta\left(\varepsilon_i^a + \varepsilon_j^b + \varepsilon_k^c + \cdots\right)} \tag{39.18}$$

Recall that in Eq. (39.18), the molecules are distinguishable and independent. Therefore, we can sum over i, j, k, etc. independently. As a result, $Q(N, \underline{V}, T)$ in Eq. (39.18) can be written as a product of partition functions associated with molecules a, b, c, etc., that is,

$$Q(N, \underline{V}, T) = \left(\sum_i e^{-\beta \varepsilon_i^a}\right)\left(\sum_j e^{-\beta \varepsilon_j^b}\right)\left(\sum_k e^{-\beta \varepsilon_k^c}\right)\cdots \tag{39.19}$$

or

$$Q(N, \underline{V}, T) = q_a(\underline{V}, T)\, q_b(\underline{V}, T)\, q_c(\underline{V}, T)\cdots \tag{39.20}$$

where each of the molecular Canonical partition functions, $q(\underline{V}, T)$, is given by:

$$q(\underline{V}, T) = \sum_i e^{-\beta \varepsilon_i} = \sum_i e^{-\varepsilon_i/k_B T} \tag{39.21}$$

and the number of molecular Canonical partition functions in Eq. (39.20) is equal to N. Because the calculation of $q(\underline{V}, T)$ in Eq. (39.20) only requires knowledge of the discrete quantum mechanical energy states of a single molecule, its calculation is often feasible, as we will show shortly.

If the quantum mechanical energy states of all the molecules are the same, then, all the molecular Canonical partition functions are equal, and Eq. (39.20) reduces to:

$$Q(N, \underline{V}, T) = [q(\underline{V}, T)]^N \tag{39.22}$$

where $q(\underline{V}, T)$ is given by Eq. (39.21).

39.6   Canonical Partition Function of a System of Independent and Indistinguishable Molecules

In the case of indistinguishable molecules, the total energy is given by:

$$E_{ijk\cdots} = \underbrace{\varepsilon_i + \varepsilon_j + \varepsilon_k + \cdots}_{\text{N Terms}} \tag{39.23}$$

where, because the N molecules are indistinguishable, no superscripts appear in the energies $\{\varepsilon_i\}$. Each molecular energy in Eq. (39.23) depends on the system volume, $\underline{V}$, and therefore should carry an underbar, which we have not included for notational simplicity. Using Eq. (39.23) in the expression for the Canonical partition function of the system, we obtain:

$$Q(N, \underline{V}, T) = \sum_{i, j, k, \cdots} e^{-\beta(\varepsilon_i + \varepsilon_j + \varepsilon_k + \cdots)} \tag{39.24}$$

Because the molecules are indistinguishable, the quantum mechanical Pauli exclusion principle applies and states that: "No two electrons in a molecule can be in the same quantum state." As a result, it is clear that the indices i, j, k, $\cdots$ in Eq. (39.24) are not independent. Therefore, the summations in Eq. (39.24) cannot be carried out independently as we did in the case of molecules which are distinguishable (see Eq. (39.19)).

Nevertheless, if the number of terms in Eq. (39.24) in which two or more indices are the same, and hence, are not independent, is small, we could carry out the summations in an unrestricted manner and obtain $[q(\underline{V}, T)]^N$, as we did in the distinguishable molecule case. Subsequently, we would need to divide this result by N! in order to account for the overcounting because the molecules are indistinguishable in this case. This, in turn, would result in $[q(\underline{V}, T)]^N / N!$

It is reasonable to assume that if the number of quantum states available to any molecule is much larger than the number of molecules, it would be very unlikely for any two molecules to be in the same quantum state. The following quantum mechanical criterion can be used to ensure that this condition is satisfied:

$$\frac{N}{\underline{V}} \left(\frac{h^2}{8mk_BT} \right)^{3/2} \ll 1 \tag{39.25}$$

The inequality in Eq. (39.25) is favored by a large molecular mass (m), a high temperature (T), and a low number density ($\rho = N/\underline{V}$). Table 39.1 lists the values of $(N/\underline{V})(h^2/8mk_BT)^{3/2}$, at a pressure of one bar and various temperatures, for several systems.

Table 39.1

System	T/K	$\frac{N}{\underline{V}}\left(\frac{h^2}{8mk_BT}\right)^{3/2}$
Liquid helium	4	1.5
Gaseous helium	4	0.11
Gaseous helium	20	1.8×10^{-3}
Gaseous helium	100	3.3×10^{-5}
Liquid hydrogen	20	0.29
Gaseous hydrogen	20	5.1×10^{-3}
Gaseous hydrogen	100	9.4×10^{-5}
Liquid neon	27	1.0×10^{-2}
Gaseous neon	27	7.8×10^{-5}
Liquid krypton	127	5.1×10^{-5}
Electrons in metals	300	1400

As Table 39.1 shows, systems for which $(N/\underline{V})(h^2/8mk_BT)^{3/2}$ is not $\ll 1$, therefore violating the criterion in Eq. (39.25), include liquid helium and liquid hydrogen (due to their small masses and low temperatures) and electrons in metals (due to their small mass).

When Eq. (39.25) is satisfied, the summations in Eq. (39.24) can be carried out independently, which after dividing by N!, yields:

$$Q(N, \underline{V}, T) = \frac{[q(\underline{V}, T)]^N}{N!} \qquad (39.26)$$

where $q(\underline{V}, T)$ is given in Eq. (39.22).

When Eqs. (39.25) and (39.26) are satisfied, the molecules obey Boltzmann statistics. As Eq. (39.25) shows, Boltzmann statistics becomes increasingly applicable as the temperature increases.

39.7   Decomposition of a Molecular Canonical Partition Function into Canonical Partition Functions for Each Degree of Freedom

We would like to express the average energy of a macroscopic system, $\langle E \rangle$, in terms of the molecular Canonical partition function, $q(\underline{V}, T)$. We begin with Eq. (39.7), where for indistinguishable molecules taking the natural logarithm of Eq. (39.26) yields:

$$\ln Q = \ln\left(\frac{q^N}{N!}\right) = N\ln q - \ln N! \qquad (39.27)$$

Using Eq. (39.27) in Eq. (39.7), while keeping N and $\underline{V}$ constant, yields:

$$\langle \underline{E} \rangle = \frac{N k_B T^2}{q} \left(\frac{\partial q}{\partial T} \right)_{\underline{V}}$$

(39.28)

Because (see Eqs. (39.22) and (39.6))

$$q = \sum_j e^{-\beta \varepsilon_j} \quad \text{and} \quad \frac{d\beta}{dT} = -\frac{1}{k_B T^2}$$

(39.29)

it follows that:

$$\left(\frac{\partial q}{\partial T} \right)_{\underline{V}} = \frac{1}{k_B T^2} \sum_j \varepsilon_j e^{-\beta \varepsilon_j}$$

(39.30)

Using Eq. (39.30) in Eq. (39.28) yields:

$$\langle \underline{E} \rangle = N \langle \varepsilon \rangle$$

(39.31)

where the average molecular energy, $\langle \varepsilon \rangle$, is given by:

$$\langle \varepsilon \rangle = \sum_j \varepsilon_j \frac{e^{-\varepsilon_j/k_B T}}{q(\underline{V}, T)}$$

(39.32)

Equation (39.32) shows that the probability that a molecule is in quantum state j, denoted by $\pi_j(\underline{V}, T)$, is given by:

$$\pi_j(\underline{V}, T) = \frac{e^{-\varepsilon_j/k_B T}}{q(\underline{V}, T)} = \frac{e^{-\varepsilon_j/k_B T}}{\sum_j e^{-\varepsilon_j/k_B T}}$$

(39.33)

Equations (39.31), (39.32), and (39.33) can be simplified even further if we assume that the energy of a molecule can be decomposed into several contributions associated with the various molecular degrees of freedom: translational (trans), rotational (rot), vibrational (vib), and electronic (elec). Specifically:

$$\varepsilon = \varepsilon_i^{trans} + \varepsilon_j^{rot} + \varepsilon_k^{vib} + \varepsilon_l^{elec}$$

(39.34)

Because the various molecular degrees of freedom are distinguishable, it follows that:

$$q(V,T) = \underbrace{\left(\sum_i e^{-\varepsilon_i^{trans}/k_B T}\right)}_{q_{trans}} \underbrace{\left(\sum_j e^{-\varepsilon_j^{rot}/k_B T}\right)}_{q_{rot}} \underbrace{\left(\sum_k e^{-\varepsilon_k^{vib}/k_B T}\right)}_{q_{vib}} \underbrace{\left(\sum_l e^{-\varepsilon_l^{elec}/k_B T}\right)}_{q_{elec}}$$

or

$$q(\underline{V},T) = q_{trans}\, q_{rot}\, q_{vib}\, q_{elec} \tag{39.35}$$

39.8 Energy States and Energy Levels

So far, we have expressed partition functions as summations over energy states. Each energy state is represented by a quantum mechanical wave function with an associated energy, that is,

$$q(\underline{V},T) = \sum_{\substack{j \\ (\text{States})}} e^{-\varepsilon_j/k_B T} \tag{39.36}$$

Alternatively, if we refer to sets of states that have the same energy as levels, we can write $q(\underline{V},T)$ as a summation over energy levels by including the degeneracy, g_j, of each energy level j, that is,

$$q(\underline{V},T) = \sum_{\substack{j \\ (\text{Levels})}} g_j\, e^{-\varepsilon_j/k_B T} \tag{39.37}$$

In Eq. (39.36), the terms representing a degenerate energy state j are repeated g_j times. On the other hand, in Eq. (39.37), degenerate energy state j is written only once and then multiplied by g_j. Including degeneracies explicitly, as we did in Eq. (39.37), is usually more convenient from a computational point of view, as we will show in the coming lectures.

Lecture 40

Translational, Vibrational, Rotational, and Electronic Contributions to the Partition Function of Monoatomic and Diatomic Ideal Gases and Sample Problem

40.1 Introduction

The material presented in this lecture is adapted from Chapter 4 in M&S. First, we will derive an expression for the Canonical partition function of a monoatomic ideal gas, including calculating the translational contribution to the partition function and its average translation energy. Second, we will discuss the Energy Equipartition Theorem. Third, we will derive an expression for the electronic contribution to the atomic partition function. Fourth, we will solve Sample Problem 40.1 to calculate the fraction of helium atoms in the first excited state at 300 K, given information about the degeneracies and energies of the excited states involved. Fifth, we will calculate the average energy, average heat capacity at constant volume, and average pressure of a monoatomic ideal gas. Finally, we will begin discussing a diatomic ideal gas, where in addition to the translational and electronic degrees of freedom encountered in the case of a monoatomic ideal gas, we will need to incorporate the vibrational and rotational degrees of freedom, which we will discuss in Lecture 41. Here, we will calculate the translational and electronic contributions to the molecular partition function of a diatomic ideal gas.

40.2 Partition Functions of Ideal Gases

We already saw that if the number of available quantum states is much larger than the number of molecules (or atoms) in the system, then, we can write the Canonical partition function of the entire system in terms of the individual Canonical molecular partition functions. Specifically,

© Springer Nature Switzerland AG 2020
D. Blankschtein, *Lectures in Classical Thermodynamics with an Introduction to Statistical Mechanics*, https://doi.org/10.1007/978-3-030-49198-7_40

$$Q(N, \underline{V}, T) = \frac{[q(\underline{V}, T)]^N}{N!} \tag{40.1}$$

Equation (40.1) for Q is particularly valid for ideal gases, because the molecules are independent in that case and the densities $(N/\underline{V})$ of gases that behave ideally are sufficiently low that the criterion, $(N/\underline{V})\left(h^2/8mk_BT\right)^{3/2} \ll 1$, required for Eq. (40.1) to be valid, is satisfied. We will discuss a monoatomic ideal gas first, followed by a diatomic ideal gas. At this introductory level, we will not discuss a polyatomic ideal gas.

40.3 Translational Partition Function of a Monoatomic Ideal Gas

The energy of an atom in a monoatomic ideal gas is the sum of its translational energy and electronic energy, that is,

$$\varepsilon_{\text{atomic}} = \varepsilon_{\text{trans}} + \varepsilon_{\text{elec}} \tag{40.2}$$

In Eq. (40.2), nuclear degrees of freedom are not included because they are not accessible at the temperatures of interest. Using Eq. (40.2) in the general expression for q involving these two degrees of freedom (see Lecture 39), we obtain:

$$q(\underline{V}, T) = q_{\text{trans}}(\underline{V}, T)q_{\text{elec}}(T) \tag{40.3}$$

To compute q_{trans} in Eq. (40.3), we need to make use of some well-known quantum mechanics results. Specifically, the solution of the Schrödinger equation for the translational energy states of an atom in a cubic container of side length, d, is given by:

$$\varepsilon_{n_x n_y n_z} = \frac{h^2}{8md^2}\left(n_x^2 + n_y^2 + n_z^2\right), \text{for } n_x, n_y, n_z = 1, 2, \ldots \tag{40.4}$$

where the translational energy states are quantized (discrete) and are labeled by the quantum numbers (n_x, n_y, and n_z) corresponding to the three spatial dimensions (x, y, and z).

Substituting Eq. (40.4) in the expression for q_{trans} derived in Lecture 39 yields:

$$q_{\text{trans}} = \sum_j e^{-\beta \varepsilon_j^{\text{trans}}} \tag{40.5}$$

with $j = n_x, n_y, n_z = 1, 2$, etc. Carrying out the summations in Eq. (40.5), we obtain:

$$q_{trans} = \sum_{n_x, n_y, n_z=1}^{\infty} e^{-\beta \epsilon_{n_x n_y n_z}}$$

$$= \left[\sum_{n_x=1}^{\infty} \exp\left(-\frac{\beta h^2 n_x^2}{8md^2} \right) \right] \left[\sum_{n_y=1}^{\infty} \exp\left(-\frac{\beta h^2 n_y^2}{8md^2} \right) \right]$$

$$\times \left[\sum_{n_z=1}^{\infty} \exp\left(\frac{-\beta h^2 n_z^2}{8md^2} \right) \right] \tag{40.6}$$

Because each summation in Eq. (40.6) is identical, we can rewrite it as follows:

$$q_{trans} = \left[\sum_{n=1}^{\infty} \exp\left(\frac{-\beta h^2 n^2}{8md^2} \right) \right]^3 \tag{40.7}$$

Recall that in Eqs. (40.4), (40.6), and (40.7), m is the mass of the atom, $\beta = 1/k_B T$, and h is Planck's constant $= 6.626 \times 10^{-34}$ Js.

To a very good approximation, the summation in Eq. (40.7) can be replaced by an integral over n. Specifically,

$$q_{trans}(\underline{V}, T) = \left(\int_0^{\infty} e^{-\beta h^2 n^2 / 8md^2} \, dn \right)^3 \tag{40.8}$$

Recall that the integral in Eq. (40.8) is a Gaussian integral. If we denote $\beta h^2 / 8md^2$ by z, then, this integral is given by:

$$\int_0^{\infty} e^{-zn^2} \, dn = \left(\frac{\pi}{4z} \right)^{1/2} \tag{40.9}$$

Using Eq. (40.9) in Eq. (40.8) yields:

$$q_{trans}(\underline{V}, T) = \left(\frac{2\pi m k_B T}{h^2} \right)^{3/2} d^3 \tag{40.10}$$

where $d^3 = \underline{V}$. In Eq. (40.10), the factor $(h^2/2\pi m k_B T)^{1/2}$ has units of length and is known as the thermal de Broglie wavelength of the atom. Denoting this length by Λ, Eq. (40.10) can be written as follows:

$$q_{trans} = \frac{V}{\Lambda^3} \qquad (40.11)$$

Recall that the criterion for the applicability of Boltzmann statistics presented in Lecture 39, $(N/\underline{V})(h^2/8mk_BT) \ll 1$, can be expressed in terms of Λ as follows:

$$\left(\frac{\pi}{4}\right)^{3/2}\left(\frac{\Lambda}{d}\right)^3 \ll 1 \qquad (40.12)$$

Equation (40.12) shows that for the number of available quantum states to be much larger than the number of atoms in the system, so that Boltzmann statistics is valid, we require that $\Lambda \ll d$, that is, that the spatial spread of the quantum mechanical wave function be much smaller than the interatomic distance. This, in turn, guarantees that the wave functions do not overlap. Clearly, quantum mechanical effects become less important as $\Lambda \ll d$, which can be attained when m is large, when T is high, or when N/V is small.

Next, we will calculate the average translational energy, $\langle\varepsilon_{trans}\rangle$, of an atom using Eq. (40.10) in the expression for $\langle\varepsilon_{trans}\rangle$ derived in Lecture 39. Specifically,

$$\langle\varepsilon_{trans}\rangle = k_BT^2\left(\frac{\partial\ln q_{trans}}{\partial T}\right)_V \qquad (40.13)$$

$$\langle\varepsilon_{trans}\rangle = k_BT^2\left[\frac{\partial}{\partial T}\left(\frac{3}{2}\ln T + (\text{Terms that do not depend on T})\right)\right]_{\underline{V}}$$

$$\langle\varepsilon_{trnas}\rangle = \frac{3}{2}k_BT \qquad (40.14)$$

Equation (40.14) indicates that each of the three atomic translational degrees of freedom contributes $\frac{1}{2}k_BT$ to the average translational energy of the atom. This result is an example of the Energy Equipartition Theorem to be discussed below.

40.4 Electronic Contribution to the Atomic Partition Function

It is convenient to write the electronic partition function of an atom as a sum over energy levels, with known degeneracies. Specifically,

$$q_{elec} = \sum_i g_{ei}e^{-\beta\varepsilon_{ei}} \qquad (40.15)$$

where g_{ei} is the degeneracy of quantum level i and ε_{ei} is the electronic energy of quantum level i. If we measure all the electronic energies relative to the ground electronic state (denoted by $i = 1$), or equivalently, if we choose $\varepsilon_{e1} = 0$, we can express Eq. (40.15) as follows:

$$q_{elec}(T) = g_{e1} + g_{e2}e^{-\beta\varepsilon_{e2}} + \cdots \tag{40.16}$$

Because q_{elec} is an atomic property, it does not depend on $\underline{V}$, and depends solely on T.

Quantum mechanics indicates that, typically,

$$\beta\varepsilon_{elec} = \frac{\varepsilon_{elec}}{k_B T} \approx \frac{10^4 K}{T} \tag{40.17}$$

which is equal to 10, even for a relatively high temperature like $T = 1000$ K. Clearly, as T decreases, $\beta\varepsilon_{elec}$ becomes even larger.

In view of Eq. (40.17), $e^{-\beta\varepsilon_{e2}}$ in Eq. (40.16) has a typical value of around 10^{-5} for most atoms at ordinary temperatures. As a result, in Eq. (40.16), only the first term, g_{e1} – the degeneracy of the ground electronic level – is $\neq 0$.

40.5 Sample Problem 40.1

Calculate the fraction of helium atoms in the first excited state at $T = 300$ K. It is known that $g_{e1} = 1$, $g_{e2} = 3$, and $\beta\varepsilon_{e2} = 767$ at 300 K. It is also known that $g_{e3} = 1$ and $\beta\varepsilon_{e3} = 797$ at 300 K.

40.5.1 Solution

The fraction of helium atoms in the first excited state ($j = 2$) is given by the probability of finding the helium atoms in that state. In other words,

$$f_2 = p_2 = \frac{g_{e2}e^{-\beta\varepsilon_{e2}}}{q_{elec}(T)}$$

where

$$q_{elec}(T) = g_{e1} + g_{e2}e^{-\beta\varepsilon_{e2}} + g_{e3}e^{-\beta\varepsilon_{e3}} + \cdots$$

Using the information provided in the Problem Statement, it follows that:

$$f_2 = \frac{3e^{-767}}{1 + 3e^{-767} + 1e^{-797}} \simeq 10^{-334}$$

which clearly shows that the first excited state is not populated. In fact, for most atoms, including the first two terms of the electronic partition function is usually sufficient, that is,

$$q_{elec}(T) = g_{e1} + g_{e2}e^{-\beta\varepsilon_{e2}} \tag{40.18}$$

In summary, for a monoatomic ideal gas:

$$Q(N, \underline{V}, T) = \frac{q^N}{N!} = \frac{(q_{trans}\, q_{elec})^N}{N!}$$

$$q_{trans} = \left(\frac{2\pi m k_B T}{h^2}\right)^{3/2} \underline{V} \tag{40.19}$$

$$q_{elec} = g_{e1} + g_{e2}e^{-\beta\varepsilon_{e2}}$$

40.6   Average Energy of a Monoatomic Ideal Gas

Denoting the average energy by $\underline{U} = \langle \underline{E} \rangle$ and using the result obtained in Lecture 39, it follows that:

$$\underline{U} = k_B T^2 \left(\frac{\partial \ln Q}{\partial T}\right)_{\underline{V},N} = N k_B T^2 \left(\frac{\partial \ln q}{\partial T}\right)_{\underline{V}} \tag{40.20}$$

Using the expression for $q = q_{trans} \cdot q_{elec}$, along with Eq. (40.19), in Eq. (40.20) yields:

$$\underline{U} = \underbrace{\frac{3}{2}N k_B T}_{\substack{\text{Average} \\ \text{Translational} \\ \text{(Kinetic)} \\ \text{Energy}}} + \overbrace{\frac{N g_{e2}\varepsilon_{e2}e^{-\beta\varepsilon_{e2}}}{q_{elec}}}^{\text{Average Electronic Energy}} \tag{40.21}$$

In most cases, the electronic contribution in Eq. (40.21) is negligible.

40.7   Average Heat Capacity at Constant Volume of a Monoatomic Ideal Gas

Neglecting the electronic contribution to $\underline{U}$, the average heat capacity at constant volume, C_V, is given by (see Part I):

$$C_V = \left(\frac{\partial U}{\partial T}\right)_V = \frac{1}{N}\left(\frac{\partial \underline{U}}{\partial T}\right)_{\underline{V},N} = \overbrace{\frac{3}{2}k_B}^{\text{Per Molecule}} = \underbrace{\frac{3}{2}R}_{\text{Per Mole}} \qquad (40.22)$$

40.8   Average Pressure of a Monoatomic Ideal Gas

As shown in Lecture 39:

$$P = k_B T\left(\frac{\partial \ln Q}{\partial \underline{V}}\right)_{T,N} = N k_B T\left(\frac{\partial \ln q}{\partial \underline{V}}\right)_T \qquad (40.23)$$

Using the expression for q in Eq. (40.19) in Eq. (40.23), we obtain:

$$P = N k_B T\left[\frac{\partial}{\partial \underline{V}}(\ln \underline{V} + \text{Terms that do not depend on } \underline{V})\right]_T$$

or

$$P = \frac{N k_B T}{\underline{V}} \qquad (40.24)$$

Equation (40.24) is, of course, the ideal gas equation of state, expressed on a per molecule basis (see Part I), that we have now derived molecularly using statistical mechanics. Further, Eq. (40.24) was obtained because $q(\underline{V}, T)$ is of the form $f(T)\underline{V}$. As a result, only the translational degrees of freedom of the atoms depend on $\underline{V}$ and contribute to P. Below, we will show that this is also the case for a diatomic ideal gas.

40.9   Diatomic Ideal Gas

In addition to the translational and electronic degrees of freedom, diatomic molecules possess vibrational and rotational degrees of freedom. As we did to model a monoatomic ideal gas, for a diatomic ideal gas, we will use solutions of the

Schrödinger equation to obtain expressions for the allowable energy levels and their degeneracies. Specifically, to describe/model the rotations and vibrations, we will utilize the rigid rotator-harmonic oscillator approximation to solve the Schrödinger equation. In this approximation, the rotational energy levels are quantized (discrete) and are given by:

$$\varepsilon_J = \frac{\hbar^2}{2I} J(J+1), \text{for } J = 0, 1, 2, \ldots \tag{40.25}$$

where

$$\hbar = h/2\pi$$
$$I = \text{Moment of Inertia}$$

with each energy level, J, having a degeneracy, g_J, given by:

$$g_J = 2J + 1 \tag{40.26}$$

The vibrational energies of the harmonic oscillator are also quantized (discrete) and are given by:

$$\varepsilon_n = h\nu\left(n + \frac{1}{2}\right), \text{for } n = 0, 1, 2, \ldots \tag{40.27}$$

where

$$\nu = \text{Vibrational frequency} = c/\lambda = c\tilde{\nu}$$

where

c – Speed of light
λ – Wavelength
$\tilde{\nu}$ – Wave number

In Eq. (40.27), each energy level, n, is non-degenerate, that is, it has:

$$g_n = 1 \tag{40.28}$$

In the rigid rotator-harmonic oscillator approximation, we can write the total energy of the diatomic molecule as a sum of its translational, rotational, vibrational, and electronic energies, that is,

$$\varepsilon = \varepsilon_{trans} + \varepsilon_{rot} + \varepsilon_{vib} + \varepsilon_{elec} \tag{40.29}$$

As in the case of a monoatomic ideal gas, the criterion for the applicability of Boltzmann's statistics, $(N/\underline{V})(h^2/8mk_BT) \ll 1$, is readily satisfied at normal temperatures. For a diatomic molecule comprising atoms 1 and 2, $m = m_1 + m_2$, which is the diatomic mass. As before, we can write:

$$Q(N, \underline{V}, T) = \frac{[q(\underline{V}, T)]^N}{N!} \tag{40.30}$$

In addition, the energy decomposition in Eq. (40.29) allows us to write:

$$q(\underline{V}, T) = q_{trans}\, q_{rot}\, q_{vib}\, q_{elec} \tag{40.31}$$

Using Eq. (40.31) in Eq. (40.30) yields:

$$Q(N, \underline{V}, T) = \frac{(q_{trans}\, q_{rot}\, q_{vib}\, q_{elec})^N}{N!} \tag{40.32}$$

As discussed above, the translational partition function of a diatomic molecule is similar to that of an atom, except that the atomic mass, m, is replaced by the diatomic mass, $m_1 + m_2$, that is:

$$q_{trans}(\underline{V}, T) = \left(\frac{2\pi(m_1 + m_2)k_BT}{h^2}\right)^{3/2} \underline{V} \tag{40.33}$$

Next, in order to calculate q_{rot}, q_{vib}, and q_{elec}, we first need to select the zero of each energy contribution. Specifically,

(i)   The zero of the rotational energy corresponds to the $J = 0$ state, where $\varepsilon_{J=0} = 0$ (see Eq. (40.25)).

(ii) The zero of the vibrational energy corresponds to the bottom of the internuclear potential well of the lowest electronic state, so that the energy of the ground vibrational state is $h\nu/2$ (see Eq. (40.27)).

(iii) The zero of the electronic energy corresponds to that of the separated atoms comprising the diatomic molecule at rest in their ground electronic states. If we denote the depth of the ground electronic energy state by $D_e(>0)$, the energy of the ground electronic state corresponds to $\varepsilon_{e1} = -D_e$ with a degeneracy g_{e1}. The energy of the first excited electronic state corresponds to ε_{e2} with a degeneracy g_{e2}. Figure 40.1 illustrates all the energies involved as a function of the internuclear separation, R.

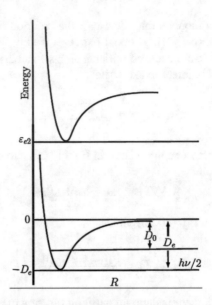

Fig. 40.1

Note that, in Fig. 40.1, $D_e = D_o + h\nu/2$, where D_o corresponds to the energy difference between the lowest vibrational state and the dissociated molecule and can be measured spectroscopically. Using g_{e1}, $\varepsilon_{e1} = -D_e$, g_{e2} and ε_{e2}, the electronic partition function can be written as follows:

$$q_{elec} = g_{e1}e^{D_e/k_B T} + g_{e2}e^{-\varepsilon_{e2}/k_B T} \tag{40.34}$$

Lecture 41

Thermodynamic Properties of Ideal Gases of Diatomic Molecules Calculated Using Partition Functions and Sample Problems

41.1 Introduction

The material presented in this lecture is adapted from Chapter 4 in M&S. In Lecture 40, we presented expressions for the translational and electronic contributions to the partition function of a diatomic molecule. In this lecture, first, we will discuss the vibrational contribution to the partition function of a diatomic molecule, including deriving expressions for the average vibrational energy and vibrational contribution to the heat capacity at constant volume of an ideal gas of diatomic molecules. Second, we will introduce the characteristic vibrational temperature, whose value relative to the system temperature will determine the probability that a given vibrational energy level is populated. Further, we will solve Sample Problem 41.1 to calculate the probability that a vibrational energy level is populated in the case of nitrogen molecules, including discussing the effect of temperature. Third, we will discuss the rotational contribution to the partition function of a diatomic molecule, including deriving expressions for the average rotational energy and rotational contribution to the heat capacity at constant volume of an ideal gas of diatomic molecules. In addition, we will introduce the characteristic rotational temperature, whose value relative to the system temperature will determine the probability that a given rotational energy level is populated. Fourth, we will emphasize that the rotational partition function of a diatomic molecule contains a symmetry number, which is equal to one when the two atoms comprising the diatomic molecule are different (referred to as the heteronuclear case) and is equal to two when the two atoms are identical (referred to as the homonuclear case). Fifth, we will present an expression for the total partition function of a diatomic molecule, which includes translational, vibrational, rotational, and electronic contributions. Finally, we will solve Sample Problem 41.2 to calculate the average molar internal energy and molar heat capacity at constant volume of an ideal gas of diatomic molecules, including identifying the various contributions to each thermodynamic property.

© Springer Nature Switzerland AG 2020
D. Blankschtein, *Lectures in Classical Thermodynamics with an Introduction to Statistical Mechanics*, https://doi.org/10.1007/978-3-030-49198-7_41

41.2   Vibrational Partition Function of a Diatomic Molecule

If we measure the vibrational levels relative to the bottom of the internuclear potential well, then, as discussed in Lecture 40, the vibrational energies are given by:

$$\varepsilon_n = \left(n + \frac{1}{2}\right)h\nu, \text{ for } n = 0, 1, 2, \ldots \tag{41.1}$$

where

$\nu(\text{Vibrational frequency}) = (K/\mu)^{1/2}/2\pi$
$\mu(\text{Reduced mass}) = (m_1 m_2)/(m_1 + m_2)$
K – Hooke spring constant

and where, $\nu = c\tilde{\nu}$, where c is the speed of light and $\tilde{\nu}$ is the wave number.

Using Eq. (41.1), the vibrational partition function of a diatomic molecule is given by:

$$q_{vib}(T) = \sum_n e^{-\beta\varepsilon_n} = \sum_{n=0}^{\infty} e^{-\beta\left(n+\frac{1}{2}\right)h\nu} \tag{41.2}$$

or

$$q_{vib}(T) = e^{-\beta h\nu/2} \sum_{n=0}^{\infty} e^{-\beta h\nu n} \tag{41.3}$$

The sum in Eq. (41.3) corresponds to a geometric series. Denoting $e^{-\beta h\nu} = x < 1$, it follows that:

$$\sum_{n=0}^{\infty} x^n = \frac{1}{1-x} \tag{41.4}$$

Using Eq. (41.4) in Eq. (41.3) yields:

$$q_{vib}(T) = \frac{e^{-\beta h\nu/2}}{1 - e^{-\beta h\nu}} \tag{41.5}$$

It is convenient to introduce a characteristic vibrational temperature, θ_{vib}, defined as follows:

$$\theta_{vib} = \frac{h\nu}{k_B} = \frac{hc\tilde{\nu}}{k_B} \tag{41.6}$$

Using Eq. (41.6) in Eq. (41.5) yields:

$$q_{vib}(T) = \frac{e^{-\theta_{vib}/2T}}{1 - e^{-\theta_{vib}/T}} \qquad (41.7)$$

As we will see, θ_{vib}/T is a central quantity whose value will determine the occupancy of the various vibrational energy levels.

Next, we will calculate the average vibrational energy of an ideal gas of N diatomic molecules using $q_{vib}(T)$. Specifically,

$$\langle \underline{E}_{vib} \rangle = N\langle \varepsilon_{vib} \rangle = Nk_B T^2 \left(\frac{d \ln q_{vib}}{dT} \right) \qquad (41.8)$$

Using Eq. (41.7) for q_{vib} in Eq. (41.8), we obtain:

$$\langle \underline{E}_{vib} \rangle = Nk_B T^2 \frac{d}{dT} \left[-\frac{\theta_{vib}}{2T} - \ln\left(1 - e^{-\theta_{vib}/T}\right) \right]$$

which, after taking the temperature derivative, yields:

$$\langle \underline{E}_{vib} \rangle = Nk_B \left(\frac{\theta_{vib}}{2} + \frac{\theta_{vib}}{e^{\theta_{vib}/T} - 1} \right) \qquad (41.9)$$

Next, we will calculate the average vibrational contribution to the molar heat capacity at constant volume of an ideal gas of diatomic molecules. To this end, we take the temperature derivative of Eq. (41.9), with $N = N_A$ (Avogadro's number) and $k_B N_A = R$ (the Gas constant). This yields:

$$C_{v,vib} = \frac{d\langle \underline{E}_{vib} \rangle}{dT} \qquad (41.10)$$

or

$$C_{v,vib} = R \frac{d}{dT} \left(\frac{\theta_{vib}}{e^{\theta_{vib}/T} - 1} \right)$$

Taking the temperature derivative in the last equation and rearranging, we obtain:

$$C_{v,vib} = R \left(\frac{\theta_{vib}}{T} \right)^2 \frac{e^{-\theta_{vib}/T}}{(1 - e^{-\theta_{vib}/T})^2} \qquad (41.11)$$

Equation (41.11) shows that the high-temperature limit of $C_{v,vib}$ (that is, $\theta_{vib} \ll T$) corresponds to $e^{-\theta_{vib}/T} \approx 1 - \theta_{vib}/T + \cdots$. In that limit, the denominator in Eq. (41.11) can be replaced by $(\theta_{vib}/T)^2$ and the numerator by $R\left(\frac{\theta_{vib}}{T}\right)^2$, which yields:

$$\text{Lim } C_{v,vib} \rightarrow R \qquad\qquad (41.12)$$
$$\scriptstyle T \rightarrow \infty$$
$$\text{or}$$
$$\scriptstyle \frac{\theta_{vib}}{T} \ll 1$$

41.3   Sample Problem 41.1

Calculate the fraction of diatomic molecules in the nth vibrational state and then use the result to calculate the fraction of $N_2(g)$ molecules in the $n = 0$ (ground vibrational state) and in the $n = 1$ (first excited vibrational state) at 300 K. It is known that $\theta_{vib} = 3374$ K.

41.3.1   Solution

First, we are asked to calculate the probability of occupancy of the nth vibrational state, which is given by $f_n = p_n = e^{-\beta h\nu\left(n+\frac{1}{2}\right)}/q_{vib}$. If we substitute the expression for q_{vib} in Eq. (41.7), we obtain:

$$f_n = e^{-n\,\theta_{vib}/T}\left(1 - e^{-\theta_{vib}/T}\right) \qquad\qquad (41.13)$$

Next, to address the second part of the problem, let us first calculate $e^{-\theta_{vib}/T}$ for $T = 300$ K and $\theta_{vib} = 3374$K. Specifically,

$$e^{-\theta_{vib}/T} = e^{-11.25} = 1.31 \times 10^{-5}$$

We can then use the last result in Eq. (41.13), for $n = 0$ and $n = 1$. This yields:

$$f_0 = 1 - e^{-\theta_{vib}/T} = 1 - 1.31 \times 10^{-5} \approx 1$$

and

$$f_1 = e^{-\theta_{vib}/T}\left(1 - e^{-\theta_{vib}/T}\right) \approx 1.31 \times 10^{-5}$$

The last two results clearly show that at 300 K, all the nitrogen molecules in the gas are in the ground vibrational state ($n = 0$). This important finding is quite typical, because $\theta_{vib}/T \gg 1$, and therefore, according to Eq. (41.13), $f_0 \approx 1$ and $f_{n>0} \approx 0$.

41.4   Rotational Partition Function of a Diatomic Molecule

As we saw in Lecture 40, the discrete energy levels of a rigid rotator are given by:

$$\varepsilon_J = \frac{\hbar^2 J(J+1)}{2I}, J = 0, 1, 2, \ldots \tag{41.14}$$

where I is the moment of inertia, given by:

$$I = \mu d^2, \text{with} \quad \mu \, (\text{Reduced mass}) = \frac{m_1 m_2}{(m_1 + m_2)} \tag{41.15}$$

In Eq. (41.15), d is the mass-to-mass separation. In Lecture 40, we saw that each rotational energy level, J, has a degeneracy, g_J, given by:

$$g_J = 2J + 1 \tag{41.16}$$

Using Eqs. (41.14) and (41.16), we can write the partition function of a rigid rotator as follows:

$$q_{rot}(T) = \sum_{J=0}^{\infty} \underbrace{(2J+1)}_{g_J} e^{-\beta \hbar^2 J(J+1)/2I} \tag{41.17}$$

For convenience, as we did in the harmonic oscillator case, we introduce the rotational temperature, θ_{rot}, as follows:

$$\theta_{rot} = \frac{\hbar^2}{2Ik_B} = \frac{hB}{k_B} \tag{41.18}$$

where

$$B = \frac{h}{8\pi^2 I}$$

Using Eq. (41.18) for θ_{rot} in Eq. (41.17) yields:

$$q_{rot}(T) = \sum_{J=0}^{\infty} (2J+1) e^{-\theta_{rot} J(J+1)/T} \tag{41.19}$$

The summation in Eq. (41.19) cannot be expressed in closed form. Nevertheless, at ordinary temperatures, we find that the value of (θ_{rot}/T) is quite small for diatomic molecules that do not contain hydrogen atoms. For example, for CO(g), $\theta_{rot} = 2.77$ K, and, therefore, $\theta_{rot}/T \approx 10^{-2}$ at room temperature. In that case, we can approximate the summation in Eq. (41.19) by an integral, because (θ_{rot}/T) is small for most diatomic molecules at ordinary temperatures. It is therefore a very good approximation to write $q_{rot}(T)$ in Eq. (41.19) as follows:

$$q_{rot}(T) = \int_0^\infty (2J + 1)e^{-\theta_{rot} J(J+1)/T} dJ \qquad (41.20)$$

The integral in Eq. (41.20) can be evaluated if we let $x = J(J + 1)$, for which $dx = (2J + 1)dJ$, so that Eq. (41.20) can be expressed as follows:

$$q_{rot}(T) = \int_0^\infty e^{-\theta_{rot}x/T} dx = \int_0^\infty e^{-\theta_{rot}x/T} d\left(\overbrace{\frac{\theta_{rot}x}{T}}^{y}\right)\left(\frac{T}{\theta_{rot}}\right) = \underbrace{\left(\int_0^\infty e^{-y} dy\right)}_{=1}\left(\frac{T}{\theta_{rot}}\right)$$

or

$$q_{rot}(T) = \frac{T}{\theta_{rot}} = \frac{8\pi^2 I k_B T}{h^2}, \text{ for } \theta_{rot} \ll T \qquad (41.21)$$

The approximation leading to Eq. (41.21) improves as T increases and is referred to as the high-temperature limit. For low temperatures, or for molecules with large values of θ_{rot}, say $H_2(g)$ with $\theta_{rot} = 85.3$ k, we can use Eq. (41.19) directly. For example, the first four terms in Eq. (41.19) are sufficient to calculate $q_{rot}(T)$ to within 0.1% for $T < 3\theta_{rot}$. In what follows, we will use the high-temperature limit, for which $(\theta_{rot}/T) \ll 1$, for most molecules at room temperature.

41.5   Average Rotational Energy of an Ideal Gas of Diatomic Molecules

The average rotational energy, $\langle E_{rot} \rangle$, of an ideal gas of diatomic molecules is given by:

$$\langle E_{rot} \rangle = N \langle \varepsilon_{rot} \rangle = N k_B T^2 \left(\frac{d \ln q_{rot}}{dT}\right) = N k_B T^2 \frac{d}{dT}\underbrace{\left(\ln T - \ln \theta_{rot} \right)}_{\text{From Eq.(41.20)}}$$

or

$$\langle E_{rot} \rangle = N k_B T \qquad (41.22)$$

41.6 Average Rotational Heat Capacity at Constant Volume of an Ideal Gas of Diatomic Molecules

For 1 mole of molecules, $N = N_A$ (Avogadro's number) and $N_A k_B = R$ (the Gas constant), the average rotational contribution to the heat capacity C_v of an ideal gas of diatomic molecules is given by:

$$C_{v,rot} = \frac{d\langle E_{rot}\rangle}{dT} = R \tag{41.23}$$

where we have used Eq. (41.22) for $\langle E_{rot}\rangle$. Equation (41.23) confirms the expected result, based on the Energy Equipartition Theorem, that a diatomic molecule has two rotational degrees of freedom, with each one contributing $R/2$ to $C_{v,rot}$.

41.7 Fraction of Diatomic Molecules in the Jth Rotational Level

The fraction of diatomic molecules in the Jth rotational level is given by the probability of occupancy of the Jth rotational level. Specifically,

$$f_J - p_J = \frac{g_J e^{-\theta_{rot} J(J+1)/T}}{q_{rot}}$$

Using Eqs. (41.16) and (41.21) in the last result, we obtain:

$$f_J = \frac{(2J + 1)e^{-\theta_{rot} J(J+1)/T}}{(T/\theta_{rot})}$$

or

$$f_J = (2J + 1)(\theta_{rot}/T)e^{-\theta_{rot} J(J+1)/T} \tag{41.24}$$

Contrary to the case for the vibrational energy levels, for most diatomic molecules, the use of Eq. (41.24) reveals that, at ordinary temperatures, the majority are in the excited rotational energy levels. This, of course, is due to the fact that $\theta_{rot}/T \ll 1$ at ordinary temperatures, so that the excited rotational energy levels are well populated.

41.8 Rotational Partition Functions of Diatomic Molecules Contain a Symmetry Number

Although it may not be apparent based on our derivation of $q_{rot}(T)$, Eqs. (41.17) and (41.21) apply only to heteronuclear diatomic molecules. This is a manifestation of the celebrated Pauli exclusion principle of quantum mechanics, which requires that the wave function of a homonuclear diatomic molecule must possess a certain symmetry with respect to the interchange of the two identical nuclei comprising the molecule. Specifically,

(i) If the two nuclei have integral spins (referred to as bosons), the molecular wave function must remain unchanged under an interchange of the two identical nuclei.

(ii) If the two nuclei have half-odd integer spins (referred to as fermions), the molecular wave function must change sign under an interchange of the two identical nuclei.

This symmetry requirement impacts greatly the population of the rotational energy levels of a homonuclear diatomic molecule. A detailed quantum mechanical analysis, which is beyond the scope of this introductory exposure to statistical mechanics, reveals that in the case of a homonuclear diatomic molecule, when the criterion $\theta_{rot}/T \ll 1$ is satisfied, we obtain:

$$q_{rot}(T) = \frac{T}{2\theta_{rot}} \tag{41.25}$$

The factor of 2 in the denominator of Eq. (41.25) reflects the fact that a homonuclear diatomic molecule has two indistinguishable orientations, that is, there is a two-fold axis of symmetry perpendicular to the internuclear axis.

Equations (41.21) and (41.25) for $q_{rot}(T)$ can be conveniently expressed as a single equation by introducing a factor, σ, referred to as the symmetry number of the molecule. Specifically, the symmetry number, σ, represents the number of indistinguishable orientations of the molecule. Using the symmetry number, we obtain:

$$q_{rot}(T) = \frac{T}{\sigma\theta_{rot}} \tag{41.26}$$

where $\sigma = 1$ for a heteronuclear diatomic molecule and $\sigma = 2$ for a homonuclear diatomic molecule.

41.9   Total Partition Function of a Diatomic Molecule

We can now combine all the molecular contributions to the partition function of a diatomic molecule using the results presented in Lecture 40 and in this lecture. Specifically, in the rigid rotator-harmonic oscillator approximation that we used, we obtain:

$$q\left(\underline{V}, T\right) = q_{trans} q_{rot} q_{vib} q_{elec}$$

$$q\left(\underline{V}, T\right) = \left[\left(\frac{2\pi M k_B T}{h^2}\right)^{3/2} \underline{V}\right] \left[\frac{T}{\sigma \theta_{rot}}\right] \left[\frac{e^{-\theta_{vib}/2T}}{(1 - e^{-\theta_{vib}/T})}\right] \left[g_{e1} e^{De/k_B T}\right] \qquad (41.27)$$

where the applicability of Eq. (41.27) requires that:

(i) $\theta_{rot}/T \ll 1$
(ii) Only the ground electronic state be populated
(iii) The zero of energy for the electronic energy be taken as the separated atoms at rest in their ground electronic states
(iv) The zero of energy for the vibrational energy be at the bottom of the internuclear potential well of the lowest electronic state

As we saw in the case of the atomic partition function, only q_{trans} in Eq. (41.27) depends on $\underline{V}$, and therefore, $q(\underline{V}, T)$ is again of the form $f(T)\underline{V}$, where $f(T)$ is a more complex function of temperature than in the atomic case.

41.10   Sample Problem 41.2

Derive an expression for the molar internal energy and heat capacity at constant volume of an ideal gas of diatomic molecules, including identifying the various contributions to both thermodynamic properties.

41.10.1   Solution

We begin with:

$$Q\left(N, \underline{V}, T\right) = \frac{[q(\underline{V}, T)]^N}{N!} \qquad (41.28)$$

Using Eq. (41.28), the internal energy of an ideal gas of diatomic molecules can be written as follows:

$$\underline{U} = k_B T^2 \left(\frac{\partial \ln Q}{\partial T} \right)_{\underline{V},N} = N k_B T^2 \left(\frac{\partial \ln q}{\partial T} \right)_{\underline{V}} \tag{41.29}$$

Using Eq. (41.27) for q($\underline{V}$,T) in Eq. (41.29), we obtain:

$$\underline{U} = N k_B T^2 \left(\frac{\partial}{\partial T} \left[\frac{3}{2} \ln T + \ln T - \frac{\theta_{vib}}{2T} - \ln \left(1 - e^{-\theta_{vib}/T} \right) + \frac{De}{k_B T} + \begin{array}{c} \text{Terms} \\ \text{independent} \\ \text{of T} \end{array} \right] \right)_{\underline{V}} \tag{41.30}$$

For 1 mole (N = N_A and $N_A k_B = R$), Eq. (41.30) yields:

$$U = \overbrace{\frac{3}{2} RT}^{\langle U_{trans} \rangle} + \underbrace{RT}_{\langle U_{rot} \rangle} + \overbrace{\frac{R\theta_{vib}}{2}}^{\substack{\text{Zero-Point} \\ \text{Vibrational} \\ \text{Energy}}} + \underbrace{\frac{R\theta_{vib} e^{-\theta_{vib}/T}}{\left(1 - e^{-\theta_{vib}/T} \right)}}_{\substack{\langle U_{vib} \rangle \text{ in Excess} \\ \text{of the Zero-Point} \\ \text{Vibrational Energy}}} - \overbrace{N_A D_e}^{\langle U_{elec} \rangle} \tag{41.31}$$

The heat capacity at constant volume of an ideal gas of diatomic molecules is obtained by differentiating U in Eq. (41.31) with respect to T. Specifically,

$$\frac{C_V}{R} = \frac{5}{2} + \left(\frac{\theta_{vib}}{T} \right)^2 \frac{e^{-\theta_{vib}/T}}{\left(1 - e^{-\theta_{vib}/T} \right)^2} \tag{41.32}$$

Equation (41.32) shows that the three translational and two rotational degrees of freedom contribute $\frac{5}{2}R$ to the heat capacity at constant volume of an ideal gas of diatomic molecules, a result which is consistent with the Energy Equipartition Theorem. The last term in Eq. (41.32) reflects the vibrational contribution.

Lecture 42

Statistical Mechanical Interpretation of Reversible Mechanical Work, Reversible Heat, and the First Law of Thermodynamics, the Micro-Canonical Ensemble and Entropy, and Sample Problem

42.1 Introduction

The material presented in this lecture is adapted from Chapter 5 in M&S. First, we will present a statistical mechanical interpretation of the First Law of Thermodynamics for a closed system, where we will identify the role of probability in determining the differential reversible mechanical work done on the system and the differential heat absorbed by the system reversibly. Second, we will introduce and discuss the Micro-Canonical ensemble where the energy, volume, and number of molecules of every component comprising the system are fixed. We will then present a celebrated result derived by Boltzmann which relates entropy to the extent of disorder in the system, quantified in terms of the number of distinct states that the system can access, consistent with its energy, volume, and number of molecules of each type. Third, we will derive an expression for the entropy in terms of the probabilities of the system being in quantum state j and a related expression in terms of the Canonical partition function. Fourth, we will solve Sample Problem 42.1 to calculate the entropy of a monoatomic ideal gas where every atom is in its ground electronic state. Finally, we will relate the statistical mechanical expression for entropy derived in this lecture to the thermodynamic definition of entropy presented in Part I. As promised, we will also show that $\beta = 1/k_B T$.

42.2 Statistical Mechanical Interpretation of Reversible Mechanical Work, Reversible Heat, and the First Law of Thermodynamics

In Lecture 39, we derived the following expression for the average energy, $\langle \underline{E} \rangle = \underline{U}$, of a macroscopic system:

© Springer Nature Switzerland AG 2020
D. Blankschtein, *Lectures in Classical Thermodynamics with an Introduction to Statistical Mechanics*, https://doi.org/10.1007/978-3-030-49198-7_42

$$\underline{U} = \sum_j p_j(N, \underline{V}, \beta) \, \underline{E}_j(N, \underline{V}) \tag{42.1}$$

where the probability, p_j, is given by:

$$p_j(N, \underline{V}, \beta) = \frac{e^{-\beta \underline{E}_j(N, \underline{V})}}{Q(N, \underline{V}, \beta)} \tag{42.2}$$

where

$$Q(N, \underline{V}, \beta) = \sum_j e^{-\beta \underline{E}_j(N, \underline{V})} \tag{42.3}$$

is the Canonical partition function. Equation (42.1) represents the average energy of an equilibrium system characterized by N, $\underline{V}$, and T (or $\beta = 1/k_B T$) as its natural thermodynamic variables in the context of the Canonical ensemble.

The differential of Eq. (42.1) is given by:

$$d\underline{U} = \sum_j p_j d\underline{E}_j + \sum_j \underline{E}_j dp_j \tag{42.4}$$

If the system is closed, N cannot change, and $\underline{E}_j$ (N, $\underline{V}$) can only change as a result of changes in the system volume, $\underline{V}$. We can therefore write:

$$d\underline{E}_j = \left(\frac{\partial \underline{E}_j}{\partial \underline{V}}\right)_N d\underline{V} \tag{42.5}$$

Substituting Eq. (42.5) in the first summation in Eq. (42.4) yields:

$$d\underline{U} = \sum_j p_j \left(\frac{\partial \underline{E}_j}{\partial \underline{V}}\right)_N d\underline{V} + \sum_j \underline{E}_j dp_j \tag{42.6}$$

The first term in Eq. (42.6) can be viewed as the average change in the system energy induced by a small change in the system volume, or in other words, as the average differential mechanical work introduced in Part I. Furthermore, if this change is carried out reversibly, such that the system remains at equilibrium throughout the process, then, p_j in Eqs. (42.4) and (42.6) is given by Eq. (42.2) throughout the entire process. Accordingly, Eq. (42.6) can be rewritten as follows:

$$d\underline{U} = \sum_j p_j(N, \underline{V}, \beta) \left(\frac{\partial \underline{E}_j}{\partial \underline{V}}\right)_N d\underline{V} + \sum_j \underline{E}_j(N, \underline{V}) dp_j(N, \underline{V}, \beta) \tag{42.7}$$

If we compare Eq. (42.7) with the First Law of Thermodynamics for a closed system undergoing a reversible process (introduced in Part I), that is, with:

$$d\underline{U} = \delta Q_{rev} + \delta W_{rev} \tag{42.8}$$

where the first term on the right-hand side of the equal sign corresponds to the differential heat absorbed by the system reversibly, and the second term on the right-hand side of the equal sign corresponds to the differential reversible mechanical work done on the system, it follows that:

$$\delta W_{rev} = \left\{ \sum_j p_j(N, \underline{V}, \beta) \left(\frac{\partial E_j}{\partial \underline{V}} \right)_N \right\} d\underline{V} \tag{42.9}$$

and

$$\delta Q_{rev} = \sum_j E_j(N, \underline{V}) dp_j(N, \underline{V}, \beta) \tag{42.10}$$

Equation (42.9) reveals that the differential reversible mechanical work done on the system, δW_{rev}, results from an infinitesimal change in the allowed energies of the system due to infinitesimal changes in the system volume, without changing the probability distribution of its quantum states, p_j. On the other hand, Eq. (42.10) reveals that the differential reversible heat, δQ_{rev}, results from a change in the probability distribution of the quantum states of the system, without changing the allowed energies of these states. In the next section, we will discuss the connection with entropy, which was first introduced thermodynamically in Part I.

If we compare Eq. (42.9) with the expression for the differential reversible P$d\underline{V}$-type work done on the system presented in Part I, that is, with:

$$\delta W_{rev} = -Pd\underline{V} \tag{42.11}$$

it follows that:

$$P = -\sum_j p_j(N, \underline{V}, \beta) \left(\frac{\partial E_j}{\partial \underline{V}} \right)_N \tag{42.12}$$

Recall that, in Lecture 39, we used Eq. (42.12), without proof, to calculate the pressure of an ideal gas.

42.3 Micro-Canonical Ensemble and Entropy

So far, we have not discussed the concept of entropy from a molecular perspective in the context of statistical mechanics.

In Part I, we showed that entropy is a thermodynamic function of state which can be viewed as a quantitative measure of the extent of disorder in a system. The concept of disorder can be described in a number of ways, and here, we will adopt the following description. Consider an ensemble of $\mathcal{A}$ isolated systems, each with energy, $\underline{E}$, volume, $\underline{V}$, and number of particles (molecules), N. Such an ensemble is known as the Micro-Canonical ensemble (see Fig. 42.1):

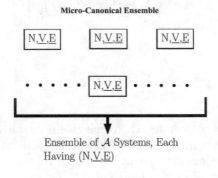

Micro-Canonical Ensemble

Ensemble of $\mathcal{A}$ Systems, Each
Having (N,$\underline{V}$,$\underline{E}$)

Fig. 42.1

Recall that each value adopted by $\underline{E}$ must be an eigenvalue of the Schrödinger equation describing the system. As we already saw, the energy, $\underline{E}$, is a function of N and $\underline{V}$, so that we can write $\underline{E} = \underline{E}(N, \underline{V})$. Although each system of the Micro-Canonical ensemble has the same energy $\underline{E}$, it may be in a different quantum state because of the degeneracy. Let the degeneracy associated with the energy $\underline{E}$ be $\Omega(\underline{E})$, such that one can label the $\Omega(\underline{E})$ degenerate quantum states as 1, 2, ..., $\Omega(\underline{E})$. Typically, the degeneracy $\Omega(\underline{E})$ is $\gg 1$.

Next, let a_j be the number of systems in the Micro-Canonical ensemble that are in quantum state j. Because the $\mathcal{A}$ systems of the Micro-Canonical ensemble are distinguishable, the number of ways of having a_1 systems in quantum state 1, a_2 systems in quantum state 2, etc., is given by the following combinatorial expression:

$$W(a_1, a_2, a_3 \ldots) = \frac{\mathcal{A}!}{a_1! a_2! a_3! \cdots} = \frac{\mathcal{A}!}{\prod_j a_j!} \tag{42.13}$$

Interestingly, Eq. (42.13) is analogous to that used to calculate the number of distinct ways in which $\mathcal{A}$ distinguishable objects (say, books on a shelf) can be divided into groups, the first containing a_1 objects, the second containing a_2 objects, etc. In Eq. (42.13), the following constraint applies:

$$\sum_j a_j = A \qquad (42.14)$$

If all the A systems are in one particular state (a totally ordered configuration), say state 1, then: $a_1 = A$, $a_2 = 0$, $a_3 = 0$, etc., and $W = 1$ (see Eq. (42.13)), which is the smallest value that W can attain. In the other extreme, when all the a_js are equal (a disordered configuration), W attains its largest value.

It then follows that W can be viewed as a quantitative measure of the disorder in a system. Further, it follows that entropy, being a quantitative measure of disorder as well, should be related to W. However, because W is multiplicative, while entropy is additive, in the sense that two systems having W_1 and W_2 will have disorder $W_{1+2} = W_1 \times W_2$, while the entropy will be $\underline{S}_{1+2} = \underline{S}_1 + \underline{S}_2$, it follows that the entropy, $\underline{S}$, should depend on $\ln W$, which will ensure the additivity of the entropy. The proportionality constant is the Boltzmann constant, in honor of Boltzmann, who first introduced the relation between $\underline{S}$ and W as follows:

$$\underline{S} = k_B \ln W \qquad (42.15)$$

Equation (42.15) is probably the most celebrated equation in statistical mechanics and provides a quantitative relation between the thermodynamic quantity, $\underline{S}$-the entropy, and the statistical quantity, W. The Boltzmann constant was chosen to obtain the appropriate units of the entropy, because $\ln W$ is a pure number. Recall that $k_B = 1.38 \times 10^{-23}$J/K. Further, recall that $\underline{S} = 0$ for a completely ordered system ($a_1 = 1, a_2 = a_3 = \cdots = 0; W = 1$) and attains its maximum value for a completely disordered system ($a_1 = a_2 = a_3 = \cdots; W_{max}$).

42.4   Relating Entropy to the Canonical Partition Function

In the previous lectures in Part III, we showed that it is possible to calculate many thermodynamic properties if we have access to the Canonical partition function, Q. For example, we saw that:

$$\underline{U} = k_B T^2 \left(\frac{\partial \ln Q}{\partial T} \right)_{\underline{V},N} = -\left(\frac{\partial \ln Q}{\partial \beta} \right)_{\underline{V},N} \qquad (42.16)$$

and that:

$$P = k_B T \left(\frac{\partial \ln Q}{\partial \underline{V}} \right)_{T,N} \qquad (42.17)$$

In the entropy case, we begin with Eq. (42.15), where we substitute the expression for W in Eq. (42.13). Specifically,

$$\underline{S} = k_B \ln\left(\frac{\mathcal{A}!}{\prod_j a_j!}\right) = k_B\left\{\ln\mathcal{A}! - \sum_j \ln a_j!\right\} \tag{42.18}$$

In order to deal with the factorials in Eq. (42.18), we utilize Stirling's approximation given by:

$$\ln z! = z\ln z - z, \text{ for } z \gg 1 \tag{42.19}$$

Using Eq. (42.19) in the last term of Eq. (42.18) yields:

$$\underline{S} = k_B\left\{\mathcal{A}\ln\mathcal{A} - \mathcal{A} - \sum_j\left(a_j\ln a_j - a_j\right)\right\}$$

$$= k_B\left\{\mathcal{A}\ln\mathcal{A} - \mathcal{A} - \sum_j a_j\ln a_j + \overbrace{\sum_j a_j}^{\mathcal{A}}\right\}$$

or

$$\underline{S} = k_B\left\{\mathcal{A}\ln\mathcal{A} - \sum_j a_j\ln a_j\right\} \tag{42.20}$$

The entropy of a typical system in an ensemble of $\mathcal{A}$ systems is given by:

$$\underline{S}_{system} = \frac{\underline{S}}{\mathcal{A}} \tag{42.21}$$

with $\underline{S}$ given by Eq. (42.20). If we use the fact that the probability of finding a system in the jth quantum state, p_j, is given by:

$$p_j = \frac{a_j}{\mathcal{A}} \tag{42.22}$$

and then substitute $a_j = \mathcal{A}p_j$ in Eq. (42.20), we obtain:

$$\underline{S} = k_B\mathcal{A}\ln\mathcal{A} - k_B\sum_j p_j\mathcal{A}\ln\left(p_j\mathcal{A}\right)$$

$$= k_B\mathcal{A}\ln\mathcal{A} - k_B\sum_j p_j\mathcal{A}\ln p_j - k_B\mathcal{A}\ln\mathcal{A}\underbrace{\left(\sum_j p_j\right)}_{=1}$$

In the last equation, after cancelling the first and last terms on the right-hand side of the equal sign, dividing the resulting expression by $\mathcal{A}$, and using Eq. (42.21), we obtain:

$$\underline{S}_{\text{system}} = -k_B \sum_j p_j \ln p_j \tag{42.23}$$

If the system is completely ordered, say, $p_i = 1$ and $p_{j \neq i} = 0$, recall that $\sum_k p_k = 1$, Eq. (42.23) shows that, as expected, $\underline{S}_{\text{system}} = -k_B(1\ln1) = 0$. In other words, according to our molecular interpretation of entropy, $\underline{S} = 0$ for a perfectly ordered system. It is also possible to show that the entropy in Eq. (42.23) attains its maximum value when all the p_js are equal. This, in turn, corresponds to the system being maximally disordered, and as stressed earlier, to $W = W_{\text{max}}$ in Eq. (42.13).

Next, to relate $\underline{S}$ to the Canonical partition function, $Q(N, \underline{V}, T)$, we substitute

$$p_j = \frac{e^{-\beta E_j(N, \underline{V})}}{Q(N, \underline{V}, T)} \tag{42.24}$$

in Eq. (42.23), including eliminating the no-longer needed subscript "system" in $\underline{S}$. This yields:

$$\underline{S} = -k_B \sum_j p_j \ln p_j$$

$$= -k_B \sum_j \left(\frac{e^{-\beta E_j}}{Q} \right) (-\beta E_j - \ln Q)$$

$$= \beta k_B \left(\sum_j \frac{E_j e^{-\beta E_j}}{Q} \right) + \frac{k_B \ln Q}{Q} \left(\sum_i e^{-\beta E_j} \right)$$

In the last equation, $\beta k_B = 1/T$, the first term in parenthesis is equal to $\langle \underline{E} \rangle = \underline{U}$, and the last term in parenthesis is equal to Q. Using these results in the last equation, we obtain:

$$\underline{S} = \frac{U}{T} + k_B \ln Q \tag{42.25}$$

Using Eq. (42.16) for $\underline{U}$ in Eq. (42.25), we can express $\underline{S}$ completely in terms of Q $(N, \underline{V}, T)$. Specifically,

$$\underline{S} = \frac{1}{T} \left[k_B T^2 \left(\frac{\partial \ln Q}{\partial T} \right)_{\underline{V}, N} \right] + k_B \ln Q$$

Rearranging the last equation we obtain:

$$\underline{S} = k_B T \left(\frac{\partial \ln Q}{\partial T} \right)_{V,N} + k_B \ln Q \tag{42.26}$$

42.5   Sample Problem 42.1

Calculate the entropy of a monoatomic ideal gas, where all the atoms are in the ground electronic state.

42.5.1   Solution

We have already shown that for a monoatomic ideal gas, Q is given by:

$$Q(N, \underline{V}, T) = \frac{1}{N!} \left[\left(\frac{2\pi m k_B T}{h^2} \right)^{3/2} \underline{V} g_{el} \right]^N \tag{42.27}$$

We can then substitute Eq. (42.27) in Eq. (42.26). This yields:

$$\underline{S} = k_B T N \underbrace{\left\{ \frac{\partial}{\partial T} \left(\ln T^{3/2} + \text{Terms that are T-independent} \right)_{V,N} \right\}}_{3/2T}$$

$$+ k_B N \ln \left[\left(\frac{2\pi m k_B T}{h^2} \right)^{3/2} \underline{V} g_{el} \right] - k_B \ln N!$$

If we utilize Stirling's approximation to express the last term in the last equation as $-k_B N \ln N + k_B N$, including rearranging the resulting expression, we obtain:

$$\underline{S} = \frac{5}{2} k_B N + k_B N \ln \left[\left(\frac{2\pi m k_B T}{h^2} \right)^{3/2} \left(\frac{\underline{V}}{N} \right) g_{el} \right] \tag{42.28}$$

Equation (42.28) can be expressed in a more compact form as follows:

$$\underline{S} = k_B N \ln \left[\left(\frac{2\pi m k_B T}{h^2} \right)^{3/2} \left(\frac{\underline{V}}{N} \right) g_{el} \, e^{5/2} \right] \tag{42.29}$$

42.6   Relating the Statistical Mechanical Relation, $\underline{S} = k_B \ln W$, to the Thermodynamic Relation, $d\underline{S} = \delta Q_{rev}/T$

Next, we will show that the statistical mechanical relations presented above to calculate the entropy molecularly, that is, $\underline{S}=k_B\ln W$ or $\underline{S}=-k_B\sum_j p_j \ln p_j$, are both consistent with the thermodynamic definition of entropy presented in Part I. In addition, as promised, we will show that $\beta = 1/k_B T$, a result which we have used in Part III without proof until now. We begin with the entropy expression in Eq. (42.23), where as explained earlier, we will delete the subscript "system." Specifically,

$$\underline{S} = -k_B \sum_j p_j \ln p_j \tag{42.30}$$

The differential of Eq. (42.30) is given by:

$$d\underline{S} = -k_B \sum_j \left(dp_j + \ln p_j dp_j \right) \tag{42.31}$$

Because $\sum_j p_j = 1$, it follows that $\sum_j dp_j = 0$ in Eq. (42.31). Accordingly, Eq. (42.31) can be expressed as follows:

$$d\underline{S} = -k_B \sum_j \ln p_j dp_j \tag{42.32}$$

We also know that:

$$p_j(N, \underline{V}, \beta) = \frac{e^{-\beta E_j(N,\underline{V})}}{Q(N, \underline{V}, \beta)} \tag{42.33}$$

Taking the ln of Eq. (42.33) yields:

$$\ln p_j = -\beta E_j - \ln Q \tag{42.34}$$

Using $\ln p_j$ in Eq. (42.34) in Eq. (42.32) then yields:

$$dS = -k_B \sum_j \{-\beta E_j - \ln Q\} dp_j$$

$$= +(k_B \beta) \sum_j E_j(N, \underline{V}) dp_j(N, \underline{V}, \beta) + k_B \ln Q \overbrace{\left(\sum_j dp_j \right)}^{=0}$$

or

$$dS = (\beta k_B) \sum_j E_j(N, \underline{V}) dp_j(N, \underline{V}, \beta) \tag{42.35}$$

Replacing the summation over j in Eq. (42.35) by Eq. (42.10) yields:

$$dS = (\beta k_B) \delta Q_{rev} \tag{42.36}$$

Equation (42.36) clearly shows that the differential of the entropy is related to the differential of the heat absorbed by the system in a reversible process, precisely as we defined the differential of the entropy thermodynamically in Part I. Most importantly, for dS in Eq. (42.36) to be identical to the thermodynamic definition, $dS = \delta Q_{rev}/T$, the following must be true:

$$\beta k_B = \frac{1}{T} \Rightarrow \beta = \frac{1}{k_B T} \tag{42.37}$$

thereby proving what we promised to show in Lecture 38.

Lecture 43

Statistical Mechanical Interpretation of the Third Law of Thermodynamics, Calculation of the Helmholtz Free Energy and Chemical Potentials Using the Canonical Partition Function, and Sample Problems

43.1 Introduction

The material presented in this lecture is adapted from Chapter 7 in M&S. First, we will discuss the Third Law of Thermodynamics, including providing a statistical mechanical interpretation to it. Second, we will derive an expression for the Helmholtz free energy of a pure ($n = 1$) material using the Canonical partition function. Third, we will solve Sample Problem 43.1 to calculate the Helmholtz free energy of a one component ($n = 1$) ideal gas. Fourth, we will solve Sample Problem 43.2 to calculate the chemical potential of a monoatomic ideal gas. Finally, we will solve Sample Problem 43.3 to calculate the standard-state chemical potential of a one component ideal gas, including relating it to Denbigh's definition of a one component ($n = 1$) ideal gas that we presented in Part II.

43.2 The Third Law of Thermodynamics and Entropy

The Third Law of Thermodynamics states that: "Every substance has a finite positive entropy, but at absolute zero (i.e., at zero degrees Kelvin), the entropy may become zero, and does so in the case of a perfectly crystalline substance."

Unlike the First and Second Laws of Thermodynamics, the Third Law of Thermodynamics is not associated with any new thermodynamic state function. Recall that the First Law of Thermodynamics is associated with the energy, and the Second Law of Thermodynamics is associated with the entropy. On the other hand, the Third Law of Thermodynamics simply provides a numerical scale for the entropy.

Although the Third Law of Thermodynamics was formulated before the full development of quantum theory and statistical mechanics, it is most readily

© Springer Nature Switzerland AG 2020
D. Blankschtein, *Lectures in Classical Thermodynamics with an Introduction to Statistical Mechanics*, https://doi.org/10.1007/978-3-030-49198-7_43

understood at the molecular level. Recall that one of the expressions for the entropy presented in Lecture 42 is given by:

$$\underline{S} = k_B \ln W \tag{43.1}$$

where W is the number of distinct ways in which the total energy of the system may be distributed over the various energy states. At $T = 0$ K, we expect that the system will be in its lowest energy state. Accordingly, $W = 1$ and $\underline{S}$ in Eq. (43.1) is equal to $k_B \ln 1 = 0$.

Another way to arrive at this result is to begin with the entropy expression in terms of probabilities derived in Lecture 42. Specifically,

$$\underline{S} = -k_B \sum_j p_j \ln p_j \tag{43.2}$$

where p_j is the probability of finding the system in the jth quantum state with energy E_j. At $T = 0K$, there is no thermal energy, and we expect the system to be in the ground state (o). Accordingly, $p_o = 1$, and all the other p_js are equal to zero. In that case, Eq. (43.2) shows that $\underline{S} = -k_B p_o \ln p_o = -k_B(1)\ln 1 = 0$. Even if the ground state has a degeneracy, $g_o = n$, then, each of the n degenerate quantum states with energy, $\underline{E}_o$, would have equal probability, $1/n$, and $\underline{S}$ in Eq. (43.2) would be:

$$\underline{S}(T = 0K) = -k_B \sum_{j=1}^{n} \left(\frac{1}{n}\right) \ln\left(\frac{1}{n}\right) = -k_B \left(\frac{1}{n}\right) \ln\left(\frac{1}{n}\right) \underbrace{\left(\sum_{j=1}^{n} 1\right)}_{n}$$

$$\underline{S}(T = 0K) = +k_B \ln(n) \tag{43.3}$$

In Eq. (43.3), even if n were as large as Avogadro's number and $N = 1$ mole, $\underline{S}/N$ in Eq. (43.3) would be equal to:

$$\underline{S}/N = S = \frac{k_B \ln(N_A)}{N}$$

where $N_A = 6.022 \times 10^{23}$ [unitless], and $k_B = 1.38 \times 10^{-23}$ J/K. Using this data in the equation for S above yields:

$$S = \frac{\left(1.38 \times 10^{-23} J/K\right) \left(\ln\left(6.022 \times 10^{23}\right)\right)}{1 \text{mole}}$$

or

$$S = (1.38)(23)\underbrace{(\ln(60.22))}_{4.1}\left(10^{-23}\right)\frac{J}{mol\,K}$$

$$\underbrace{\phantom{(1.38)(23)(\ln(60.22))(10^{-23})}}_{130}$$

$$S = 13\times10^{-23}\,J/mol\,K \tag{43.4}$$

Clearly, the value of S in Eq. (43.4) is well below a measurable value. Recall that:

$$\left(\frac{\partial S}{\partial T}\right)_P = \frac{C_P}{T} \tag{43.5}$$

Integrating Eq. (43.5) from $T = 0K$ to some T yields:

$$S(T) = S(0K) + \int_0^T (C_P/T')\,dT', \quad P = const.$$

Using the fact that $S(0K) = 0$, it follows that:

$$S(T) = \int_0^T \frac{(C_P/T')\,dT'}{T'}, \quad P = const. \tag{43.6}$$

43.3   Calculation of the Helmholtz Free Energy of a Pure Material Using the Canonical Partition Function

We recently showed that:

$$\underline{U} = k_B T^2\left(\frac{\partial \ln Q}{\partial T}\right)_{V,N} \tag{43.7}$$

and

$$\underline{S} = k_B T\left(\frac{\partial \ln Q}{\partial T}\right)_{V,N} + k_B \ln Q \tag{43.8}$$

Recall that the Helmholtz free energy, $\underline{A}$, is given by:

$$\underline{A} = \underline{U} - T\underline{S} \tag{43.9}$$

Using Eqs. (43.7) and (43.8) in Eq. (43.9) yields a very simple expression for $\underline{A}$ in terms of Q, that is,

$$\underline{A} = k_B T^2 \left(\frac{\partial \ln Q}{\partial T}\right)_{\underline{V},N} - k_B T^2 \left(\frac{\partial \ln Q}{\partial T}\right)_{\underline{V},N} - k_B T \ln Q$$

or

$$\underline{A} = -k_B T \ln Q \qquad (43.10)$$

Equation (43.10) can be expressed as follows:

$$Q = e^{-\underline{A}/k_B T} \qquad (43.11)$$

Equation (43.11) shows that the Helmholtz free energy, $\underline{A}$, is the thermodynamic function which generates Q, because its natural independent thermodynamic variables are T, $\underline{V}$, and N. Recall that these variables characterize the Canonical ensemble and the Canonical partition function, $Q(T, \underline{V}, N)$.

43.4   Sample Problem 43.1

Calculate the Helmholtz free energy of a monoatomic ideal gas using Eq. (43.10). Assume that only the ground electronic level is populated, with a degeneracy g_{el}.

43.4.1   Solution

For a monoatomic ideal gas, we know that:

$$Q(T, \underline{V}, N) = \frac{[q(T, \underline{V})]^N}{N!} \qquad (43.12)$$

Taking the natural logarithm of Eq. (43.12) yields:

$$\ln Q = N \ln q - \ln N! \qquad (43.13)$$

Using Stirling's approximation on $\ln N!$ in Eq. (43.13) yields:

$$\ln Q = N \ln q - [N \ln N - N] \qquad (43.14)$$

$$\ln Q = N\{\underbrace{1}_{\ln e} + \ln q - \ln N\} \tag{43.15}$$

or

$$\ln Q = N \ln[q(T\underline{V})e/N] \tag{43.16}$$

We also know that for a monoatomic ideal gas:

$$q(T, \underline{V}) = \left(\frac{2\pi m k_B T}{h^2}\right)^{3/2} \cdot \underline{V} \cdot g_{el} \tag{43.17}$$

Using Eq. (43.17) for $q(T, \underline{V})$ in Eq. (43.16) for $\ln Q$ and then using the resulting $\ln Q$ expression in Eq. (43.10) yields the desired result for $\underline{A}$. Specifically,

$$\underline{A} = -Nk_B\ln\left[\left(\frac{2\pi m k_B T}{h^2}\right)^{3/2} \cdot \left(\frac{\underline{V}}{N}\right) \cdot e \cdot g_{el}\right] \tag{43.18}$$

As expected, $\underline{A}$ in Eq. (43.18) is extensive.

43.5 Sample Problem 43.2

Calculate the chemical potential of a monoatomic ideal gas. Assume that only the ground electronic level is populated, with a degeneracy g_{el1}.

43.5.1 Solution

In Part I, we saw that for a monoatomic gas:

$$\underline{A} = \underline{A}(T, \underline{V}, N) \tag{43.19}$$

We also saw that the differential of $\underline{A}$ is given by:

$$d\underline{A} = -\underline{S}dT - Pd\underline{V} + \mu dN \tag{43.20}$$

As we discussed in Part I, Eq. (43.20) indicates that:

$$\mu = \left(\frac{\partial \underline{A}}{\partial N}\right)_{T,\underline{V}} \tag{43.21}$$

Using Eq. (43.10) for $\underline{A}$ in Eq. (43.21) for μ yields:

$$\mu = -k_B T \left(\frac{\partial \ln Q}{\partial N}\right)_{T,\underline{V}} \tag{43.22}$$

Equation (43.22) is a general result for a pure ($n = 1$) gas. When the gas is ideal, $\ln Q$ is given by Eq. (43.14), where the dependence on N is clearly shown. Using Eq. (43.14) for $\ln Q$ in Eq. (43.22) for μ yields:

$$\mu = -k_B T \left\{\frac{\partial}{\partial N}\left(N\ln q - N\ln N + N\right)\right\}_{T,\underline{V}} \tag{43.23}$$

or

$$\mu = -k_B T \left\{\ln q(T, \underline{V}) - \ln N - \underbrace{(N/N)}_{1} + 1\right\} \tag{43.24}$$

Cancelling the last two terms in Eq. (43.24) and rearranging yields:

$$\mu = -k_B T \ln\left[\frac{q(T, \underline{V})}{N}\right] \tag{43.25}$$

Equation (43.25) is general for a one component ideal gas and can be used to calculate the chemical potential of any type of molecule (monoatomic or diatomic). To this end, one needs to use the expressions for q corresponding to a monoatomic or diatomic ideal gas derived in Part III. In the Sample Problem below, we will calculate μ for a monoatomic ideal gas using Eq. (43.17) for q in Eq. (43.25) for μ.

43.6   Sample Problem 43.3

Calculate the standard-state chemical potential, $\mu^\circ(T)$, or the function, $\lambda(T)$, introduced in Part II, for a monoatomic ideal gas using Eq. (43.25). Assume that only the ground electronic level is populated, with a degeneracy g_{e1}.

43.6.1 Solution

In order to use Eq. (43.25) for μ, we use Eq. (43.17) for $q(T,\underline{V})$ which we divide by N. This yields:

$$\frac{q(T,\underline{V})}{N} = \left(\frac{2\pi m k_B T}{h^2}\right)^{3/2} \cdot \left(\frac{\underline{V}}{N}\right) \cdot g_{el} \tag{43.26}$$

Because the gas is ideal, it follows that the ideal gas EOS applies (used here on a per molecule basis). Specifically,

$$\frac{\underline{V}}{N} = \frac{k_B T}{P} \tag{43.27}$$

Using Eq. (43.27) in Eq. (43.26) yields:

$$\frac{q(T,P)}{N} = \left(\frac{2\pi m k_B T}{h^2}\right)^{3/2} \cdot \left(\frac{k_B T}{P}\right) \cdot g_{el} \tag{43.28}$$

Using Eq. (43.28) in Eq. (43.25) yields:

$$\mu = -k_B T \ln\left[\left(\frac{2\pi m k_B T}{h^2}\right)^{3/2} \cdot \left(\frac{k_B T}{P}\right) \cdot g_{el}\right] \tag{43.29}$$

Next, we can expand the ln term in Eq. (43.29) to extract the P dependence. Specifically,

$$\mu(T,P) = -k_B T \underbrace{\left[\left(\frac{2\pi m k_B T}{h^2}\right)^{3/2} \cdot (k_B T) \cdot g_{el}\right]}_{\text{Unit of Pressure}} + k_B T \ln P \tag{43.30}$$

Let us add and subtract $k_B T \ln P_o$ to the right-hand side of the equal sign in Eq. (43.30), where P_o is the standard-state pressure (typically, 1 bar). This yields:

$$\mu(T,P) = -k_B T \ln \underbrace{\left[\left(\frac{2\pi m k_B T}{h^2}\right)^{3/2} \cdot \left(\frac{k_B T}{P_o}\right) \cdot g_{el}\right]}_{\text{Dimensionless!}} + k_B T \ln \underbrace{\left(\frac{P}{P_o}\right)}_{\text{Dimensionless!}} \tag{43.31}$$

We can rewrite Eq. (43.31) as follows:

$$\mu(T, P) = \mu_o(T) + k_B T[\ln(P/P_o)]$$

or, as expressed in Part II, as follows:

$$\mu(T, P) = \lambda(T) + k_B T[\ln(P/P_o)] \tag{43.32}$$

As promised in Part II when we discussed the Denbigh definition of a pure ($n = 1$) ideal gas in terms of its chemical potential, in this lecture, we used statistical mechanics to calculate $\mu_o(T)$, or equivalently, $\lambda(T)$.

Lecture 44

Grand-Canonical Ensemble, Statistical Fluctuations, and Sample Problems

44.1 Introduction

The material presented in this lecture is adapted from Chapter 3 in McQuarrie. First, we will discuss the Grand-Canonical ensemble, where the variables $\underline{V}$, T, and μ are fixed. After defining the Grand-Canonical partition function, we will derive expressions for the Grand-Canonical ensemble-averaged energy, pressure, and number of molecules. Second, we will show that the Grand Potential, $-P\underline{V}$, is the generator of the Grand-Canonical partition function. Third, we will discuss statistical fluctuations (or, in short, fluctuations) using concepts from probability theory. Specifically, we will calculate the standard deviations of the energy and number of molecules around their average or mean values. Fourth, we will solve Sample Problem 44.1 to calculate the relative spread in the energy of a one component ideal gas. Fifth, we will solve Sample Problem 44.2 to show that the distribution of the energies of the system is Gaussian and centered around the average energy, including calculating the spread of the distribution. We will see that in the thermodynamic limit (N ≫ 1), the fluctuations in the energy tend to zero, and the Gaussian distribution reduces to a delta function. Sixth, we will solve Sample Problem 44.3 to calculate the relative spread of the Gaussian distribution of the number of molecules for a one component ideal gas. Finally, we will show that in the thermodynamic limit (N ≫ 1), the Micro-Canonical, Canonical, and Grand-Canonical ensembles are all equivalent, including predicting the same thermodynamic properties, which correspond to those that are observed experimentally.

44.2 Grand-Canonical Ensemble

So far in Part III, we discussed two ensembles:

© Springer Nature Switzerland AG 2020
D. Blankschtein, *Lectures in Classical Thermodynamics with an Introduction to Statistical Mechanics*, https://doi.org/10.1007/978-3-030-49198-7_44

1. The Canonical ensemble, in which each system of the ensemble has fixed values of T, $\underline{V}$, and N

2. The Micro-Canonical ensemble, in which each system of the ensemble has fixed values of $\underline{E}$, $\underline{V}$, and N

We turn next to a discussion of a third ensemble, the Grand-Canonical ensemble, where each system of the ensemble is enclosed by walls that are rigid, heat conducting (so far, the same as in the Canonical ensemble), but permeable. As a result, the number of molecules, N, in each system can vary. We construct a Grand-Canonical ensemble by placing a large collection of such systems in a large heat bath at temperature T, and a large reservoir of molecules at chemical potential μ, until equilibrium is attained, and then isolate it from its surroundings (see Fig. 44.1). As a result, each system of the Grand-Canonical ensemble has fixed values of $\underline{V}$, T, and μ.

Grand-Canonical Ensemble

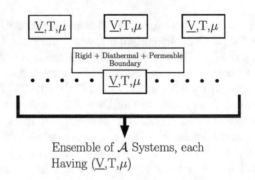

Ensemble of $\mathcal{A}$ Systems, each
Having $(\underline{V}, T, \mu)$

Fig. 44.1

Similar to the approach that we pursued when we discussed the Canonical ensemble, the probability that any randomly chosen system in the Grand-Canonical ensemble contains N molecules and is in the jth quantum state, with energy $\underline{E}_{Nj}(\underline{V})$, is given by:

$$p_{Nj}(\underline{V}, \beta, \gamma) = \frac{e^{-\beta \underline{E}_{Nj}(\underline{V})} e^{-\gamma N}}{\Xi(\underline{V}, \beta, \gamma)} \tag{44.1}$$

where $\beta = 1/k_B T$, $\gamma = -\mu/k_B T$, and Ξ is the Grand-Canonical partition function, given by:

$$\Xi(\underline{V}, \beta, \gamma) = \sum_N \sum_j e^{-\beta E_{Nj}(\underline{V})} e^{-\gamma N} \tag{44.2}$$

Using Eqs. (44.1) and (44.2), we can calculate all the average thermodynamic properties of the system in the context of the Grand-Canonical ensemble. Specifically, the internal energy of the system, $\underline{U}$, is given by:

$$\underline{U} = \langle \underline{E}(\underline{V}, \beta, \gamma) \rangle = \sum_N \sum_j \underline{E}_{Nj}(\underline{V}) \underbrace{\left[\frac{e^{-\beta E_{Nj}(\underline{V})} e^{-\gamma N}}{\Xi} \right]}_{P_{Nj}(\underline{V})}$$

or

$$\underline{U} = -\left(\frac{\partial \ln \Xi}{\partial \beta} \right)_{\underline{V}, \gamma} \tag{44.3}$$

The pressure of the system, P, is given by:

$$P = \langle P \rangle = \sum_N \sum_j \overbrace{-\left(\frac{\partial \underline{E}_{Nj}}{\partial \underline{V}} \right)_{\beta, \gamma}}^{P_{Nj}} \underbrace{\left[\frac{e^{-\beta E_{Nj}(\underline{V})} e^{-\gamma N}}{\Xi} \right]}_{P_{Nj}(\underline{V})}$$

or

$$P = \frac{1}{\beta} \left(\frac{\partial \ln \Xi}{\partial \underline{V}} \right)_{\beta, \gamma} \tag{44.4}$$

The number of molecules in the system, N, is given by:

$$N = \langle N \rangle = \sum_{N'} \sum_j N' \underbrace{\left[\frac{e^{-\beta E_{N'j}(\underline{V})} e^{-\gamma N'}}{\Xi} \right]}_{P_{N'j}(\underline{V})}$$

or

$$N = -\left(\frac{\partial \ln \Xi}{\partial \gamma} \right)_{\underline{V}, \beta} = k_B T \left(\frac{\partial \ln \Xi}{\partial \mu} \right)_{\underline{V}, T} \tag{44.5}$$

where to derive the last term, we used the relation:

$$\frac{\partial}{\partial \gamma} = -k_B T \frac{\partial}{\partial \mu}$$

We can also express Eqs. (44.1)–(44.5) directly in terms of T and μ. For example, Eq. (44.2) can be expressed as follows:

$$\Xi(\underline{V}, T, \mu) = \sum_{N} \sum_{j} e^{-E_{Nj}(\underline{V})/k_B T} e^{+\mu N/k_B T} \tag{44.6}$$

By summing Eq. (44.6) over j, for fixed N, and recalling that the Canonical partition function, $Q(N, \underline{V}, T)$, is given by:

$$Q(N, \underline{V}, T) = \sum_{j} e^{-E_{Nj}(\underline{V})/k_B T} \tag{44.7}$$

it follows that:

$$\Xi(\underline{V}, T, \mu) = \sum_{N} Q(N, \underline{V}, T) e^{N\mu/k_B T} \tag{44.8}$$

Note that the term $e^{\mu/k_B T}$ in Eq. (44.8) is often denoted as λ and is essentially the fugacity, f, introduced in Part II (recall that $\mu \sim k_B T \ln f$).

Because the number of systems in the Grand-Canonical ensemble is as large as desired, the number of molecules in the ensemble is also as large as desired. Accordingly, $\sum_{N}$ in Eq. (44.8) is given by:

$$\sum_{N} \ldots \equiv \sum_{N=0}^{\infty} \ldots \tag{44.9}$$

In Part I, we saw that the following equation is valid for a one component (n = 1) system:

$$\underline{S} = \frac{U}{T} + \frac{PV}{T} - \frac{\mu N}{T} \tag{44.10}$$

Therefore, we can use Eq. (44.10) to calculate $\underline{S}$, where we substitute $\underline{U}$ from Eq. (44.3), P from Eq. (44.4), and N from Eq. (44.5).

We can derive another useful result starting with Eq. (44.4) for P and then multiplying both sides by $d\underline{V}$. This yields:

$$PdV = k_BT\left(\frac{\partial \ln\Xi}{\partial V}\right)_{T,\mu} dV \qquad (44.11)$$

In Part I, we derived the Gibbs-Duhem equation for a one component ($n = 1$) system, which is given by:

$$SdT - VdP + Nd\mu = 0 \qquad (44.12)$$

Using the fact that:

$$-VdP = PdV - d(PV) \qquad (44.13)$$

in Eq. (44.12), including rearranging, we obtain:

$$PdV = d(PV) - SdT - Nd\mu \qquad (44.14)$$

Substituting PdV from Eq. (44.14) in Eq. (44.11) yields:

$$\overbrace{d(PV) - SdT - Nd\mu}^{d(PV)_{T,\mu}} = k_BT\left(\frac{\partial \ln\Xi}{\partial V}\right)_{T,\mu} dV = k_BTd(\ln\Xi)_{T,\mu}$$

or

$$PV = k_BT\ln\Xi(V, T, \mu) \qquad (44.15)$$

Equation (44.15) can be expressed as follows:

$$\Xi(V, T, \mu) = e^{(PV)/k_BT} \qquad (44.16)$$

Equation (44.16) shows that -PV, referred to as the Grand Potential, is the generator of the Grand-Canonical partition function. This is analogous to A, the Helmholtz free energy, being the generator of the Canonical partition function (see Lecture 43).

44.3   Statistical Fluctuations

So far in Part III, we calculated ensemble averages of mechanical properties, including energy and pressure, and then equated these with the experimentally observed thermodynamic properties. However, it is also important to calculate the expected deviation of a mechanical property from its mean (or average) value, referred to as a statistical fluctuation, or in short, as a fluctuation. This will allow us to determine to what extent we expect to observe deviations from the mean value that we calculate using statistical mechanics. If the spread about the mean value is large, then, experimentally, we would observe a range of values, whose mean (or average) is given by the value calculated using statistical mechanics. However, we will show that the probability of observing any other value, except the mean value, is extremely unlikely.

Let us first consider fluctuations in the Canonical ensemble, where T, $\underline{V}$, and N are constant. We can investigate fluctuations in the energy, pressure, and related properties, because these are the properties that vary from system to system in the Canonical ensemble. First, we will consider fluctuations in the energy. Probability theory indicates that the variance of a variable X with respect to its mean (or average) value, $\langle X \rangle$, is given by:

$$\sigma_X^2 = \left\langle (X - \langle X \rangle)^2 \right\rangle = \langle X^2 \rangle - \langle X \rangle^2 \qquad (44.17)$$

where

$$\langle X^2 \rangle = \sum_j X_j^2 p_j, \quad \langle X \rangle = \sum_j X_j p_j \qquad (44.18)$$

where p_j is the probability that the system is in quantum state j. In the case of the energy, $X = \underline{E}$ and Eq. (44.17) becomes:

$$\sigma_{\underline{E}}^2 = \langle \underline{E}^2 \rangle - \langle \underline{E} \rangle^2 = \sum_j E_j^2 p_j - \left(\sum_j E_j p_j \right)^2 \qquad (44.19)$$

where

$$p_j = \frac{e^{-\beta E_j}}{Q(\beta, \underline{V}, N)} \qquad (44.20)$$

We can rewrite Eq. (44.19) in a more convenient form by recognizing that the first summation can be expressed as follows:

$$\sum_j E_j^2 p_j = \frac{1}{Q} \sum_j E_j^2 e^{-\beta E_j} = -\frac{1}{Q} \frac{\partial}{\partial \beta} \underbrace{\left(\sum_j E_j e^{-\beta E_j} \right)}_{Q \langle \underline{E} \rangle} \qquad (44.21)$$

or

$$\sum_j E_j^2 p_j = -\frac{1}{Q}\frac{\partial}{\partial\beta}\left(\langle E\rangle Q\right) = \frac{1}{Q}\cdot Q\left(-\frac{\partial\langle E\rangle}{\partial\beta}\right) - \langle E\rangle\underbrace{\frac{1}{Q}\left(\frac{\partial Q}{\partial\beta}\right)}_{\left(\frac{\partial\ln Q}{\partial\beta}\right)} \tag{44.22}$$

or

$$\sum_j E_j^2 p_j = -\left(\frac{\partial\langle E\rangle}{\partial\beta}\right) - \langle E\rangle\left(\frac{\partial\ln Q}{\partial\beta}\right) \tag{44.23}$$

Because $Q = \sum_j e^{-\beta E_j}$, it follows that:

$$\left(\frac{\partial\ln Q}{\partial\beta}\right) = \frac{-\sum_j E_j\,e^{-\beta E_j}}{Q} = -\langle E\rangle \tag{44.24}$$

Substituting Eq. (44.24) in Eq. (44.23), and recalling that $1/\partial\beta = -k_B T^2/\partial T$, it follows that:

$$\sum_j E_j^2 p_j = k_B T^2\left(\frac{\partial\langle E\rangle}{\partial T}\right)_{N,\underline{V}} + \langle E\rangle^2 \tag{44.25}$$

Substituting Eq. (44.25) in Eq. (44.19), including cancelling the $\langle\underline{E}\rangle^2$ terms, yields:

$$\sigma_{\underline{E}}^2 = k_B T^2\left(\frac{\partial\,\langle E\rangle}{\partial T}\right)_{N,\underline{V}} \tag{44.26}$$

Next, if we equate $\langle E\rangle$ with the thermodynamic energy, $\underline{U}$, we find that:

$$\sigma_{\underline{E}}^2 = \sigma_{\underline{U}}^2 = k_B T^2\left(\frac{\partial\underline{U}}{\partial T}\right)_{N,\underline{V}} = Nk_B T^2\underbrace{\left(\frac{\partial U}{\partial T}\right)_V}_{C_V}$$

or

$$\sigma_{\underline{U}}^2 = k_B T^2 C_V N \tag{44.27}$$

The square root of $\sigma_{\underline{U}}^2$, that is, $\sqrt{\sigma_{\underline{U}}^2} = \sigma_{\underline{U}}$, is referred to as the standard deviation, and is given by (see Eq. (44.27)):

$$\sigma_{\underline{U}} = \left(k_B T^2 C_V\right)^{1/2} \sqrt{N} \tag{44.28}$$

To quantify the relative magnitude of the spread in $\underline{U}$, we examine the quantity:

$$\frac{\sigma_{\underline{U}}}{\underline{U}} = \frac{\sqrt{N}\left(k_B T^2 C_V\right)^{1/2}}{NU}$$

or

$$\frac{\sigma_{\underline{U}}}{\underline{U}} = \overbrace{\left[\frac{\left(k_B T^2 C_V\right)^{1/2}}{U}\right]}^{O(1)} \cdot \frac{1}{\sqrt{N}} \tag{44.29}$$

Equation (44.29) shows that in a typical macroscopic system, where $N \gg 1$, the relative deviations from the mean energy, $\underline{U}$, are extremely small (of order $1/\sqrt{N} \to 0$). In fact, as we will show shortly, when $N \gg 1$, the probability distribution of the energy is a Gaussian distribution which approaches a delta function centered around the mean (or average) energy.

44.4　Sample Problem 44.1

Calculate $\sigma_{\underline{U}}/\underline{U}$ for a monoatomic ideal gas.

44.4.1　Solution

As discussed in Part I, for a monoatomic ideal gas:

$$U = \frac{3}{2} k_B T \quad \text{and} \quad C_V = \frac{3}{2} k_B$$

Using the two results above in Eq. (44.29) yields:

$$\frac{\sigma_{\underline{U}}}{\underline{U}} = \frac{\left[k_B T^2 \left(\frac{3}{2} k_B\right)\right]^{1/2}}{\left(\frac{3}{2} k_B T\right)} \cdot \frac{1}{N^{1/2}}$$

or

$$\frac{\sigma_U}{U} = \left(\frac{3}{2}N\right)^{-1/2}$$ (44.30)

If N is of order 10^{23}, then, Eq. (44.30) shows that σ_U/U is of order 10^{-12}, which implies an extremely small spread of U around its mean (or average) value. Although outside the scope of this introductory exposure to statistical mechanics, we would like to stress that fluctuations become extremely large (in fact, diverge) near critical points, which results in very interesting phenomena, known collectively as critical phenomena (see below for a specific illustration).

44.5 Sample Problem 44.2

Show that the probability of observing a particular value of E is Gaussian, with the maximum of the distribution centered at $E = \langle E \rangle$, and its width determined by σ_E.

44.5.1 Solution

Using the statistical mechanical description based on quantum levels and the degeneracy of these quantum levels, the probability, $p(E)$, of observing energy level E in the Canonical ensemble (T, V, N fixed) can be written as follows:

$$p(E) = C\Omega(E)e^{-\beta E}$$ (44.31)

where $\Omega(E)$ is the degeneracy of energy level E, and C is a normalization factor that is independent of E (essentially, it is Q^{-1}). Because $\Omega(E)$ is an increasing function of E (see below), and e^{-E/k_BT} is a decreasing function of E, their product in Eq. (44.31) should peak at some value of E, referred to hereafter as E^*. However, we have just shown that the spread about the maximum value is extremely small, so that E^* and $\langle E \rangle$ are basically the same. The width of $p(E)$ is of $O(N^{-1/2})$, so that E^* and $\langle E \rangle$ also differ by $O(N^{-1/2})$. In the limit of a macroscopic system ($N \gg 1$), the difference tends to 0.

We can expand $p(E)$ in a Taylor series about $E^* = \langle E \rangle$. To this end, it is convenient to work with $\ln p(E)$. The Taylor series expansion of $\ln p(E)$ around $\ln p(\langle E \rangle)$, to quadratic order, is given by:

$$\ln(\underline{E}) = \ln(\langle\underline{E}\rangle) + \left(\frac{\partial\ln(\underline{E})}{\partial\underline{E}}\right)_{\underline{E}=\langle\underline{E}\rangle}(\underline{E} - \langle\underline{E}\rangle)$$

$$+ \frac{1}{2}\left(\frac{\partial^2\ln(\underline{E})}{\partial\underline{E}^2}\right)_{\underline{E}=\langle\underline{E}\rangle}(\underline{E} - \langle\underline{E}\rangle)^2 \qquad (44.32)$$

Because $p(\underline{E})$ exhibits a maximum at $\underline{E} = \langle\underline{E}\rangle$, it follows that:

$$\left(\frac{\partial\ln(\underline{E})}{\partial\underline{E}}\right)_{\underline{E}=\langle\underline{E}\rangle} = 0 \qquad (44.33)$$

Substituting $p(\underline{E})$ from Eq. (44.31) in Eq. (44.33), we obtain:

$$\left(\frac{\partial\ln p(\underline{E})}{\partial\underline{E}}\right)_{\underline{E}=\langle\underline{E}\rangle} = \left[\frac{\partial}{\partial\underline{E}}(\ln C + \ln\Omega(\underline{E}) - \beta\underline{E})\right]_{\underline{E}=\langle\underline{E}\rangle}$$

As indicated earlier, C is independent of $\underline{E}$, and therefore, we can express the last equation as follows:

$$\left(\frac{\partial\ln p(\underline{E})}{\partial\underline{E}}\right)_{\underline{E}=\langle\underline{E}\rangle} = \left(\frac{\partial\ln\Omega(\underline{E})}{\partial\underline{E}}\right)_{\underline{E}=\langle\underline{E}\rangle} - \beta \qquad (44.34)$$

Taking the second derivative of Eq. (44.34) with respect to $\underline{E}$, without holding $\underline{E} = \langle\underline{E}\rangle$ in the two first-order derivatives, the partial derivative of β with respect to $\underline{E}$ is zero, because the quantum mechanical $\underline{E}$s do not depend on β or on T. As a result, the second derivative of $\ln p(\underline{E})$ with respect to $\underline{E}$ in Eq. (44.34) is equal to the second derivative of $\ln\Omega(\underline{E})$ with respect to $\underline{E}$. Using Eq. (44.34) in Eq. (44.33), including rearranging, yields:

$$\left(\frac{\partial\ln\Omega(\underline{E})}{\partial\underline{E}}\right)_{\underline{E}=\langle\underline{E}\rangle} = \beta \qquad (44.35)$$

Because β is nonnegative, Eq. (44.35) shows that $\Omega(\underline{E})$ increases with $\underline{E}$ only when $\underline{E} = \langle\underline{E}\rangle$. To show that this behavior is general for all $\underline{E}$s, as assumed earlier when we discussed Eq. (44.31), requires additional analysis which is beyond the scope of this lecture.

Next, we need to compute the second derivative of $\ln p(\underline{E})$ in Eq. (44.32). To this end, we make use of Eq. (44.35), that is:

$$\left(\frac{\partial^2 \ln p(\underline{E})}{\partial \underline{E}^2}\right)_{\underline{E}=\langle \underline{E}\rangle} = \frac{\partial \beta}{\partial \langle \underline{E}\rangle} \tag{44.36}$$

The full derivation of Eq. (44.36) is beyond the scope of this lecture. For additional details, the interested reader is referred to Chapter 3 in McQuarrie. Recalling that $\beta = 1/k_B T$, the last term in Eq. (44.36) can be expressed as follows:

$$\frac{\partial \beta}{\partial \langle \underline{E}\rangle} = -\frac{1}{k_B T^2} \overbrace{\left(\frac{\partial T}{\partial \langle \underline{E}\rangle}\right)}^{(1/C_V N)} \tag{44.37}$$

Using Eq. (44.37) in Eq. (44.36) then yields:

$$\left(\frac{\partial \ln p(\underline{E})}{\partial \underline{E}^2}\right)_{\underline{E}=\langle \underline{E}\rangle} = -\frac{1}{k_B T^2 C_V N} \tag{44.38}$$

Next, we can use Eqs. (44.33) and (44.38) in Eq. (44.32), which yields:

$$\ln p(\underline{E}) = \ln P(\langle \underline{E}\rangle) - \frac{(\underline{E} - \langle \underline{E}\rangle)^2}{2 k_B T^2 C_V N}$$

or

$$p(\underline{E}) = p(\langle \underline{E}\rangle) e^{-(\underline{E}-\langle \underline{E}\rangle)^2 / (2 k_B T^2 C_V N)} \tag{44.39}$$

As we will show next, Eq. (44.39) corresponds to a Gaussian distribution of $\underline{E}$ around $\langle \underline{E}\rangle$. Indeed, recall that a Gaussian distribution of the variable, $-\infty \leq X \leq +\infty$, is given by:

$$p(X) = \frac{1}{(2\pi\sigma^2)^{1/2}} \exp\left\{-\frac{(X - \langle X\rangle)^2}{2\sigma^2}\right\} \tag{44.40}$$

where σ^2 is referred to as the variance of the distribution, and determines the width of the Gaussian distribution. Specifically, a smaller value of σ in Eq. (44.40) corresponds to a narrower Gaussian distribution. In the limit $\sigma \to 0$, Eq. (44.40) reduces to a delta function. A comparison of Eqs. (44.39) and (44.40) reveals that:

$$p(\langle \underline{E}\rangle) = \left(2\pi\sigma_{\underline{E}}^2\right)^{-1/2} \tag{44.41}$$

and

$$\sigma_{\underline{E}}^2 = k_B T^2 C_V N \qquad (44.42)$$

Equation (44.42) is identical to Eq. (44.27), with the energy $\underline{U}$ replaced by $\underline{E}$. Equation (44.39) can be rewritten as follows:

$$\frac{p(\underline{E})}{p(\langle E \rangle)} = e^{-\underbrace{\left[\frac{(E - \langle E \rangle)^2}{2k_B T^2 C_V}\right]}_{O(1)} N} \qquad (44.43)$$

Equation (44.43) is a very important result which shows that the probability of observing $\underline{E} \neq \langle E \rangle$ is of order e^{-N}, which when N is macroscopic ($\gg 1$), approaches 0.

44.6 Fluctuations in the Number of Molecules

To calculate the fluctuations in the number of molecules, N, it is convenient to use the Grand-Canonical ensemble and follow a procedure similar to that used to calculate the fluctuations in the energy using the Canonical ensemble. If σ_N^2 is the variance in the number of molecules, N, then:

$$\begin{aligned} \sigma_N^2 &= \left\langle (N - \langle N \rangle)^2 \right\rangle = \langle N^2 \rangle - \langle N \rangle^2 \\ &= \sum_N \sum_j N^2 p_{Nj} - \langle N \rangle^2 \end{aligned} \qquad (44.44)$$

where

$$p_{Nj} = \frac{e^{-\beta E_{Nj}} e^{-\gamma N}}{\Xi(\underline{V}, \beta, \gamma)} \qquad (44.45)$$

As we did with $\sigma_{\underline{E}}^2$, we first consider the double summation in Eq. (44.44). Specifically,

$$\sum_N \sum_j N^2 p_{Nj} = \frac{1}{\Xi} \sum_N \sum_j N^2 e^{-\beta E_{Nj}} e^{-\gamma N} = -\frac{1}{\Xi} \frac{\partial}{\partial \gamma} \underbrace{\left(\sum_N \sum_j e^{-\beta E_{Nj}} e^{-\gamma N} \right)}_{(\langle N \rangle \Xi)}$$

$$= -\frac{1}{\Xi} \frac{\partial}{\partial \gamma} (\langle N \rangle \Xi)$$

or

$$\sum_N \sum_j N^2 p_{Nj} = -\left(\frac{\partial \langle N \rangle}{\partial \gamma}\right) - \langle N \rangle \left(\frac{\partial \ln \Xi}{\partial \gamma}\right) \tag{44.46}$$

The second partial derivative on the right-hand side of Eq. (44.46) can be expressed as follows:

$$\left(\frac{\partial \ln \Xi}{\partial \gamma}\right) = \frac{-\sum_N \sum_j N e^{-\beta E_{Nj}} e^{-\gamma N}}{\Xi} = -\langle N \rangle \tag{44.47}$$

Substituting Eq. (44.47) in Eq. (44.46) then yields:

$$\sum_N \sum_j N^2 p_{Nj} = -\left(\frac{\partial \langle N \rangle}{\partial \gamma}\right) + \langle N \rangle^2 \tag{44.48}$$

Because $\gamma = -\mu/k_B T$ and $\left(\frac{\partial}{\partial \gamma}\right) = -k_B T \left(\frac{\partial}{\partial \mu}\right)$, Eq. (44.48) can be expressed as follows:

$$\sum_N \sum_j N^2 p_{Nj} = k_B T \left(\frac{\partial \langle N \rangle}{\partial \mu}\right)_{V,T} + \langle N \rangle^2 \tag{44.49}$$

Substituting Eq. (44.49) in Eq. (44.44), and cancelling the $\langle N \rangle^2$ terms, yields:

$$\sigma_N^2 = k_B T \left(\frac{\partial \langle N \rangle}{\partial \mu}\right)_{V,T} \tag{44.50}$$

As shown in Chapter 3 of McQuarrie, the following thermodynamic relation is valid:

$$\left(\frac{\partial \mu}{\partial \langle N \rangle}\right)_{V,T} = \frac{-V^2}{\langle N \rangle^2} \left(\frac{\partial P}{\partial V}\right)_{N,T}$$

Rearranging the last equation, we obtain:

$$\left(\frac{\partial \langle N \rangle}{\partial \mu}\right)_{V,T} = -\left(\frac{\partial V}{\partial P}\right)_{N,T} \frac{\langle N \rangle^2}{V^2} \tag{44.51}$$

In Part I, we showed that the isothermal compressibility, κ_T, is given by:

$$\kappa_T = -\frac{1}{\underline{V}}\left(\frac{\partial \underline{V}}{\partial P}\right)_{N,T} \tag{44.52}$$

Using Eq. (44.52) in Eq. (44.51) yields:

$$\left(\frac{\partial \langle N \rangle}{\partial \mu}\right)_{\underline{V},T} = \frac{\kappa_T \langle N \rangle^2}{\underline{V}} \tag{44.53}$$

Substituting Eq. (44.53) in Eq. (44.50), we obtain:

$$\sigma_N^2 = \frac{k_B T \kappa_T \langle N \rangle^2}{\underline{V}} \tag{44.54}$$

Taking the square root of both terms in Eq. (44.54), the standard deviation, σ_N, is given by:

$$\sigma_N = \left(\frac{k_B T \kappa_T}{\underline{V} N}\right)^{1/2} \langle N \rangle \tag{44.55}$$

Using Eq. (44.55), σ_N relative to $\langle N \rangle$ is then given by:

$$\frac{\sigma_N}{\langle N \rangle} = \left(\frac{k_B T \kappa_T}{\underline{V}}\right) N^{-1/2} \tag{44.56}$$

Equation (44.56) again shows that in the macroscopic limit ($N \gg 1$), $\sigma_N/\langle N \rangle$ tends to zero as $N^{-1/2}$. Accordingly, in the macroscopic limit, the fluctuations in N become vanishingly small, the distribution of N becomes sharply peaked around $\langle N \rangle$, and approaches a delta function.

44.7   Sample Problem 44.3

Calculate $\sigma_N/\langle N \rangle$ for a monoatomic ideal gas.

44.7.1   Solution

To solve this problem, we need to utilize Eq. (44.56). To this end, we need to compute the isothermal compressibility, discussed in Part I, and given in Eq. (44.52), which we repeat below for completeness:

$$\kappa_T = -\frac{1}{\underline{V}}\left(\frac{\partial \underline{V}}{\partial P}\right)_{N,T} = -\frac{1}{V}\left(\frac{\partial V}{\partial P}\right)_T \tag{44.57}$$

In Part I, we showed that for a monoatomic ideal gas (on a per atom basis):

$$V = \frac{k_B T}{P} \Rightarrow \left(\frac{\partial V}{\partial P}\right)_T = -\frac{k_B T}{P^2} = -\frac{V}{P} \tag{44.58}$$

Using Eq. (44.58) in Eq. (44.57) yields:

$$\kappa_T = \frac{1}{P} \tag{44.59}$$

Equation (44.58) shows that $(k_B T/V) = P$. Using this last result and Eq. (44.59) in Eq. (44.56) yields:

$$\sigma_N/\langle N \rangle = \frac{1}{\sqrt{N}} \tag{44.60}$$

Equation (44.60) again shows that in the macroscopic limit $(N \gg 1)$, the fluctuations in N become vanishingly small.

Important Comment Because $\underline{V}$ is fixed in the Grand-Canonical ensemble, the fluctuations in the number of molecules are proportional to the fluctuations in the number density, $\rho = N/\underline{V}$. In other words, we can rewrite Eq. (44.56) as follows:

$$\frac{\sigma_N}{\langle N \rangle} = \frac{(\sigma_N/\underline{V})}{(\langle N \rangle/\underline{V})} = \frac{\sigma_\rho}{\langle \rho \rangle} = \left(\frac{k_B T \kappa_T}{V}\right)^{1/2}\frac{1}{\sqrt{N}} \tag{44.61}$$

Equation (44.61) shows that, in the thermodynamic limit $(N \gg 1)$, $(\sigma_\rho/\langle \rho \rangle) \to 0$. However, near the critical point, density fluctuations are not negligible, because κ_T diverges. This is observed experimentally by the onset of critical opalescence when a pure fluid $(n = 1)$ becomes turbid as it approaches the critical point.

As shown in Chapter 3 of McQuarrie, the distribution of N around $\langle N \rangle$ is Gaussian and exhibits a maximum at $N = \langle N \rangle$, where:

$$p(N) = p(\langle N \rangle)\exp\left[-\frac{(N - \langle N \rangle)^2}{2\sigma_N^2}\right] \tag{44.62}$$

The interested reader is encouraged to derive Eq. (44.62) in order to further crystallize the material taught.

44.8   Equivalence of All the Ensembles in the Thermodynamic Limit

Next, we would like to show that in the thermodynamic limit ($N \gg 1$), when fluctuations are negligible, all the ensembles are equivalent. Consider first the Canonical partition function, given by:

$$Q(N, \underline{V}, T) = \sum_E \underbrace{\Omega(N, \underline{V}, E)}_{\substack{\text{Degeneracy of} \\ \text{Energy Level, E}}} e^{-E/k_B T} \qquad (44.63)$$

In this lecture, we showed that out of all the terms in the sum in Eq. (44.63), the only one which really matters corresponds to $\underline{E} = \langle \underline{E} \rangle$. Accordingly, to an excellent approximation, when $N \gg 1$, Eq. (44.63) reduces to:

$$Q(N, \underline{V}, T) = \Omega(N, \underline{V}, \langle \underline{E} \rangle) e^{-\langle \underline{E} \rangle / k_B T} \qquad (44.64)$$

Therefore, although the systems in the Canonical ensemble can, in principle, assume any value of $\underline{E}$ (as long as it is an eigenvalue of the N-molecule Schrödinger equation), in effect, every system in the Canonical ensemble is most likely to be found with the average energy, $\langle \underline{E} \rangle$. Accordingly, when $N \gg 1$, a Canonical ensemble reduces to a Micro-Canonical ensemble. To see this more clearly, we begin by taking the ln of Eq. (44.64). Specifically,

$$\ln Q = \ln \Omega - \langle \underline{E} \rangle / k_B T$$

or

$$-k_B T \ln Q = \langle \underline{E} \rangle - k_B T \ln \Omega \qquad (44.65)$$

However, in Lecture 43, we showed that:

$$-k_B T \ln Q = \underline{A} \qquad (44.66)$$

Using Eq. (44.66) in Eq. (44.65), we obtain:

$$\underline{A} = \langle \underline{E} \rangle - k_B T \ln \Omega(N, \underline{V}, \langle \underline{E} \rangle) \qquad (44.67)$$

In Part I, we saw that:

$$\underline{A} = \underline{U} - T\underline{S} \qquad (44.68)$$

In addition, in Part III, we saw that $\langle \underline{E} \rangle = \underline{U}$. Accordingly, a comparison of Eqs. (44.67) and (44.68) shows that:

$$\underline{S} = k_B \ln \Omega(N, \underline{V}, \langle \underline{E} \rangle) \tag{44.69}$$

Equation (44.69) is identical to the expression for $\underline{S}$ that we derived in Lecture 42 in the context of the Micro-Canonical ensemble. However, here, we derived Eq. (44.69) starting from the Canonical ensemble.

A similar argument can be made in the case of the Grand-Canonical ensemble. Indeed, although N and $\underline{E}$ could fluctuate in this case as well, for all practical purposes, every system in the ensemble, when $N \gg 1$, possesses $N = \langle N \rangle$ and $\underline{E} = \langle \underline{E} \rangle$, so that the Grand-Canonical ensemble reduces to a Micro-Canonical ensemble. Accordingly, in the thermodynamic limit, $N \gg 1$, the Micro-Canonical, Canonical, and Grand-Canonical ensembles are all equivalent. This very important result is rewarding because it shows that, irrespective of the specific ensemble used, when $N \gg 1$, statistical mechanics will predict the same thermodynamic properties, which correspond to those observed experimentally.

Lecture 45

Classical Statistical Mechanics and Sample Problem

45.1 Introduction

The material presented in this lecture is adapted from Chapter 7 in McQuarrie. First, we will motivate the use of classical statistical mechanics, including discussing generalized momenta and positions, the Hamiltonian, Hamilton's equations of motion, and phase space. Second, we will introduce the classical molecular partition function. Third, we will discuss the classical partition function of an atom in an ideal gas. Fourth, we will discuss the classical partition function of a rigid rotor. Fifth, we will discuss the classical partition function of a system consisting of N independent and indistinguishable molecules. Sixth, we will discuss the molecular partition function of a system consisting of N interacting and indistinguishable molecules. Seventh, we will solve Sample Problem 45.1, including deriving an expression for the configurational integral. Eight, we will show how to incorporate both classical and quantum mechanical degrees of freedom into the partition function. Finally, we will discuss the equipartition of energy in the context of statistical mechanics, as reflected in the celebrated Equipartition Theorem discussed in Part I.

45.2 Classical Statistical Mechanics

In classical mechanics, given a set of N molecules, we can uniquely determine the state of the system and its energy if we know the positions and momenta (or velocities) of each of the N molecules comprising the system. The total energy of the system is given by the Hamiltonian, H, such that:

$$H(\vec{q}, \vec{p}) = K(\vec{p}) + \Phi(\vec{q}) \tag{45.1}$$

© Springer Nature Switzerland AG 2020
D. Blankschtein, *Lectures in Classical Thermodynamics with an Introduction to Statistical Mechanics*, https://doi.org/10.1007/978-3-030-49198-7_45

In three-dimensional space, $\vec{p}$ and $\vec{q}$ in Eq. (45.1) are vectors each having 3N components, $(p_{x1}, p_{y1}, p_{z1}, p_{x2}, p_{y2}, p_{z2}, \ldots, p_{xN}, p_{yN}, p_{zN};$ $q_{x1}, q_{y1}, q_{z1}, q_{x2}, q_{y2}, q_{z2}, \ldots, q_{xN}, q_{yN}, q_{zN})$ and representing the generalized momenta and positions describing the system, respectively. In Eq. (45.1), $K(\vec{p})$ is the kinetic energy of the system and depends solely on the momenta, $\vec{p}$, and $\Phi(\vec{q})$ is the potential energy of the system and depends solely on the positions, $\vec{q}$. Unlike quantum mechanics, where the energy is quantized (discrete), in classical mechanics, the Hamiltonian and the kinetic and potential energies are not restricted to assume discrete values.

Given an initial set of coordinates $\vec{p}$ and $\vec{q}$, we can determine the evolution of the system by integrating Newton's equations of motion. Alternatively, the same information can be obtained by integrating Hamilton's equations of motion, given by:

$$\frac{\partial H}{\partial p_i} = \dot{q}_i$$

$$\frac{\partial H}{\partial q_i} = -\dot{p}_i$$

(45.2)

where p_i and q_i are the ith components of $\vec{p}$ and $\vec{q}$, respectively, and the dots denote time derivatives.

As the classical system evolves through time, it traverses a sequence of 6N-dimensional configurations, or states, $(\vec{p}, \vec{q})$. The set of all allowed configurations of $(\vec{p}, \vec{q})$ is referred to as phase space. Equations (45.2) define the trajectory of a classical system through phase space.

45.3   Classical Molecular Partition Function

As we discussed in Part III, the molecular Canonical partition function is given by:

$$q(\underline{V}, T) = \sum_j e^{-\beta E_j}$$

(45.3)

where j denotes the quantized, discrete state j having energy $\underline{E}_j$. Recall that Eq. (45.3) is a sum over all the available quantum states.

It is reasonable to conjecture that the corresponding classical expression of Eq. (45.3) is a similar sum, or because the energy in the classical limit is a continuous function of $\vec{p}$ and $\vec{q}$ (see Eq. (45.1)), the sum should become an integral over all the possible "classical states" of the system. Because the classical energy is the Hamiltonian function, $H(\vec{p}, \vec{q})$, we conjecture that the classical molecular partition function, $q_{class}(\underline{V}, T)$, is given by:

$$q_{class}(\underline{V}, T) \sim \int \cdots \int e^{-\beta H(\vec{p}, \vec{q})} d\vec{p} d\vec{q} \qquad (45.4)$$

where $d\vec{p}$ stands for $dp_1 dp_2 \ldots dp_s$, and $d\vec{q}$ stands for $dq_1 dq_2 \ldots dq_s$, where s is the number of momenta or position coordinates required to completely specify the motion or the position of the molecule, respectively. In essence, s represents the number of degrees of freedom of the molecule. It is important to recognize that the set of coordinates, $\{q_i\}$, is not necessarily a set of Cartesian coordinates, but instead, it represents a set of generalized coordinates that specifies the position of the molecule in a convenient manner. For example, for a mass point, the generalized coordinates may be x, y, and z. For a rigid rotor, one may choose the two angles, θ and ϕ, required to specify the orientation of the molecule. At this stage, Eq. (45.4) is a plausible conjecture.

45.4   Classical Partition Function of an Atom in an Ideal Gas

Let us next test the conjecture proposed above in the case of an atom in an ideal gas. Earlier in Part III, we saw that the quantum mechanical (QM) atomic translational (trans) partition function is given by:

$$q_{QM}^{trans}(\underline{V}, T) = \left(\frac{2\pi m k_B T}{h^2}\right)^{3/2} \underline{V} \qquad (45.5)$$

where m is the mass of the atom, T is the temperature in degrees Kelvin, $\underline{V}$ is the volume of the container, and h is Planck's constant. The classical Hamiltonian of an atom in an ideal gas is simply the kinetic energy, $K(\vec{p})$, in Eq. (45.1) (recall that $\Phi(\vec{q}) = 0$ for an ideal (non-interacting) gas). Specifically:

$$H = K(\vec{p}) = \frac{1}{2m}(p_x^2 + p_y^2 + p_z^2) \qquad (45.6)$$

where we used the fact that $\vec{p} = m\vec{v}$, so that $\frac{1}{2}m\vec{v}^2 = \vec{p}^2/2m$.

According to Eq. (45.4), for an atom in an ideal gas for which Eq. (45.6) applies, it follows that:

$$q_{class}^{trans} \sim \int \cdots \int \exp\left\{\frac{-\beta\left(p_x^2 + p_y^2 + p_z^2\right)}{2m}\right\} dp_x dp_y dp_z \, dxdydz \qquad (45.7)$$

where the integral over dxdydz simply yields the volume of the container, $\underline{V}$, such that Eq. (45.7) can be expressed as follows:

$$q_{class}^{trans} \sim \underline{V}\left\{\underbrace{\int\limits_{-\infty}^{+\infty} e^{-\beta p^2/2m}dp}_{(2\pi mk_BT)^{1/2}}\right\}^3$$

or

$$q_{class}^{trans}(\underline{V}, T) \sim (2\pi mk_BT)^{3/2}\underline{V} \qquad (45.8)$$

Equation (45.8) is identical to Eq. (45.5), except for the factor h^3 in the denominator. Of course, it is not possible to derive a purely classical expression where h appears, but nevertheless, our conjecture in Eq. (45.4) appears reasonable.

45.5   Classical Partition Function of a Rigid Rotor

Next, we can also check how the conjecture in Eq. (45.4) works in other cases that we have considered quantum mechanically. For example, as we saw earlier in Part III, for the rigid rotor, the classical Hamiltonian is given by:

$$H = \frac{1}{2I}\left(p_\theta^2 + p_\phi^2/\sin^2\theta\right) \qquad (45.9)$$

where I is the moment of inertia of the molecule. The generalized coordinates in this case are the two angular coordinates, θ and ϕ, and the two associated momenta coordinates, p_θ and p_ϕ. Using Eq. (45.9) in Eq. (45.4) yields:

$$q_{class}^{rot} \sim \int\limits_{-\infty}^{+\infty}\int\limits_{-\infty}^{+\infty} dp_\theta dp_\phi \int\limits_0^{2\pi} d\phi \int\limits_0^\pi d\theta\, e^{-\beta H} = 8\pi^2 Ik_BT \qquad (45.10)$$

where we have used the appropriate limits of integration for the momenta and angular coordinates. Recall that, as we saw earlier in Part III, the QM version of Eq. (45.10) is given by:

$$q_{QM}^{rot} = \frac{8\pi^2 Ik_BT}{h^2} \qquad (45.11)$$

A comparison of Eqs. (45.11) and (45.10) shows that the factor h^2 is missing in the denominator of Eq. (45.10), but nevertheless, our conjecture in Eq. (45.4) again appears reasonable.

45.6   Classical Partition Function of a System Consisting of N Independent and Indistinguishable Molecules

The examples involving translations and rotations presented above demonstrate that q_{class}^{trans} is incorrect by a factor of h^3, while q_{class}^{rot} is incorrect by a factor of h^2. Therefore, a factor of h appears for each product $(dp_i dq_i)$ in the expression for q_{class}, that is, (i) 3 in the translational case: $(dp_x dq_x)(dp_y dq_y)(dp_z dq_z)$ in Eq. (45.7), and (ii) 2 in the rotational case: $(dp_\theta d\theta)(dp_\phi d\phi)$ in Eq. (45.10). Because the partition function is dimensionless, dividing q_{class} by h^s ensures that q_{class} be dimensionless as well. Recall that s is the number of degrees of freedom of the molecule. We will therefore generalize our original conjecture in Eq. (45.4) as follows:

$$q_{QM} = \sum_j e^{-\beta E_j} \rightarrow q_{class} = \frac{1}{h^s} \int \cdots \int e^{-\beta H} \prod_{j=1}^{s} dp_j dq_j \qquad (45.12)$$

Next, we will extend Eq. (45.12) from a single molecule to N independent and indistinguishable molecules, which as discussed earlier in Part III, is a very good description at high temperatures. For N independent and indistinguishable molecules, the Canonical partition function, Q, is given by:

$$Q = \frac{q^N}{N!} \qquad (45.13)$$

Using Eq. (45.12) in Eq. (45.13), it follows that:

$$Q = \frac{1}{N!} \prod_{j=1}^{N} \left\{ \frac{1}{h^s} \int \cdots \int e^{-\beta H_j} \prod_{i=1}^{s} dp_{ji} dq_{ji} \right\} \qquad (45.14)$$

where H_j is the Hamiltonian of the jth molecule and is a function of $p_{j1}, p_{j2}, \ldots, p_{js}$ and $q_{j1}, q_{j2}, \ldots, q_{js}$.

Because every j term in Eq. (45.14) is independent, we can carry out the multiplication over j as follows:

$$Q = \frac{1}{N!}\left(\frac{1}{h^s}\int\cdots\int e^{-\beta H_1}\prod_{i=1}^{s}dp_{1i}dq_{1i}\right)\left(\frac{1}{h^s}\int\cdots\int e^{-\beta H_2}\prod_{i=1}^{s}dp_{2i}dq_{2i}\right)\cdots$$

$$\cdots\left(\frac{1}{h^s}\int\cdots\int e^{-\beta H_N}\prod_{i=1}^{s}dp_{Ni}dq_{Ni}\right)$$

or

$$Q = \frac{1}{N!h^{sN}}\int\cdots\int e^{-\beta\,\overbrace{(H_1 + H_2 + \ldots + H_N)}^{\sum_j H_j = H}}\times\underbrace{\left(\prod_{i=1}^{s}dp_{1i}dq_{1i}\right)\cdots\left(\prod_{i=1}^{s}dp_{Ni}dq_{Ni}\right)}_{\underbrace{\prod_{i=1}^{s}(dp_{1i}dp_{2i}\ldots dp_{Ni})(dq_{1i}dq_{2i}\ldots dq_{Ni})}_{\prod_{i=1}^{sN}dp_i dq_i}}$$

or

$$Q = \frac{1}{N!h^{sN}}\int\cdots\int e^{-\beta H}\prod_{i=1}^{Ns}dp_i dq_i \qquad (45.15)$$

Again, recall that Eq. (45.15) is valid for N independent and indistinguishable molecules.

45.7 Classical Partition Function of a System Consisting of N Interacting and Indistinguishable Molecules

As we did in the single-molecule case, based on Eq. (45.15), we conjecture that the classical limit of Q for a system of N interacting and indistinguishable molecules is given by:

$$Q = \frac{1}{N!h^{sN}}\int\cdots\int e^{-\beta H(\vec{p},\vec{q})}d\vec{p}\,d\vec{q} \qquad (45.16)$$

where $H(\vec{p},\vec{q})$ is the classical N-body Hamiltonian for interacting molecules and is not given by $H = \sum_j H_j$. In Eq. (45.16), the notation $(\vec{p},\vec{q})$ represents the set

$\{p_j, q_j\}$ that describes the entire system, and $\mathrm{d}\vec{p}\,\mathrm{d}\vec{q}$ stands for $\prod\limits_{j=1}^{sN} \mathrm{d}p_j\,\mathrm{d}q_j$.

To summarize, we have conjectured that the classical limit of $Q(N, \underline{V}, T)$ is given by:

$$Q = \sum_j e^{-\beta E_j} \rightarrow \frac{1}{N!h^{sN}} \int \cdots \int e^{-\beta H(\vec{p},\vec{q})}\,\mathrm{d}\vec{p}\,\mathrm{d}\vec{q} \tag{45.17}$$

45.8 Sample Problem 45.1

Calculate Q_{class} for a monoatomic, non-ideal gas consisting of N molecules, including identifying the configurational integral.

45.8.1 Solution

The classical Hamiltonian for this system is given by:

$$H_{\text{class}}(\vec{p}, \vec{q}) = \underbrace{\frac{1}{2m}\sum_{j=1}^{N}(p_{jx}^2 + p_{jy}^2 + p_{jz}^2)}_{\text{Kinetic Energy}} + \underbrace{\Phi(x_1, y_1, z_1, \ldots, x_N, y_N, z_N)}_{\text{Potential Energy}} \tag{45.18}$$

If we substitute Eq. (45.18) in Eq. (45.16), we can decouple the momentum integrations from the spatial integrations to obtain (with $s = 3$ in this case):

$$Q_{\text{class}} = \frac{1}{N!h^{3N}} \underbrace{\left(\int_{-\infty}^{+\infty} e^{-\frac{\beta p^2}{2m}}\,\mathrm{d}p\right)^{3N}}_{(2\pi m k_B T)^{3N/2}} \left(\int_{\underline{V}} e^{-\Phi(x_1,\ldots,z_N)/k_B T}\,\mathrm{d}x_1 \ldots \mathrm{d}z_N\right)$$

or

$$Q_{\text{class}} = \frac{1}{N!}\left(\frac{2\pi m k_B T}{h^2}\right)^{3N/2} Z_N \tag{45.19}$$

where

$$Z_N = \int_{\underline{V}} e^{-\Phi(x_1,\ldots,z_N)/k_BT} dx_1.\ldots dz_N \qquad (45.20)$$

and is referred to as the configurational integral. Because intermolecular forces depend on the relative distances between molecules, in most cases, it is very challenging to carry out the integrations in Eq. (45.20). However, in the absence of intermolecular forces, Φ in Eq. (45.20) is zero, and as a result, $Z_N = \underline{V}^N$. Equations (45.19) and (45.20) are the basis for studying monoatomic, classical interacting gases and liquids.

When $\Phi = 0$ and $Z_N = \underline{V}^N$, Q_{class} in Eq. (45.19) reduces to the expression for Q of a monoatomic ideal gas derived earlier in Part III in the context of quantum mechanics, that is,

$$Q_{class} = \frac{q_{class}^N}{N!} ; q_{class} = \left(\frac{2\pi m k_B T}{h^2}\right)^{3/2} \underline{V} \qquad (45.21)$$

45.9   Simultaneous Treatment of Classical and Quantum Mechanical Degrees of Freedom

It often happens that not all the degrees of freedom of a molecule can be treated classically. We have already seen that the spacings between the translational and the rotational energy levels are small enough that these energies are almost continuous. As a result, we have seen that the sum over states, or levels, can be replaced by an integral. In other words, the translational and the rotational degrees of freedom can be treated classically. On the other hand, as discussed earlier in Part III, the vibrational degrees of freedom must be treated quantum mechanically. We can therefore write the Hamiltonian as follows:

$$H = H_{class} + H_{QM} \qquad (45.22)$$

where H_{class} refers to the s degrees of freedom that can be treated classically, and H_{QM} refers to those degrees of freedom that must be treated quantum mechanically. Accordingly, the molecular partition function can be written as follows:

$$q = q_{class} q_{QM} \qquad (45.23)$$

where

$$q_{class} = \frac{1}{h^s} \int e^{-\beta H_{class}(\vec{p}, \vec{q})} dp_1 dq_1 \ldots dp_s dq_s \tag{45.24}$$

Equations (45.22) and (45.24) can be generalized to a system of interacting molecules. If the Hamiltonian of the entire system is separable into a classical part and a quantum mechanical part, it follows that:

$$H = H_{class} + H_{QM}$$
$$Q = Q_{class} Q_{QM} = \frac{Q_{QM}}{N! h^{sN}} \int e^{-H_{class}/k_B T} d\vec{p}_{class} d\vec{q}_{class} \tag{45.25}$$

45.10   Equipartition of Energy

We have seen that classical statistical mechanics is valid when the temperature is sufficiently high to allow replacement of the QM summations by an integral. Under these conditions, it is not necessary to know the energy eigenvalues of the quantum mechanical problem. Indeed, only the classical Hamiltonian is required.

There is an interesting theorem of classical statistical mechanics which we will discuss next. Consider the expression for the average energy of a molecule in a system of independent molecules. Specifically,

$$\langle \varepsilon \rangle = \frac{\int H e^{-\beta H} dp_1 dq_1 \ldots dp_s dq_s}{\int e^{-\beta H} dp_1 dq_1 \ldots dp_s dq_s} \tag{45.26}$$

The integrals in Eq. (45.26) can be evaluated for any known dependence of H on the p's and the q's.

As we have shown earlier, for N independent molecules, it follows that (i) $\langle \underline{E} \rangle = N \langle \varepsilon \rangle$, and (ii) the heat capacity at constant volume is given by:

$$C_V = \left(\frac{\partial \langle \varepsilon \rangle}{\partial T} \right)_V \tag{45.27}$$

where $\langle \varepsilon \rangle$ is given in Eq. (45.26).

If the Hamiltonian is of the form:

$$H(p_1, p_2, \ldots, p_s, q_1, q_2, \ldots, q_s) = \sum_{j=1}^{m} a_j p_j^2 + \sum_{j=1}^{n} b_j q_j^2 \tag{45.28}$$
$$+ H(p_{m+1}, \ldots, p_s, q_{n+1}, \ldots, q_s)$$

where a_j and b_j are constants, it is possible to show, using Gaussian integrals, that each of the quadratic terms (p_j^2 and q_j^2) in Eq. (45.28) contributes $\frac{1}{2}k_BT$ to $\langle \varepsilon \rangle$ and $\frac{1}{2}k_B$ to C_V. This important result is known as the Principle of Equipartition of Energy and results from the quadratic form of the terms in H given in Eq. (45.28).

For an atom in an ideals gas, the Hamiltonian is given by:

$$H = \frac{p_x^2 + p_y^2 + p_z^2}{2m} \tag{45.29}$$

In Eq. (45.29), there are three quadratic terms in H, and therefore, the Equipartition Theorem indicates that the three translational degrees of freedom of the atom contribute $3(k_BT/2)$ to the total energy of the atom and $3(k_B/2)$ to C_V of the atom. Of course, these results are consistent with the results presented in Part I.

Lecture 46

Configurational Integral and Statistical Mechanical Derivation of the Virial Equation of State

46.1 Introduction

The material presented in this lecture is adapted from Chapter 12 in McQuarrie. First, we will use classical statistical mechanics to derive expressions for the Hamiltonian, the partition function, and the configurational integral of a non-ideal gas, including examining the low number density limit where the interaction potential is zero, the configuration integral is equal to $\underline{V}^N$, and the gas behaves ideally. Second, we will use classical statistical mechanics to model gases at higher number densities where the interaction potential is no longer zero and the configurational integral is no longer equal to $\underline{V}^N$. Third, we will use the Grand-Canonical ensemble to derive the virial expansion of the pressure in powers of the number density, truncated at cubic order. Finally, we will calculate the second and third virial coefficients in terms of the intermolecular potential.

46.2 Modeling Gases at Number Densities Approaching Zero

As we saw in Part I, in the limit of low number densities, $\rho = N/\underline{V} \ll 1$, all gases approach ideal gas behavior. In that limit, gases obey the well-known ideal gas equation of state, which we derived in a recent lecture using statistical mechanics. Specifically, on a per molecule basis:

$$P = \frac{Nk_BT}{\underline{V}} = \rho k_BT, \text{ for } \rho \ll 1 \tag{46.1}$$

© Springer Nature Switzerland AG 2020
D. Blankschtein, *Lectures in Classical Thermodynamics with an Introduction to Statistical Mechanics*, https://doi.org/10.1007/978-3-030-49198-7_46

Physically, Eq. (46.1) indicates that, on average, the molecules are far away from each other in the volume that they occupy and as a result do not interact with each other.

To understand how Eq. (46.1) can be derived using the classical Canonical partition function approach that we discussed in Lecture 45, we consider N monoatomic molecules contained in a volume, $\underline{V}$, at temperature, T. It then follows that:

$$Q = \frac{1}{N! h^{3N}} \int \ldots \int e^{-\beta H} d\vec{p}_1 \ldots d\vec{p}_N d\vec{r}_1 \ldots d\vec{r}_N \qquad (46.2)$$

Because the Hamiltonian, H, is of the form:

$$H = \frac{1}{2m} \sum_{n=1}^{N} \left(p_{nx}^2 + p_{ny}^2 + p_{nz}^2 \right) + \Phi_N(x_1, y_1, z_1, \ldots, x_N, y_N, z_N) \qquad (46.3)$$

we can carry out the integration over the momenta in Eq. (46.2) using Eq. (46.3) for H. This yields:

$$Q = \frac{1}{N!} \left(\frac{2\pi m k_B T}{h^2} \right)^{3N/2} Z_N \qquad (46.4)$$

where Z_N is the configurational integral introduced in Lecture 45 and given by:

$$Z_N = \int \ldots \int e^{-\Phi_N/k_B T} d\vec{r}_1 d\vec{r}_2 \cdots d\vec{r}_N \qquad (46.5)$$

For an ideal gas, $\Phi_N = 0$, and Eq. (46.5) reduces to:

$$Z_N = \left(\int d\vec{r}_1 \right) \left(\int d\vec{r}_2 \right) \cdots \left(\int d\vec{r}_N \right) = (\underline{V})(\underline{V}) \cdots (\underline{V}) = \underline{V}^N \qquad (46.6)$$

Combining Eqs. (46.4) and (46.6) then yields (for a monoatomic ideal gas):

$$Q = \frac{q^N}{N!}, \text{ where}$$
$$q = \left(\frac{2\pi m k_B T}{h^2} \right)^{3/2} \underline{V} \qquad (46.7)$$

In Eq. (46.7), the atomic partition function, q, is of the form $f(T)\underline{V}$, which results directly in the ideal gas equation of state, as emphasized recently. Note that the same result holds true for polyatomic ideal gases, albeit with different expressions for the function $f(T)$.

46.3   Modeling Gases at Higher Number Densities

As the number density of the gas increases, on average, the molecules get closer to each other. As a result, the molecules begin to increasingly interact with each other, that is, Φ_N is no longer zero. In that case, Z_N in Eq. (46.5) is no longer equal to $\underline{V}^N$, and the ideal gas EOS no longer describes the volumetric behavior of the gas.

As we saw in Parts I and II, many empirical and semiempirical equations of state have been developed to describe deviations from the simple ideal gas EOS. The most fundamental of these, because it has a solid molecular foundation, is the virial equation of state. Next, we will derive the virial EOS using statistical mechanics.

Recall that the virial EOS expresses deviations from the ideal behavior as an infinite power series in the number density, $\rho = N/\underline{V}$, that is:

$$\frac{P}{k_B T} = \rho + B_2(T)\rho^2 + B_3(T)\rho^3 + \cdots \qquad (46.8)$$

The parameters, $B_2(T)$, $B_3(T)$, ..., in Eq. (46.8) are known as the second, third, etc. virial coefficients, respectively, and depend only on the temperature, T, and on the particular gas under consideration but are independent of ρ or P. Our central goal in this lecture is to use statistical mechanics to derive expressions for the second and third virial coefficients in Eq. (46.8).

46.4   Derivation of the Virial Equation of State Using the Grand-Canonical Partition Function

As shown in Lecture 44, the Grand-Canonical partition function is given by:

$$\Xi(\underline{V}, T, \mu) = \sum_{N=0}^{\infty} Q(N, \underline{V}, T)\lambda^N \qquad (46.9)$$

where

$$\lambda - e^{\beta\mu} \qquad (46.10)$$

When $N = 0$, the system has only one state with $\underline{E} = 0$, and therefore:

$$Q(N = 0, \underline{V}, T) = 1 \qquad (46.11)$$

Using Eq. (46.11) in Eq. (46.9) yields:

$$\Xi(\underline{V}, T, \mu) = 1 + \sum_{N=1}^{\infty} \underbrace{Q(N, \underline{V}, T)}_{Q_N(\underline{V}, T)} \lambda^N \qquad (46.12)$$

We recently showed that the characteristic thermodynamic function associated with Ξ is $-P\underline{V}$, according to the relation:

$$P\underline{V} = k_B T \ln\Xi \qquad (46.13)$$

In the context of the Grand-Canonical ensemble, the number of molecules in the system is given by:

$$N = \langle N \rangle = k_B T \left(\frac{\partial \ln\Xi}{\partial \mu}\right)_{\underline{V}, T} = \lambda \left(\frac{\partial \ln\Xi}{\partial \lambda}\right)_{\underline{V}, T} \qquad (46.14)$$

Equation (46.13) yields the pressure, P, in terms of Ξ, and Eq. (46.14) yields the number of molecules, N, in terms of Ξ. Our goal is to obtain a relation between P and N (or ρ). This can be done by eliminating Ξ between these two quantities in Eqs. (46.14) and (46.13). A convenient way of doing this is to derive a power series of $\ln\Xi$ as a function of λ (see Eq. (46.10)) and then to eliminate λ between Eqs. (46.13) and (46.14). As we will show below, this will involve making use of several power series expansions.

It turns out that it is more convenient, although not necessary, to introduce a new variable, z, which is proportional to λ, such that $z \to \rho$ as $\rho \to 0$. In the limit $\lambda \to 0$, Eq. (46.12) becomes:

$$\Xi = 1 + Q_1 \lambda + O(\lambda^2) \qquad (46.15a)$$

where $Q_1 \equiv Q(N = 1, \underline{V}, T)$. Using Eq. (46.15a) in Eq. (46.14) then yields:

$$N = \lambda \left(\frac{\partial \ln\Xi}{\partial \lambda}\right)_{\underline{V}, T} = \frac{\lambda}{\Xi} \left(\frac{\partial \Xi}{\partial \lambda}\right)_{\underline{V}, T} \qquad (46.15b)$$

where when $\lambda \to 0$, Eq. (46.15b) reduces to (note that $\Xi \to 1$ and $(\partial\Xi/\partial\lambda)_{\underline{V}, T} \to Q_1$):

$$N = \lambda Q_1 (\lambda \to 0) \qquad (46.15c)$$

Because $N/\underline{V} = \rho$, Eq. (46.15c) shows that as $\lambda \to 0$, the density $\rho = N/\underline{V} \to \lambda Q_1/\underline{V}$, and therefore, as $\rho \to 0$, the new variable $z = \lambda Q_1/\underline{V}$ or $\lambda = z\underline{V}/Q_1$. In terms of the new variable, z, Eq. (46.12) can be expressed as follows:

$$\Xi = 1 + \sum_{N=1}^{\infty} \left(\frac{Q_N V^N}{Q_1^N} \right) z^N \tag{46.16}$$

It is now convenient to define a quantity, Z_N, as follows:

$$Z_N = N! \left(\frac{V}{Q_1} \right)^N Q_N \tag{46.17}$$

We will show that the classical limit of Z_N, as defined in Eq. (46.17), is in fact the configurational integral given in Eq. (46.5). Using Eq. (46.17) in Eq. (46.16) yields an expression for Ξ as a power series in z. Specifically,

$$\Xi(\underline{V}, T, \mu) = 1 + \sum_{N=1}^{\infty} \frac{Z_N(\underline{V}, T) z^N}{N!} \tag{46.18}$$

Next, we assume that the pressure can also be expanded as a power series in z according to:

$$P = k_B T \sum_{j=1}^{\infty} b_j z^j \tag{46.19}$$

where we would like to determine the unknown coefficients b_j in terms of Z_N in Eq. (46.17). To do that, we substitute Eq. (46.19) directly into the exponentiated form of Eq. (46.13), given by:

$$\Xi = e^{P\underline{V}/k_B T} \tag{46.20}$$

which yields:

$$\Xi = \exp \left[\underline{V} \sum_{j=1}^{\infty} b_j z^j \right] \tag{46.21}$$

If we now expand the exponential in Eq. (46.21) in a power series in z, we obtain:

$$\Xi = 1 + \underline{V} \sum_{j=1}^{\infty} b_j z^j + \frac{\underline{V}^2}{2!} \left(\sum_{j=1}^{\infty} b_j z^j \right)^2 + \frac{\underline{V}^3}{3!} \left(\sum_{j=1}^{\infty} b_j z^j \right)^3 + \dots \tag{46.22}$$

or (to order z^3):

$$\Xi = 1 + z[b_1 \underline{V}] + z^2[b_2\underline{V} + b_1^2\underline{V}^2/2!]$$
$$+ z^3[b_3\underline{V} + 2b_1b_2\underline{V}^2/2! + b_1^3\underline{V}/3!] \qquad (46.23)$$

We can also expand Ξ in Eq. (46.18) as a power series in z to order z^3, which yields:

$$\Xi = 1 + Z_1 z + Z_2 z^2/2! + Z_3 z^3/3! \qquad (46.24)$$

A term-by-term comparison of Eqs. (46.23) and (46.24) yields:

(i) Order z:

$$b_1\underline{V} = Z_1 \Rightarrow b_1 = \frac{Z_1}{\underline{V}} \qquad (46.25)$$

However, Eq. (46.17) shows that:

$$Z_1 = \left(\frac{\underline{V}}{Q_1}\right)Q_1 = \underline{V} \qquad (46.26)$$

Using Eq. (46.26) in Eq. (46.25) then yields:

$$b_1 = 1 \qquad (46.27)$$

(ii) Order z^2:

$$b_2\underline{V} + b_1\underline{V}^2/2! = Z_2/2! \Rightarrow b_2\underline{V} = \frac{Z_2}{2!} - \frac{b_1\underline{V}^2}{2!}$$

Because $b_1 = 1$ (see Eq. (46.27)), and $Z_1 = \underline{V}$ (see Eq. (46.26)), the equation to the right of the arrow above, including dividing by $\underline{V}$ and rearranging, yields:

$$b_2 = \frac{1}{2!\underline{V}}\left\{Z_2 - Z_1^2\right\} \qquad (46.28)$$

(iii) Order z^3:

$$b_3\underline{V} + \frac{2\,b_1b_2\underline{V}^2}{2!} + \frac{(b_1\underline{V})^3}{3!} = \frac{Z_3}{3!} \Rightarrow$$

or

$$b_3 \underline{V} = \frac{Z_3}{3!} - \overbrace{\frac{(b_1 \underline{V})^3}{3!}}^{Z_1^3} - \underbrace{\frac{2 \overbrace{(b_1 \underline{V})}^{Z_1} \overbrace{(b_2 \underline{V})}^{(Z_2 - Z_1^2)/2!}}{2!}}_{\frac{(Z_1 Z_2 - Z_1^3)\left(\frac{3!}{3!}\right)}{2}}$$

or

$$b_3 = \frac{1}{3!\underline{V}} \left\{ Z_3 - 3Z_1 Z_2 + 2Z_1^3 \right\} \tag{46.29}$$

We have now determined the b_js ($j = 1$, 2, and 3) that appear in Eq. (46.19) for P, in terms of Z_1, Z_2, and Z_3 (see Eqs. (46.27), (46.28), and (46.29)). The calculation of b_2 involves the evaluation of Z_2 and Z_1 only, that is, dealing with at most two molecules. Similarly, the calculation of b_3 involves evaluating Z_1, Z_2, and Z_3 only, that is, dealing with at most three molecules. Therefore, the original N-body problem was simplified into a series of few-body problems.

However, recall that our goal is to express the pressure, P, as a function of the number density, ρ, and not of the variable, z (as in Eq. (46.19)). To make this connection, we need to express $z = z(\rho)$, and then use this result in Eq. (46.19) for P. To this end, we note from Eq. (46.14) that:

$$\rho = \frac{N}{\underline{V}} = \frac{\lambda}{\underline{V}} \left(\frac{\partial \ln \Xi}{\partial \lambda} \right)_{\underline{V},T} = \frac{\lambda}{\underline{V}} \left(\frac{\partial \ln \Xi}{\partial z} \right)_{\underline{V},T} \left(\frac{\partial z}{\partial \lambda} \right)_{\underline{V},T}$$

Recalling that $z = \lambda Q_1/\underline{V}$, it follows that $(\partial z/\partial \lambda)_{\underline{V},T} = Q_1/\underline{V}$, so that we obtain:

$$\rho = \frac{N}{\underline{V}} = \overbrace{\left(\frac{\lambda Q_1}{\underline{V}} \right)}^{z} \left(\frac{1}{\underline{V}} \right) \left(\frac{\partial \ln \Xi}{\partial z} \right)_{\underline{V},T}$$

or

$$\rho = \frac{N}{\underline{V}} = \frac{z}{\underline{V}} \left(\frac{\partial \ln \Xi}{\partial z} \right)_{\underline{V},T} \tag{46.30}$$

Because according to Eq. (46.13), $\ln \Xi = P\underline{V}/k_B T$, it follows that:

$$\left(\frac{\partial \ln \Xi}{\partial z} \right)_{\underline{V},T} = \frac{\underline{V}}{k_B T} \left(\frac{\partial P}{\partial z} \right)_{\underline{V},T} \tag{46.31}$$

Using Eq. (46.19) for $P = P(z)$ in Eq. (46.31) and then the resulting Eq. (46.31) in Eq. (46.30) yields:

$$\rho = \frac{z}{k_B T}\left(\frac{\partial P}{\partial z}\right)_{\underline{V},T} = \frac{z}{k_B T}\left(\frac{\partial}{\partial z}\left[k_B T \sum_{j=1}^{\infty} b_j z^j\right]\right)_{\underline{V},T}$$

or

$$\rho = \sum_{j=1}^{\infty} j b_j z^j \qquad (46.32)$$

Equations (46.32) and (46.19) express the number density, ρ, and the pressure, P, respectively, as a series expansion in z. Accordingly, we need to eliminate z between the two equations to obtain $P = P(\rho)$. To this end, we will simply determine the first few terms in an algebraic way. Specifically, we simply invert z as a series expansion in ρ as follows:

$$z = a_1 \rho + a_2 \rho^2 + a_3 \rho^3 + \cdots \qquad (46.33)$$

and then substitute Eq. (46.33) in Eq. (46.32). This yields:

$$\rho = b_1 z + 2 b_2 z^2 + 3 b_3 z^3 + \cdots$$
$$\rho = b_1 \left(a_1 \rho + a_2 \rho^2 + a_3 \rho^3\right) + 2 b_2 \left(a_1 \rho + a_2 \rho^2 + a_3 \rho^3\right)^2$$
$$+ 3 b_3 \left(a_1 \rho + a_2 \rho^2 + a_3 \rho^3\right)^3 + \cdots$$

or

$$[a_1 b_1 - 1]\rho + \left[b_1 a_2 + 2 b_2 a_1^2\right]\rho^2 + \left[b_1 a_3 + 4 b_2 a_1 a_2 + 3 b_3 a_1^3\right]\rho^3 + \cdots = 0 \quad (46.34)$$

Each term in Eq. (46.34) must vanish independently, so that:

(i) Order ρ:

$$a_1 b_1 = 1 \Rightarrow a_1 = 1/b_1 \Rightarrow$$
$$a_1 = 1 (\text{Because } b_1 = 1) \qquad (46.35)$$

(ii) Order ρ^2:

$$b_1 a_2 = -2b_2 a_1^2 \Rightarrow$$

$$a_2 = -2b_2 \, (\text{Because } b_1 = 1 \text{ and } a_1 = 1) \tag{46.36}$$

(iii) Order ρ^3:

$$a_3 = -\frac{(4b_2 a_1 a_2 + 3b_3 a_1^3)}{b_1} \Rightarrow$$

$$a_3 = -3b_3 + 8b_2^2 \, (\text{Because } a_1 = 1 \text{ and } a_2 = -2b_2) \tag{46.37}$$

Now that we know z as a power series in ρ, we can substitute this information in Eq. (46.29) to obtain P as a function of ρ. Specifically,

$$\frac{P}{k_B T} = \sum_{j=1}^{\infty} b_j z^j = b_1 z + b_2 z^2 + b_3 z^3 + \cdots$$

$$\frac{P}{k_B T} = b_1 \left(a_1 \rho + a_2 \rho^2 + a_3 \rho^3 \right) + b_2 \left(a_1 \rho + a_2 \rho^2 + a_3 \rho^3 \right)^2$$

$$+ b_3 \left(a_1 \rho + a_2 \rho^2 + a_3 \rho^3 \right)^3 + \cdots$$

Recalling that $b_1 = 1$, $a_1 = 1$, $a_2 = -2b_2$, and $a_3 = -3b_3 + 8b_2^2$, we obtain:

$$\frac{P}{k_B T} = \rho + B_2(T)\rho^2 + B_3(T)\rho^3 + \cdots \tag{46.38}$$

where (see Eq. (46.28)):

$$B_2(T) = -b_2 = -\frac{1}{2!\underline{V}} \left\{ Z_2 - Z_1^2 \right\} \tag{46.39}$$

and (see Eqs. (46.28) and (46.29)):

$$B_3(T) = 4b_2^2 - 2b_3$$

or

$$B_3(T) = 4 \left[\frac{1}{2!\underline{V}} \left(Z_2 - Z_1^2 \right) \right]^2 +$$

$$-2 \left[\frac{1}{3!\underline{V}} \left(Z_3 - 3 Z_2 Z_1 + 2 Z_1^3 \right) \right]$$

or

$$B_3(T) = -\frac{1}{3\underline{V}^2}\left\{\underline{V}(Z_3 - 3Z_2Z_1 + 2Z_1^3) - 3(Z_2 - Z_1^2)^2\right\} \tag{46.40}$$

Equations (46.39) and (46.40) allow us to calculate the second and third virial coefficients if we know the intermolecular potential from which we can calculate the configurational integrals (see Eq. (46.5)) which appear in these two equations.

Lecture 47

Virial Coefficients in the Classical Limit, Statistical Mechanical Derivation of the van der Waals Equation of State, and Sample Problem

47.1 Introduction

The material presented in this lecture is adapted from Chapter 12 in McQuarrie. First, we will review material discussed in Lecture 46, including referring to specific equations, in order to derive general expressions for the second and third virial coefficients in terms of the two-body and three-body interaction potentials, respectively. Subsequently, we will specialize on spherically symmetric interaction potentials. Second, we will discuss the spatial dependence of the two-body interaction potential, including its long-range asymptotic behavior. Third, we will solve Sample Problem 47.1 to calculate the second virial coefficient corresponding to the hard-sphere interaction potential. Fourth, we will calculate the second virial coefficient corresponding to an interaction potential consisting of a hard-sphere repulsion and a van der Waals attraction. Fifth, we will make several important remarks about the behavior of interaction potentials. Finally, we will derive the van der Waals equation of state using statistical mechanics, in the context of the virial expansion of the pressure versus number density truncated at second order, thereby connecting with material presented in Part I.

47.2 Virial Coefficients in the Classical Limit

To avoid dealing with internal degrees of freedom to simplify the derivation, we will consider a monoatomic gas. As we saw in Lecture 46, in the classical limit, $Q(N, \underline{V}, T) \equiv Q_N(\underline{V}, T)$ is given by Eq. (46.2). In particular, for $N = 1$, in the absence of interactions between the molecules, the interaction potential $\Phi = 0$, and therefore, $Q(N = 1, \underline{V}, T) \equiv Q_1(\underline{V}, T) = q(\underline{V}, T)$ of a monoatomic ideal gas, that is,

© Springer Nature Switzerland AG 2020
D. Blankschtein, *Lectures in Classical Thermodynamics with an Introduction to Statistical Mechanics*, https://doi.org/10.1007/978-3-030-49198-7_47

$$Q_1(\underline{V}, T) = \left(\frac{2\pi m k_B T}{h^2}\right)^{3/2} \underline{V} = \frac{\underline{V}}{\Lambda^3} \tag{47.1}$$

where

$$\Lambda = \left(\frac{h^2}{2\pi m k_B T}\right)^{1/2}$$

For $N > 1$, we can still integrate over the momenta in Eq. (46.2), as we did to obtain Eq. (46.4). This yields:

$$Q_N = \frac{Z_N}{N! \Lambda^{3N}} \tag{47.2}$$

where Z_N is the configurational integral given in Eq. (46.5). When $N = 1$, a comparison of Eqs. (47.2) and (47.1) shows that $Z_1 = \underline{V}$ and, hence, that $Q_1/\underline{V} = 1/\Lambda^3$ in Eq. (47.2). As a result, Eq. (47.2) can be expressed as follows:

$$Q_N = \frac{1}{N!} \left(\frac{Q_1}{\underline{V}}\right)^N Z_N \tag{47.3}$$

Equation (47.3) is identical to Eq. (46.17), and hence, the quantity Z_N in Eq. (47.3) is indeed the configurational integral, as we stated without proof in Lecture 46.

To calculate $B_2(T)$ using Eq. (46.39), we first need to evaluate Z_1 and Z_2. To calculate $B_3(T)$ using Eq. (46.40), we first need to evaluate Z_1, Z_2, and Z_3. These are given by (see Eq. (46.5)):

$$Z_1 = \int d\vec{r}_1 = \underline{V} \tag{47.4}$$

$$Z_2 = \int\int e^{-\Phi_2/k_B T} d\vec{r}_1 d\vec{r}_2 \tag{47.5}$$

$$Z_3 = \int\int\int e^{-\Phi_3/k_B T} d\vec{r}_1 d\vec{r}_2 d\vec{r}_3 \tag{47.6}$$

Next, we will illustrate the calculation of the second virial coefficient, $B_2(T)$, for a monoatomic gas. As Eq. (46.39) shows, in order to calculate $B_2(T)$, we first need to evaluate Z_2 using Eq. (47.5) and Z_1 (which is equal to $\underline{V}$; see Eq. (47.4)). To calculate Z_2 in Eq. (47.5), we require the two-body interaction potential, Φ_2. For monoatomic molecules positioned at $\vec{r}_1$ and $\vec{r}_2$, it is reasonable to assume that $\Phi_2(\vec{r}_1, \vec{r}_2)$ depends only on the distance of separation between the centers of the two molecules, $r_{12} = |\vec{r}_2 - \vec{r}_1|$, that is, that $\Phi_2 = \Phi_2(r_{12})$. Figure 47.1 illustrates two spherical molecules of equal radius separated by a center-to-center distance r_{12}.

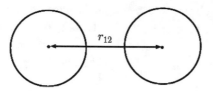

Fig. 47.1

The corresponding interaction potential between the two spherical molecules depicted in Fig. 47.1 is shown in Fig. 47.2.

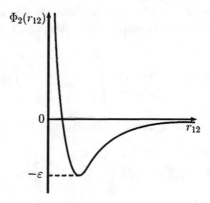

Fig. 47.2

We can calculate $B_2(T)$ in terms of $\Phi_2(r_{12})$ by substituting Z_2 and Z_1 from Eqs. (47.5) and (47.4), respectively, in Eq. (46.39). Specifically,

$$B_2(T) = -\frac{1}{2\underline{V}} \left(Z_2 - Z_1^2 \right)$$

$$B_2(T) = -\frac{1}{2\underline{V}} \int \int \left[e^{-\beta\Phi_2(r_{12})} - 1 \right] d\vec{r_1} d\vec{r_2} \tag{47.7}$$

where we have used the fact that:

$$Z_1^2 = \underline{V}^2 = \int \int 1 \, d\vec{r_1} d\vec{r_2} = \overbrace{\left(\int d\vec{r_1} \right)}^{\underline{V}} \overbrace{\left(\int d\vec{r_2} \right)}^{\underline{V}} \tag{47.8}$$

For two neutral molecules in their ground electronic states, $\Phi_2(r_{12})$ decays to zero quite rapidly, say, within a few molecular diameters. As a result, the integrand, $\left(e^{-\beta\Phi_2} - 1 \right)$, in Eq. (47.7) is essentially zero, unless $d\vec{r_1}$ and $d\vec{r_2}$ are close to each other. Consequently, it is possible to change the integration variables to $\vec{r}_1$ and to the relative coordinate, $\vec{r}_{12} = \vec{r}_2 - \vec{r}_1$, and express Eq. (47.7) as follows:

$$B_2(T) = -\frac{1}{2\underline{V}} \int d\vec{r}_1 \int \left[e^{-\beta \Phi_2(r_{12})} - 1 \right] d\vec{r}_{12} \tag{47.9}$$

In Eq. (47.9), the integration over $\vec{r}_{12}$ is independent of the location of the pair of molecules (1,2) in the volume $\underline{V}$, except in cases when the pair is located near the walls of the container. However, for macroscopic systems for which both $\underline{V}$ and $N \to \infty$, this wall, or surface, effect is negligible. Therefore, we can carry out the integration over $d\vec{r}_1$ separately to yield a factor of $\underline{V}$, and take advantage of the spherical symmetry of Φ_2 which only depends on r_{12}, by expressing $d\vec{r}_{12} = 4\pi r_{12}^2 dr_{12}$. Using these coordinate transformations in Eq. (47.9) yields:

$$B_2(T) = -\frac{1}{2\underline{V}} (4\pi)(\underline{V}) \int_0^\infty \left[e^{-\beta \Phi_2(r_{12})} - 1 \right] r_{12}^2 \, dr_{12}$$

Denoting $r_{12} = r$, the last equation can be expressed as follows:

$$B_2(T) = -2\pi \int_0^\infty \left[e^{-\beta \Phi_2(r)} - 1 \right] r^2 dr \tag{47.10}$$

In the last two equations, we have extended the upper limits of integration to ∞, because $\Phi_2(r)$, and therefore the integrands, decay to zero quite rapidly.

47.3   Spatial Dependence of the Two-Body Interaction Potential Including Its Long-Range Asymptotic Behavior

Equation (47.10) shows that when $\Phi_2(r)$ is known, $B_2(T)$ can be calculated at the molecular level. In principle, $\Phi_2(r)$ can be obtained using quantum mechanics, but this involves a challenging numerical problem. Instead, using perturbation theory, it is possible to show that asymptotically:

$$\Phi_2(r) \to -C_6 r^{-6} (\text{for } r \to \infty) \tag{47.11}$$

Typically, we utilize analytical expressions for $\Phi_2(r)$, with adjustable parameters, that decay asymptotically as r^{-6}. The adjustable parameters in the model for

$\Phi_2(r)$ are optimized to fit experimental data. Probably, the best-known expression for the two-body interaction potential is given by:

$$\Phi_2(r) = \frac{n\varepsilon}{(n-6)} \left(\frac{n}{6}\right)^{6/(n-6)} \left\{\left(\frac{\sigma}{r}\right)^n - \left(\frac{\sigma}{r}\right)^6\right\}, 9 \leq n \leq 15 \qquad (47.12)$$

In Eq. (47.12), σ is the distance at which $\Phi_2(r) = 0$, and ε is the depth of the potential well. In addition, the exponent n is chosen to be an integer between 9 and 15, where for historical reasons, $n = 12$ is the most popular value, which corresponds to the well-known (12-6) Lennard-Jones potential. It is noteworthy that the r^{-6} term in Eq. (47.12) ensures the correct asymptotic behavior $(\Phi_2(r) \to r^{-6})$ for $r \to \infty$. The actual Lennard-Jones potential is given by:

$$\Phi_2(r) = 4\varepsilon \left(\left(\frac{\sigma}{r}\right)^{12} - \left(\frac{\sigma}{r}\right)^6\right) \qquad (47.13)$$

It is not possible to integrate $B_2(T)$ analytically using Eq. (47.10) if Eq. (47.13), or any other realistic expression for $\Phi_2(r)$, is used. Therefore, in Sample Problem 47.1, we will discuss a simpler model interaction potential that enables an analytical integration of $B_2(T)$ using Eq. (47.10).

47.4 Sample Problem 47.1: Calculate the Second Virial Coefficient Corresponding to the Hard-Sphere Interaction Potential

47.4.1 Solution

Mathematically, the hard-sphere interaction potential can be expressed as follows (see Fig. 47.3):

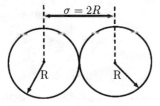

Fig. 47.3

$$\Phi_2(r) = \begin{cases} +\infty, & \text{for} \quad r \leq \sigma \\ 0, & \text{for} \quad r > \sigma \end{cases} \tag{47.14}$$

where r is the center-to-center separation distance and the distance of closest approach between the centers of the two spheres of equal radius R is $\sigma = 2R$.

The interaction potential in Eq. (47.14) has no attractive part (it is purely repulsive) and is able to model the steep, repulsive part of realistic intermolecular potentials. In fact, this is the simplest intermolecular potential used and is the only one for which the first seven virial coefficients have been evaluated.

A collection of molecules interacting via the interaction potential in Eq. (47.14) is known as a hard-sphere system and is the basis for many statistical mechanical calculations of interacting fluids. A hard-sphere system behaves like a collection of molecular size "billiard balls."

Using Eq. (47.14) for $\Phi_2(r)$ in Eq. (47.10) for $B_2(T)$ yields:

$$B_2(T) = -2\pi\{\int_0^\sigma [e^{\overbrace{-\Phi_2(r)/k_BT}^{-\infty}} - 1]r^2dr + \int_\sigma^\infty [e^{\overbrace{-\Phi_2(r)/k_BT}^{0}} - 1]r^2dr\}$$

or

$$B_2 = -2\pi\{\int_0^\sigma [\underbrace{0 - 1}_{-1}]r^2dr + \int_\sigma^\infty [\underbrace{1 - 1}_{0}]r^2dr\}$$

or

$$B_2 = 2\pi\left\{\int_0^\sigma \overbrace{r^2dr}^{\frac{1}{3}r^3\big|_0^\sigma = \frac{\sigma^3}{3}} + 0\right\}$$

or

$$B_2 = \frac{2\pi\sigma^3}{3} \tag{47.15}$$

As shown in Fig. 47.3, for two spheres of equal radius R, $\sigma = 2R$, so that: $B_2 = \frac{2\pi}{3}(2R)^3 = 4\left(\frac{4\pi}{3}R^3\right) = 4\,(\text{sphere volume})$. Further, for the hard-sphere potential, B_2 does not depend on T and, as expected, is always positive, which is indicative of repulsive interactions.

47.5 Calculating the Second Virial Coefficient Corresponding to an Interaction Potential Consisting of a Hard-Sphere Repulsion and a van der Waals Attraction

Next, we will consider a new interaction potential consisting of a hard-sphere repulsion at short distances and a van der Waals attraction $(\sim -r^{-6})$ at large distances. In essence, the new interaction potential approximates the Lennard-Jones potential, albeit with an infinite repulsion at contact. Specifically,

$$\Phi_2(r) = \begin{cases} +\infty, & \text{for } r \leq \sigma \\ -\varepsilon\left(\dfrac{\sigma}{r}\right)^6, & \text{for } r > \sigma \end{cases} \tag{47.16}$$

Using Eq. (47.16) for $\Phi_2(r)$ in Eq. (47.10) for $B_2(T)$ yields:

$$B_2(T) = -2\pi\left\{ \int_0^\sigma \left[e^{\overbrace{-\Phi_2(r)/k_BT}^{-\infty}} - 1 \right] r^2 dr + \int_\sigma^\infty \left[e^{\overbrace{-\Phi_2(r)/k_BT}^{+\varepsilon(\sigma/r)^6/k_BT}} - 1 \right] r^2 dr \right\}$$

$$\underbrace{}_{[0-1]=-1}$$

or

$$B_2(T) = -2\pi\left\{ \underbrace{-\int_0^\sigma r^2 dr}_{=-\frac{\sigma^3}{3}} + \underbrace{\int_\sigma^\infty \left[e^{(\varepsilon/k_BT)\left(\frac{\sigma}{r}\right)^6} - 1 \right] r^2 dr}_{I} \right\}$$

In the last equation, the integral denoted as I cannot be evaluated analytically. However, we can approximate the integrand in I by assuming that $(\varepsilon/k_BT) \ll 1$ (high-temperature limit) and then expanding the exponent to order $(\varepsilon/k_BT)^2$. Specifically,

$$\text{Integrand} = e^{\left(\frac{\varepsilon}{k_BT}\right)\left(\frac{\sigma}{r}\right)^6} - 1 = 1 + \left(\frac{\varepsilon}{k_BT}\right)\left(\frac{\sigma}{r}\right)^6 \quad 1 + O\left(\left(\frac{\varepsilon}{k_BT}\right)^2\right)$$

$$\text{Integrand} = \left(\frac{\varepsilon}{k_BT}\right)\left(\frac{\sigma}{r}\right)^6 + O\left(\left(\frac{\varepsilon}{k_BT}\right)^2\right) \tag{47.17}$$

Using Eq. (47.17) in the integral I yields:

$$I = \left(\frac{\varepsilon\sigma^6}{k_BT}\right) \int\limits_{\sigma}^{\infty} (r^{-6})r^2 dr = \left(\frac{\varepsilon\sigma^6}{k_BT}\right) \underbrace{\int\limits_{\sigma}^{\infty} r^{-4} dr}_{\frac{-r^{-3}}{3}\Big|_{\sigma}^{\infty} = \frac{1}{3\sigma^3}}$$

or

$$I = \frac{\sigma^3}{3}\left(\frac{\varepsilon}{k_BT}\right) \tag{47.18}$$

Accordingly, the second virial coefficient corresponding to this interaction potential is given by:

$$B_2(T) = -2\pi\left\{ -\frac{\sigma^3}{3} + \frac{\sigma^3}{3}\left(\frac{\varepsilon}{k_BT}\right) \right\}$$

or

$$B_2(T) = \frac{2\pi\sigma^3}{3}\left\{ 1 - \left(\frac{\varepsilon}{k_BT}\right) \right\} \tag{47.19}$$

We can define an excluded volume $V_{excl} = 4\pi\sigma^3/3$, which is eight times the molecular volume, because the distance of closest approach for two spheres of equal radius, R, is $\sigma = 2R$ (see Fig. 47.3), and therefore, $V_{excl} = 4\frac{\pi}{3}(2R)^3 = 8\left(4\frac{\pi}{3}R^3\right) = 8V_{sphere}$. We can therefore rewrite Eq. (47.19) as follows:

$$B_2(T) = \frac{V_{excl}}{2}\left(1 - \frac{\varepsilon}{k_BT}\right) \tag{47.20}$$

Because in the high-temperature limit $\frac{\varepsilon}{k_BT}$ in Eq. (47.20) is $\ll 1$, it follows that the second virial coefficient is always positive, which is indicative of a repulsive interaction.

47.6   Important Remarks About the Behavior of Interaction Potentials

1. The tail of the attractive van der Waals interaction potential ($\sim r^{-6}$) extends to very long separations. Nevertheless, its integral is dominated by contributions from the short distances σ. As a result, the van der Waals interaction potential is short-ranged and results in corrections to the ideal gas behavior that are analytical in the number density, ρ, leading to the virial series.

2. On the other hand, interaction potentials that decay with distance as r^{-3} or slower are long-ranged. In that case, the integral that appears in the calculation of $B_2(T)$ is dominated by the long distances and, therefore, diverges. As a result, corrections to the ideal gas behavior cannot be expressed in the form of a virial series and are, in fact, nonanalytic.

3. The second virial coefficient, $B_2(T)$, has dimensions of volume and, for short-ranged interaction potentials, is proportional to the excluded volume, V_{excl} (see Eq. (47.20)). Because the virial expansion of the pressure in powers of the number density is given by:

$$\frac{P}{k_BT} = \rho + B_2(T)\rho^2 + \cdots \tag{47.21}$$

it will break down at high number densities, as well as at low temperatures. Indeed, due to the attractive interactions, as $T \to 0$, the molecules in the gas can lower the system energy by condensing into a liquid.

47.7 Derivation of the van der Waals Equation of State Using Statistical Mechanics

Using the expression for $B_2(T)$ given in Eq. (47.20) in the truncated virial expansion, Eq. (47.21), we obtain:

$$\frac{P}{k_BT} = \rho + \frac{V_{excl}}{2}\left(1 - \frac{\varepsilon}{k_BT}\right)\rho^2 + \cdots \tag{47.22}$$

Equation (47.22) can be rearranged as follows:

$$\frac{1}{k_BT}\left(P + \frac{\varepsilon V_{excl}}{2}\rho^2\right) = \rho + \rho^2\frac{V_{excl}}{2} = \rho\left(1 + \rho\frac{V_{excl}}{2}\right) \approx \frac{\rho}{1 - \rho\frac{V_{excl}}{2}}$$

or

$$P + \frac{\varepsilon V_{excl}}{2}\rho^2 = \frac{k_BT\rho}{1 - \rho\frac{V_{excl}}{2}} \tag{47.23}$$

Because $\rho = N/\underline{V}$, using this result in Eq. (47.23) yields:

$$P + \left(\frac{\varepsilon V_{excl}}{2}\right)\left(\frac{N}{\underline{V}}\right)^2 = \frac{k_B T\left(\frac{N}{\underline{V}}\right)}{1 - \left(\frac{N}{\underline{V}}\right)\left(\frac{V_{excl}}{2}\right)} = \frac{k_B TN}{\underline{V} - N\left(\frac{V_{excl}}{2}\right)}$$

or

$$\left[P + \left(\frac{\varepsilon V_{excl}}{2}\right)\left(\frac{N}{\underline{V}}\right)^2\right]\left[\underline{V} - N\left(\frac{V_{excl}}{2}\right)\right] = Nk_B T \qquad (47.24)$$

If we identify:

$$b = \frac{V_{excl}}{2} \qquad (47.25)$$

$$a = \frac{\varepsilon V_{excl}}{2} \qquad (47.26)$$

then, Eq. (47.24) can be written as follows:

$$\left[P + a\left(\frac{N}{\underline{V}}\right)^2\right][\underline{V} - Nb] = Nk_B T \qquad (47.27)$$

Equation (47.27) is the celebrated van der Waals equation of state first introduced in Part I and derived here using statistical mechanics.

Lecture 48

Statistical Mechanical Treatment of Chemical Reaction Equilibria and Sample Problem

48.1 Introduction

The material presented in this lecture is adapted from Chapter 9 in McQuarrie. First, we will show how to express the equilibrium constant of a gas-phase chemical reaction in terms of the partition functions of the various species participating in the chemical reaction. Second, we will discuss an interesting relation between the pressure-based and the number density-based equilibrium constants for ideal gas mixtures. Finally, we will solve Sample Problem 48.1 to calculate, at the molecular level, the equilibrium constant of a specific gas-phase chemical reaction using statistical mechanics, thereby establishing contact with material presented in Part II using thermodynamic considerations.

48.2 Expressing the Equilibrium Constant Using Partition Functions

Consider the following gas-phase chemical reaction taking place at equilibrium in a closed and insulated vessel:

$$|\upsilon_A|A + |\nu_B|B \leftrightarrows |\upsilon_C|C + |\upsilon_D|D \tag{48.1}$$

As discussed in Part II, in Eq. (48.1), the υs are stoichiometric numbers (> 0 for products, < 0 for reactants, and $= 0$ for inert species), A and B are reactants, and C and D are products.

First, we will review the derivation of the condition of chemical reaction equilibria presented in Part II. For this purpose, it is convenient to introduce the extent of reaction, ξ, to express the differential change in the number of moles of component j participating in the chemical reaction. Specifically:

© Springer Nature Switzerland AG 2020
D. Blankschtein, *Lectures in Classical Thermodynamics with an Introduction to Statistical Mechanics*, https://doi.org/10.1007/978-3-030-49198-7_48

$$dN_j = \upsilon_j d\xi, \quad j = A, B, C, \text{ and } D \tag{48.2}$$

Because the chemical reaction in Eq. (48.1) takes place in a vessel of fixed volume, $\underline{V}$, at fixed temperature, T, where the total number of species is conserved, as discussed in Part II, the relevant fundamental equation to use in order to describe the thermodynamic behavior of the system is the Helmholtz free energy, $\underline{A}$, whose natural variables are T, $\underline{V}$, and N. In particular, we know that:

$$d\underline{A} = -\underline{S}dT - Pd\underline{V} + \sum_j \mu_j dN_j \tag{48.3}$$

Substituting dN_j from Eq. (48.2) in Eq. (48.3) yields:

$$d\underline{A} = -\underline{S}dT - Pd\underline{V} + \left(\sum_j \nu_j \mu_j\right) d\xi \tag{48.4}$$

For a system at equilibrium, at fixed volume, $\underline{V}$, and temperature, T, the Helmholtz free energy, $\underline{A}$, in Eq. (48.4) must be a minimum with respect to all possible changes, $d\xi$, and therefore:

$$\left(\frac{\partial \underline{A}}{\partial \xi}\right)_{T,\underline{V}} = \sum_j \nu_j \mu_j = 0 \tag{48.5}$$

or

$$\sum_j \nu_j \mu_j = 0 \tag{48.6}$$

Equation (48.6) is the condition of chemical reaction equilibria, which we first derived in Part II by minimizing the Gibbs free energy, $\underline{G}$, as a function of its natural variables, T, P, and N. Using Eq. (48.6) for the chemical reaction in Eq. (48.1) yields:

$$|\nu_C|\mu_C + |\nu_D|\mu_D - |\nu_A|\mu_A - |\nu_B|\mu_B = 0 \tag{48.7}$$

Next, we will use statistical mechanics to relate the various chemical potentials in Eq. (48.7) to partition functions. This, in turn, will result in an expression for the equilibrium constant for the chemical reaction in Eq. (48.1) expressed in terms of partition functions. Let us consider a mixture of ideal gases A, B, C, and D, where the various species are independent and distinguishable. As shown earlier in Part III, the Canonical partition function of the mixture, $Q(N_A, N_B, N_C, N_D, \underline{V}, T)$, can be expressed as the product of the Canonical partition functions of the individual species (A, B, C, and D). Specifically:

$$Q(N_A, N_B, N_C, N_D, \underline{V}, T) = Q_A(N_A, \underline{V}, T) \cdot Q_B(N_B, \underline{V}, T) \cdot Q_C(N_C, \underline{V}, T)$$
$$\cdot \, Q_D(N_D, \underline{V}, T) \tag{48.8}$$

Because the species in each gas are indistinguishable, it follows that:

$$Q_j(N_j, \underline{V}, T) = \frac{[q_j(\underline{V}, T)]^{N_j}}{N_j!} \tag{48.9}$$
$$j = A, B, C, \text{ and } D$$

where q_j is the Canonical partition function of species of type j (A, B, C, and D). Using Eq. (48.9) for j = A, B, C, and D in Eq. (48.8) yields:

$$Q(N_A, N_B, N_C, \underline{V}, T) = \prod_{j=A,B,C,D} \frac{[q_j(\underline{V}, T)]^{N_j}}{N_j!} \tag{48.10}$$

As discussed in Part II, the chemical potential of species j (A, B, C, and D) is given by (see Eq. (48.3)):

$$\mu_j = \left(\frac{\partial \underline{A}}{\partial N_j}\right)_{T,\underline{V},N_{i \neq j}} \tag{48.11}$$

In Part III, we recently showed that:

$$\underline{A} = -k_B T \ln Q \tag{48.12}$$

Using Eq. (48.12) in Eq. (48.11) yields:

$$\mu_j = -k_B T \left(\frac{\partial \ln Q}{\partial N_j}\right)_{T,\underline{V},N_{i \neq j}} \tag{48.13}$$

Taking the ln of Q in Eq. (48.9) yields:

$$\ln Q = \sum_{j=A,B,C,D} \left(N_j \ln q_j - \ln N_j!\right) \tag{48.14}$$

Substituting lnQ in Eq. (48.14) in Eq. (48.13), including using Sterling's approximation for $\ln N_j!$, yields:

$$\mu_j = -k_B T \left[\ln q_j - \ln N_j\right]$$

or

$$\mu_j = -k_BTln\left[\frac{q_j(\underline{V}, T)}{N_j}\right]$$

(48.15)

$$\text{for } j = A, B, C, \text{and } D$$

Equation (48.15) shows that μ_j of species j is determined solely by the Canonical partition function, q_j, and by N_j of species j. This result reflects the ideality of the gas mixture, where the various species do not interact with each other.

Substituting Eq. (48.15), for j = A, B, C, and D, in the condition of chemical reaction equilibria (Eq. (48.7)) yields:

$$|\nu_C|\ln\left(\frac{q_C}{N_C}\right) + |\nu_D|\ln\left(\frac{q_D}{N_D}\right) - |\nu_A|\ln\left(\frac{q_A}{N_A}\right) - |\nu_B|\ln\left(\frac{q_B}{N_B}\right) = 0$$

or

$$\ln\left[\frac{q_C^{|\nu_C|} q_D^{|\nu_D|}}{N_C^{|\nu_C|} N_D^{|\nu_D|}}\right] = \ln\left[\frac{q_A^{|\nu_A|} q_B^{|\nu_B|}}{N_A^{|\nu_A|} N_B^{|\nu_B|}}\right]$$

or

$$\frac{N_C^{|\nu_C|} N_D^{|\nu_D|}}{N_A^{|\nu_A|} N_B^{|\nu_B|}} = \frac{q_C^{|\nu_C|} q_D^{|\nu_D|}}{q_A^{|\nu_A|} q_B^{|\nu_B|}}$$

(48.16)

Earlier in Part III, we showed that for an ideal gas (like gases A, B, C, and D), q is of the form $q = f(T)\underline{V}$. Therefore, $q/\underline{V} = f(T)$ is a function of temperature only. If we divide each factor on both sides of Eq. (48.16) by $\underline{V}^{|\nu_j|}$ and recall that $N_j/\underline{V} = \rho_j$, the number density of species j, we obtain the following expression for the equilibrium constant, K(T), in terms of the Canonical partition functions of species A, B, C, and D:

$$\frac{\rho_C^{|\nu_C|} \rho_D^{|\nu_D|}}{\rho_A^{|\nu_A|} \rho_B^{|\nu_B|}} \equiv \underbrace{K(T)}_{\substack{\text{Equilibrium} \\ \text{Constant}}} = \frac{(q_C/\underline{V})^{|\nu_C|}(q_D/\underline{V})^{|\nu_D|}}{(q_A/\underline{V})^{|\nu_A|}(q_B/\underline{V})^{|\nu_B|}}$$

(48.17)

Using Eq. (48.17) for K(T) and the expressions for the Canonical partition functions of species A, B, C, and D participating in gas-phase chemical reactions like that in Eq. (48.1), we can calculate equilibrium constants at the molecular level in the context of statistical mechanics, thereby fulfilling a goal that we set to accomplish in Part II.

48.3 Relating the Pressure-Based and the Number Density-Based Equilibrium Constants for Ideal Gas Mixtures

In Part II, we saw that, in general, for an ideal gas mixture (IGM), the equilibrium constant can be expressed in terms of the partial pressures, P_i, of every species i participating in the chemical reaction. Specifically, the equilibrium constant of the IGM corresponds to the pressure-based equilibrium constant which is given by:

$$K^{IGM} \equiv K_P^{IGM} = \prod_{i=1}^{n} P_i^{\nu_i} = \prod_{i=1}^{n} (y_i P)^{\nu_i} \tag{48.18}$$

For species i in an IGM, we can express its partial pressure (on a per molecule basis) as follows:

$$P_i = \frac{k_B N_i T}{V} = \rho_i (k_B T) \tag{48.19}$$

where ρ_i is the number density of species i. Replacing P_i in Eq. (48.18) by Eq. (48.19) yields:

$$K_P^{IGM} = (k_B T)^{\Delta \nu} \cdot \prod_{i=1}^{n} \rho_i^{\nu_i} \tag{48.20}$$

where $\Delta \nu = \sum_{i=1}^{n} \nu_i$ and k_B is the Boltzmann constant.

If we introduce a new equilibrium constant, K_ρ, expressed in terms of the number densities of the various species i participating in the chemical reaction, we obtain:

$$K_\rho \equiv \prod_{i=1}^{n} \rho_i^{\nu_i} \tag{48.21}$$

where

$$\left(K_P^{IGM} / K_\rho^{IGM} \right) = (k_B T)^{\Delta \nu} \tag{48.22}$$

48.4 Sample Problem 48.1

Consider the following gas-phase chemical reaction:

$$H_2(g) + I_2(g) \leftrightarrows 2HI(g) \qquad (48.23)$$

and evaluate the equilibrium constant, K(T), at the molecular level using statistical mechanics.

48.4.1 Solution

For the chemical reaction in Eq. (48.23), we know that $|\nu_{HI}|=2$, $|\nu_{H_2}|=1$, and $|\nu_{I_2}|=1$. Following an approach similar to the one that we used above to derive Eq. (48.17), we can express the equilibrium constant associated with Eq. (48.23) as follows:

$$K(T) = \frac{(q_{HI}/\underline{V})^2}{(q_{H_2}/\underline{V})(q_{I_2}/\underline{V})} = \frac{q_{HI}^2}{q_{H_2} q_{I_2}} \qquad (48.24)$$

In Part III, we recently showed that for a diatomic molecule:

$$q(\underline{V}, T) = q_{trans}\, q_{rot}\, q_{vib}\, q_{elec}$$

$$q(\underline{V}, T) = \left[\left(\frac{2\pi M k_B T}{h^2}\right)^{3/2} \underline{V}\right]\left[\frac{T}{\sigma\theta_{rot}}\right]\left[\frac{e^{-\theta_{vib}/2T}}{(1 - e^{-\theta_{vib}/T})}\right]\left[g_{el}\, e^{D_e/k_B T}\right] \qquad (48.25)$$

In Eq. (48.25), we can combine the exponential in the numerator of the third term on the right of the equal sign in Eq. (48.25) and the exponential in the last term in Eq. (48.25) as follows (recall that $\theta_{vib}/2T = h\nu/2k_B T$ and $D_e = D_o + h\nu/2 \Rightarrow D_e/k_B T = D_o/k_B T + h\nu/2k_B T$):

$$e^{-\theta_{vib}/2T}\, e^{D_e/k_B T} = e^{D_o/k_B T} \qquad (48.26)$$

Using Eq. (48.26) in Eq. (48.25) yields:

$$q(\underline{V}, T) = \left[\left(\frac{2\pi M k_B T}{h^2}\right)^{3/2} \underline{V}\right]\left[\frac{T}{\sigma\theta_{rot}}\right]\left[\frac{1}{(1 - e^{-\theta_{vib}/T})}\right]\left[g_{el}\, e^{D_e/k_B T}\right] \qquad (48.27)$$

Next, if we use Eq. (47.27) for H_2, I_2, and HI participating in the chemical reaction in Eq. (48.23) and factor in that $\sigma_{HI} = 1$, $\sigma_{H_2} = 2$, $\sigma_{I_2} = 2$, and that all the g_{el}s are equal, K(T) in Eq. (48.24) can be expressed as follows:

$$K(T) = \left(\frac{M_{HI}^2}{M_{H_2} M_{I_2}}\right)^{3/2} \left(\frac{4 \theta_{rot}^{H_2} \theta_{rot}^{I_2}}{\left(\theta_{rot}^{HI}\right)^2}\right) \left(\frac{\left(1 - e^{-\theta_{vib}^{H_2}/T}\right)\left(1 - e^{-\theta_{vib}^{I_2}/T}\right)}{\left(1 - e^{-\theta_{vib}^{HI}/T}\right)^2}\right)$$

$$\times \left(e^{\left(2 D_o^{HI} - D_o^{H_2} - D_o^{I_2}\right)/k_B T}\right) \tag{48.28}$$

Equation (48.28) enables us to calculate $K(T)$ molecularly in the context of statistical mechanics, thereby fulfilling another goal that we set to accomplish in Part III, including establishing contact with Part II.

Lecture 49

Statistical Mechanical Treatment of Binary Liquid Mixtures

49.1 Introduction

The material presented in this lecture is adapted from my lecture notes. First, we will discuss modeling binary liquid mixtures using statistical mechanics. Second, we will show how to calculate the entropy of mixing of a binary liquid mixture in the context of the Micro-Canonical ensemble using a lattice description. Third, we will discuss the range of validity of the lattice theory description to model binary liquid mixtures. Fourth, we will implement the lattice theory description. Fifth, we will derive expressions for the chemical potentials of the two components comprising the binary liquid mixture. Finally, we will discuss several molecular characteristics of ideal solutions, including what happens when deviations from these characteristics occur.

49.2 Modeling Binary Liquid Mixtures Using a Statistical Mechanical Approach

For simplicity, we consider a binary liquid mixture containing N_1 molecules of liquid 1 and N_2 molecules of liquid 2. An extension to additional liquids is conceptually similar. We will also assume that the molecules of liquid 1 and liquid 2, while distinguishable, are nearly identical to each other in size and interaction energy. In other words, we will assume that we can replace a molecule of liquid 1 in the mixture by one of liquid 2, with no volume of mixing or enthalpy of mixing changes. Accordingly, for the mixing process considered here, we assume that $\Delta \underline{V}_{mix} = 0$ and $\Delta \underline{H}_{mix} = 0$ (see Part II).

We would like to calculate the Gibbs free energy of mixing, $\Delta \underline{G}_{mix}$, for this binary liquid mixture at the molecular level. In addition, we would like to calculate the chemical potentials μ_1 and μ_2 of molecules of liquid 1 and 2, respectively. In Part II, we showed that:

© Springer Nature Switzerland AG 2020
D. Blankschtein, *Lectures in Classical Thermodynamics with an Introduction to Statistical Mechanics*, https://doi.org/10.1007/978-3-030-49198-7_49

$$\underline{G} = \underline{H} - T\underline{S}$$

and that:

$$\Delta \underline{G}_{mix} = \Delta \underline{H}_{mix} - T\Delta \underline{S}_{mix} \qquad (49.1)$$

However, because we assumed that $\Delta \underline{H}_{mix} = 0$, Eq. (49.1) reduces to:

$$\Delta \underline{G}_{mix} = -T\Delta \underline{S}_{mix} \qquad (49.2)$$

Equation (49.2) shows that calculating $\Delta \underline{G}_{mix}$ requires calculating $\Delta \underline{S}_{mix}$. As shown in Part II:

$$\Delta \underline{S}_{mix} = \underline{S}_{1+2} - \underline{S}_1^o - \underline{S}_2^o \qquad (49.3)$$

where $\Delta \underline{S}_{1+2}$ is the entropy of the binary mixture of liquids 1 and 2 and $\underline{S}_1^o$ and $\underline{S}_2^o$ are the entropies of liquid 1 and liquid 2, respectively.

49.3 Calculating $\Delta \underline{S}_{mix}$ Using the Micro-Canonical Ensemble

We will use the Micro-Canonical ensemble definition of entropy presented recently in Part III, that is,

$$\underline{S} = k_B \ln \Omega \qquad (49.4)$$

where Ω is the number of distinct available configurations. Using Eq. (49.4) in Eq. (49.3), it follows that:

$$\Delta \underline{S}_{mix} = \underbrace{k_B \ln \Omega_{1+2}}_{\underline{S}_{1+2}} - \underbrace{k_B \ln \Omega_1^o}_{\underline{S}_1^o} - \underbrace{k_B \ln \Omega_2^o}_{\underline{S}_2^o}$$

where Ω_{1+2}, Ω_1^o, and Ω_2^o are the numbers of distinct available configurations in the binary liquid mixture, in liquid 1, and in liquid 2, respectively. Simplifying the last equation yields:

$$\Delta \underline{S}_{mix} = k_B \ln\left(\frac{\Omega_{1+2}}{\Omega_1^0 \, \Omega_2^0}\right)$$ (49.5)

In order to enumerate (count) the three Ωs in Eq. (49.5), we will utilize a lattice model, statistical mechanical approach to describe the binary liquid mixture (1+2), as well as the two liquids (1 and 2). Specifically, we will model the binary liquid mixture of N_1 molecules of liquid 1 and N_2 molecules of liquid 2 as a cubic lattice of cells, where each cell can accommodate a molecule of liquid 1 or a molecule of liquid 2. Recall that this is possible because we assumed that the molecules of liquid 1 and liquid 2 have nearly the same volumes.

49.4   Range of Validity of the Lattice Description of Binary Liquid Mixtures

The description of the binary mixture of liquids 1 and 2 using a lattice is reasonable. In fact, the lattice description would be precise for a regular crystal. This is because in both liquids 1 and 2 and in the binary liquid mixture of liquids 1 and 2, the molecules are assumed to be sufficiently ordered in space to justify the approximate representation using a lattice. Indeed, in a simple liquid consisting of nearly spherical molecules, the first neighbors of a given molecule in the liquid are positioned at a distance from its center which is fairly well defined, albeit not as precisely as in a crystal. The second neighbors are positioned at less accurately specified distances, and so on. However, because we will typically be interested in the first coordination shell around a given molecule, the lattice representation is a reasonable description of the short-range order which is characteristic of a liquid. Therefore, the use of a lattice description is a valid idealization. On the other hand, the additional assumption that the same lattice may be used to describe the configurations of liquids 1 and 2, and of the binary liquid mixture of 1 and 2, is a much more serious assumption when modeling real binary liquid mixtures. Indeed, this demands that the geometry of the molecules of liquid 1 and liquid 2 be nearly identical.

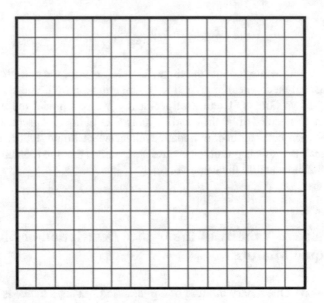

Fig. 49.1

49.5   Lattice Theory Calculation

With the above in mind, in order to accommodate $(N_1 + N_2)$ molecules in the cubic lattice, we will require a total of $(N_1 + N_2)$ cubic cells, where each cubic cell can accommodate both a molecule of liquid 1 and a molecule of liquid 2. A sample two-dimensional square lattice is shown in Fig. 49.1.

In Fig. 49.1, the number of lattice cells is equal to $(N_1 + N_2) = N$, where each lattice cell can accommodate a molecule of liquid 1 or a molecule of liquid 2. The number of ways in which the $(N_1 + N_2)$ molecules can be positioned in the lattice of $N = (N_1 + N_2)$ lattice cells, one at a time, is given by:

$$N(N-1)(N-2)\cdots 1 = N! \qquad (49.6)$$

Because interchanging any of the N_1 identical (indistinguishable) molecules of liquid 1, or any of the N_2 identical (indistinguishable) molecules of liquid 2, makes no difference, we need to divide Eq. (49.6) by the number of such interchanges, which is $N_1!$ for the molecules of liquid 1 and $N_2!$ for the molecules of liquid 2. Therefore, the total number of distinct configurations available in the binary liquid mixture is given by:

$$\Omega_{1+2} = \frac{N!}{N_1!N_2!} = \frac{(N_1 + N_2)!}{N_1!N_2!} \tag{49.7}$$

For liquid 1, $N_2 = 0$, and Eq. (49.7) shows that:

$$\Omega_1^o = \frac{N_1!}{N_1!} = 1 \tag{49.8}$$

Similarly, for liquid 2, $N_1 = 0$, and Eq. (49.7) shows that:

$$\Omega_2^o = \frac{N_2!}{N_2!} = 1 \tag{49.9}$$

Using Eqs. (49.7), (49.8), and (49.9) in Eq. (49.5) yields:

$$\Delta \underline{S}_{mix} = k_B \ln \left[\frac{(N_1 + N_2)!}{N_1!N_2!} \right] \tag{49.10}$$

or

$$\Delta \underline{S}_{mix} = k_B \{ \ln[(N_1 + N_2)!] - \ln[N_1!] - \ln[N_2!] \} \tag{49.11}$$

Next, to simplify Eq. (49.11), we can use Stirling's approximation, $\ln[z!] = z\ln z - z$ (for $z \gg 1$). This yields:

$$\Delta \underline{S}_{mix} = k_B \{ (N_1 + N_2) \ln(N_1 + N_2) - (N_1 + N_2) - [N_1 \ln N_1 - N_1] - [N_2 \ln N_2 - N_2] \} \tag{49.12}$$

Rearranging Eq. (49.12), including cancelling the equal terms, yields:

$$\Delta \underline{S}_{mix} = -k_B \left\{ N_1 \ln \left(\frac{N_1}{N_1 + N_2} \right) + N_2 \ln \left(\frac{N_2}{N_1 + N_2} \right) \right\} \tag{49.13}$$

Because

$$\frac{N_1}{N_1 + N_2} = x_1 \text{ (Mole fraction of 1)}$$
$$\frac{N_2}{N_1 + N_2} = x_2 \text{ (Mole fraction of 2)} \tag{49.14}$$

Equation (49.13) can be expressed as follows:

$$\Delta \underline{S}_{mix} = -k_B\{N_1 \ln x_1 + N_2 \ln x_2\} \tag{49.15}$$

Equation (49.15) is expressed in terms of the numbers of molecules, N_1 and N_2, and the Boltzmann constant k_B. This is due to the fact that in the lattice description, we position individual molecules in the available lattice cells. However, once derived, Eq. (49.15) can be expressed in terms of the numbers of moles, n_1 and n_2. This is simply done by recalling that $k_B N_A = R$, $N_1/N_A = n_1$ and $N_2/N_A = n_2$, where N_A is Avogadro's number.

We are now ready to calculate the Gibbs free energy of mixing, $\Delta \underline{G}_{mix}$, by substituting Eqs. (49.13) or (49.15) in Eq. (49.2). This yields:

$$\Delta \underline{G}_{mix} = k_B T\left\{N_1 \ln\left(\frac{N_1}{N_1 + N_2}\right) + N_2 \ln\left(\frac{N_2}{N_1 + N_2}\right)\right\} \tag{49.16}$$

or

$$\Delta \underline{G}_{mix} = k_B T\{N_1 \ln x_1 + N_2 \ln x_2\} \tag{49.17}$$

49.6 Calculation of Chemical Potentials

Next, we can calculate the chemical potentials μ_1 and μ_2. Specifically, in Part II, we saw that:

$$\begin{aligned}
\Delta \mu_1 = \mu_1 - \mu_1^o = \left(\frac{\partial \Delta \underline{G}_{mix}}{\partial N_1}\right)_{T,P,N_2} \\
\Delta \mu_2 = \mu_2 - \mu_2^o = \left(\frac{\partial \Delta \underline{G}_{mix}}{\partial N_2}\right)_{T,P,N_1}
\end{aligned} \tag{49.18}$$

where μ_1^o and μ_2^o are the chemical potentials of pure component 1 and pure component 2, respectively.

Differentiating Eq. (49.16) with respect to N_1, at constant T, P, N_2, yields μ_1. Similarly, differentiating Eq. (49.16) with respect to N_2, at constant T, P, N_1, yields μ_2. In both cases, recall that x_1 and x_2 depend on N_1 and N_2 (see Eq. (49.14)). This yields:

$$\begin{aligned}
\mu_1 = \mu_1^o(T, P) + k_B T \ln x_1 \\
\mu_2 = \mu_2^o(T, P) + k_B T \ln x_2
\end{aligned} \tag{49.19}$$

In Part II, we also showed that:

$$\mu_1 = \mu_1^o(T, P) + k_B T \ln a_1$$
$$\mu_2 = \mu_2^o(T, P) + k_B T \ln a_2$$

(49.20)

where a_1 and a_2 are the activities of component 1 and component 2, respectively.
   A comparison of Eqs. (49.20) and (49.19) shows that:

$$a_1 = x_1$$
$$a_2 = x_2$$

(49.21)

   Equation (49.21) shows that the binary mixture of liquids 1 and 2 considered here is ideal over the entire composition range.

49.7 Molecular Characteristics of Ideal Solutions

The equations derived above using statistical mechanics pertain to ideal solutions and connect nicely with material derived thermodynamically in Part II. In addition, the lattice theory approach that we used provides further insights about the molecular characteristics associated with the ideality of the binary liquid mixture considered here. Specifically:

1. The interaction energy between a pair of molecules of liquid 1, [1-1], a pair of molecules of liquid 2, [2-2], and a pair of molecules of liquids 1 and 2, [1-2] must satisfy $\varepsilon_{12} = \frac{1}{2}(\varepsilon_{11} + \varepsilon_{22})$, where ε_{11} is the interaction energy of 1 and 1 and ε_{22} is the interaction energy of 2 and 2. As discussed in Part II, this will result in $\Delta \underline{H}_{mix} = 0$, for which the mixing of liquids 1 and 2 is athermal.

2. The molecules of liquid 1 and liquid 2 must be of the same size, and therefore, the volume of mixing, $\Delta \underline{V}_{mix}$, is zero.

3. Solutions deviate from ideality because they fail to satisfy one, or both, of requirements 1 and 2 above. For example, in the case of a high-molecular-weight liquid mixed with a low-molecular-weight liquid, the non-idealities of mixing are primarily a result of not satisfying requirement (2), irrespective of whether the mixing process is athermal or not (requirement 1).

   Finally, if the binary liquid mixture of 1 and 2 is not athermal, one can use the lattice theory approach to calculate $\Delta \underline{H}_{mix}$. This calculation is beyond the scope of this lecture.

Lecture 50

Review of Part III and Sample Problem

50.1 Introduction

In this final lecture, we will first present a comprehensive review of the material covered in Part III, and following that, we will solve Sample Problem 50.1, an interesting problem dealing with a lattice theory calculation.

50.2 Canonical Ensemble

Canonical partition function:

$$Q(N, \underline{V}, T) = \sum_j \exp(-\beta \underline{E}_j(N, \underline{V})) \tag{50.1}$$

where the summation is carried out over the different quantum states, and:

$$\beta = \frac{1}{k_B T} \tag{50.2}$$

Probability of finding a system in energy state j:

$$p_j(N, \underline{V}, T) = \frac{\exp(-\beta \underline{E}_j(N, \underline{V}))}{Q} \tag{50.3}$$

Probability of finding a system in energy level i:

© Springer Nature Switzerland AG 2020
D. Blankschtein, *Lectures in Classical Thermodynamics with an Introduction to Statistical Mechanics*, https://doi.org/10.1007/978-3-030-49198-7_50

$$p_i(N, \underline{V}, T) = \frac{g_i \exp(-\beta \underline{E}_i(N, \underline{V}))}{Q} \tag{50.4}$$

where g_i is the degeneracy of energy level i.

50.3   Average Properties in the Canonical Ensemble

Averaging properties:

$$<X> = \sum_j X_j p(X_j) \tag{50.5}$$

Average energy $<\underline{E}>$:

$$<\underline{E}> = \underline{U}(N, \underline{V}, T) = -\left[\frac{\partial \ln Q(N, \underline{V}, \beta)}{\partial \beta}\right]_{N, \underline{V}}$$

$$= k_B T^2 \left[\frac{\partial \ln Q(N, \underline{V}, T)}{\partial T}\right]_{N, \underline{V}} \tag{50.6}$$

Relative deviations from the average energy ($\underline{U}$):

$$\frac{\sigma_{\underline{U}}}{\underline{U}} = \left[\frac{\sqrt{k_B T^2 C_V}}{U}\right] \cdot \frac{1}{\sqrt{N}} \tag{50.7}$$

Average pressure:

$$<P> = P = k_B T \left[\frac{\partial \ln Q(N, \underline{V}, \beta)}{\partial \underline{V}}\right]_{N, \beta} \tag{50.8}$$

Average entropy $\underline{S}(N, \underline{V}, T)$:

$$\underline{S} = -k_B \sum_j p_j \ln p_j \tag{50.9}$$

$$\underline{S} = k_B T \left[\frac{\partial \ln Q(N, \underline{V}, T)}{\partial T}\right]_{N, \underline{V}} + k_B \ln Q(N, \underline{V}, T) \tag{50.10}$$

Average Helmholtz free energy $A(N, \underline{V}, T)$:

$$\underline{A} = -k_B T \ln Q \tag{50.11}$$

Average chemical potential $\mu(N, \underline{V}, T)$:

$$\mu = -k_B T \left[\frac{\partial \ln Q(N, \underline{V}, T)}{\partial N} \right]_{T, \underline{V}} \tag{50.12}$$

50.4 Calculation of the Canonical Partition Function

In general, use the definition:

$$Q(N, \underline{V}, T) = \sum_j \exp\left(-\beta \underline{E}_j(N, \underline{V})\right) \tag{50.13}$$

For independent and distinguishable molecules (or atoms) with identical energy states:

$$Q(N, \underline{V}, T) = [q(\underline{V}, T)]^N \tag{50.14}$$

where

$$q(\underline{V}, T) = \sum_j \exp\left(-\beta \epsilon_j\right) \tag{50.15}$$

where ϵ_j is the energy of state j.

For independent and indistinguishable molecules (or atoms) with identical energy states:

$$Q(N, \underline{V}, T) = \frac{[q(\underline{V}, T)]^N}{N!} \tag{50.16}$$

where

$$q(\underline{V}, T) = \sum_j \exp\left(-\beta \epsilon_j\right) \tag{50.17}$$

Equation (50.16) is valid only when the condition for the applicability of Boltzmann's statistics is satisfied, namely, when:

$$\frac{N}{\underline{V}} \left(\frac{h^2}{8mk_BT} \right)^{3/2} \ll 1 \tag{50.18}$$

Determining $q(\underline{V}, T)$:

$$q(\underline{V}, T) = q_{trans} \cdot q_{rot} \cdot q_{vib} \cdot q_{elec} \tag{50.19}$$

where

$$q_{trans}(\underline{V}, T) = \sum_j \exp\left(-\beta \epsilon_j^{trans}\right) \tag{50.20}$$

$$q_{rot}(T) = \sum_j \exp\left(-\beta \epsilon_j^{rot}\right) \tag{50.21}$$

$$q_{vib}(T) = \sum_j \exp\left(-\beta \epsilon_j^{vib}\right) \tag{50.22}$$

$$q_{elec}(T) = \sum_j \exp\left(-\beta \epsilon_j^{elec}\right) \tag{50.23}$$

and j refers to a quantum state.

50.5   Molecular Partition Functions of Ideal Gases

Monoatomic ideal gas:

Translation:

$$q_{trans}(\underline{V}, T) = \left(\frac{2\pi mk_BT}{h^2} \right)^{3/2} \underline{V} \tag{50.24}$$

Electronic:

$$q_{elec}(T) = g_{e1} + g_{e2}e^{-\beta \epsilon_{e2}} \tag{50.25}$$

Diatomic ideal gas:

Translation:

$$q_{\text{trans}}(\underline{V}, T) = \left(\frac{2\pi(m_1 + m_2)k_BT}{h^2}\right)^{3/2} \underline{V} \tag{50.26}$$

Electronic:

$$q_{\text{elec}}(T) = g_{e1}e^{+\beta D_e} + g_{e2}e^{-\beta\epsilon_{e2}} \tag{50.27}$$

In Eq. (50.27), $D_e = D_0 + h\nu/2$.

Vibration:

$$q_{\text{vib}}(T) = \frac{\exp\left(-\frac{\theta_{\text{vib}}}{2T}\right)}{1 - \exp\left(-\frac{\theta_{\text{vib}}}{T}\right)} \tag{50.28}$$

where

$$\theta_{\text{vib}} = \frac{h\nu}{k_B} = \frac{hc\tilde{\nu}}{k_B} \tag{50.29}$$

Rotation:

$$q_{\text{rot}}(T) = \frac{T}{\sigma\theta_{\text{rot}}} \quad \text{for} \quad \theta_{\text{rot}} \ll T \tag{50.30}$$

where

$$\theta_{\text{rot}} = \frac{h^2}{8\pi^2 I k_B} = \frac{hB}{k_B} \quad \text{and} \quad B = \frac{h}{8\pi^2 I} \tag{50.31}$$

where

$$I = \frac{m_1 m_2}{m_1 + m_2}d^2 \tag{50.32}$$

$\sigma = 1$, for a heteronuclear diatomic molecule

$\sigma = 2$, for a homonuclear diatomic molecule

50.6   Summary of Thermodynamic Functions of Ideal Gases

Below, we present expressions for some thermodynamic functions of monoatomic and diatomic ideal gases.

Monoatomic ideal gas:

$$\frac{A}{k_B T} = -\ln\left[\left(\frac{2\pi m k_B T}{h^2}\right)^{3/2} \frac{V}{N} e g_{e1}\right] \tag{50.33}$$

$$\frac{U}{k_B T} = \frac{3}{2} \tag{50.34}$$

$$\frac{C_V}{k_B} = \frac{3}{2} \tag{50.35}$$

$$\frac{S}{k_B} = \ln\left[\left(\frac{2\pi m k_B T}{h^2}\right)^{3/2} \frac{V}{N} e^{5/2} g_{e1}\right] \tag{50.36}$$

Diatomic ideal gas:

$$\frac{U}{k_B T} = \frac{3}{2} + \frac{2}{2} + \left[\frac{\theta_{vib}}{2T} + \frac{\theta_{vib}/T}{e^{\theta_{vib}/T} - 1}\right] - \frac{D_e}{k_B T} \tag{50.37}$$

$$\frac{C_V}{k_B} = \frac{3}{2} + \frac{2}{2} + \left[\left(\frac{\theta_{vib}}{T}\right)^2 \frac{e^{\theta_{vib}/T}}{(e^{\theta_{vib}/T} - 1)^2}\right] \tag{50.38}$$

50.7   Grand-Canonical Ensemble

Grand-Canonical partition function:

$$\Xi(\underline{V}, T, \mu) = \sum_N \sum_j e^{-\frac{E_{Nj}}{k_B T}} e^{+\frac{\mu N}{k_B T}} \tag{50.39}$$

Probability of finding the system with N molecules in state j:

$$p_{Nj}(\underline{V}, T, \mu) = \frac{e^{-\frac{E_{Nj}}{k_B T}} e^{+\frac{\mu N}{k_B T}}}{\Xi(\underline{V}, T, \mu)} \tag{50.40}$$

Relation between Ξ and Q:

$$\Xi(\underline{V}, T, \mu) = \sum_N Q(N, \underline{V}, T) e^{+\frac{\mu N}{k_B T}} \tag{50.41}$$

Equation (50.41) shows that we can use the Canonical partition function Q given in Eq. (50.1) to calculate the Grand-Canonical partition function Ξ.

50.8   Average Properties in the Grand-Canonical Ensemble

Average energy:

$$<\underline{E}> = \underline{U} = -\left(\frac{\partial \ln \Xi}{\partial \beta}\right)_{\underline{V}, \mu} \tag{50.42}$$

Average pressure:

$$<P> = P(\underline{V}, T, \mu) = k_B T \left(\frac{\partial \ln \Xi}{\partial \underline{V}}\right)_{T, \mu} \tag{50.43}$$

Average number of molecules:

$$<N> = N = k_B T \left(\frac{\partial \ln \Xi}{\partial \mu}\right)_{\underline{V}, T} \tag{50.44}$$

Relative deviations from the average number of molecules $<N>$:

$$\frac{\sigma_N}{<N>} = \frac{1}{\sqrt{N}} \sqrt{\frac{k_B T \kappa_T}{V}} \tag{50.45}$$

where κ_T is the isothermal compressibility:

$$\kappa_T = -\frac{1}{\underline{V}} \left(\frac{\partial \underline{V}}{\partial P}\right)_{N, T} \tag{50.46}$$

50.9   Micro-Canonical Ensemble

Probability of finding a system in state j:

$$p_j = \frac{1}{W} \tag{50.47}$$

where $W(N, \underline{V}, \underline{E})$ is the degeneracy of the system characterized by $(N, \underline{V}, \underline{E})$, that is,

$$W = \frac{A!}{\prod_j a_j!} \tag{50.48}$$

50.10   Average Entropy in the Micro-Canonical Ensemble

$$\underline{S} = -k_B \sum_j p_j \ln p_j = -k_B \sum_{j=1}^{W} \left(\frac{1}{W}\right) \ln\left(\frac{1}{W}\right) = k_B \ln W(N, \underline{V}, \underline{E}) \tag{50.49}$$

50.11   Classical Statistical Mechanics

Definition of the Hamiltonian:

$$H(\vec{q}, \vec{p}) = K(\vec{p}) + \Phi(\vec{q}) \tag{50.50}$$

where $K(\vec{p})$ is the molecular kinetic energy and $\Phi(\vec{q})$ is the intermolecular potential energy.

Classical molecular partition function:

$$q_{class}(\underline{V}, T) = \frac{1}{h^s} \int \cdots \int e^{-\beta H} \prod_{j=1}^{s} dp_j dq_j \tag{50.51}$$

where s denotes the number of degrees of freedom of the molecule.

Classical Canonical partition function for N independent and indistinguishable molecules (or atoms):

$$Q_{class} = \frac{1}{N! h^{sN}} \int \cdots \int e^{-\beta H_{class}(\vec{q},\vec{p})} d\vec{p} d\vec{q} \tag{50.52}$$

where $H_{class}(\vec{q},\vec{p})$ is the classical N-body Hamiltonian for interacting particles and is given by:

$$H_{class}(\vec{q},\vec{p}) = \frac{1}{2m} \sum_{j=1}^{N} (p_{jx}^2 + p_{jy}^2 + p_{jz}^2) + \Phi(x_1, y_1, z_1, \ldots, x_N, y_N, z_N) \tag{50.53}$$

Using Eq. (50.53) in Eq. (50.52) and integrating over the momenta coordinates, we obtain:

$$Q_{class} = \frac{1}{N!} \left(\frac{2\pi m k_B T}{h^2} \right)^{3N/2} Z_N \tag{50.54}$$

where the classical configurational integral, Z_N, in Eq. (50.54) is given by:

$$Z_N = \int_{\underline{V}} e^{-\beta \Phi(x_1, y_1, z_1, \ldots, x_N, y_N, z_N)} dx_1 dy_1 dz_1 \ldots dz_N \tag{50.55}$$

Canonical ensemble partition function for interacting molecules (or atoms):

$$Q = Q_{class} Q_{QM} \tag{50.56}$$

50.12 Calculation of Virial Coefficients

Virial equation of state (EOS):

$$\frac{P}{k_B T} = \rho + B_2(T)\rho^2 + B_3(T)\rho^3 + \ldots \tag{50.57}$$

50.13   Statistical Mechanical Treatment of Chemical Reaction Equilibria

Chemical potential μ_j:

$$\mu_j = -k_B T \ln\left[\frac{q_j(T, \underline{V})}{N_j}\right] \tag{50.58}$$

Equilibrium constant $K(T)$:

For a general homogeneous gas-phase chemical reaction:

$$|\nu_A|A + |\nu_B|B \rightleftarrows |\nu_C|C + |\nu_D|D \tag{50.59}$$

$$\frac{\rho_C^{|\nu_C|} \rho_D^{|\nu_D|}}{\rho_A^{|\nu_A|} \rho_B^{|\nu_B|}} = K(T) = \frac{(q_C/\underline{V})^{|\nu_C|}(q_D/\underline{V})^{|\nu_D|}}{(q_A/\underline{V})^{|\nu_A|}(q_B/\underline{V})^{|\nu_B|}} \tag{50.60}$$

50.14   Statistical Mechanical Treatment of Binary Liquid Mixtures

Refer to Lecture 49 for details.

50.15   Useful Constants in Statistical Mechanics

Plank's constant: $h = 6.626 \times 10^{-34}$ Js
Boltzmann's constant: $k_B = 1.381 \times 10^{-23}$ J K^{-1}

50.16   Useful Relations in Statistical Mechanics

Useful conversions:

$$\frac{\partial}{\partial \beta} = -k_B T^2 \frac{\partial}{\partial T} \tag{50.61}$$

$$\frac{\partial}{\partial\gamma} = -k_BT\frac{\partial}{\partial\mu}$$

(50.62)

Stirling's approximation: For large N,

$$\ln N! = N\ln N - N$$

(50.63)

The Gaussian integral:

$$\int_{-\infty}^{\infty} e^{-\alpha^2 x^2}\,dx = \frac{\sqrt{\pi}}{\alpha}$$

(50.64)

Infinite sum for e:

$$e^x = \sum_{n=0}^{\infty} \frac{x^n}{n!}$$

(50.65)

Statistical fluctuations:

$$\sigma_x^2 = <(x - <x>)^2> = <x^2> - <x>^2$$

(50.66)

50.17   Sample Problem 50.1

Consider a lattice consisting of 10,000 lattice cells which are occupied by 3,000 indistinguishable particles at temperature, T. It is known that only one particle can reside in each lattice cell. It is also known that the particles do not interact. The initial state is such that the 3,000 particles occupy 30% of the lattice, which is separated by an impermeable partition from the rest of the lattice. Calculate the change in the system Gibbs free energy after the partition is removed and the system reaches equilibrium.

50.17.1   Solution

To compute the change in the system Gibbs free energy for the given process, we recall that:

$$\underline{G} = \underline{H} - T\underline{S}$$

(50.67)

Because the particles do not interact (e.g., there is no bond breakage or formation in the lattice representation), the enthalpy change associated with this process is zero! Accordingly, for the constant temperature process considered here, Eq. (50.67) indicates that:

$$\Delta \underline{G}_{i \to f} = \underline{G}_f - \underline{G}_i = (\underline{H}_f - \underline{H}_i) - T(\underline{S}_f - \underline{S}_i)$$

$$\Delta \underline{G}_{i \to f} = 0 - T(\underline{S}_f - \underline{S}_i) = -T\Delta \underline{S}_{i \to f} \qquad (50.68)$$

Equation (50.68) shows that for the process considered here, the evaluation of $\Delta \underline{G}_{i \to f}$ is equivalent to the evaluation of $-T\Delta \underline{S}_{i \to f}$!

Because the energy of all the states is the same, we can use the entropy definition in the Micro-Canonical ensemble ($\underline{E}$, $\underline{V}$, N). Note that for indistinguishable particles where only one particle can reside at each lattice cell, it follows that:

$$\underline{S} = k_B \ln \Omega \qquad (50.69)$$

where Ω is the number of distinct available configurations, or the degeneracy, of the system. Note that Ω in Eq. (50.69) is the same as W in Eq. (50.48).

We can therefore write that:

$$\Delta \underline{S}_{i \to f} = \underline{S}_f - \underline{S}_i = k_B \ln \Omega_f - k_B \ln \Omega_i \qquad (50.70)$$

or that:

$$\Delta \underline{S}_{i \to f} = k_B \ln \left(\frac{\Omega_f}{\Omega_i} \right) \qquad (50.71)$$

We are told that the lattice consists of M = 10,000 lattice cells, which are occupied by N = 3,000 indistinguishable particles. We are also told that in the initial state, i, only 30% of the lattice cells are occupied. In other words, initially, only 3,000 lattice cells are occupied.

This implies that in the initial state, there are only $(3{,}000!/3{,}000!) = 1$ distinct configurations available to arrange the 3,000 indistinguishable particles in the 3,000 lattice cells. That is:

$$\Omega_i = 1 \Rightarrow \underline{S}_i = k_B \ln \Omega_i = 0 \qquad (50.72)$$

In the final state, f, after the partition is removed, the entire lattice comprising M = 10,000 lattice cells becomes available to the N = 3,000 indistinguishable particles. Therefore, we need to compute the number of distinct configurations where N particles and (M-N) empty lattice cells are arranged in a lattice comprising M lattice cells. The answer is:

$$\Omega_f = \frac{M!}{N!(M-N)!} \tag{50.73}$$

Using Eq. (50.73) in Eq. (50.69), it follows that:

$$\underline{S}_f = k_B \ln \Omega_f = k_B \ln \left[\frac{M!}{N!(M-N)!} \right] \tag{50.74}$$

Expanding the natural logarithm in Eq. (50.74) yields:

$$\underline{S}_f = k_B \{ \ln M! - \ln N! - \ln(M-N)! \} \tag{50.75}$$

Using Stirling's approximation for each ln(factorial) in Eq. (50.75) yields:

$$\underline{S}_f = k_B \{ (M \ln M - M) - (N \ln N - N) - [(M-N)\ln(M-N) - (M-N)] \} \tag{50.76}$$

Manipulation of Eq. (50.76) yields:

$$\underline{S}_f = k_B \{ M \ln M - N \ln N - (M-N) + (M-N) - (M-N)\ln(M-N) \} \tag{50.77}$$

or

$$\underline{S}_f = k_B \{ M \ln M - N \ln N - (M-N)\ln(M-N) \} \tag{50.78}$$

Using $M = 10{,}000$ and $N = 3{,}000$ in Eq. (50.78) yields:

$$\underline{S}_f = 6{,}109 \, k_B \tag{50.79}$$

Using $\underline{S}_i = 0$, $\underline{S}_f = 6{,}109 \, k_B$, in Eq. (50.68) for $\Delta \underline{G}_{i \to f}$, yields:

$$\Delta \underline{G}_{i \to f} = -T(\underline{S}_f - \underline{S}_i) = -T \underline{S}_f \tag{50.80}$$

or

$$\Delta G_{i \to f} = -6{,}109 \, k_B T \tag{50.81}$$

The negative sign in Eq. (50.81) indicates that the process considered is favored and, therefore, occurs spontaneously.

Solved Problems for Part I

© Springer Nature Switzerland AG 2020
D. Blankschtein, *Lectures in Classical Thermodynamics with an Introduction to Statistical Mechanics*, https://doi.org/10.1007/978-3-030-49198-7

Problem 1

Problem 3.4 in Tester and Modell

Two cylinders are attached as shown in the following figure. Both cylinders and pistons are adiabatic and have walls of negligible heat capacity. The connecting rod is nonconducting.

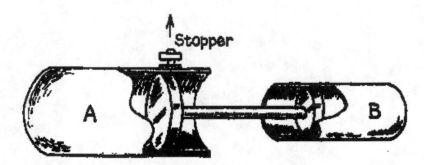

The initial conditions and pertinent dimensions are as follows:

	Cylinder A	Cylinder B
Initial pressure (bar)	10	1
Initial temperature (K)	300	300
Initial volume (m³)	6.28×10^{-3}	1.96×10^{-3}
Piston area (m²)	3.14×10^{-3}	1.96×10^{-3}

The pistons are, initially, prevented from moving by a stop on the outer face of piston A. When the stop is removed, the pistons move and finally reach an end state characterized by a balance of forces on the connecting rod. There is some friction in both piston and cylinders during this process. Gases A and B are ideal and have constant values of $C_v = 20.9$ J/mol K.

What are the final pressures in both A and B? Consider two cases, one where the ambient pressure is 0 bar and another where it is 1 bar.

Solution to Problem 1

Solution Strategy

To solve this problem, we will use the four-step strategy discussed in Part I, where we:

1. Draw the pertinent problem configuration
2. Summarize the given information
3. Identify critical issues
4. Make physically reasonable approximations

1. Draw the pertinent problem configuration

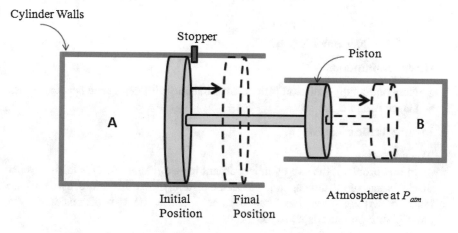

Fig. 1

2. Summarize the given information

Two cylinders, each containing a known quantity of an ideal gas, are connected to each other via a double piston connected by a single rod (see Fig. 1). The entire assembly of cylinders, pistons, and rod is adiabatic and nonconducting. At time zero, a stopper that prevents the piston assembly from moving is removed. We would like to determine the pressure in each of the cylinders as a function of the atmospheric pressure for the cases, $p_{atm} = 0$ bar and $p_{atm} = 1$ bar, when the system reaches its final equilibrium state.

In addition to the information provided above, we know the initial pressure, temperature, and volume in each cylinder, as well as the areas of the pistons in each cylinder. We are also told that the heat capacities at constant volume of gases A and B are constant and equal to 20.9 J/mol K. Finally, we are told that at equilibrium, all the forces acting on the pistons are balanced.

- **Data provided**

  - Initial conditions of gas A:

    - $p_{AI} = 10$ bar
    - $\underline{V}_{AI} = 6.28 \times 10^{-3}$ m^3
    - $T_{AI} = 300$ K

  - Initial conditions of gas B:

    - $p_{BI} = 1$ bar
    - $\underline{V}_{BI} = 1.96 \times 10^{-3}$ m^3
    - $T_{BI} = 300$ K

- Physical dimensions of the pistons:

  - $A_A = 3.14 \times 10^{-3}$ m^2
  - $A_B = 1.96 \times 10^{-3}$ m^2

- Other:

  - $C_V = 20.9$ J/mol K

- **Given assumptions**

  - Nonconducting walls and pistons are adiabatic with negligible heat capacities
  - Gases A and B are ideal

3. **Identify critical issues**

- **The geometry**

  In this problem, the two cylinders do not have equal areas. This leads to a net displacement of the atmospheric gas as the pistons move. Therefore, care must be taken in writing down the force balance between the two pistons when the atmospheric pressure is nonzero.

- **Friction**

  There is some friction associated with the motion of the pistons. However, we do not have enough information to uniquely specify the friction. Therefore, we anticipate that the system of equations may be underspecified. Consequently, we may have to make additional assumptions in order to obtain a reasonable range for the solution.

- **System boundary**

  Many system boundaries are possible and if solved correctly will lead to the same solution. Because the pistons and the walls have negligible heat capacities and the pistons have negligible masses, it is advantageous to include the pistons and the walls within the system boundary. This will ensure that the pressure at the system boundary is the well-defined atmospheric pressure, rather than the gas pressures which vary during the process. Alternatively, one could choose a system which includes the atmosphere (but not the pistons or the chambers) and then relate the work done on the atmosphere to the work done on the gas/pistons system.

  The least desirable boundaries would involve the gas in the cylinders, but not the pistons or the walls. In this case, the system is simple, but the assumption of an adiabatic system is not valid. In Fig. 2, we show a schematic depicting the boundary used in the solution presented here, which contains the cylinder walls, the pistons, and the gas in the chambers, but none of the atmosphere (see the dashed black line).

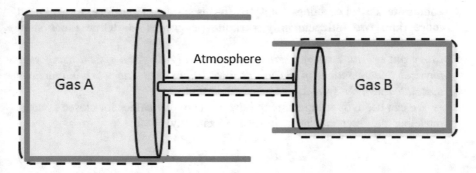

Fig. 2

4. Make physically reasonable assumptions

- **Quasi-static process**

  If sufficient friction is present, we can assume a slow quasi-static process

- **Negligible gravity**

  Due to the horizontal configuration and the low density of the gases, we can assume that gravity has a negligible effect on the system

Solving the Problem

There are a number of steps which, taken together, provide a good method to solve problems involving the First Law and Second Law of Thermodynamics. Below, we first outline these steps and then use them to develop our solution.

1. *Define the system, keeping in mind the need to implement the First Law of Thermodynamics to the system.*

   A convenient choice for the system is the entirety of the piston/cylinder assembly including the gases within. We will not include any of the gas at atmospheric pressure outside of the system (see Fig. 2).

2. *Determine the properties of the system boundaries (i.e., adiabatic vs. diathermal, permeable vs. impermeable, movable vs. rigid).*

   The Problem Statement indicates that the walls are nonconducting, and based on our choice of system boundary, we can further conclude that the boundary is impermeable (no gas enters or leaves). Finally, because the pistons can move, the boundary is movable.

   It is important at this point to consider the complexity of the system in question. Is it simple or composite? The presence of an adiabatic and impermeable piston between the two chambers represents an internal boundary. Systems with internal boundaries are, by definition, composite. **Because the system is**

composite, Postulate I does not apply to this system. Therefore, we will need more than two independently variable properties to define the system completely.

3. *Carry out a First Law of Thermodynamics analysis of the system using your physical understanding of the problem to deduce heat and work interactions associated with the system.*

   We can begin by using the First Law of Thermodynamics for closed systems modeling a differential change in state. Specifically:

$$d\underline{U} = \delta Q + \delta W \tag{1}$$

Note: We have replaced $\underline{E}$ with $\underline{U}$ because the pistons are horizontally aligned (no potential energy effect) and stationary at the initial and final positions (no kinetic energy effect).

Remember that we have defined the system to include both gases, A and B, and therefore, recognizing that the walls and the pistons have negligible heat capacities, it follows that:

$$d\underline{U} = d\underline{U}_A + d\underline{U}_B \tag{2}$$

All the walls in the system are adiabatic and nonconducting. Therefore, we can set δQ in Eq. (1) to zero. Furthermore, we can define the work done on the system by adding up the works done by the environment on both pistons. Following the definition, we can define a pressure/volume work interaction as follows (recall that $Pd\underline{V}$ is the **differential work done by the system** and $-Pd\underline{V}$ is the **differential work done on the system**):

$$\delta W = -Pd\underline{V} \tag{3}$$

Applying Eq. (3) to our specific system and inserting it into Eq. (1), we obtain:

$$d\underline{U} = -P_{atm}d\underline{V}_A - P_{atm}d\underline{V}_B \tag{4}$$

4. *Replace the internal energy term using the ideal gas law.*

   We can also expand the $d\underline{U}$ term by expressing the internal energy of an ideal gas in terms of the temperature and the gas heat capacity at constant volume. Specifically:

$$d\underline{U} = d\underline{U}_A + d\underline{U}_B = N_A C_V dT_A + N_B C_V dT_B \tag{5}$$

Combining Eqs. (4) and (5), we obtain:

$$N_A C_V dT_A + N_B C_V dT_B = -P_{atm}d\underline{V}_A - P_{atm}d\underline{V}_B \tag{6}$$

We can then integrate Eq. (6) from the initial state (state 1) to the final state (state 2):

$$N_A C_V \int_{T_{A1}}^{T_{A2}} dT_A + N_B C_V \int_{T_{B1}}^{T_{B2}} dT_B = -P_{atm} \int_{V_{A1}}^{V_{A2}} d\underline{V}_A - P_{atm} \int_{V_{B1}}^{V_{B2}} d\underline{V}_B \qquad (7)$$

Because P_{atm}, N_A, N_B, and C_V are all constant, all the integrals in Eq. (7) are simple to carry out and yields:

$$N_A C_V (T_{A2} - T_{A1}) + N_B C_V (T_{B2} - T_{B1})$$
$$= -P_{atm}(\underline{V}_{A2} - \underline{V}_{A1}) - P_{atm}(\underline{V}_{B2} - \underline{V}_{B1}) \qquad (7a)$$

Next, we can use the ideal gas law to express each of the four temperatures, T_{A1}, T_{A2}, T_{B1}, and T_{B2}, in Eq. (7a) in terms of pressures, volumes, and number of moles. For example:

$$\frac{P_{A1}\underline{V}_{A1}}{N_A R} = T_{A1} \qquad (7b)$$

Substituting the four temperatures in Eq. (7a), using expressions similar to that in Eq. (7b), we obtain:

$$\frac{C_V}{R}[(P_{A2}\underline{V}_{A2} - P_{A1}\underline{V}_{A1}) + (P_{B2}\underline{V}_{B2} - P_{B1}\underline{V}_{B1})]$$
$$= -P_{atm}[(\underline{V}_{A2} - \underline{V}_{A1}) + (\underline{V}_{B2} - \underline{V}_{B1})] \qquad (8)$$

Note that there are **four** unknowns in Eq. (8): P_{A2}, P_{B2}, V_{A2}, and V_{B2}. Therefore, we recognize that three additional equations are required to obtain a unique solution.
5. *Use your physical understanding of the problem to define the internal constraints imposed on the system.*

A second equation can be derived by carrying out a force balance on the pistons, because according to the Problem Statement, the pistons are balanced, and therefore:

$$\sum F = 0 = (P_{A2} - P_{atm})A_A - (P_{B2} - P_{atm})A_B \qquad (9)$$

A third equation can be derived based on the geometry of the pistons to relate the changes in volume in cylinders A and B. Specifically:

$$\frac{V_{A2} - V_{A1}}{V_{B2} - V_{B1}} = -\frac{A_A}{A_B} \qquad (10)$$

At this point, we have used all the provided information but still are one equation short of fully defining the problem. This is because we are told that there is **some** friction, but are not given a friction coefficient. As we will see, the problem is still solvable when P_{atm} equals zero. However, when P_{atm} equals 1 bar, we will need to make further assumptions to obtain reasonable ranges rather than exact solutions. In this respect, a good engineering approximation can be made by using the solutions when (1) there is no friction in cylinder A and (2) there is no friction in cylinder B. This will give us the upper/lower bounds of the final pressures and temperatures in the two cylinders.

6. *Do the math.*

To simplify the math, we can begin by rearranging Eqs. (9) and (10) to eliminate the dependence on the area in one of the equations. Specifically:

$$\frac{\underline{V}_{A2} - \underline{V}_{A1}}{\underline{V}_{B2} - \underline{V}_{B1}} = -\frac{P_{B2} - P_{atm}}{P_{A2} - P_{atm}} \tag{11}$$

Equation (11) will be one of our three linearly independent equations (replacing Eq. (10)).

For $P_{atm} = 0$

Combining Eq. (8) and Eq. (11) when P_{atm} equals zero leads to the cancelation of both $\underline{V}_{A2}$ and $\underline{V}_{B2}$ from the system of equations, that is:

$$\underline{V}_{A1}(P_{A2} - P_{A1}) + \underline{V}_{B1}(P_{B2} - P_{B1}) = 0 \tag{12}$$

Note that combining only two equations led to the cancelation of **two** variables rather than just one. We can now combine Eq. (12) with Eq. (9) to solve explicitly for the values of P_{A2} and P_{B2}. Specifically, we find that:

$$\boxed{\begin{aligned}
P_{A2} &= \frac{P_{A1}\underline{V}_{A1} + P_{B1}\underline{V}_{B1}}{\underline{V}_{A1} + \underline{V}_{B1}\dfrac{A_A}{A_B}} = 6.87\,\text{bar} \\[2ex]
P_{B2} &= \frac{P_{A1}\underline{V}_{A1} + P_{B1}\underline{V}_{B1}}{\underline{V}_{A1}\dfrac{A_B}{A_A} + \underline{V}_{B1}} = 11.01\,\text{bar}
\end{aligned}} \tag{13}$$

For $P_{atm} = 1$

Unfortunately, in this case, the favorable cancelation does not occur, and our system is still incompletely defined. To proceed, we must make additional reasonable approximations to study the limiting behaviors of the system. These limiting cases will yields a **range** for the possible system pressures rather than yielding a single solution.

First, we will assume that friction is only present in cylinder A and, subsequently, that it is only present in cylinder B. These two cases represent the extremes of the possible processes and, hence, should give us the full range of feasible solutions.

If no friction is present in cylinder A, we can assume an adiabatic expansion of the gas in that compartment modeled by:

$$\left(\frac{P_{A2}}{P_{A1}}\right) = \left(\frac{V_{A1}}{V_{A2}}\right)^{\frac{c_V + R}{c_V}} \tag{14}$$

Combining this result with Eqs. (8), (9), and (11), we have four equations and four unknowns. Using a numerical solving technique (e.g., MATLAB), it is possible to solve for the system pressures. Doing this, we find:

$$\boxed{\begin{aligned} P_{A2} &= 6.89\,\text{bar} \\ P_{B2} &= 10.44\,\text{bar} \end{aligned}} \tag{15}$$

This procedure can be repeated to solve for the case of no friction present in cylinder B. Doing this, we find:

$$\left(\frac{P_{B2}}{P_{B1}}\right) = \left(\frac{V_{B1}}{V_{B2}}\right)^{\frac{c_V + R}{c_V}} \tag{16}$$

By solving Eqs. (8), (9), (11), and (16), we find:

$$\boxed{\begin{aligned} P_{A2} &= 6.86\,\text{bar} \\ P_{B2} &= 10.38\,\text{bar} \end{aligned}} \tag{17}$$

Equations (15) and (17) show that the final pressure in cylinder A must be between 6.86 bar and 6.89 bar and that the final pressure in cylinder B must be between 10.38 bar and 10.44 bar.

We can compare the calculated final pressures in chambers A and B for the two different P_{atm} scenarios. Interestingly, the values are very close. We find that P_{B2} is slightly lower and that P_{A2} is slightly higher when the atmospheric pressure is 1 bar as opposed to 0 bar. This makes physical sense because when the atmospheric pressure is nonzero, the atmosphere will exert a larger force on the larger piston in chamber A than on the smaller piston in chamber B. This will provide extra resistance against the movement of gas A, preventing the pistons from moving as far to the right (toward chamber B) than in the case with vacuum. The relatively small difference between the pressures calculated with and without atmosphere shows that the atmosphere yields only a small contribution to the change in the energy of the system when compared to the gases in the cylinders.

Other Possible Solution Strategies

There are, of course, many other solution strategies to arrive at the same answers. Initially, as stated earlier, several different boundary conditions are possible and should all eventually lead to an expression analogous to Eq. (8). In addition, while unnecessary, it is valid to solve the $P_{atm} = 0$ bar case by making the same friction assumptions used in the $P_{atm} = 1$ case. We encourage the interested reader to try other solution strategies to further penetrate this interesting and challenging problem.

Problem 2

Problem 3.8 in Tester and Modell

Bottles of compressed gases are commonly found in chemistry and chemical engineering laboratories. They present a serious safety hazard unless they are properly handled and stored. Oxygen cylinders are particularly dangerous. Pressure regulators for oxygen must be kept scrupulously clean, and no oil or grease should ever be applied to any threads or on moving parts within the regulator. The rationale for this rule comes from the fact that if oil *were* present – and if it *were* to ignite in the oxygen atmosphere – this "hot" spot could lead to ignition of the metal tubing and regulator and cause a disastrous fire and failure of the pressure container. Yet it is hard to see how a trace of heavy oil or grease could become ignited even in pure, compressed oxygen since ignition points probably are over 800 K if "nonflammable" synthetic greases are employed.

Let us model the simple act of opening an oxygen cylinder that is connected to a closed regulator (see the following figure). Assume that the sum of the volumes of the connecting line and the interior of the regulator is V_R. V_R is negligible compared to the bottle volume. Opening valve A pressurizes V_R from some initial pressure to full bottle pressure.

Presumably, the temperature in V_R also changes. The question we would like to raise is: Can the temperature in V_R ever rise to a sufficiently high value to ignite any traces of oil or grease in the line or regulator?

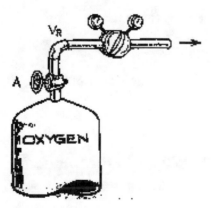

Data: The oxygen cylinder is at 15.17 MPa and 311 K. The connecting line to the regulator and the regulator interior (V_R) are initially at 0.101 MPa, 311 K, and contain pure oxygen.

Assume no heat transfer to the metal tubing or regulator during the operation. Oxygen is essentially an ideal gas. $C_P = 29.3$ J/mol K, $C_V = 20.9$ J/mol K, and both are independent of pressure or temperature.

(a) If gas entering V_R mixes completely with the initial gas, what is the final temperature in V_R?

(b) An alternative model assumes that there is *no* mixing between the gas originally in V_R and that which enters from the bottle. In this case, after the pressures are equalized, we would have two identifiable gas slugs which presumably are at different temperatures. Assuming no axial heat transfer between the gas slugs, what is the final temperature of each?

(c) Comment on your assessment of the hazard of this simple operation of bottle opening. Which of the models in (a) and (b) is more realistic? Can you suggest other improved models?

Solution to Problem 2

Solution Strategy

To solve this problem, we will use the following information:

1. Properties of an ideal gas:

$$P\underline{V} = NRT \tag{1}$$

$$C_P - C_V = R \tag{2}$$

2. The First Law of Thermodynamics for:

(i) Closed systems:

$$d\underline{E} = \delta Q + \delta W \tag{3}$$

(ii) Open systems:

$$d\underline{E} = \delta Q + \delta W + H_{in}\delta N_{in} - H_{out}\delta N_{out} \tag{4}$$

Part (a)

We are told that upon opening an oxygen cylinder connected to a closed regulator, the entering oxygen from the cylinder mixes with the already existing gas, thereby raising its pressure. Let the gas in the connecting tubes and interior of the regulator be our system (see the colored region in Fig. 1). Based on the Problem Statement, we recognize that the initial state of this system is characterized by:

$$P_i = 0.101\,\text{MPa}$$

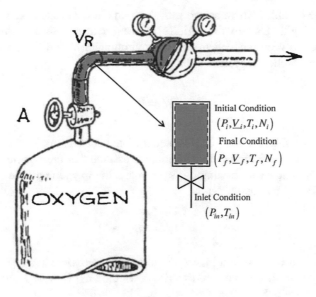

Initial Condition
$(P_i, \underline{V}_i, T_i, N_i)$
Final Condition
$(P_f, \underline{V}_f, T_f, N_f)$
Inlet Condition
(P_{in}, T_{in})

Fig. 1

$$\underline{V}_i = V_R$$
$$T_i = 311\,\mathrm{K}$$
$$N_i = ?$$

The final state of this system is characterized by:

$$P_f = 15.17\,\mathrm{MPa}$$
$$\underline{V}_f = V_R$$
$$T_f = ?$$
$$N_f = ?$$

Next, we proceed to determine the characteristics of the system boundary:

1. **Permeable or Impermeable?** Oxygen from the tank enters our system. There-fore, the system boundary is permeable, indicating that the chosen system is open. The conditions of the inlet (*in*) stream are given by:

$$P_{in} = 15.17 \; \mathrm{MPa} = 15.17 \times 10^6 \; \mathrm{Pa}$$
$$T_{in} = 311 \; \mathrm{K}$$

2. **Rigid or Movable?** There are no movable parts in the system boundary. Therefore, it is rigid. Due to the rigidity of the boundary, there is no P$\underline{V}$ work interaction between the system and its environment. In addition, no other work is done on the chosen system. As a result:

$$\delta W = 0$$

3. **Adiabatic or Diathermal?** The Problem Statement tells us to neglect any heat transfer to the metal tubing. In addition, we assume that there is no heat transfer across the boundary separating our system and the oxygen cylinder. As a result, the operation is adiabatic, that is:

$$\delta Q = 0$$

Because the system does not have any internal boundaries, it is a simple system. We can use the information above to write down the First Law of Thermodynamics for our open system. Specifically:

$$d\underline{E} = d\underline{U} = \cancel{\delta Q} + \cancel{\delta W} + H_{in}\delta N_{in} - H_{out}\cancel{\delta N_{out}} \tag{5}$$

From this point on, we can either follow the steps presented in Lecture 6 or directly integrate Eq. (5) as is. Either way, we will rewrite the internal energy and enthalpy in terms of a reference state and be able to cancel out several terms. Below, we directly integrate Eq. (5) between the initial (i) and final (f) states. This yields:

$$\underline{U}_f - \underline{U}_i = H_{in}N_{in} \tag{6}$$

Using the definitions of $\underline{U}$ and H for ideal gases presented in Part I (see Eq. (7a)), Eq. (6) can be simplified to yields Eq. (7b):

$$\begin{aligned} \underline{U}_j &= N_j C_v\left(T_j - T_o\right) + N_j U_o, \text{for } j = f, i \\ H_{in} &= C_p(T_{in} - T_o) + H_o \end{aligned} \tag{7a}$$

$$\begin{aligned} & N_f C_V\left(T_f - T_0\right) + N_f U_0 - \left(N_i C_V(T_i - T_0) + N_i U_0\right) = N_{in} C_P(T_{in} - T_0) + N_{in} H_0 \\ \Rightarrow & N_f C_V T_f - N_i C_V T_i = N_{in} C_P T_{in} + \left(N_{in} H_0 - \left(N_f - N_i\right) U_0\right) - N_{in} C_P T_0 \\ & + \left(N_f - N_i\right) C_V T_0 \\ \Rightarrow & N_f C_V T_f - N_i C_V T_i = N_{in} C_P T_{in} + \left(N_{in}(U_0 + RT_0) - \left(N_f - N_i\right) U_0\right) \\ & - N_{in}(C_V + R)T_0 + \left(N_f - N_i\right) C_V T_0 \end{aligned} \tag{7b}$$

Using a mole balance, we can eliminate N_{in} in favor of N_f and N_i to obtain:

$$N_f - N_i = N_{in} \tag{8}$$

Substituting Eq. (8) on the right-hand side of the equality in the last line of Eq. (7b) and cancelling the equal terms, we obtain:

$$N_f C_V T_f - N_i C_V T_i = (Nf - Ni)C_P T_{in} + \left((N_f - N_i)(U_0 + RT_0) - (N_f - N_i)U_0\right)$$
$$- (N_f - N_i)(C_V + R)T_0 + (N_f - N_i)C_V T_0$$
$$\Rightarrow N_f C_V T_f - N_i C_V T_i = (Nf - Ni)C_P T_{in} + (N_f - N_i)RT_0 - (N_f - N_i)(C_V + R - C_V)T_0$$
$$\Rightarrow N_f C_V T_f - N_i C_V T_i = (Nf - Ni)C_P T_{in} + (N_f - N_i)RT_0 - (N_f - N_i)RT_0$$
$$\Rightarrow N_f C_V T_f - N_i C_V T_i = (N_f - N_i)C_P T_{in} \tag{9}$$

Using the ideal gas law in the last line of Eq. (9), we can express N_f and N_i in terms of P_f, V_f, T_f and P_i, V_i, and T_i, respectively. After cancelling terms and rearranging, we obtain Eq. (10) below:

$$\left(\frac{P_f V_f}{RT_f}\right)C_V T_f - \left(\frac{P_i V_i}{RT_i}\right)C_V T_i = \left(\left(\frac{P_f V_f}{RT_f}\right) - \left(\frac{P_i V_i}{RT_i}\right)\right)C_P T_{in}$$

$$\Rightarrow P_f V_R C_V - P_i V_R C_V = \left(\left(\frac{P_f V_R}{T_f}\right) - \left(\frac{P_i V_R}{T_i}\right)\right)C_P T_{in}$$

$$\Rightarrow P_f C_V - P_i C_V = \left(\frac{P_f}{T_f} - \frac{P_i}{T_i}\right)C_P T_{in} \Rightarrow \frac{(P_f C_V - P_i C_V)}{C_P T_{in}} + \frac{P_i}{T_i} = \frac{P_f}{T_f}$$

$$\Rightarrow T_f = \frac{P_f}{\left(\frac{C_V(P_f - P_i)}{C_P T_{in}} + \frac{P_i}{T_i}\right)} = \frac{\kappa T_{in}}{\left(1 + \frac{P_i}{P_f}\left(\kappa \frac{T_{in}}{T_i} - 1\right)\right)}, \text{ where } \kappa = C_p/C_v \tag{10}$$

Note that Eq. (10) is identical to Eq. (6.14) presented in Lecture 6 where $P = P_f$. Substituting numerical values of the different variables in Eq. (10) yields:

$$T_f = \frac{(29.3/20.9)(311K)}{\left(1 + \frac{0.101}{15.17}\left[\left(\frac{29.3}{20.9}\right)\frac{311}{311} - 1\right]\right)} = 434.8K \tag{11}$$

Part (b)

In this case, the volume V_R houses two subsystems (see colored regions A and B in Fig. 2). Subsystem A contains the gas that occupied the volume V_R before the valve was opened and that gets compressed as the gas enters from the oxygen tank. Subsystem B contains the gas that enters from the oxygen tank. Let us first consider subsystem A. The initial and final conditions for subsystem A are given by:

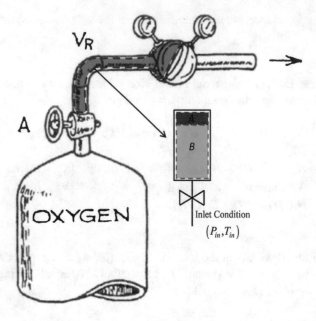

Fig. 2

Initial state:

$$P_i = 0.101 \ \text{MPa}$$
$$\underline{V}_i = V_R$$
$$T_i = 311 \ \text{K}$$
$$N_i = ?$$

Final state:

$$P_f = 15.17 \ \text{MPa}$$
$$\underline{V}_f = ?$$
$$T_f = ?$$
$$N_f = N_i$$

Next, we proceed to determine the characteristics of the boundary of this system:

1. **Permeable or Impermeable?** There is no mixing between the incoming gas and the gas already present in the connecting pipe. As a result, the boundary of the system is impermeable and the system is closed.

2. **Rigid or Movable?** The bottom boundary of subsystem A is moved by the gas which enters into system B. Therefore, the bottom boundary of subsystem A is movable and undergoes PV work ($\delta W = -Pd\underline{V}$).

3. **Adiabatic or Diathermal?** The Problem Statement asks us to neglect any heat transfer to the metal tubing or between the two gas slugs A and B. As a result, the operation is adiabatic ($\delta Q = 0$).

Next, we can use the information above to write down the First Law of Thermodynamics for our closed system. This yields:

$$d\underline{E} = d\underline{U} = \cancel{\delta Q} + \delta W \tag{12}$$

Equation (12) can be solved by writing $d\underline{U}$ and δW in terms of P, $\underline{V}$, T, and N, including rearranging as needed. This yields:

$$NC_V dT = -Pd\underline{V} = -\frac{NRT}{\underline{V}}d\underline{V}$$

$$\Rightarrow \cancel{N}C_V dT = -\frac{\cancel{N}RT}{\underline{V}}d\underline{V}$$

$$\Rightarrow \frac{C_V}{T}dT = -\frac{R}{\underline{V}}d\underline{V} \tag{13}$$

$$\Rightarrow \frac{C_V}{R}\ln\frac{T_f}{T_i} = -\ln\frac{\underline{V}_f}{\underline{V}_i}$$

Using the ideal gas law, we can write the two $\underline{V}$s in Eq. (13) in terms of P, N, and T. Rearranging as needed and finally integrating from the initial (i) to the final (f) condition, we obtain:

$$\frac{C_V}{R}\ln\frac{T_f}{T_i} = -\ln\left[\left(\frac{N_f \cancel{R} T_f}{P_f}\right)\left(\frac{P_i}{N_i \cancel{R} T_i}\right)\right]$$

$$\Rightarrow \frac{C_V}{R}\ln\frac{T_f}{T_i} = -\ln\left[\left(\frac{\cancel{N}_i T_f}{P_f}\right)\left(\frac{P_i}{\cancel{N}_i T_i}\right)\right]$$

$$\Rightarrow \frac{C_V}{R}\ln\frac{T_f}{T_i} = -\ln\left[\left(\frac{T_f}{T_i}\right)\left(\frac{P_i}{P_f}\right)\right] \tag{14}$$

$$\Rightarrow \left(\frac{C_V}{R}+1\right)\ln\frac{T_f}{T_i} = -\ln\left(\frac{P_i}{P_f}\right)$$

$$\Rightarrow \boxed{T_f = T_i\left(\frac{P_f}{P_i}\right)^{(R/(C_v+R))}}$$

Substituting numerical values of the different variables in the last expression in Eq. (14) yields:

$$T_f = 311T_f \left(\frac{15.17}{0.101}\right)^{\left(\frac{8.314}{20.9+8.314}\right)} = 1294.8 \ \text{K} \qquad (15)$$

To determine the temperature of the gas in subsystem B, we consider the entire volume V_R, consisting of subsystems A and B, as our new system (see the shaded region in Fig. 1). Because the system is a composite system, to carry out any analysis, we need to determine the initial and final conditions for both subsystem A and subsystem B. We already know the conditions for subsystem A. Those for subsystem B are given by:

$$P_{i,B} = \text{n}/\text{a}$$
$$\underline{V}_{i,B} = 0$$
$$T_i = \text{n}/\text{a}$$
$$N_{i,B} = 0$$
$$P_{f,B} = 15.17 \ \text{MPa}$$
$$\underline{V}_{f,B} = V_R - \underline{V}_f$$
$$T_{f,B} = ?$$
$$N_{f,B} = ?$$

Again, we proceed to determine the characteristics of the boundary of our composite system (A + B):

1. **Permeable or Impermeable?** Oxygen from the tank enters our system. As a result, the boundary is permeable, and the system is open. The conditions for the inlet stream are given by:

$$P_{in} = 15.17 \ \text{MPa}$$
$$T_{in} = 311 \ \text{K}$$

2. **Rigid or Movable?** Similar to Part (a), there are no movable parts in the system boundary. As a result, the boundary is rigid, and no $P\underline{V}$ work is incurred, that is:

$$\delta W = 0$$

Note that the movable boundary between subsystems A and B is internal to our chosen system. Therefore, any $P\underline{V}$ work done across this boundary does not appear in our analysis.

3. **Adiabatic or Diathermal?** Based on an argument identical to that made in Part (a), we can conclude that the system boundary is adiabatic. Therefore,

$$\delta Q = 0$$

Next, we can use the information above to write down the First Law of Thermodynamics for our open system. Although the system is a composite system with an adiabatic internal boundary, it is comprised of two simple subsystems (A and B). Therefore, the sum of the internal energies of these two simple subsystems is equal to the energy of the composite system, that is:

$$dE = d\underline{U}_A + d\underline{U}_B = \cancel{\delta Q} + \cancel{\delta W} + H_{in}\delta N_{in} - H_{out}\cancel{\delta N}_{out} \tag{16}$$

Integrating Eq. (16) between the initial (i) and final (f) conditions yields:

$$\underline{U}_{f,A} - \underline{U}_{i,A} + \underline{U}_{f,B} - \underline{U}_{i,B} = H_{in}N_{in} \tag{17}$$

Because our system is comprised of two subsystems, the internal energy change of the composite system can be written as the sum of the internal energy changes of the two subsystems comprising the composite system. Using a derivation similar to the derivation in Part (a), Eq. (17) can be further simplified to yields (note that $N_{f,A}$ is the same as N_f, because we continue to use the same notation as in the first section of Part (b)):

$$\Rightarrow \left(N_{f,B}C_V\left(T_{f,B} - T_0\right) + N_{f,B}U_0 + N_f C_V\left(T_f - T_0\right) + N_f U_0\right)$$
$$-(N_{i,B}C_V(T_{i,B} - T_0) + N_{i,B}U_0 + N_iC_V(T_i - T_0) + N_iU_0) = N_{in}C_P(T_{in} - T_0) + N_{in}H_0$$
$$\Rightarrow \left(N_{f,B}C_V T_{f,B} + N_iC_V T_f\right) - (N_iC_V T_i) = N_{in}C_P T_{in}$$
$$+\left(N_{f,B}C_V T_0 + N_f C_V T_0 - N_{i,B}C_V T_0 - N_iC_V T_0 - N_{in}C_P T_0\right)$$
$$+\left(-N_{f,B}U_0 - N_f U_0 + N_{i,B}U_0 + N_iU_0 + N_{in}H_0\right)$$
$$\Rightarrow \left(N_{f,B}C_V T_{f,B} + N_iC_V T_f\right) - (N_iC_V T_i) = N_{in}C_P T_{in}$$
$$+\left(N_{f,B}C_V T_0 + N_iC_V T_0 - N_iC_V T_0 - N_{in}(C_V + R)T_0\right)$$
$$+\left(-N_{f,B}U_0 - N_iU_0 + N_iU_0 + N_{in}H_0\right)$$
$$\tag{18}$$

Using a mole balance, we conclude that:

$$N_{f,B} = N_{in} \tag{19}$$

Substituting Eq. (19) into Eq. (18) as needed, including rearranging, yields:

$$\left(N_{f,B}C_V T_{f,B} + N_i C_V T_f\right) - \left(N_i C_V T_i\right) = N_{f,B}C_P T_{in}$$
$$+\left(N_{f,B}C_V T_0 - N_{f,B}(C_V + R)T_0\right) + \left(-N_{f,B}U_0 + N_{f,B}H_0\right) \qquad (20)$$
$$\Rightarrow \left(N_{f,B}C_V T_{f,B} + N_i C_V T_f\right) - \left(N_i C_V T_i\right) = N_{f,B}C_P T_{in}$$

Using the ideal gas law in the last expression in Eq. (20), we can eliminate $N_{f,\,B}$ and N_i in terms of $P_{f,\,B}$, $\underline{V}_{f,B}$, $T_{f,\,B}$, and P_i, $\underline{V}_i$, and T_i, respectively. This yields:

$$\left(\left(\frac{P_{f,B}\underline{V}_{f,B}}{R T_{f,B}}\right)C_V T_{f,B} + \left(\frac{P_f \underline{V}_f}{R T_f}\right)C_V T_f\right) - \left(\frac{P_i \underline{V}_i}{R T_i}\right)C_V T_i = \left(\frac{P_{f,B}\underline{V}_{f,B}}{R T_{f,B}}\right)C_P T_{in}$$

$$\Rightarrow \left(P_{f,B}\left(V_R - \underline{V}_f\right)C_V + P_f \underline{V}_f C_V\right) - P_i V_R C_V = P_{f,B}\left(V_R - \underline{V}_f\right)C_P\left(\frac{T_{in}}{T_{f,B}}\right)$$

$$\Rightarrow \left(P_{f,B}\left(1 - \frac{\underline{V}_f}{V_R}\right)C_V + P_f \frac{\underline{V}_f}{V_R}C_V\right) - P_i C_V = P_{f,B}\left(1 - \frac{\underline{V}_f}{V_R}\right)C_P\left(\frac{T_{in}}{T_{f,B}}\right)$$
$$(21)$$

Note that in the derivation above, we used the fact that $N_i = N_f$. To further simplify the last expression in Eq. (21), we can use the ideal gas law to express $\underline{V}_f/V_R$ in terms of P_i, T_i, P_f, and T_f. This yields:

$$\left(P_{f,B}\left(1 - \left(\frac{N_f R T_f}{P_f}\right)\left(\frac{P_i}{N_i R T_i}\right)\right)C_V + P_f\left(N_f R T_f P_f\right)\left(\frac{P_i}{N_i R T_i}\right)C_V\right) - P_i C_V$$

$$= P_{f,B}\left(1 - \left(\frac{N_f R T_f}{P_f}\right)\left(\frac{P_i}{N_i R T_i}\right)\right)C_P\left(\frac{T_{in}}{T_{f,B}}\right)$$

$$\Rightarrow \left(P_{f,B}\left(1 - \left(\frac{N_i T_f}{P_f}\right)\left(\frac{P_i}{N_i T_i}\right)\right)C_V + \left(N_i T_f\right)\left(\frac{P_i}{N_i T_i}\right)C_V\right) - P_i C_V$$

$$= P_{f,B}\left(1 - \left(\frac{N_i T_f}{P_f}\right)\left(\frac{P_i}{N_i T_i}\right)\right)C_P\left(\frac{T_{in}}{T_{f,B}}\right)$$

$$\Rightarrow \left(P_{f,B}\left(1 - \left(\frac{T_f}{P_f}\right)\left(\frac{P_i}{T_i}\right)\right)C_V + T_f\left(\frac{P_i}{T_i}\right)C_V\right) - P_i C_V$$

$$= P_{f,B}\left(1 - \left(\frac{T_f}{P_f}\right)\left(\frac{P_i}{T_i}\right)\right)C_P\left(\frac{T_{in}}{T_{f,B}}\right)$$

$$\Rightarrow \boxed{T_{f,B} = \frac{T_{in}P_{f,B}C_P\left(1 - \left(\frac{T_f}{T_i}\right)\left(\frac{P_i}{P_f}\right)\right)}{\left(P_{f,B}\left(1 - \left(\frac{T_f}{T_i}\right)\left(\frac{P_i}{P_f}\right)\right)C_V + P_i\left(\frac{T_f}{T_i}\right)C_V\right) - P_i C_V}}$$

$$(22)$$

Substituting numerical values of the different variables in the last expression in Eq. (22) yields:

$$T_{f,B} = \frac{311 \times 15.17 \times 10^6 \times 29.3 \left(1 - \left(\frac{1294.84}{311}\right)\left(\frac{0.101 \times 10^6}{15.17 \times 10^6}\right)\right)}{\left(15.17 \times 10^6 \left(1 - \left(\frac{1294.84}{311}\right)\left(\frac{0.101 \times 10^6}{15.17 \times 10^6}\right)\right)\right)20.9 + 0.101 \times 10^6 \left(\frac{1294.84}{311}\right)20.9\right) - 0.101 \times 10^6 \times 20.9}$$

$$= 426.75 \text{ K}$$

$$(23)$$

Part (c)

From Parts (a) and (b), we conclude that there is a possibility for the temperature in V_R to reach a value higher than 800 K if the system behaves similar to that in Part (b), i.e., the incoming gas and the already existing gas do not mix with each other. A comparison of the solutions to Parts (a) and (b) suggests that, in Part (b), a very small fraction of the gas in V_R is at a temperature of 1294.84 K. This follows because the solution to Part (a) should be the average temperature of the gas in Part (b), because the state of the gas in Part (a) can be attained by going through the state of the gas in Part (b) and then going through an additional step of mixing the two gas slugs. The final temperature obtained in Part (a) is very close to $T_{f,\,B}$ obtained in Part (b), which suggests that most of the gas is at a temperature of $T_{f,\,B}$.

Quantitatively, it can be shown that the gas at a higher temperature in Part (b) occupies only 2.8% of the total volume and constitutes only 0.93% of the total mass. Therefore, it is highly unlikely that such a small amount of gas will not get mixed with the gas coming in from the valve. In addition, it is also highly unlikely that a gas slug at a temperature of over 1000 K will not transfer heat to the metal tubing and the adjacent gas which are at a significantly lower temperature. Therefore, the probability of the gas catching fire is very low. However, such precautions are warranted when handling potential safety hazards.

A more realistic stratified model for the system would be one where heat transfer between the two gas slugs is allowed, i.e., where the boundary between systems A and B in Fig. 2 is not adiabatic.

Problem 3

Problem 4.11 in Tester and Modell

In several parts of the world, there exist ocean currents of different temperatures that come into contact. An example of this takes place off the coast of Southern Africa, where the warm Agulhas and the cold Benguela currents meet at Cape Point, near Cape Town (see the figure below). It has been proposed that work may be obtained from these ocean currents by operating a heat engine between the warm current as a source and the cold current as a sink (shown schematically in the illustration). This renewable form of energy production has been called *ocean thermal*, and the Department of Energy needs your help in evaluating the proposal.

Assume that the system may be simplified into two channels of water in contact and flowing cocurrently. Furthermore, assume that no mixing occurs between the streams and that heat transfer between the stratified streams under natural conditions is negligible. Finally, assume that the two streams have equal specific heat capacities at constant pressure equal to 4.186 J/gK and mass densities of 1000 kg/m^3.

(a) Derive a general expression for the maximum amount of power that could be obtained from the system. Express your result in terms of temperatures, flow rates, and physical properties of the streams.

(b) What is the pinch temperature, and where does it occur? The pinch is defined as the limiting condition where the stream temperatures approach each other.

(c) Repeat part (a) if the two streams flow countercurrent rather than cocurrent.

(d) It has been estimated that the Benguela current is 16×10^6 m^3/s and its initial temperature is 278 K. The Agulhas current is 20×10^6 m^3/s and its initial temperature is 300 K. Calculate and compare the power obtainable from these two currents assuming that they flow (1) cocurrently and (2) countercurrently.

Solution to Problem 3

Solution Strategy

To solve this problem, we will use the following four-step strategy discussed in Part I:

1. Draw the pertinent problem configuration
2. Summarize the given information
3. Identify critical issues
4. Make physically reasonable approximations

1. **Draw the pertinent problem configuration (see Fig. 1)**

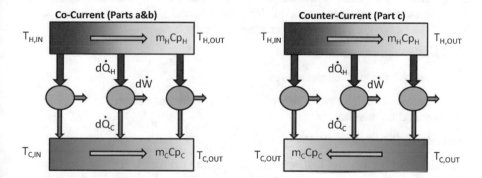

Fig. 1

2. **Summarize the given information**

Two ocean streams at different temperatures are used to power a series of differential Carnot engines. The streams each have given mass flow rates and inlet temperatures. Because the heat capacities of these streams are finite, their temperatures will change throughout the heat interaction. The goal is to extract the maximum amount of work from the two streams. To extract the work, the streams can be contacted either cocurrently or countercurrently. It is necessary to obtain an algebraic solution for the maximum work attainable in both cases in terms of the temperatures, the flow rates, and the stream properties. Using the given information about the two streams in question, a numerical solution is also required. Because no information is given about the specific heat capacities and densities of the streams, we will assume these to be equal for both streams.

Given Data:

- Cold Stream (Benguela Current):

  - $T_{C,IN} = 278$ K
  - $F_C = 16 \times 10^6$ m^3/s

- Hot Stream (Agulhas Current):

  - $T_{H,IN} = 300$ K
  - $F_H = 20 \times 10^6$ m^3/s

- Assumptions:

  - No heat transfer between the stratified layers
  - No mixing between the streams
  - The specific heat capacities at constant pressure and mass densities of the streams are equal ($C_P = 4.186$ J/g K, $\rho = 1000$ kg/m^3)
  - The heat engine operates at an ideal (Carnot) efficiency

3. **Identify critical issues**

Constantly Varying Temperatures

The most difficult aspect of this problem is that the temperatures of both streams vary throughout the process.

Finding the Maximum Work

When solving this problem, we are asked to find the maximum work that can be extracted. Although it is possible to make assumptions of what the end conditions should be, it is more rigorous to define the amount of work in terms of the outlet temperatures and then to optimize the result using calculus.

Defining the System

Another challenging part of this problem is defining an appropriate system to analyze. This problem does not immediately suggest a logical system. While many systems may be possible, for the analysis presented here, we will consider a composite system of differential slices of each stream which are connected by a differential Carnot engine. Figure 2 shows a diagram where the areas surrounded by the dashed lines represent our system.

Defining the Boundaries

The boundaries of this system are open because the hot and cold streams are fluxing in and out of the differential volume. The boundaries are adiabatic because the Problem Statement specifically indicates that the stratified layers do not interact. Finally, the boundary is rigid. It should be noted, however, that although the boundary does not move, we will be integrating over the entirety of both streams.

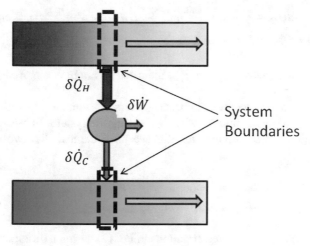

Fig. 2

Therefore, we will expand our knowledge of this one generic differential element to the full process in order to calculate the total work.

4. **Make physically reasonable assumptions**

In general, if possible, it is best to keep the number of assumptions to a minimum. As will be stressed below, it is easy to make intuitive, but incorrect, assumptions for this problem that will lead to erroneous solutions.

Cocurrent Streams Reaching Thermal Equilibrium

One assumption that can be made (although is not required) is that the streams, when in a cocurrent configuration, will reach a final equilibrated temperature where heat transfer will no longer occur. This assumption does simplify the analysis and understanding of the problem.

System Steady State

Because we are dealing with an open system where the supply of the two streams should be constant, we can assume that there is no accumulation (steady state) in the differential system that we have defined.

Solving the Problem

Parts (a and b): Cocurrent Flow
Apply the First Law of Thermodynamics to the System

To apply the First Law of Thermodynamics to the system, we will first consider each differential element separately. Specifically, we will use the First Law of Thermodynamics for the hot (H) and cold (C) streams, each an open system. Specifically:

$$dE_H = dU_H = -\delta Q_H + \delta W_H + \sum H_{H,in}\delta n_{H,in} - \sum H_{H,out}\delta n_{H,out}$$
$$dE_C = dU_C = \delta Q_C + \delta W_C + \sum H_{C,in}\delta n_{C,in} - \sum H_{C,out}\delta n_{C,out}$$

(1)

In Eq. (1), the negative sign in δQ_H indicates that δQ_H is removed from system "H" (see also Fig. 1).

Assuming no mass accumulation in the system, the number of moles entering the system must equal to the number of moles leaving the system. Because there are no work interactions in the boundaries that we have defined, the two equations in Eq. (1) reduce to:

$$dU_H = -\delta Q_H + \delta n_H(H_{H,in} - H_{H,out})$$
$$dU_C = \delta Q_C + \delta n_C(H_{C,in} - H_{C,out})$$

(2)

Next, we can express the two equations in Eq. (2) during a differential change in time. We recognize that the change in internal energy per time is zero because the system is at steady state. In other words, $\frac{dU_H}{dt} = 0$ and $\frac{dU_C}{dt} = 0$. Instead of heats or molar quantities, it is convenient to use heat flow rates ($\delta \dot{Q}_i$) and volumetric flow rates (F_i) (molar or mass flow rates would also be acceptable). We also use the mass density, ρ, to match our heat capacity definition which is on a per mass basis. Specifically, the two equations in Eq. (2) can be expressed as follows:

$$\frac{dU_H}{dt} = 0 = -\delta \dot{Q}_H + F_H\rho(H_{H,in} - H_{H,out})$$
$$\frac{dU_C}{dt} = 0 = \delta \dot{Q}_C + F_C\rho(H_{C,in} - H_{C,out})$$

(3)

Next, we can rewrite the difference in enthalpies as simply the difference in temperatures multiplied by the specific heat capacity (C_P). Doing that and then using the results in Eq. (3) yields the last two expressions in Eq. (4), that is:

$$H_{H,in} - H_{H,out} = C_p\left(T_{H,in} - \cancel{T_0}\right) + \cancel{C_p H_0} - \left[C_p\left(T_{H,out} - \cancel{T_0}\right) + \cancel{C_p H_0}\right] = -C_P dT_H$$
$$H_{C,in} - H_{C,out} = C_p\left(T_{C,in} - \cancel{T_0}\right) + \cancel{C_p H_0} - \left[C_p\left(T_{C,out} - \cancel{T_0}\right) + \cancel{C_p H_0}\right] = -C_P dT_C$$
$$0 = -\delta \dot{Q}_H - F_H\rho C_P dT_H$$
$$0 = \delta \dot{Q}_C - F_C\rho C_P dT_C$$

(4)

Note that we changed the sign of the second terms in the last two expressions in Eq. (4) to be consistent (the derivative refers to **out minus in** rather than to in minus out). Our First Law of Thermodynamics analysis is nearly complete, and next we will use the Second Law of Thermodynamics in order to derive another equation relating the differential heat flows.

Apply the Second Law of Thermodynamics to the System of the Carnot Engine

As indicated above, application of the Second Law of Thermodynamics will provide us another equation to relate the differential heat flows. This relation will be based on the efficiency of the Carnot engine, which is a function of the temperatures of the two streams in our system. Using the Second Law of Thermodynamics and the fact that the Carnot engine undergoes a reversible (δS_{gen}, the generated entropy is zero), cyclic process ($d\underline{S} = 0$), it follows that:

$$
d\underline{S} = \sum \frac{\delta Q_i}{T_i} + \delta \underline{S}_{gen}
$$

$$
0 = \frac{\delta \dot{Q}_H}{T_H} + \frac{-\delta \dot{Q}_C}{T_C} \tag{5}
$$

$$
\Rightarrow \frac{\delta \dot{Q}_H}{\delta \dot{Q}_C} = \frac{T_H}{T_C}
$$

Combining the First Law and the Second Law of Thermodynamics

By combining our results from the First Law of Thermodynamics in Eq. (4) and our results from the Second Law of Thermodynamics in Eq. (5), we obtain:

$$
\frac{\delta \dot{Q}_H}{\delta \dot{Q}_C} = \frac{F_H dT_H}{F_C dT_C} = \frac{T_H}{T_C} \tag{6}
$$

Rearranging Eq. (6), we obtain:

$$
\frac{dT_H}{T_H} = -\frac{F_C}{F_H} \frac{dT_C}{T_C} \tag{7}
$$

Through integration of Eq. (7) from the inlet to the outlet (noting that the flow rates are constant), it follows that:

$$
\int_{T_{H,IN}}^{T_{H,OUT}} \frac{dT_H}{T_H} = -\frac{F_C}{F_H} \int_{T_{C,IN}}^{T_{C,OUT}} \frac{dT_C}{T_C} \tag{7a}
$$

$$
\ln\left(\frac{T_{H,OUT}}{T_{H,IN}}\right) = -\frac{F_C}{F_H} \ln\left(\frac{T_{C,OUT}}{T_{C,IN}}\right) \tag{7b}
$$

$$
T_{H,OUT} = T_{H,IN} \left(\frac{T_{C,IN}}{T_{C,OUT}}\right)^\alpha \tag{8}
$$

where $\alpha = F_C/F_H$. If we define the pinch (P) temperature, T_P, to be the equilibrated final temperatures of the hot and cold streams, it follows that:

$$T_P = T_{H,IN}\left(\frac{T_{C,IN}}{T_P}\right)^{\alpha} \tag{8a}$$

$$\boxed{T_P = T_{H,OUT} = T_{C,OUT} = \left(T_{H,IN}T_{C,IN}^{\alpha}\right)^{\frac{1}{1+\alpha}}} \tag{9}$$

The pinch temperature will be found at the outlet of the process after the streams have had ample time to equilibrate via the heat engines. Through our understanding of thermodynamics, we can assume that when the outlets reach the pinch temperatures, the maximum possible quantity of work would have been extracted. As will be shown in the countercurrent section, this assumption is true, but not necessary to solve the problem. In making use of this assumption, we can define the total fluxes of heat transfer and work generation as follows:

$$\boxed{\begin{aligned}
\dot{Q}_H &= -F_H\rho C_P(T_{H,IN} - T_P) \\
\dot{Q}_C &= F_C\rho C_P(T_{C,IN} - T_P) \\
\dot{W} &= -(-Q_H + Q_C) = F_H\rho C_P(T_{H,IN} - T_P) + F_C\rho C_P(T_{C,IN} - T_P) \\
\dot{W} &= F_H\rho C_P\left[T_{H,IN} - \left(T_{H,IN}T_{C,IN}^{\alpha}\right)^{\frac{1}{1+\alpha}}\right] + F_C\rho C_P\left[T_{C,IN} - \left(T_{H,IN}T_{C,IN}^{\alpha}\right)^{\frac{1}{1+\alpha}}\right]
\end{aligned}} \tag{10}$$

Part (c): Countercurrent Flow

The solution strategy for the countercurrent flow is very similar to that for the cocurrent flow. What we find is that the change in the flow direction of one of the streams leads to a change in the last expression in Eq. (4), resulting in (this can be explained by switching the order of the enthalpies in Eq. (3) to **out minus in**):

$$\begin{aligned}
0 &= -\delta\dot{Q}_H - F_H\rho C_P dT_H \\
0 &= \delta\dot{Q}_C + F_C\rho C_P dT_C
\end{aligned} \tag{11}$$

(**Note:** It would have been equally valid to change the sign in the hot stream term in Eq. (11), but either way, the end result is identical). Propagating the change caused by the sign change through to Eq. (7), we find that (note the sign change relative to that in Eq. (7) in the cocurrent case):

$$\frac{dT_H}{T_H} = \frac{F_C}{F_H}\frac{dT_C}{T_C} \tag{12}$$

As we consider Eq. (12), we recognize that we must be very careful about the integration limits that we use. While in Part (a) we simply integrated both sides from the inlet to the outlet temperatures, for the countercurrent case, one of the integrals must be flipped. Therefore, we obtain:

$$\int_{T_{H,IN}}^{T_{H,OUT}} \frac{dT_H}{T_H} = \frac{F_C}{F_H} \int_{T_{C,OUT}}^{T_{C,IN}} \frac{dT_C}{T_C} = -\frac{F_C}{F_H} \int_{T_{C,IN}}^{T_{C,OUT}} \frac{dT_C}{T_C} \tag{13}$$

Integrating Eq. (13) and rearranging leads to the following result:

$$T_{H,OUT} = T_{H,IN} \left(\frac{T_{C,IN}}{T_{C,OUT}} \right)^\alpha \tag{14}$$

Note that Eq. (14) is identical to Eq. (8)! In other words, even with the changing of the flow direction, the basic equations relating the inlet and outlet temperatures are identical. However, it is unclear if the total amount of work will be equal, because the final temperatures of the process could be different. It is tempting to assume that the outlet temperature of the hot stream should be equal to the inlet temperature of the cold stream (which may be the case in a pure heat exchanger). It can be shown, however, that if that were doable, it would lead to a pure conversion of heat into work in violation of the Second Law of Thermodynamics (see the comment section at the end of the solution).

To solve for the optimal outlet temperatures, we can setup a simple optimization to maximize the work with respect to the outlet cold temperature. To this end, we will set the derivative of the work with respect to the outlet cold temperature equal to zero and then solve to find a maximum. This yields:

$$\dot{W}(T_{C,OUT}) = -(-Q_H + Q_C)$$

$$= F_H \rho C_P \left[T_{H,IN} - T_{H,IN} \left(\frac{T_{C,IN}}{T_{C,OUT}} \right)^\alpha \right]$$

$$+ F_C \rho C_P [T_{C,IN} - T_{C,OUT}] \tag{15}$$

$$\frac{d\dot{W}}{dT_{C,OUT}} = -(-\alpha) \frac{F_H \rho C_P T_{H,IN} T_{C,IN}^\alpha}{T_{C,OUT}^{(\alpha+1)}} - F_C \rho C_P = 0 \tag{16}$$

Solving for the outlet cold stream temperature in Eq. (16), we find that:

$$T_{C,OUT} = \left(T_{H,IN} T_{C,IN}^\alpha \right)^{\frac{1}{1+\alpha}} \tag{17}$$

Note that Eq. (17) is identical to the expression for the pinch temperature in Eq. (9)! Using Eq. (14), we can show that the outlet temperature of the hot stream must be equal to the outlet temperature of the cold stream, which equals the pinch temperature of the cocurrent process. We can check to make sure that this is indeed a maximum by checking the sign of the second derivative:

$$\frac{d^2\dot{W}}{dT^2_{C,OUT}} = -(\alpha)(\alpha+1)\frac{F_H\rho C_P T_{H,IN}T^\alpha_{C,IN}}{T^{(\alpha+2)}_{C,OUT}} \tag{18}$$

Equation (18) clearly shows that the second derivative is always negative, ensuring that our solution is indeed a maximum. We can define the total work done using Eq. (10), and obtain the exact same solution for the countercurrent and cocurrent cases:

$$\boxed{\dot{W}_{counter} = \dot{W}_{co} = F_H\rho C_P\left[T_{H,IN} - \left(T_{H,IN}T^\alpha_{C,IN}\right)^{\frac{1}{1+\alpha}}\right] + F_C\rho C_P\left[T_{C,IN} - \left(T_{H,IN}T^\alpha_{C,IN}\right)^{\frac{1}{1+\alpha}}\right]}$$
$$\tag{19}$$

Part (d): Plugging in the Numbers

To calculate the total amounts of work, we simply use Eq. (19) with the values given in the Problem Statement. Specifically:

$$F_H = 20 \times 10^6 \text{ m}^3/\text{s}$$

$$F_C = 16 \times 10^6 \text{ m}^3/\text{s}$$

$$C_P = 4.186 \text{ J/gK}$$

$$\rho = 1000 \text{ kg/m}^3$$

$$\alpha = F_C/F_H = 0.8$$

$$T_{H,IN} = 300 \text{ K}$$

$$T_{C,IN} = 278 \text{ K}$$

Results

$$\boxed{T_p = 290.01 \text{ K}}$$

$$\boxed{\dot{W}_{co} = \dot{W}_{counter} = 3.12 \times 10^{13} \text{ J/s}}$$

Additional Comments

Why Is the $T_{H,OUT} = T_{C,IN}$ Assumption Invalid for the Countercurrent Flow?

Several logical arguments exist which show why the final condition of $T_{H,OUT} = T_{C,IN}$ will not lead to more work production than using the final condition presented above. Below is just one example:

Consider the system as a black box (see Fig. 3).

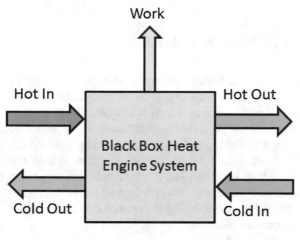

Fig. 3

Now, if the flow rates of the hot and cold streams are equal, then, the energy of the exiting hot stream is equal to that of the entering cold stream. We can then redirect the hot outlet stream to the cold inlet stream and redraw the diagram (see Fig. 4):

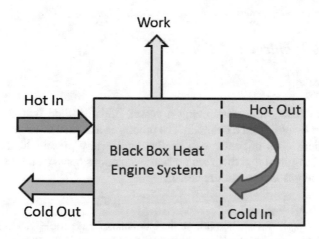

Fig. 4

Examining Fig. 4, it is clear that heat (being taken from the hot inlet stream) is being converted directly into work in violation of the Second Law of Thermodynamics (unless $T_{C,OUT} = T_{H,IN}$, which would be predicted by Eqs. (8) and (14), thus confirming their validity)!

Problem 4

Problem 4.29 in Tester and Modell

We are given a cylindrical vessel with an initial volume of 2 m³ that is filled with helium at 1 bar, 300 K. We plan to pressurize this vessel to 10 bar using a large external source of helium gas maintained at 300 bar, 300 K.

Our model assumes that the initial helium present in the vessel does not mix with the entering helium and that the initial gas is layered (layer A) and compressed by the entering gas (layer B) from 1 bar to 10 bar. The final system then contains two "layers" of helium, both at 10 bar, but presumably at different temperatures.

You can assume that (1) there is no heat transfer between the gas layers, (2) there is no heat transfer to the vessel during the operation, and (3) helium behaves as an ideal gas with a constant value of $C_p = 20.9$ J/mol K.

(a) Calculate the final temperature in layer B.
(b) Calculate the entropy change of the universe after pressurization.
(c) If the system consisting of layers A and B was thermally isolated from the environment and no additional gas allowed to enter, what is the maximum work that one could obtain for the *reversible* mixing of A and B (in the same vessel) to some final homogeneous temperature T_f? What would be the final temperature and pressure?

Solution to Problem 4

Solution Strategy

This problem deals with pressurizing a vessel, initially filled with helium, using high-pressure helium from a cylinder. The process is assumed to occur such that the entering helium does not mix with the helium already present in the vessel. In addition, it is assumed that there are no heat interactions between the two gas layers or between the gas layers and the vessel.

Part (a)

The figure below shows a schematic of the elements of this problem, where system A denotes the gas already present in the vessel and system B represents the gas that enters the vessel from the helium cylinder. Note that determining the final state of the two systems is identical to what we did in the Solution to Problem 2, Part (b). Therefore, the relevant equations are directly reproduced here without a derivation. For a detailed derivation, please refer to the Solution to Problem 2.

To solve the rest of the problem below, we list the initial and final states of the two systems.

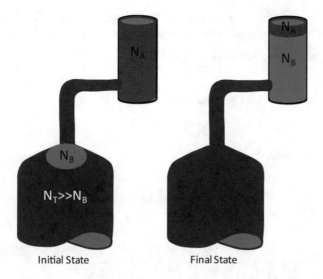

Initial State   Final State

Initial state of system A:

$$P_{i,A} = 1 \text{ bar} = 1 \times 10^5 \text{ Pa}$$

$$V_{i,A} = 2 \text{ m}^3$$

$$T_{i,A} = 300 \text{ K}$$

$$N_{i,A} = ? = \frac{P_{i,A}V_{i,A}}{RT_{i,A}} = \frac{1 \times 10^5 \times 2}{8.314 \times 300} = 80.19 \text{ moles}$$

Final state of system A:

$$P_{i,A} = 10 \text{ bar} = 10 \times 10^5 \text{ Pa}$$

$$N_{f,A} = N_{i,A} = 80.19 \text{ moles}$$

$$T_{f,A} = ?$$

$$= T_{i,A}\left(\frac{P_{f,A}}{P_{i,A}}\right)^{R/(C_v+R)} = T_{i,A}\left(\frac{P_{f,A}}{P_{i,A}}\right)^{R/C_p}$$

$$= 300\left(\frac{10}{1}\right)^{8.314/20.9} = 749.76 \text{ K}$$

See Eq. (14) in the Solution to Problem 2, Part (b).

$$V_{f,A} = ? = \frac{N_{f,A}RT_{f,A}}{P_{f,A}} = \frac{80.19 \times 8.314 \times 749.76}{10 \times 10^5} = 0.5 \text{ m}^3$$

Note that to be consistent with this problem, in the equations above, we added the subscript "A" to the gas system that was originally in the cylinder.

Inlet conditions of the helium stream:

$$P_{in} = 300 \text{ bar} = 300 \times 10^5 \text{ Pa}$$

$$T_{in} = 300 \text{ K}$$

Final state of system B:

$$P_f = 10 \text{ bar} = 10 \times 10^5 \text{ Pa}$$

$T_{f,B} = ?$

$$= \frac{T_{in} P_{f,B} C_P \left(1 - \left(\frac{T_{f,A}}{T_{i,A}}\right)\left(\frac{P_{i,A}}{P_{f,A}}\right)\right)}{\left(P_{f,B}\left(1 - \left(\frac{T_{f,A}}{T_{i,A}}\right)\left(\frac{P_{i,A}}{P_{f,A}}\right)\right)C_V + P_{i,A}\left(\frac{T_{f,A}}{T_{i,A}}\right)C_V\right) - P_{i,A}C_V}$$

(see Eq. (22) in the Solution to Problem 2, Part (b))

$$= \frac{300 \times 10 \times 10^5 \times 20.9\left(1 - \left(\frac{749.76}{300}\right)\left(\frac{1 \times 10^5}{10 \times 10^5}\right)\right)}{(20.9 - 8.314)\left(10 \times 10^5\left(1 - \left(\frac{749.76}{300}\right)\left(\frac{1 \times 10^5}{10 \times 10^5}\right)\right) + 1 \times 10^5\left(\frac{749.76}{300}\right) - 1 \times 10^5\right)}$$

$= 415.19 \text{ K}$

Note that to be consistent with this problem, in the equation above, we again added the subscript "A" to the gas system that was originally in the cylinder:

$$V_{f,B} = ? = V_{i,A} - V_{f,A} = 2 - 0.5 = 1.5 \text{ m}^3$$

$$N_{f,B} = N_B = \frac{P_{f,B} V_{f,B}}{RT_{f,B}} = \frac{10 \times 10^5 \times 1.5}{8.314 \times 415.19} = 434.54 \text{ moles}$$

Part (b)

To find the change in the entropy of the universe, let us first identify the important components of the universe that are undergoing a change in state. The universe consists of the vessel, the helium cylinder, and the ambient in which these two are kept. There are no interactions between the vessel and the ambient or between the helium cylinder and the ambient. Therefore, the entropy change of the universe is purely due to the entropy change of the vessel (consisting of systems A and B) and the helium cylinder. Specifically:

$$\Delta \underline{S}_{universe} = \Delta \underline{S}_{vessel} + \Delta \underline{S}_{cylinder}$$
$$= \Delta \underline{S}_A + \Delta \underline{S}_B + \Delta \underline{S}_{cylinder} \tag{1}$$

We begin the calculation of $\Delta \underline{S}_{universe}$ by first computing the entropy change of system A, $\Delta \underline{S}_A$. The gas in system A is undergoing an adiabatic, reversible compression, and therefore, the entropy change of system A is 0, that is:

$$\Delta \underline{S}_A = \left(\frac{\delta Q}{T} \right)_{reversible} = 0 \tag{2}$$

To calculate $\Delta \underline{S}_B$, let us consider the N_B moles of helium that were initially in the helium cylinder at a temperature and pressure of 300 K and 300 bar, respectively, and finally occupy the vessel at a temperature and pressure of 415.19 K and 10 bar, respectively. Although we know nothing about the path taken by the N_B moles of gas during this process, we know that entropy is a state function, and therefore, we can construct a hypothetical reversible path between the initial and final states to calculate the entropy change undergone by the N_B moles of gas.

For a simple system following a reversible path, we know that:

$$d\underline{U} = \delta Q + \delta W = Td\underline{S} - Pd\underline{V}$$
$$\Rightarrow d\underline{S} = \frac{d\underline{U}}{T} + \frac{P}{T}d\underline{V} \tag{3}$$

Because the working fluid is an ideal gas, we can simplify Eq. (3), including integrating it from the initial state i to the final state f. This yields:

$$d\underline{S} = \frac{NC_V dT}{T} + \frac{NR}{\underline{V}}d\underline{V} = N\left(\frac{C_V dT}{T} + \frac{R}{\underline{V}}d\underline{V} \right)$$
$$\Rightarrow \int_i^f d\underline{S} = \Delta \underline{S} = \int_i^f N\left(C_V \frac{dT}{T} + R\frac{d\underline{V}}{\underline{V}} \right) = N\left(C_V \ln \frac{T_f}{T_i} + R \ln \frac{\underline{V}_f}{\underline{V}_i} \right)$$
$$= N\left(C_V \ln \frac{T_f}{T_i} + R \ln \frac{T_f P_i}{T_i P_f} \right) \tag{4}$$
$$\Rightarrow \Delta \underline{S} = N\left((C_V + R) \ln \frac{T_f}{T_i} + R \ln \frac{P_i}{P_f} \right)$$
$$\Rightarrow \Delta \underline{S} = N\left(C_p \ln \frac{T_f}{T_i} + R \ln \frac{P_i}{P_f} \right)$$

For system B, the last expression in Eq. (4) can be written as follows:

$$\Delta \underline{S}_B = N_f \left(C_p \ln \frac{T_{f,B}}{T_{in}} + R \ln \frac{P_{in}}{P_{f,B}} \right) \tag{5}$$

Substituting values of $T_{f,B}$, T_{in}, $P_{f,B}$, and P_{in} in Eq. (5) yields:

$$\Delta \underline{S}_B = 434.54 \left((20.9) \ln \frac{415.19}{300} + 8.314 \ln \frac{300}{10} \right) = 15238.92 \text{ J/K} \qquad (6)$$

The last contribution that we need to evaluate is the entropy change of the moles of helium that never exited the cylinder. Here, we assume that the helium cylinder behaves like an infinite reservoir, such that even after removing N_B moles of gas from the cylinder, the temperature and pressure of the cylinder do not change appreciably. As a result, the entropy change of the cylinder is negligible compared to that of system B, that is:

$$\Delta \underline{S}_{cylinder} = 0 \qquad (7)$$

Combining Eqs. (1), (2), (6), and (7), we obtain:

$$\Delta \underline{S}_{universe} = \Delta \underline{S}_A + \Delta \underline{S}_B + \Delta \underline{S}_{cylinder} = 0 + 15238.92 + 0$$
$$= 15238.92 \text{ J/K} \qquad (8)$$

As expected, $\Delta \underline{S}_{universe} > 0$ because the overall process is irreversible. Indeed, the overall process is irreversible because it involves the rapid expansion followed by compression of gas as it transfers from the gas reservoir to the cylinder. There is no way to return the system *and* the environment back to their original state.

Part (c)

To solve this part of the problem, let us first understand why the system is considered to be unmixed. A solution is considered well mixed when the pressure, temperature, and composition of the solution are uniform throughout. Because the vessel consists of two identical gas layers (see the figure below), the only reason that it is considered unmixed is because the two helium layers in the vessel are at different temperatures. Therefore, the process of mixing can be envisioned as one that would ultimately result in the vessel having a uniform pressure and temperature. Furthermore, we seek to carry out this process reversibly to extract useful work out of it. Therefore, let us run a Carnot engine such that the hot helium layer A acts as the hot source and the cold helium layer B acts as the cold sink. At any point during the process, the Carnot engine absorbs heat equal to δQ_H from layer A, rejects heat equal to δQ_C to layer B, and converts the rest to work, δW. To calculate δQ_H, let us carry out a First Law of Thermodynamics analysis of layer A. Specifically:

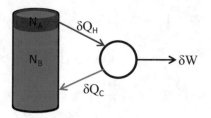

$$dE_A = dU_A = N_A C_V dT_A = \delta Q_A + \delta W_A = -|\delta Q_H| - P_A \underline{V}_A$$
$$\Rightarrow |\delta Q_H| = -N_A C_V dT_A - \frac{N_A R T_A}{\underline{V}_A} d\underline{V}_A \tag{9}$$

Similarly, a First Law of Thermodynamics analysis of layer B yields:

$$dE_B = dU_B = N_B C_V dT_B = \delta Q_B + \delta W_B = |\delta Q_C| - P_B \underline{V}_B$$
$$\Rightarrow |\delta Q_C| = N_B C_V dT_B + \frac{N_B R T_B}{\underline{V}_B} d\underline{V}_B \tag{10}$$

Based on the efficiency of the Carnot engine, we know that:

$$\left| \frac{\delta Q_C}{\delta Q_H} \right| = \frac{T_C}{T_H} \tag{11}$$

Here, we recognize that T_C is equal to T_B and T_H is equal to T_A. Substituting Eqs. (9) and (10) in Eq. (11) yields:

$$\frac{N_B C_V dT_B + \dfrac{N R T_B}{\underline{V}_B} d\underline{V}_B}{-N_A C_V dT_A - \dfrac{N_A R T_A}{\underline{V}_A} d\underline{V}_A} = \frac{T_B}{T_A}$$

$$\Rightarrow \frac{N_B C_V \dfrac{dT_B}{T_B} + N_B R \dfrac{d\underline{V}_B}{\underline{V}_B}}{-N_A C_V \dfrac{dT_A}{T_A} + N_A R \dfrac{d\underline{V}_A}{\underline{V}_A}} = 1 \tag{12}$$

$$\Rightarrow N_B C_V \frac{dT_B}{T_B} + N_B R \frac{d\underline{V}_B}{\underline{V}_B} = -N_A C_V \frac{dT_A}{T_A} + N_A R \frac{d\underline{V}_A}{\underline{V}_A}$$

As stated earlier, the gas layers are considered mixed when the temperature and pressure of both gas layers are equal. Let us denote this temperature and pressure by T and P, respectively. Integrating Eq. (12) between $\left(T_{B,f}, T_{A,f}, P_{B,f}, P_{A,f}, \underline{V}_{B,f}, \underline{V}_{A,f} \right)$ and $\left(T, T, P, P, \underline{V}_{B,m}, \underline{V}_{A,m} \right)$ yields:

$$N_B\left(C_V \ln \frac{T_{B,f}}{T} + R \ln \frac{V_{B,f}}{\underline{V}_{B,m}}\right) = -N_A\left(C_V \ln \frac{T_{A,f}}{T} + R \ln \frac{V_{A,f}}{\underline{V}_{A,m}}\right)$$

$$\Rightarrow N_B\left(C_V \ln \frac{T_{B,f}}{T} + R \ln \left(\frac{T_{B,f}}{T}\frac{P}{P_{B,f}}\right)\right) = -N_A\left(C_V \ln \frac{T_{A,f}}{T} + R \ln \left(\frac{T_{A,f}}{T}\frac{P}{P_{A,f}}\right)\right)$$

$$\Rightarrow N_B\left(C_V \ln \frac{T_{B,f}}{T} + R \ln \frac{T_{B,f}}{T} + R \ln \frac{P}{P_{B,f}}\right)$$

$$= -N_A\left(C_V \ln \frac{T_{A,f}}{T} + R \ln \frac{T_{A,f}}{T} + N_A R \ln \frac{P}{P_{A,f}}\right)$$

$$\Rightarrow N_B\left((C_V + R) \ln T_{B,f} - (C_V + R) \ln T + R \ln P - R \ln P_{B,f}\right)$$

$$= -N_A\left((C_V + R) \ln T_{A,f} - (C_V + R) \ln T + R \ln P - R \ln P_{A,f}\right)$$

$$\Rightarrow C_P\left(N_B \ln T_{B,f} + N_A \ln T_{A,f}\right) - R\left(N_B \ln P_{B,f} + N_A \ln P_{A,f}\right)$$

$$= (N_A + N_B)(C_P \ln T - R \ln P)$$

$$\Rightarrow C_P\left(N_B \ln T_{B,f} + N_A \ln T_{A,f}\right) - R\left(N_B \ln P_{B,f} + N_A \ln P_{A,f}\right)$$

$$= (N_A + N_B)\left(C_P \ln T - R \ln \left(\frac{(N_A + N_B)RT}{(\underline{V}_{A,m} + \underline{V}_{B,m})}\right)\right)$$

$$\Rightarrow C_P\left(N_B \ln T_{B,f} + N_A \ln T_{A,f}\right) - R\left(N_B \ln P_{B,f} + N_A \ln P_{A,f}\right)$$

$$= (N_A + N_B)R \ln \left(\frac{(N_A + N_B)R}{(\underline{V}_{A,m} + \underline{V}_{B,m})}\right) = (N_A + N_B)((C_P - R) \ln T)$$

$$\Rightarrow T = T_{B,f}^{\alpha_B \frac{C_P}{C_V}} T_{A,f}^{\alpha_A \frac{C_P}{C_V}} P_{B,f}^{-\alpha_B \frac{R}{C_V}} P_{A,f}^{-\alpha_A \frac{R}{C_V}}\left(\frac{(N_A + N_B)R}{(\underline{V}_{A,m} + \underline{V}_{B,m})}\right)^{\frac{R}{C_V}}$$

$$(13)$$

where $\alpha_i = N_i/(N_A + N_B)$ for $i=$ A or B. Note that $\underline{V}_{A,m}$ and $\underline{V}_{B,m}$ represent the volumes occupied by layers A and B at the end of the mixing process, such that $\underline{V}_A + \underline{V}_B = 2 \text{ m}^3$. Substituting the values of the known variables yields:

$$T = (415.19)^{\left(\frac{434.54}{80.19+434.54}\right)\left(\frac{20.9}{20.9-8.314}\right)} (749.76)^{\left(\frac{80.19}{80.19+434.54}\right)\left(\frac{20.9}{20.9-8.314}\right)} (10 \times 10^5)^{-\left(\frac{80.19}{80.19+434.54}\right)\left(\frac{8.314}{20.9-8.314}\right)}$$

$$(10 \times 10^5)^{-\left(\frac{434.54}{80.19+434.54}\right)\left(\frac{8.314}{20.9-8.314}\right)} \left(\frac{(80.19 + 434.54)8.314}{2}\right)^{\left(\frac{8.314}{20.9-8.314}\right)}$$

$$= 447.40 \text{ K}$$

$$(14)$$

Using the ideal gas law, we can calculate the final pressure:

$$P = \frac{(N_A + N_B)RT}{\underline{V}_A + \underline{V}_B} = \frac{(80.19 + 434.54) \times 8.314 \times 447.40}{2} = 957316.37 \text{ Pa}$$

$$= 9.57 \text{ bar}$$

$$(15)$$

Finally, we can calculate the maximum available work using a First Law of Thermodynamics analysis of the Carnot engine. Specifically:

$$W = \int \delta W = \int |\delta Q_H| - \int |\delta Q_C| \tag{16}$$

Substituting the expressions for δQ_H and δQ_C from Eqs. (9) and (10), respectively, in Eq. (16) yields:

$$W = \int (-N_A C_V dT_A - P_A d\underline{V}_A) - \int (N_B C_V dT_B - P_B d\underline{V}_B) \tag{17}$$

To maintain mechanical equilibrium at the interface separating the two helium layers, it is necessary that the pressures on either side of the interface be equal, i.e., $P_A = P_B$. In addition, we know that the volume of the vessel is fixed. Therefore, $d\underline{V}_A + d\underline{V}_B = 0 \Rightarrow d\underline{V}_A = -d\underline{V}_B$. It may seem odd to maintain the identity of $\underline{V}_A$ and $\underline{V}_B$ in this mixing process as one could expect that, upon mixing, both gas layers would occupy the entire vessel and not some part of the vessel. However, we can assume that we carry out the process of equilibrating the temperature and pressure of the vessel by modeling the interface between the two gas layers as impermeable, and at the end of the process, we make the interface permeable to allow "mixing" between the two identical gases. However, once the temperatures and pressures are equilibrated, we realize that the states of the two gas layers are identical to each other. Therefore, even after we remove the interface (or make it permeable), it is not going to change the state of the system. Consequently, we cannot obtain any more useful work. Therefore, the thought process presented here to obtain useful work indeed gives the maximum possible work that can be done by the two gas layers in the isolated vessel. Note that if the two gas layers consisted of different gases (say helium and argon instead of helium and helium), then, due to the presence of the impermeable interface, the two gas layers would be at a different composition (but at the same temperature and pressure). Therefore, upon making the interface permeable, one would need to carry out an additional step of mixing to obtain the final fully mixed state (uniform temperature, pressure, and composition).

Returning to calculating the work, we substitute the relations between P_A, P_B, $\underline{V}_A$, and $\underline{V}_B$ in Eq. (17) to obtain:

$$W = \int (-N_A C_V dT_A - P_A dV_A) - \int (N_B C_V dT_B - P_A d\underline{V}_A)$$

$$\Rightarrow \int (-N_A C_V dT_A - N_B C_V dT_B) = -N_A C_V (T - T_{Af}) - N_B C_V (T - T_{Bf}) \tag{18}$$

Substituting values for the different variables in Eq. (18) yields:

$$W = -(20.9 - 8.314) \times (80.19 \times (447.40 - 749.76)) + 434.54$$
$$\times (447.40 - 415.19)$$
$$= 129002.91 \text{ J}$$
$$= 129.00 \text{ kJ}$$

(19)

Other Possible Solution Strategies

Alternate Way of Solving for the Final Temperature in Part (c)

From the Second Law of Thermodynamics, we know that for any reversible process, the entropy change of the universe is 0. Our system (the vessel) is an isolated system. Therefore, the mixing process is going to be reversible only when the entropy change of the system is equal to 0, that is:

$$\Delta \underline{S}_A + \Delta \underline{S}_B = 0 \tag{20}$$

Like in Part (b), we know the initial and final states of the constituents of the vessel (both helium layers attain a temperature and pressure of T and P, respectively). Therefore, we can make use of Eq. (4) to evaluate the entropy change of systems A and B. Specifically:

$$\Delta \underline{S}_A = N_A \left(C_p \ln \frac{T}{T_{f,A}} + R \ln \frac{P_{f,A}}{P} \right) \tag{21}$$

$$\Delta \underline{S}_B = N_B \left(C_p \ln \frac{T}{T_{f,B}} + R \ln \frac{P_{f,B}}{P} \right) \tag{22}$$

Substituting Eqs. (21) and (22) in Eq. (20) then yields:

$$N_A \left(C_p \ln \frac{T}{T_{f,A}} + R \ln \frac{P_{f,A}}{P} \right) + N_B \left(C_p \ln \frac{T}{T_{f,B}} + R \ln \frac{P_{f,B}}{P} \right) = 0$$
$$\Rightarrow (N_A + N_B)(C_p \ln T - R \ln P) = C_p \left(N_B \ln T_{B,f} + N_A \ln T_{A,f} \right)$$
$$- R \left(N_B \ln P_{B,f} + N_A \ln P_{A,f} \right)$$
$$\Rightarrow (N_A + N_B) \left(C_p \ln T - R \ln \left(\frac{(N_A + N_B)RT}{(\underline{V}_A + \underline{V}_B)} \right) \right) = C_p \left(N_B \ln T_{B,f} + N_A \ln T_{A,f} \right)$$
$$- R \left(N_B \ln P_{B,f} + N_A \ln P_{A,f} \right)$$
$$\Rightarrow (N_A + N_B)((C_p - R) \ln T) = C_p \left(N_B \ln T_{B,f} + N_A \ln T_{A,f} \right)$$
$$- R \left(N_B \ln P_{B,f} + N_A \ln P_{A,f} \right) + (N_A + N_B)R \ln \left(\frac{(N_A + N_B)R}{(\underline{V}_A + \underline{V}_B)} \right)$$
$$\Rightarrow T = T_{B,f}^{\alpha_B \frac{C_p}{C_V}} T_{A,f}^{\alpha_A \frac{C_p}{C_V}} P_{B,f}^{-\alpha_B \frac{R}{C_V}} P_{A,f}^{-\alpha_A \frac{R}{C_V}} \left(\frac{(N_A + N_B)R}{(\underline{V}_A + \underline{V}_B)} \right)^{\frac{R}{C_V}}$$

(23)

Note that, as expected, the expression for T in Eq. (23) is identical to the expression for T in Eq. (13)!

Alternate Way of Solving for the Total Work in Part (c)

Instead of carrying out a First Law of Thermodynamics analysis on the Carnot engine, we can select a system that includes gas A, gas B, and the Carnot engine. This yields:

$$\Delta \underline{E} = W_{\text{on sys}} + Q_{\text{into sys}} \tag{24}$$

However, during the final work extraction process, there is no heat interaction between the composite system and the environment, and therefore, $Q_{\text{into sys}} = 0$. Furthermore, the work done on the system is related to the total work as follows: $W = -W_{\text{on sys}}$. Finally, the total energy of the system can be decomposed into the sum of the energies of the three subsystems, that is:

$$\Delta \underline{E} = \Delta \underline{E}_A + \Delta \underline{E}_B + \Delta \underline{E}_{engine} \tag{25}$$

However, because the engine operates in a cyclic manner, its energy is constant over the process. Further, because the two gaseous subsystems are simple ideal gases, their energies are simply their internal energies, that is:

$$\Delta \underline{E} = \Delta \underline{E}_A + \Delta \underline{E}_B = \Delta \underline{U}_A + \Delta \underline{U}_B = N_A C_V (T - T_{A,f}) + N_B C_V (T - T_{B,f}) \tag{26}$$

Combining Eqs. (24), (25), and (26), we obtain:

$$W = -\Delta \underline{E} = -N_A C_V (T - T_{A,f}) - N_B C_V (T - T_{B,f}) \tag{27}$$

Note that, as expected, the expression for W in Eq. (27) is identical to the expression for W in Eq. (18)!

Problem 5

Problems 5.17 + 5.27 in Tester and Modell

1. Problem 5.17

Show that:

$$\kappa = \frac{C_p}{C_v} = \left[\frac{\partial V}{\partial P}\right]_T \left[\frac{\partial P}{\partial V}\right]_S$$

and check to see if the relation holds for an ideal gas.

2. Problem 5.27

A non-ideal gas of constant heat capacity $C_v = 12.56$ J/mol K undergoes a *reversible adiabatic* expansion. The gas is described by the van der Waals equation of state:

$$(P + a/V^2)(V - b) = RT$$

where $a = 0.1362$ Jm^3/mol^2 and $b = 3.215 \times 10^{-5}$ m^3/mol. Derive an expression for the temperature variation of the gas internal energy, and calculate its value when the gas volume of 400 moles is 0.1 m^3 and its temperature is 294 K.

Solution to Problem 5

Solution to Problem 5.17

Solution Strategy

1. Summarize what we know

From Part I, we already know the definitions of both the constant pressure and the constant volume heat capacities. Specifically:

$$C_P \equiv \left(\frac{\partial H}{\partial T}\right)_P \tag{1}$$

$$C_V \equiv \left(\frac{\partial U}{\partial T}\right)_V \tag{2}$$

Therefore, to solve the problem, we must show that:

$$\frac{C_P}{C_V} = \frac{\left(\frac{\partial H}{\partial T}\right)_P}{\left(\frac{\partial U}{\partial T}\right)_V} = \left(\frac{\partial V}{\partial P}\right)_T \left(\frac{\partial P}{\partial V}\right)_S \tag{3}$$

2. Select Possible Solution Strategies

1. Use the differential forms of the fundamental equations which include heat capacities.
2. Standard calculus strategies:

   (i) Derivative inversion
  (ii) Triple product rule
 (iii) Chain rule expansion
 (iv) Maxwell reciprocity relationships

3. Advanced calculus strategies:

   (i) Jacobian transformations

3. Logically analyze the problem

What we should recognize from Eq. (3) is that the variables that appear in the heat capacity definitions (H, U, T, P, V) are somewhat different than the variables in the desired proof (V, P, T, S). Therefore, we anticipate that it will be important through our calculus transitions to remove the U and H dependencies and to somehow add S terms. The simplest way to accomplish these goals is to replace the standard heat capacity definitions in Eqs. (1) and (2) with their other, well-known, entropy-based definitions. Specifically, as shown in Part I:

$$C_P \equiv \left(\frac{\partial H}{\partial T}\right)_P = T\left(\frac{\partial S}{\partial T}\right)_P \tag{4}$$

$$C_V \equiv \left(\frac{\partial U}{\partial T}\right)_V = T\left(\frac{\partial S}{\partial T}\right)_V \tag{5}$$

For completeness, we will first show that Eqs. (4) and (5) are valid representations. The simplest way to do is to consider the molar enthalpy and molar internal energy fundamental equations, that is:

$$dH = TdS - VdP \tag{6}$$

$$dU = TdS - PdV \tag{7}$$

By taking the derivative with respect to temperature of both sides of Eqs. (6) and (7), with the appropriate variables held constant (P for Eq. (6) and V for Eq. (7)), the results in Eqs. (4) and (5) are readily obtained. The use of the chain rule expansion is also possible:

$$\left(\frac{\partial H}{\partial T}\right)_P = \left(\frac{\partial H}{\partial S}\right)_P \left(\frac{\partial S}{\partial T}\right)_P = T\left(\frac{\partial S}{\partial T}\right)_P \tag{8}$$

$$\left(\frac{\partial U}{\partial T}\right)_V = \left(\frac{\partial U}{\partial S}\right)_V \left(\frac{\partial S}{\partial T}\right)_V = T\left(\frac{\partial S}{\partial T}\right)_V \tag{9}$$

Using Eqs. (4) and (5) in our Problem Statement yields:

$$\frac{T\left(\frac{\partial S}{\partial T}\right)_P}{T\left(\frac{\partial S}{\partial T}\right)_V} = \frac{\left(\frac{\partial S}{\partial T}\right)_P}{\left(\frac{\partial S}{\partial T}\right)_V} = \left(\frac{\partial V}{\partial P}\right)_T \left(\frac{\partial P}{\partial V}\right)_S \tag{10}$$

We must prove Eq. (10) in order to solve the problem. At this point, the problem may still seem challenging, because the variables that are held constant in our derivation (P, V) differ from the variables that are held constant in the proposed solution (T, S). Many of our strategies, including derivative inversion and chain rule expansion, do not change the variables that are held constant. As a result, they will not be useful to us. We could try to use the Maxwell reciprocity relationships, but eventually, this would lead us to a dead end (as would be found by trial and error).

Our remaining strategies would be to use the fundamental equation (unfortunately, leading again to a dead end) or the triple product rule. Interestingly, the triple product rule may turn out to be an effective strategy. Indeed, we know that using the triple product rule on both the top and the bottom of the left-hand side of Eq. (10) will result in partial derivatives where T and S are held constant, similar to the proposed solution that we are trying to prove (the right-hand side of Eq. (10)).

Utilizing the triple product rule on the numerator and the denominator of the left-hand side of Eq. (10), we obtain:

$$T\left(\frac{\partial S}{\partial T}\right)_P = \frac{-T}{\left(\frac{\partial T}{\partial P}\right)_S \left(\frac{\partial P}{\partial S}\right)_T} \tag{11}$$

and

$$T\left(\frac{\partial S}{\partial T}\right)_V = \frac{-T}{\left(\frac{\partial T}{\partial V}\right)_S \left(\frac{\partial V}{\partial S}\right)_T} \tag{12}$$

Dividing Eq. (11) by Eq. (12), including cancelling the two Ts and rearranging, we obtain:

$$\frac{\left(\frac{\partial S}{\partial T}\right)_P}{\left(\frac{\partial S}{\partial T}\right)_V} = \frac{-\left(\frac{\partial T}{\partial V}\right)_S \left(\frac{\partial V}{\partial S}\right)_T}{-\left(\frac{\partial T}{\partial P}\right)_S \left(\frac{\partial P}{\partial S}\right)_T} = \left(\frac{\partial P}{\partial V}\right)_S \left(\frac{\partial V}{\partial P}\right)_T \tag{13}$$

Equation (13) shows that $\kappa = \frac{C_P}{C_V} = \left(\frac{\partial V}{\partial P}\right)_T \left(\frac{\partial P}{\partial V}\right)_S$ is correct!

4. Validate the proof for an ideal gas

Recall that, for an ideal gas, the equation of state is given by:

$$PV = RT \tag{14}$$

and that U is only a function of T, where:

$$dU = C_V dT \tag{15}$$

Accordingly, the molar internal energy form of the fundamental equation can be written as follows:

$$C_V dT = TdS - PdV \tag{16}$$

To obtain $\left(\frac{\partial V}{\partial P}\right)_T$, we use the ideal gas law. This yields:

$$\left(\frac{\partial V}{\partial P}\right)_T = \left(\frac{\partial \frac{RT}{P}}{\partial P}\right)_T = -\frac{RT}{P^2} = \frac{-V}{P} \tag{17}$$

To obtain $\left(\frac{\partial P}{\partial V}\right)_S$, we use Eq. (16) at constant entropy, which yields:

$$C_V dT = -PdV \tag{18}$$

From the ideal gas law:

$$d(PV) = RdT \tag{19}$$

$$VdP + PdV = RdT \tag{20}$$

$$dT = \frac{V}{R}dP + \frac{P}{R}dV \tag{21}$$

Substituting Eq. (21) in Eq. (18) and simplifying yields:

$$C_V \frac{V}{R} dP + C_V \frac{P}{R} dV = -PdV \tag{22}$$

$$C_V \frac{V}{R} dP = -\left(P + C_V \frac{P}{R}\right) dV \tag{23}$$

$$C_V V dP = -(R + C_V) PdV \tag{24}$$

Recall that $C_V + R = C_P$, and therefore, Eq. (24) can be expressed as follows:

$$C_V V dP = -C_P PdV \tag{25}$$

Rearranging Eq. (25) yields Eq. (26), which upon integration from an initial state i to a final state f yields Eq. (27):

$$\frac{dP}{P} = -\frac{C_P}{C_V} \frac{dV}{V} \tag{26}$$

$$ln\left(\frac{P_f}{P_i}\right) = \frac{C_P}{C_V} ln\left(\frac{V_i}{V_f}\right) \tag{27}$$

Recall that $\kappa = \frac{C_P}{C_V}$, and therefore, Eq. (27) can be simplified as follows:

$$\left(\frac{P_f}{P_i}\right) = \left(\frac{V_i}{V_f}\right)^{\kappa} \tag{28}$$

Therefore, at constant entropy, Eq. (28) yields:

$$P_f V_f^{\kappa} = P_i V_i^{\kappa} = const. \tag{29}$$

Note: Equation (29) is the same expression as that for the reversible adiabatic expansion, or compression, of an ideal gas $\Rightarrow$ constant entropy!

Accordingly, for an ideal gas:

$$\left(\frac{\partial P}{\partial V}\right)_S = \left(\frac{\partial \alpha / V^{\kappa}}{\partial V}\right)_S = \frac{-\kappa \alpha}{V^{\kappa+1}}, \text{ where } \alpha = PV^{\kappa}(\text{a constant}) \tag{30}$$

Combining Eqs. (17) and (30), we obtain:

$$\left(\frac{\partial P}{\partial V}\right)_S \left(\frac{\partial V}{\partial P}\right)_T = \left(\frac{-\kappa\alpha}{V^{\kappa+1}}\right)\left(\frac{-V}{P}\right) = \frac{\kappa\alpha}{PV^\kappa} = \frac{\kappa\alpha}{\alpha} = \kappa \qquad (31)$$

Equation (31) shows that, as expected, the general relationship also holds for an ideal gas!

5. **Alternative solution using Jacobian transformations**

We begin with Eq. (3), which we repeat below for completeness:

$$\frac{C_P}{C_V} = \left(\frac{\partial H}{\partial T}\right)_P \Big/ \left(\frac{\partial U}{\partial T}\right)_V \qquad (32)$$

We then express the two partial derivatives in Eq. (32) using the appropriate Jacobians. This yields:

$$\frac{C_P}{C_V} = \left(\frac{\partial H}{\partial T}\right)_P \Big/ \left(\frac{\partial U}{\partial T}\right)_V = \frac{\frac{\partial(H,P)}{\partial(T,P)}}{\frac{\partial(U,V)}{\partial(T,V)}} \qquad (33)$$

Next, we carry out a chain rule expansion on the top and bottom derivatives in Eq. (33) such that we end up with the derivative for entropy, that is:

$$\frac{C_P}{C_V} = \frac{\frac{\partial(H,P)}{\partial(S,P)}\frac{\partial(S,P)}{\partial(T,P)}}{\frac{\partial(U,V)}{\partial(S,V)}\frac{\partial(S,V)}{\partial(T,V)}} \qquad (34)$$

$$\frac{C_P}{C_V} = \frac{\left(\frac{\partial H}{\partial S}\right)_P \frac{\partial(S,P)}{\partial(T,P)}}{\left(\frac{\partial U}{\partial S}\right)_V \frac{\partial(S,V)}{\partial(T,V)}} \qquad (35)$$

$$\frac{C_P}{C_V} = \frac{T\,\frac{\partial(S,P)}{\partial(T,P)}}{T\,\frac{\partial(S,V)}{\partial(T,V)}} \qquad (36)$$

$$\frac{C_P}{C_V} = \frac{\partial(S,P)}{\partial(T,P)}\frac{\partial(T,V)}{\partial(S,V)} \qquad (37)$$

Finally, we can rearrange Eq. (37) to obtain the desired result, that is:

$$\frac{C_P}{C_V} = \frac{\partial(T,V)}{\partial(T,P)}\frac{\partial(S,P)}{\partial(S,V)} \tag{38}$$

$$\frac{C_P}{C_V} = \left(\frac{\partial V}{\partial P}\right)_T \left(\frac{\partial P}{\partial V}\right)_S \tag{39}$$

Solution to Problem 5.27

We first rephrase the problem in a clearer manner. Basically, we are asked to derive an expression for $\left(\partial U/\partial T\right)_S$ and then to evaluate it at the conditions given below:

- $V = 0.1\,\text{m}^3/400\,\text{mol} = 0.25 \times 10^{-3}\,\text{m}^3/\text{mol}$
- $T = 294\,\text{K}$
- $a = 0.1362\,\text{Jm}^3/\text{mol}$
- $b = 3.25 \times 10^{-5}\,\text{m}^3/\text{mol}$
- $C_V = 12.56\,\text{J/molK}$
- $R = 8.314\,\text{J/molK}$

We are told that the volumetric behavior of the gas is described by the van der Waals equation of state (EOS), given by:

$$\left(P + \frac{a}{V^2}\right)(V - b) = RT$$

We are also told that the process under consideration involves the reversible expansion of a non-ideal gas.

Solution Strategy

1. Summarize what we know

In considering the problem, we should recognize several key points:

(a) The system is closed, and the process is adiabatic and reversible (i.e., $dS = 0$).
(b) The heat capacity at constant volume is given, although at the moment there is no apparent use for it.
(c) The cubic van der Waals EOS is given, which has explicit solutions for $T(P,V)$ and $P(T,V)$, but not for $V(T,P)$.

2. Use the internal energy fundamental equation

While the initial solution strategy is not entirely clear based on the Problem Statement, beginning with fundamental equations is always a good option. Because the internal energy appears in the partial derivative that we need to calculate, we begin with the molar internal energy fundamental equation, given by:

$$dU = TdS - PdV \tag{1}$$

Because the entropy term in Eq. (1) is equal to zero (recall that the process is adiabatic and reversible), it follows that:

$$dU = -PdV \tag{2}$$

Because our desired solution is $(\partial U/\partial T)_S$, we can take the partial derivative of Eq. (2) with respect to temperature, at constant entropy, to obtain:

$$\left(\frac{\partial U}{\partial T}\right)_S = -P\left(\frac{\partial V}{\partial T}\right)_S \tag{3}$$

The derivative on the right-hand side of Eq. (3) cannot yet be evaluated using the given information because the van der Waals EOS is pressure explicit, that is, of the form $P(T,V)$, with no information given about S. Again, we need to use our calculus techniques to find a relationship that we can evaluate with the given information.

As usual, it is not immediately obvious which strategy will lead to a useful solution. If we consider the different strategies, we find that:

- Using derivative inversion would not get us very far (try it out).
- Using the triple product rule could be useful (our technique of choice; see below).
- Using a chain rule expansion would not remove the difficult constant entropy constraint (try it out).
- Using Maxwell's reciprocity theorem would only move the entropic term inside the derivative, and would not really help (try it out and you will find that the variable kept constant will always end up inside).

Therefore, the best option among the standard techniques appears to be using the triple product rule:

$$\left(\frac{\partial U}{\partial T}\right)_S = -P\left(\frac{\partial V}{\partial T}\right)_S = P\frac{\left(\frac{\partial S}{\partial T}\right)_V}{\left(\frac{\partial S}{\partial V}\right)_T} \tag{4}$$

where Eq. (3) was used. The partial derivative in the numerator of Eq. (4), as shown in the solution of the previous problem, is equal to C_V/T. The partial derivative in the denominator of Eq. (4), again, cannot be immediately evaluated. However, using Maxwell's reciprocity relationships will be useful to replace the entropy term in the partial derivative. Specifically:

$$\left(\frac{\partial S}{\partial V}\right)_T = \left(\frac{\partial \left(\frac{\partial A}{\partial T}\right)_V}{dV}\right)_T = \left(\frac{\partial \left(\frac{\partial A}{\partial V}\right)_T}{dT}\right)_V = \left(\frac{\partial P}{\partial T}\right)_V \tag{5}$$

Equation (5) is an expression that we can readily evaluate based on the pressure-explicit EOS given in the Problem Statement. Combining all our results, we obtain:

$$\left(\frac{\partial V}{\partial T}\right)_S = \frac{-C_V}{T\left(\frac{\partial P}{\partial T}\right)_V} = \frac{-C_V}{T}\left(\frac{\partial T}{\partial P}\right)_V \tag{6}$$

Using Eq. (6) in Eq. (3), we obtain:

$$\left(\frac{\partial U}{\partial T}\right)_S = P\frac{C_V}{T}\left(\frac{\partial T}{\partial P}\right)_V \tag{7}$$

Using the van der Waals EOS, we can expand both the pressure and the partial derivative with respect to P terms in Eq. (7) as follows:

$$\left(\frac{\partial U}{\partial T}\right)_S = \left(\frac{RT}{V-b} - \frac{a}{V^2}\right)\frac{C_V}{T}\frac{(V-b)}{R} = C_V\left(1 - \frac{a(V-b)}{RV^2T}\right) \tag{8}$$

Using the variable values given in the Problem Statement in Eq. (8) yields:

$$\left(\frac{\partial U}{\partial T}\right)_S = 12.56\,\text{J/molK}\left(1 - \frac{0.1362\,\text{Jm}^3/\text{mol}\left(0.25 \times 10^{-3}\text{m}^3/\text{mol} - 3.25 \times 10^{-5}\text{m}^3/\text{mol}\right)}{(8.314\,\text{J/molK})\left(0.25 \times 10^{-3}\text{m}^3/\text{mol}\right)^2(294\,\text{K})}\right) \tag{9}$$

This leads us to our final result of:

$$\boxed{\left(\frac{\partial U}{\partial T}\right)_S = 10.12\,\text{J/molK}}$$

3. Alternative solution using Jacobian transformations

An alternative method of getting from Eq. (3) to Eq. (7) above is to utilize Jacobian transformations. Specifically, we again begin with Eq. (3), which we repeat below for completeness:

$$\left(\frac{\partial U}{\partial T}\right)_S = -P\left(\frac{\partial V}{\partial T}\right)_S \tag{10}$$

Then, we use a Maxwell relationship to obtain:

$$\left(\frac{\partial U}{\partial T}\right)_S = -P\left(\frac{\partial V}{\partial T}\right)_S = P\left(\frac{\partial S}{\partial P}\right)_V \tag{11}$$

Expanding Eq. (11) using Jacobians yields:

$$\left(\frac{\partial U}{\partial T}\right)_S = P\frac{\partial(S,V)}{\partial(P,V)} \tag{12}$$

Subsequently, carrying out a chain rule expansion, including adding in $d(T,V)$ to obtain one derivative in terms of T, V, and P, yields:

$$\left(\frac{\partial U}{\partial T}\right)_S = P\frac{\partial(S,V)}{\partial(T,V)}\frac{\partial(T,V)}{\partial(P,V)} \tag{13}$$

Carrying out another chain rule expansion, including adding in $d(U,V)$ to obtain an expression for $1/T$, yields:

$$\left(\frac{\partial U}{\partial T}\right)_S = P\frac{\partial(S,V)}{\partial(U,V)}\frac{\partial(U,V)}{\partial(T,V)}\frac{\partial(T,V)}{\partial(P,V)} \tag{14}$$

$$\left(\frac{\partial U}{\partial T}\right)_S = P\left(\frac{\partial S}{\partial U}\right)_V\left(\frac{\partial U}{\partial T}\right)_V\left(\frac{\partial T}{\partial P}\right)_V \tag{15}$$

$$\left(\frac{\partial U}{\partial T}\right)_S = P\frac{1}{T}C_V\left(\frac{\partial T}{\partial P}\right)_V \tag{16}$$

Equation (16) is identical to Eq. (7)!

Problem 6

Problem 5.28 in Tester and Modell

The basic thermodynamic relationships for an axially stressed bar can be written as follows:

$$dQ_{rev} = Td\underline{S}, \quad dW_{rev} = -\tau d\underline{\varepsilon}, \quad \underline{\varepsilon} = N\varepsilon$$

where τ is the stress and ε is the strain. Derive the fundamental equation for a one-component bar and show that:

$$\left(\frac{\partial \mu}{\partial T}\right)_{\tau,N} = -\left(\frac{\partial \underline{S}}{\partial N}\right)_{T,\tau}$$

Solution to Problem 6

Solution Strategy

Obtaining the Fundamental Equation

Let us assume that the axially stressed bar forms a closed, simple system. Carrying out a First Law of Thermodynamics analysis of the bar yields:

$$dE = d\underline{U} = \delta Q_{rev} + \delta W_{rev} = Td\underline{S} - \tau d\underline{\varepsilon}$$
$$\Rightarrow d\underline{U} = Td\underline{S} - \tau d\underline{\varepsilon} \tag{1}$$

From Postulate I, we know that the state of a closed, simple system can be characterized by two independently variable properties in addition to the masses of the components comprising the system. From Eq. (1), we see that, in this system, the two independently variable properties are the entropy of the system, $\underline{S}$, and the strain, $\underline{\varepsilon}$. If the mass of the bar is denoted by N, we can conclude that:

$$\underline{U} = \underline{U}(\underline{S}, \underline{\varepsilon}, N)$$
$$\Rightarrow d\underline{U} = \left(\frac{\partial \underline{U}}{\partial \underline{S}}\right)_{\underline{\varepsilon},N} d\underline{S} + \left(\frac{\partial \underline{U}}{\partial \underline{\varepsilon}}\right)_{N,\underline{S}} d\underline{\varepsilon} + \left(\frac{\partial \underline{U}}{\partial N}\right)_{\underline{S},\underline{\varepsilon}} dN \tag{2}$$

where

$$\left(\frac{\partial \underline{U}}{\partial \underline{S}}\right)_{\underline{\varepsilon},N} = T, \quad \left(\frac{\partial \underline{U}}{\partial \underline{\varepsilon}}\right)_{N,\underline{S}} = -\tau, \quad \text{and} \quad \left(\frac{\partial \underline{U}}{\partial N}\right)_{\underline{S},\underline{\varepsilon}} = \mu$$

Therefore, the differential form of the internal energy fundamental equation can be written as follows:

$$dU = TdS - \tau d\varepsilon + \mu dN \tag{3}$$

Euler integrating Eq. (3), we obtain:

$$U = TS - \tau \varepsilon + \mu N \tag{4}$$

Partial Derivative of the Chemical Potential

In the second half of the problem, we want to establish the following relationship:

$$\left(\frac{\partial \mu}{\partial T}\right)_{\tau,N} = -\left(\frac{\partial S}{\partial N}\right)_{T,\tau} \tag{4a}$$

Because the relationship in Eq. (4a) contains many variables (μ, T, N, S, τ), let us first try to create the two partial derivatives separately and then device a way to prove their equality. To create the partial derivative on the left-hand side of Eq. (4a), we begin with Eq. (3) and manipulate it such that we can obtain an equation for $d\mu$. To do that, let us separate the intensive variables from the extensive ones as follows:

$$d(NU) = Td(NS) - \tau d(N\varepsilon) + \mu dN$$
$$\Rightarrow NdU + UdN = NTdS + TSdN - N\tau d\varepsilon - \tau\varepsilon dN + \mu dN \tag{5}$$
$$\Rightarrow N(dU - TdS + \tau d\varepsilon) + (U - TS + \tau\varepsilon - \mu)dN = 0$$

For any simple system, we can choose the size of the system and vary it arbitrarily while maintaining all other intensive properties at a fixed value. Consequently, for Eq. (5) to be true, it is necessary that the coefficients of N and dN be both zero, that is:

$$U = TS - \tau\varepsilon + \mu \tag{6}$$
$$dU = TdS - \tau d\varepsilon \tag{7}$$

Note that Eq. (6) is simply the integrated version of the fundamental equation, which is already given in Eq. (4). Equation (7) shows that U is a function of the two variables S and ε. Equivalently, we can also choose U to be a function of T and $-\tau$ (the conjugate variables of S and ε). Similarly, from the Corollary to Postulate I, we can conclude that S is a function of T and $-\tau$, that is:

$$S = S(T, \tau) \tag{8}$$

We will use Eq. (8) later in the proof.

Returning back to the problem of finding an expression for $d\mu$, we next proceed to derive an equation which is analogous to the Gibbs-Duhem equation, by taking the total differential of Eq. (6). This yields:

$$
\begin{aligned}
dU &= TdS + SdT - \tau d\varepsilon - \varepsilon d\tau + d\mu \\
&\Rightarrow dU - TdS + \tau d\varepsilon = SdT - \varepsilon d\tau + d\mu \\
&\Rightarrow 0 = SdT - \varepsilon d\tau + d\mu \\
&\Rightarrow d\mu = \varepsilon d\tau - SdT
\end{aligned}
\tag{9}
$$

To derive Eq. (9), we made use of Eq. (7). We can obtain the left-hand side of Eq. (4a) by differentiating Eq. (9) with respect to T at constant τ and N. This yields:

$$
\left(\frac{\partial \mu}{\partial T}\right)_{\tau,N} = 0 - S\left(\frac{\partial T}{\partial T}\right)_{\tau,N} = -S
\tag{10}
$$

Because Eq. (9) already provides a relationship between the left-hand side of Eq. (4a) in terms of the entropy of the system, next, let us consider the right-hand side of Eq. (4a) and try to simplify it. Specifically:

$$
\begin{aligned}
-\left(\frac{\partial \underline{\underline{S}}}{\partial N}\right)_{T,\tau} &= -\left(\frac{\partial (NS)}{\partial N}\right)_{T,\tau} \\
&= -S\left(\frac{\partial N}{\partial N}\right)_{T,\tau} - N\left(\frac{\partial S}{\partial N}\right)_{T,\tau} \\
&= -S - N\left(\frac{\partial S}{\partial N}\right)_{T,\tau}
\end{aligned}
\tag{11}
$$

From Eq. (8), we know that S is a function of T and τ. Accordingly, $(\partial S/\partial N)_{T,\tau} = 0$. Substituting this result in Eq. (11) yields:

$$
-\left(\frac{\partial \underline{\underline{S}}}{\partial N}\right)_{T,\tau} = -S
\tag{12}
$$

Combining Eqs. (12) and (10) yields:

$$
\begin{aligned}
\left(\frac{\partial \mu}{\partial T}\right)_{\tau,N} &= -S = -\left(\frac{\partial \underline{\underline{S}}}{\partial N}\right)_{T,\tau} \\
&\Rightarrow \left(\frac{\partial \mu}{\partial T}\right)_{\tau,N} = -\left(\frac{\partial \underline{\underline{S}}}{\partial N}\right)_{T,\tau}
\end{aligned}
\tag{13}
$$

Equation (13) is precisely what we set out to prove!

Other Possible Solution Strategies

Alternate proof for $\left(\frac{\partial \mu}{\partial T}\right)_{\tau,N} = -\left(\frac{\partial S}{\partial N}\right)_{T,\tau}$

Method 1

We begin with the internal energy fundamental equation given in Eq. (3), which we repeat below for completeness:

$$dU = TdS - \tau d\varepsilon + \mu dN \tag{14}$$

By inspection, we recognize that direct differentiation of Eq. (14) will yields the desired derivative that contains S. Therefore, we differentiate Eq. (14) with respect to N, at constant T and τ, which will give us μ in terms of the other derivatives, including $(\partial S / \partial N)_{T,\tau}$. Specifically:

$$
\begin{aligned}
&\left(\frac{\partial U}{\partial N}\right)_{T,\tau} = T\left(\frac{\partial S}{\partial N}\right)_{T,\tau} - \tau\left(\frac{\partial \varepsilon}{\partial N}\right)_{T,\tau} + \mu\left(\frac{\partial N}{\partial N}\right)_{T,\tau} \\
&\Rightarrow \left(\frac{\partial U}{\partial N}\right)_{T,\tau} = T\left(\frac{\partial S}{\partial N}\right)_{T,\tau} - \tau\left(\frac{\partial \varepsilon}{\partial N}\right)_{T,\tau} + \mu \\
&\Rightarrow \mu = \left(\frac{\partial U}{\partial N}\right)_{T,\tau} - T\left(\frac{\partial S}{\partial N}\right)_{T,\tau} + \tau\left(\frac{\partial \varepsilon}{\partial N}\right)_{T,\tau}
\end{aligned}
\tag{15}
$$

Next, we need to obtain the partial derivative of μ with respect to T, at constant N and τ. Therefore, we differentiate the last expression in Eq. (15) with respect to T, at constant N and τ. This yields:

$$
\begin{aligned}
&\left(\frac{\partial \mu}{\partial T}\right)_{N,\tau} = \left[\frac{\partial}{\partial T}\left(\frac{\partial U}{\partial N}\right)_{T,\tau}\right]_{N,\tau} - \left[\frac{\partial}{\partial T}\left(T\left(\frac{\partial S}{\partial N}\right)_{T,\tau}\right)\right]_{N,\tau} + \left[\frac{\partial}{\partial T}\left(\tau\left(\frac{\partial \varepsilon}{\partial N}\right)_{T,\tau}\right)\right]_{N,\tau} \\
&\Rightarrow \left(\frac{\partial \mu}{\partial T}\right)_{N,\tau} = \left[\frac{\partial}{\partial T}\left(\frac{\partial U}{\partial N}\right)_{T,\tau}\right]_{N,\tau} - \left(\frac{\partial S}{\partial N}\right)_{T,\tau} - T\left[\frac{\partial}{\partial T}\left(\frac{\partial S}{\partial N}\right)_{T,\tau}\right]_{N,\tau} \\
&\quad + \tau\left[\frac{\partial}{\partial T}\left(\frac{\partial \varepsilon}{\partial N}\right)_{T,\tau}\right]_{N,\tau} \\
&\Rightarrow \left(\frac{\partial \mu}{\partial T}\right)_{N,\tau} = -\left(\frac{\partial S}{\partial N}\right)_{T,\tau} + \left[\frac{\partial}{\partial T}\left(\frac{\partial U}{\partial N}\right)_{T,\tau}\right]_{N,\tau} - T\left[\frac{\partial}{\partial T}\left(\frac{\partial S}{\partial N}\right)_{T,\tau}\right]_{N,\tau} \\
&\quad + \tau\left[\frac{\partial}{\partial T}\left(\frac{\partial \varepsilon}{\partial N}\right)_{T,\tau}\right]_{N,\tau}
\end{aligned}
\tag{16}
$$

To complete the proof, we need to show that the sum of the last three terms on the right-hand side of Eq. (16) is 0, that is:

$$\left[\frac{\partial}{\partial T}\left(\frac{\partial U}{\partial N}\right)_{T,\tau}\right]_{N,\tau} - T\left[\frac{\partial}{\partial T}\left(\frac{\partial \underline{S}}{\partial N}\right)_{T,\tau}\right]_{N,\tau} + \tau\left[\frac{\partial}{\partial T}\left(\frac{\partial \underline{\varepsilon}}{\partial N}\right)_{T,\tau}\right]_{N,\tau} = 0 \qquad (17)$$

Assuming that all the physical observables behave smoothly, the order of differentiation of every term in Eq. (17) can be switched. For example, $\left[\frac{\partial}{\partial T}\left(\frac{\partial U}{\partial N}\right)_{T,\tau}\right]_{N,\tau}$ can be expressed as $\left[\frac{\partial}{\partial N}\left(\frac{\partial U}{\partial T}\right)_{N,\tau}\right]_{T,\tau}$. Next, after switching the order of differentiation in every term in Eq. (17), we would like to show that:

$$0 = \left[\frac{\partial}{\partial N}\left(\frac{\partial U}{\partial T}\right)_{N,\tau}\right]_{T,\tau} - T\left[\frac{\partial}{\partial N}\left(\frac{\partial \underline{S}}{\partial T}\right)_{N,\tau}\right]_{T,\tau} + \tau\left[\frac{\partial}{\partial N}\left(\frac{\partial \underline{\varepsilon}}{\partial T}\right)_{N,\tau}\right]_{T,\tau}$$

$$= \left[\frac{\partial}{\partial N}\left(\left(\frac{\partial U}{\partial T}\right)_{N,\tau} - T\left(\frac{\partial \underline{S}}{\partial T}\right)_{N,\tau} + \tau\left(\frac{\partial \underline{\varepsilon}}{\partial T}\right)_{N,\tau}\right)\right]_{T,\tau} \qquad (18)$$

Because T and τ are held constant in the partial derivatives with respect to N, moving T and τ terms into the derivative is allowed. However, it can be readily observed that the terms within the outmost derivative sum up to zero, because:

$$d\underline{U} = Td\underline{S} - \tau d\underline{\varepsilon} + \mu dN$$

$$\Rightarrow \left(\frac{\partial U}{\partial T}\right)_{N,\tau} = T\left(\frac{\partial \underline{S}}{\partial T}\right)_{N,\tau} - \tau\left(\frac{\partial \underline{\varepsilon}}{\partial T}\right)_{N,\tau} + 0 \qquad (18a)$$

$$\Rightarrow \left(\frac{\partial U}{\partial T}\right)_{N,\tau} - T\left(\frac{\partial \underline{S}}{\partial T}\right)_{N,\tau} + \tau\left(\frac{\partial \underline{\varepsilon}}{\partial T}\right)_{N,\tau} = 0$$

$$\left[\frac{\partial}{\partial T}\left(\frac{\partial U}{\partial N}\right)_{T,\tau}\right]_{N,\tau} - T\left[\frac{\partial}{\partial T}\left(\frac{\partial \underline{S}}{\partial N}\right)_{T,\tau}\right]_{N,\tau} + \tau\left[\frac{\partial}{\partial T}\left(\frac{\partial \underline{\varepsilon}}{\partial N}\right)_{T,\tau}\right]_{N,\tau} = 0 \qquad (19)$$

Therefore, we have shown that Eq. (16) is equivalent to the relationship that we set out to prove, that is:

$$\left(\frac{\partial \mu}{\partial T}\right)_{\tau,N} = -\left(\frac{\partial \underline{S}}{\partial N}\right)_{T,\tau} \qquad (20)$$

Method 2

A simple way to prove that $\left(\frac{\partial \mu}{\partial T}\right)_{\tau,N} = -\left(\frac{\partial \underline{S}}{\partial N}\right)_{T,\tau}$ is to use a suitable Maxwell relation. Because we need to calculate the derivatives of μ and $\underline{S}$ with respect to

T and N, respectively, under fixed τ, we recognize that we are searching for a Maxwell relation established by a thermodynamic function which is naturally described by T, τ, and N. It turns out that this corresponds to the Gibbs free energy, $\underline{G}$. To see this, we begin by writing the differential form of the internal energy fundamental equation given by:

$$d\underline{U} = Td\underline{S} - \tau d\underline{\varepsilon} + \mu dN \tag{21}$$

However, Eq. (21) shows that $\underline{U}$ is a function of $\underline{S}$, $\underline{\varepsilon}$, and N. To change the dependence to T, τ, and N, we carry out the following variable transformation:

$$\begin{aligned} d\underline{U} - d(T\underline{S}) + d(\tau\underline{\varepsilon}) &= Td\underline{S} - \tau d\underline{\varepsilon} + \mu dN - Td\underline{S} - \underline{S}dT + \tau d\underline{\varepsilon} + \underline{\varepsilon}d\tau \\ \Rightarrow d(\underline{U} - T\underline{S} + \tau\underline{\varepsilon}) &= -\underline{S}dT + \underline{\varepsilon}d\tau + \mu dN \end{aligned} \tag{22}$$

From the analogy between $P\underline{V}$ and $\tau\underline{\varepsilon}$, we recognize that the Gibbs free energy has the form $\underline{G} = \underline{U} - T\underline{S} + \tau\underline{\varepsilon}$ in this case, with $d\underline{G}$ given by:

$$d\underline{G} = -\underline{S}dT + \underline{\varepsilon}d\tau + \mu dN \tag{23}$$

Because the order of differentiation does not matter, the following second derivatives associated with Eq. (23) are identical:

$$\left(\frac{\partial}{\partial N}\left(\frac{\partial\underline{G}}{\partial T}\right)_{\tau,N}\right)_{\tau,T} = \left(\frac{\partial}{\partial T}\left(\frac{\partial\underline{G}}{\partial N}\right)_{\tau,T}\right)_{\tau,N} \tag{24}$$

In addition, as shown in Part I, we have:

$$\left(\frac{\partial\underline{G}}{\partial T}\right)_{\tau,N} = -\underline{S}, \qquad \left(\frac{\partial\underline{G}}{\partial N}\right)_{\tau,T} = \mu \tag{25}$$

Substituting Eq. (25) in Eq. (24) yields:

$$-\left(\frac{\partial\underline{S}}{\partial N}\right)_{\tau,T} = \left(\frac{\partial\mu}{\partial T}\right)_{\tau,N} \tag{26}$$

which is precisely what we set out to prove.

Problem 7

Problem 8.2 in Tester and Modell

Our research laboratory has synthesized a new material, and the properties of this material are being studied under conditions where it is always a vapor. Two sets of experiments have been carried out, and they are described below. Given the results of these experiments, the relationship $P\underline{V} = NCT$ is proposed to describe the $P\underline{V}T$ properties of the material (C is constant). If you agree with this proposal, show a rigorous proof. If you do not agree, either prove that the relationship cannot be applicable or describe clearly what additional experiments you would recommend, and demonstrate how you would use these data to show whether or not $P\underline{V} = NCT$.

Experiment A: A rigid and well-insulated container consists of compartment I separated from compartment II by an impermeable partition. Compartment I is initially filled with gas, and compartment II is initially evacuated. When the experiment begins, the partition between compartments I and II is broken, and gas fills both compartments I and II. Over a wide range of initial temperatures, pressures, and volumes of compartment I, the final temperature, after expansion, equals the initial temperature.

Experiment B: Gas flows in an insulated pipe and through an insulated throttling valve wherein the pressure is reduced. Over a wide range of upstream temperatures, pressures, as well as downstream pressures, the temperature of the gas does not change when the gas flows through the valve.

Solution to Problem 7

Solution Strategy

We need to determine the $P\underline{V}TN$ equation of state (EOS) of a new pure material and then compare it to the given $P\underline{V} = NCT$ EOS to determine if they are equivalent.

The only sources of information that we have are the two experiments A and B. Because both experiments provide information about the temperature and its change under different scenarios, it is convenient to express the EOS (i.e., the relation between P, $\underline{V}$, N, and T) as follows:

$$T = f(P, \underline{V}, N) \tag{1}$$

Taking the differential of Eq. (1), we obtain:

$$dT = \left(\frac{\partial T}{\partial P}\right)_{\underline{V},N} dP + \left(\frac{\partial T}{\partial \underline{V}}\right)_{P,N} d\underline{V} + \left(\frac{\partial T}{\partial N}\right)_{P,\underline{V}} dN \tag{2}$$

In order to determine the function f in Eq. (1) and therefore the equation of state, we first need to determine the three partial derivatives in Eq. (2), that is: $\left(\frac{\partial T}{\partial P}\right)_{\underline{V},N}$, $\left(\frac{\partial T}{\partial \underline{V}}\right)_{P,N}$, and $\left(\frac{\partial T}{\partial N}\right)_{P,\underline{V}}$ utilizing the information provided in the Problem Statement.

1. Experiment A

Experiment A describes the *adiabatic* expansion of the gas from an initial $P\underline{V}T$ state to a final $P\underline{V}T$ state. We are told that the final temperature is always equal to the initial one.

Let us consider the system to be the gas occupying the total volume of the container (consisting of compartments I and II). The system is simple, closed, rigid, and adiabatic. We can therefore use the First Law of Thermodynamics as follows:

$$dU = \delta Q + \delta W \tag{3}$$

Because $\delta Q = 0$ (adiabatic) and $\delta W = 0$ (rigid), Eq. (3) yields:

$$d\underline{U} = 0 \text{ or } \underline{U} = \text{constant} \tag{4}$$

Note: We would obtain the same result if we considered the gas in compartment I as our system. Assuming a quasi-static expansion, the system is closed and adiabatic but with a movable boundary. However, the gas in compartment I is expanding against vacuum, and therefore, $\delta W = 0$. Using the First Law of Thermodynamics in this case, we would obtain the same result: the internal energy of the gas remains constant!

Therefore, experiment A provides us with information about the change in the temperature of the gas with respect to the volume, when the total number of moles, N, and the internal energy of the system, $\underline{U}$, both remain constant. We learn that, in this case, there is no change in the temperature of the gas. Mathematically, then:

$$\left(\frac{\partial T}{\partial \underline{V}}\right)_{\underline{U},N} = 0 \tag{5}$$

However, note that the partial derivative in Eq. (5), while similar, is not equal to the partial derivative multiplying $d\underline{V}$ in Eq. (2). Nevertheless, as shown next, it can be transformed into something useful. Indeed, using the triple product rule to move $\underline{U}$ into the partial derivative, we obtain:

$$\left(\frac{\partial T}{\partial \underline{V}}\right)_{\underline{U},N} \left(\frac{\partial \underline{V}}{\partial \underline{U}}\right)_{T,N} \left(\frac{\partial \underline{U}}{\partial T}\right)_{\underline{V},N} = -1 \tag{6}$$

or

$$\left(\frac{\partial T}{\partial \underline{V}}\right)_{\underline{U},N} = -\frac{\left(\frac{\partial U}{\partial \underline{V}}\right)_{T,N}}{\left(\frac{\partial U}{\partial T}\right)_{\underline{V},N}} \tag{7}$$

According to Eqs. (5) and (7), it follows that:

$$\left(\frac{\partial \underline{U}}{\partial \underline{V}}\right)_{T,N} = 0 \tag{8}$$

We can then use Eq. (8) to obtain one of the partial derivatives that we need in Eq. (2). To do so, we would like to express the derivative in Eq. (8) as a function of P, $\underline{V}$, T, and N. For this purpose, we write the differential form of $\underline{U} = f(\underline{S}, \underline{V}, N)$ as follows:

$$d\underline{U} = Td\underline{S} - Pd\underline{V} + \mu dN \tag{9}$$

Taking the derivative of Eq. (9) with respect to $\underline{V}$, at constant T and N, yields:

$$\left(\frac{\partial \underline{U}}{\partial \underline{V}}\right)_{T,N} = T\left(\frac{\partial \underline{S}}{\partial \underline{V}}\right)_{T,N} - P\left(\frac{\partial \underline{V}}{\partial \underline{V}}\right)_{T,N} + \mu\left(\frac{\partial N}{\partial \underline{V}}\right)_{T,N} \tag{10}$$

In Eq. (10), the third term is equal to $-P$, and the last term is equal to 0, which leads to:

$$\left(\frac{\partial \underline{U}}{\partial \underline{V}}\right)_{T,N} = T\left(\frac{\partial \underline{S}}{\partial \underline{V}}\right)_{T,N} - P \tag{11}$$

Using a Maxwell relation, it follows that:

$$-\left(\frac{\partial}{\partial \underline{V}}\left(\frac{\partial \underline{A}}{\partial T}\right)_{\underline{V},N}\right)_{T,N} = \left(\frac{\partial \underline{S}}{\partial \underline{V}}\right)_{T,N} = \left(\frac{\partial P}{\partial T}\right)_{\underline{V},N} = -\left(\frac{\partial}{\partial T}\left(\frac{\partial \underline{A}}{\partial \underline{V}}\right)_{T,N}\right)_{\underline{V},N} \tag{12}$$

Combining Eqs. (8), (11), and (12), we obtain:

$$T\left(\frac{\partial P}{\partial T}\right)_{\underline{V},N} - P = 0$$

or

$$\left(\frac{\partial P}{\partial T}\right)_{\underline{V},N} = \frac{P}{T}$$

or

$$\boxed{\left(\frac{\partial T}{\partial P}\right)_{V,N} = \frac{T}{P}} \tag{13}$$

Equation (13) is the first partial derivative on the right-hand side of Eq. (1).

2. Experiment B

Experiment B describes the flow of gas in an insulated pipe and through an insulated throttling valve. Considering the gas going through the valve as our system, we can use the First Law of Thermodynamics and describe the process as we did with Experiment A.

In this case, the system is open, adiabatic, and simple. Using the First Law of Thermodynamics on the open system, we obtain:

$$dU = \delta Q + \delta W + H_{in}\delta n_{in} - H_{out}\delta n_{out} \tag{14}$$

We know that:

$$dU = 0 \quad \text{(Steady state)} \tag{15}$$

$$\delta Q = 0 \quad \text{(Adiabatic)} \tag{16}$$

$$\delta W = 0 \quad \text{(Rigid)} \tag{17}$$

$$\delta n_{in} = \delta n_{out} = dN \quad \text{(Steady state)} \tag{18}$$

Using Eqs. (15), (16), (17), and (18) in Eq. (14) yields:

$$(H_{in} - H_{out})dN = 0$$

or

$$H_{in} = H_{out}, \text{ or } dH = 0 \tag{19}$$

Therefore, Experiment B describes the change in the temperature of the gas with respect to pressure when the molar enthalpy remains constant. This change can be described mathematically as the derivative of T with respect to P when H remains constant. From the Problem Statement, we know that there is no change in the temperature. Therefore, it follows that:

$$\left(\frac{\partial T}{\partial P}\right)_H = 0 \tag{20}$$

Because $\underline{H} = N\underline{H}$, Eq. (20) is equivalent to:

$$\left(\frac{\partial T}{\partial P}\right)_{\underline{H},N} = 0 \tag{21}$$

Note: When we deal with a throttling valve, the molar enthalpy remains constant.

Next, we follow the same procedure as the one in Experiment A, in order to express the partial derivative in Eq. (21) in terms of measurable quantities. Using the triple product rule (to move $\underline{H}$ into the derivative), we obtain:

$$\left(\frac{\partial T}{\partial P}\right)_{\underline{H},N} \left(\frac{\partial P}{\partial \underline{H}}\right)_{T,N} \left(\frac{\partial \underline{H}}{\partial T}\right)_{P,N} = -1$$

or

$$\left(\frac{\partial T}{\partial P}\right)_{\underline{H},N} = -\frac{\left(\frac{\partial \underline{H}}{\partial P}\right)_{T,N}}{\left(\frac{\partial \underline{H}}{\partial T}\right)_{P,N}} \tag{22}$$

Combining Eqs. (21) and (22), we obtain:

$$\left(\frac{\partial \underline{H}}{\partial P}\right)_{T,N} = 0 \tag{23}$$

Again, we would like to make use of Eq. (23) by converting the partial derivative to one of the partial derivatives in Eq. (2). To this end, we write the differential form of $\underline{H}$ as follows:

$$d\underline{H} = Td\underline{S} + \underline{V}dP + \mu dN \quad \text{(For a one component system)} \tag{24}$$

Taking the partial derivative of Eq. (24) with respect to P, at constant T and N, yields:

$$\left(\frac{\partial \underline{H}}{\partial P}\right)_{T,N} = T\left(\frac{\partial \underline{S}}{\partial P}\right)_{T,N} + \underline{V}\left(\frac{\partial P}{\partial P}\right)_{T,N} + \mu\left(\frac{\partial N}{\partial P}\right)_{T,N}$$

In the last equation, the third term is equal to $\underline{V}$ and the last term is equal to 0. As a result, we obtain:

$$\left(\frac{\partial \underline{H}}{\partial P}\right)_{T,N} = T\left(\frac{\partial \underline{S}}{\partial P}\right)_{T,N} + \underline{V} \tag{25}$$

Using one of the Maxwell relations, it follows that:

$$-\left(\frac{\partial}{\partial P}\left(\frac{\partial \underline{G}}{\partial T}\right)_{P,N}\right)_{T,N} = \left(\frac{\partial \underline{S}}{\partial P}\right)_{T,N} = -\left(\frac{\partial \underline{V}}{\partial T}\right)_{P,N} = -\left(\frac{\partial}{\partial T}\left(\frac{\partial \underline{G}}{\partial P}\right)_{T,N}\right)_{P,N} \quad (26)$$

Combining Eqs. (25) and (26), we obtain:

$$\left(\frac{\partial \underline{H}}{\partial P}\right)_{T,N} = -T\left(\frac{\partial \underline{V}}{\partial T}\right)_{P,N} + \underline{V} \quad (27)$$

Using Eq. (23) in Eq. (27), we obtain:

$$\left(\frac{\partial \underline{V}}{\partial T}\right)_{P,N} = \frac{\underline{V}}{T} \quad (28)$$

Equation (28) can be rewritten as follows:

$$\boxed{\left(\frac{\partial T}{\partial \underline{V}}\right)_{P,N} = \frac{T}{\underline{V}}} \quad (29)$$

Equation (29) is the second partial derivative on the right-hand side of Eq. (1). To summarize, below, we repeat the main results derived so far:

$$\boxed{\left(\frac{\partial T}{\partial P}\right)_{\underline{V},N} = \frac{T}{P}} \quad (13)$$

$$\boxed{\left(\frac{\partial T}{\partial \underline{V}}\right)_{P,N} = \frac{T}{\underline{V}}} \quad (29)$$

Finally, we need to determine $\left(\frac{\partial T}{\partial N}\right)_{P,\underline{V}}$ in Eq. (1) in order to construct the complete differential of dT. Recall that for a one-component ($n = 1$) system, the number of independent first-order partial derivatives is $n + 1 = 2$. Accordingly, because we have already determined two independent ones (see Eqs. (13) and (29)), any other one can be expressed as a function of those two. We can actually obtain the third partial derivative by first using the triple product rule as follows:

$$\left(\frac{\partial T}{\partial N}\right)_{\underline{V},P}\left(\frac{\partial \underline{V}}{\partial T}\right)_{N,P}\left(\frac{\partial N}{\partial \underline{V}}\right)_{T,P} = -1$$

or

$$\left(\frac{\partial T}{\partial N}\right)_{\underline{V},P} = -\frac{\left(\frac{\partial \underline{V}}{\partial N}\right)_{T,P}}{\left(\frac{\partial \underline{V}}{\partial T}\right)_{N,P}} \tag{30}$$

or

$$\left(\frac{\partial \underline{V}}{\partial N}\right)_{T,P} = \left(\frac{\partial NV}{\partial N}\right)_{T,P} = V + N\left(\frac{\partial V}{\partial N}\right)_{T,P} = V \tag{31}$$

Combining Eqs. (29), (30), and (31), we obtain:

$$\left(\frac{\partial T}{\partial N}\right)_{\underline{V},P} = -\frac{V}{\frac{\underline{V}}{T}} = -\frac{V}{\frac{NV}{T}} = -\frac{T}{N} \tag{32}$$

Euler integrating Eq. (2), recognizing that T and P are intensive variables, it follows that:

$$0 = 0 + \left(\frac{\partial T}{\partial \underline{V}}\right)_{P,N} \underline{V} + \left(\frac{\partial T}{\partial N}\right)_{P,\underline{V}} N$$

or

$$\left(\frac{\partial T}{\partial N}\right)_{P,\underline{V}} = -\left(\frac{\partial T}{\partial \underline{V}}\right)_{P,N} \frac{\underline{V}}{N}$$

Using Eq. (29) in the last equation yields the third required partial derivative in Eq. (1), that is:

$$\boxed{\left(\frac{\partial T}{\partial N}\right)_{P,\underline{V}} = -\frac{T}{\underline{V}} \frac{\underline{V}}{N} = -\frac{T}{N}} \tag{33}$$

Substituting Eqs. (13), (29), and (33) in Eq. (2), we finally obtain:

$$dT = \frac{T}{P}dP + \frac{T}{\underline{V}}d\underline{V} - \frac{T}{N}dN$$

or

$$\frac{dT}{T} = \frac{dP}{P} + \frac{d\underline{V}}{\underline{V}} - \frac{dN}{N} \tag{34}$$

Integrating Eq. (34) yields:

$$\ln T = \ln P + \ln \underline{V} - \ln N + C'$$ (35)

where C' is the constant of integration.

We can rewrite Eq. (35) as follows:

$$\ln T + \ln N = \ln P + \ln \underline{V} + \ln C_1$$ (36)

where $\ln C_1 = C'$. Combining the logarithmic terms in Eq. (36) yields:

$$\ln (NT) = \ln (P\underline{V}C_1)$$

or

$$\boxed{CNT = P\underline{V}}$$ (37)

where $C = \frac{1}{C_1}$ is a constant.

We therefore conclude that the EOS in Eq. (37) does describe the volumetric behavior of the new synthesized material.

Problem 8

Problem 8.4 in Tester and Modell

We have a constant volume, closed vessel filled with dichlorodifluoromethane gas. We plan to heat the gas and would like to know how the molar entropy of the gas varies with pressure. Derive a general relation to calculate the desired derivative, $(\partial S/\partial P)_V$, assuming that we know the total volume of the vessel, the moles of gas, C_P as a function of T and P, and have access to a pressure-explicit equation of state.

Illustrate your result at the start of the heating process, where $T = 365.8$ K, $P = 16.5$ bar, and the total volume $= 1.51 \times 10^{-3}$ m^3. Assume that at this condition, $C_P = 94.9$ J/mol K and the pressure-explicit Redlich-Kwong equation of state is applicable, with $a = 20.839$ J m^3 K$^{1/2}$/mol^2 and $b = 6.725 \times 10^{-5}$ m^3/mol.

Solution to Problem 8

Solution Strategy

Summarize What We Know

We are asked to find a general relation to calculate how the molar entropy of dichlorodifluoromethane gas varies with pressure at constant molar volume using solely:

- The heat capacity at constant pressure
- A general, pressure-explicit equation of state
- The volume and number of moles (or, equivalently, the intensive volume)

Following that, we are asked to choose the pressure-explicit Redlich-Kwong (RK) equation of state, along with all the information provided in the Problem Statement, to obtain a numerical value for the desired partial derivative.

Select Possible Solution Strategies

As discussed in Part I, because we are given a general pressure-explicit EOS, where the variables T, P, and V appear, as well as heat capacity data, we should be able to calculate any thermodynamic property of a one-component ($n = 1$) system, including the partial derivative of the molar entropy with respect to pressure at constant molar volume. Recall that the heat capacities at constant pressure and volume can be expressed as follows:

$$C_P \equiv \left(\frac{\partial H}{\partial T}\right)_P = T\left(\frac{\partial S}{\partial T}\right)_P \tag{1}$$

$$C_V \equiv \left(\frac{\partial U}{\partial T}\right)_V = T\left(\frac{\partial S}{\partial T}\right)_V \tag{2}$$

Below, we will make use of Eqs. (1) and (2), along with the general pressure-explicit EOS.

Strategy I: Using Eq. (1) to Obtain the Desired Result

Choosing T and P as the two independent intensive variables, according to the Corollary to Postulate I, the molar entropy can be expressed as $S = S(T,P)$. The differential of S is then given by:

$$dS = \left(\frac{\partial S}{\partial T}\right)_P dT + \left(\frac{\partial S}{\partial P}\right)_T dP \tag{3}$$

Using Eq. (1) to simplify the first partial derivative in Eq. (3) and using a Maxwell relationship to transform the second partial derivative in Eq. (3), we obtain:

$$dS = \left(\frac{C_P}{T}\right)dT - \left(\frac{\partial V}{\partial T}\right)_P dP \tag{4}$$

Next, if we take the derivative of Eq. (4) with respect to pressure, at constant molar volume, we obtain:

$$\left(\frac{\partial S}{\partial P}\right)_V - \left(\frac{C_P}{T}\right)\left(\frac{\partial T}{\partial P}\right)_V - \left(\frac{\partial V}{\partial T}\right)_P\left(\frac{\partial P}{\partial P}\right)_V = \left(\frac{C_P}{T}\right)\left(\frac{\partial T}{\partial P}\right)_V - \left(\frac{\partial V}{\partial T}\right)_P \tag{5}$$

An examination of Eq. (5) indicates that we are quite close to the desired solution, because everything that appears on the right-hand side of Eq. (5) is expressed in terms of T, P, and V, as well as the heat capacity at constant pressure. However, in order to use the general pressure-explicit equation of state, we need to modify our derivatives slightly. Specifically, using a derivative inversion of the $(\partial T/\partial P)_V$ term and the triple-product rule on the $(\partial V/\partial T)_P$ term, we obtain:

$$\left(\frac{\partial S}{\partial P}\right)_V = \left(\frac{C_P}{T}\right)\left(\frac{\partial P}{\partial T}\right)_V^{-1} + \left(\frac{\partial P}{\partial T}\right)_V\left(\frac{\partial P}{\partial V}\right)_T^{-1} \tag{6}$$

We can now evaluate Eq. (6) using the given pressure-explicit EOS and heat capacity data. We will do so after we consider Strategy II.

Strategy II: Using Eq. (2) to Obtain the Desired Result

Considering Eq. (2), we recognize that we can use the heat capacity definition by a quick chain rule expansion:

$$\left(\frac{\partial S}{\partial P}\right)_V = \left(\frac{\partial S}{\partial T}\right)_V \left(\frac{\partial T}{\partial P}\right)_V = \left(\frac{C_V}{T}\right)\left(\frac{\partial T}{\partial P}\right)_V \tag{7}$$

Unfortunately, neither of the last two terms in Eq. (7) can be evaluated directly using the information provided. However, we can use transformations to obtain useful results. The constant volume heat capacity can be expanded as shown in Part I, that is:

$$C_V = C_P - T\left(\frac{\partial V}{\partial T}\right)_P \left(\frac{\partial P}{\partial T}\right)_V \tag{8}$$

Using Eq. (8) in Eq. (7), we obtain:

$$\left(\frac{\partial S}{\partial P}\right)_V = \left[\left(\frac{C_P}{T}\right) - \left(\frac{\partial V}{\partial T}\right)_P \left(\frac{\partial P}{\partial T}\right)_V\right]\left(\frac{\partial T}{\partial P}\right)_V = \left(\frac{C_P}{T}\right)\left(\frac{\partial T}{\partial P}\right)_V - \left(\frac{\partial V}{\partial T}\right)_P \tag{9}$$

Equation (9) is identical to Eq. (5)!

Evaluate the Desired Partial Derivative

To solve for the desired partial derivative, we will evaluate all the terms in Eq. (6) independently and then combine them all to obtain the desired result. The first term, C_P/T, requires no further simplification because it is given. The other terms can be readily calculated using the RK EOS as follows:

$$P = \left[\frac{RT}{(V-b)}\right] - \left[\frac{a}{T^{1/2}V(V+b)}\right] \tag{10}$$

$$\left(\frac{\partial P}{\partial T}\right)_V = \left[\frac{R}{V-b}\right] + \left[\frac{a}{2T^{3/2}V(V+b)}\right] \tag{11}$$

$$\left(\frac{\partial P}{\partial V}\right)_T = -\left[\frac{RT}{(V-b)^2}\right] + \left[\frac{a(2V+b)}{T^{1/2}V^2(V+b)^2}\right] \tag{12}$$

Using the variable values provided in the Problem Statement, it is possible via a numerical solver to evaluate the volume at the specified temperature and pressure. Doing so, we obtain:

$$\boxed{V = 1.51 \times 10^{-3} \text{ m}^3/\text{mol}} \tag{13}$$

(**Note:** If we use this result to find the number of moles in the system, we find $N = 1$ mole). Next, we can evaluate the two derivatives based on Eqs. (11) and (12):

$$\left(\frac{\partial P}{\partial T}\right)_V = 6.387 \times 10^3 \frac{N}{m^2 \cdot K} \tag{14}$$

$$\left(\frac{\partial P}{\partial V}\right)_T = -8.674 \times 10^8 \frac{N \cdot mol}{m^5} \tag{15}$$

Using the results in Eqs. (13), (14), and (15), as well as the C_P and T information given in the Problem Statement, we obtain our final result:

$$\left(\frac{\partial S}{\partial P}\right)_V = 3.33 \times 10^{-5} \frac{J}{mol \cdot K \cdot Pa} \tag{16}$$

Problem 9

Problem 8.6 in Tester and Modell

Sulfur dioxide gas at 520 K and 100 bar fills one-half of a rigid, adiabatic cylinder. The other half is evacuated, and the two halves are separated by a metal diaphragm. If this should rupture, what would be the final temperature and pressure? Assume that the gas is well mixed and that expansion is sufficiently rapid so that negligible heat transfer occurs between the walls and the SO_2 gas.

For SO_2, $T_c = 430.8$ K, $P_c = 78.8$ bar, $V_c = 1.22 \times 10^{-4}$ m^3/mol, $\omega = 0.251$, and C_P^0 (J/mol K) is given by:

$$C_P^0 = 23.852 + 6.699 \times 10^{-2}T - 4.961 \times 10^{-5}T^2 + 1.328 \times 10^{-8}T^3$$

with T in degrees Kelvin.

Solution to Problem 9

Solution Strategy

1. **Summarize what we know**

The problem asks us to find the final temperature and pressure of a gas undergoing an expansion against vacuum. To find the final state of the system (the gas), we can either use the First Law of Thermodynamics to find the change in the internal energy of the gas, and then relate it to the final state of the gas or we can use the Second Law of Thermodynamics to calculate the entropy change of the gas and then relate it to the final state of the gas.

The most convenient choice here is to use the First Law of Thermodynamics to model the expansion of the gas. However, the gas expansion is not reversible (the gas is initially at a pressure of 100 bar and expands against a pressure of 0 bar). Consequently, although the gas expands adiabatically, we cannot claim that the entropy is conserved, and as a result, do not have enough information to calculate the entropy change directly. All that we know is that the change in entropy will be positive because the expansion of the gas is irreversible. However, because of the rigid, adiabatic nature of the cylinder, the gas is an isolated system, and because it expands against vacuum, no mechanical work is done by the gas as it expands. As a result, a First Law of Thermodynamics analysis of the system (the gas) yields:

$$dU = \delta Q + \delta W = 0 + 0$$
$$\Rightarrow dU = 0 \tag{1}$$

Because the number of moles of gas does not change throughout the expansion, Eq. (1) can be rewritten in terms of the molar (intensive) internal energy as follows:

$$dU = 0$$
$$\Rightarrow U_f = U_i \tag{2}$$

Therefore, in this problem, we need to find the final state of the gas following its expansion at constant molar internal energy. We will proceed using solely the following information provided in the Problem Statement:

- Heat capacity
- Critical parameters
- Initial T and P of the gas

2. Make reasonable approximations

If we assume that the SO_2 gas behaves like an ideal gas, Eq. (2) implies that the final temperature of the gas is equal to its initial temperature. This is because the molar internal energy of an ideal gas is only a function of temperature. In that case, the pressure of the ideal gas would decrease by a factor of two due to the doubling of the volume occupied by the ideal gas at the end of its expansion.

However, at pressures as high as 100 bar, it is evident that the gas would not behave ideally. As a result, its molar internal energy would not necessarily be a function of temperature only. From the Corollary to Postulate I introduced in Part I, we know that the molar internal energy for any pure (one-component, $n = 1$) fluid can be written in terms of $(n + 1) = 2$ independent intensive variables, for example, T and V, T and P, or P and V. One of the key steps in solving this problem is to select the correct set of variables that would make the analysis simple. Because we have been given heat capacity data, it is natural to select T as one of the two variables. In order to decide between P and V, we note that we know the final value of V ($V_{final} = 2V_{initial}$). Moreover, if we use a pressure-explicit equation of state, it would be easier for us to eliminate P in favor of T and V. Therefore, we choose T and V as our two independent intensive variables.

From Part I, we know that the differential of $U(T, V)$ is given by:

$$dU = C_V dT + \left[T \left(\frac{\partial P}{\partial T} \right)_V - P \right] dV \tag{3}$$

Combining Eqs. (2) and (3), including integrating from the initial (i) to the final (f) states, yields:

$$\int\limits_{U_i}^{U_f} dU = U_f - U_i = 0 = \int\limits_{T_i}^{T_f} C_V dT + \int\limits_{V_i}^{V_f} \left[T\left(\frac{\partial P}{\partial T}\right)_V - P \right] dV$$

$$\Rightarrow \int\limits_{T_i}^{T_f} C_V dT + \int\limits_{V_i}^{V_f} \left[T\left(\frac{\partial P}{\partial T}\right)_V - P \right] dV = 0 \tag{4}$$

We know that the internal energy is a state function, and therefore, to use Eq. (4), it would be beneficial to construct a convenient path that allows us to calculate the change in internal energy using the limited information available to us. The Problem Statement provides us with heat capacity data for the SO_2 gas when it behaves ideally. Therefore, we want to carry out the temperature integration in Eq. (4) when the gas behaves ideally (i.e., at a very large V value in this case). Accordingly, for the purpose of integration, the gas first traverses an isothermal path at T_i from a volume V_i to a volume $V^* \to \infty$. Next, the gas traverses an isochoric path at a volume V^* from a temperature T_i to a temperature T_f. Finally, to reach the final state, the gas again traverses an isothermal path at a temperature T_f from a volume V^* to a volume V_f. This three-step integration method is what we referred to in Part I as the attenuated-state approximation. Mathematically, this can be written as follows:

$$\int\limits_{T_i}^{T_f} C_V dT + \int\limits_{V_i}^{V_f} \left[T\left(\frac{\partial P}{\partial T}\right)_V - P \right] dV = \int\limits_{V_i}^{V^*} \left[T\left(\frac{\partial P}{\partial T}\right)_V - P \right]_{T=T_i} dV$$

$$+ \int\limits_{T_i}^{T_f} C_V|_{V=V^*} dT + \int\limits_{V^*}^{V_f} \left[T\left(\frac{\partial P}{\partial T}\right)_V - P \right]_{T=T_f} dV \tag{5}$$

3. Choose a suitable EOS and solve

So far, our derivation is general. However, to proceed and evaluate the first and third integrals on the right-hand side of Eq. (5), we need to choose a suitable equation of state. Because we have chosen T and V as our two independent intensive variables, it is advantageous to use a pressure-explicit equation of state for which P is a function of T and V.

We need to use a pressure-explicit equation of state that utilizes all the information provided in the Problem Statement. With some reflection, this gives us a choice between the Peng-Robinson (PR) equation of state and the Redlich-Kwong-Soave (RKS) equation of state. From the Problem Statement, we know that the critical compressibility factor Z_C is $P_C V_C/RT_C = 0.268$, which is close to the Z_C for the Peng-Robinson equation of state discussed in Part I. Therefore, we assume that the volumetric behavior of the SO_2 gas can be accurately modeled using the Peng-Robinson equation of state. For the Peng-Robinson equation of state, the integrands of the first and third integrals on the right-hand side of Eq. (5) can be evaluated as follows:

$$P = \frac{RT}{(V-b)} - \frac{a(\omega, T_r)}{V(V+b) + b(V-b)}$$

$$\Rightarrow \left(\frac{\partial P}{\partial T}\right)_V = \frac{R}{(V-b)} - \frac{da/dT}{V(V+b) + b(V-b)}$$

$$\Rightarrow T\left(\frac{\partial P}{\partial T}\right)_V - P = \frac{RT}{(V-b)} - \frac{T(da/dT)}{V(V+b) + b(V-b)} - \frac{RT}{(V-b)}$$

$$+ \frac{a(\omega, T_r)}{V(V+b) + b(V-b)}$$

$$\Rightarrow T\left(\frac{\partial P}{\partial T}\right)_V - P = \frac{a(\omega, T_r) - T(da/dT)}{V(V+b) + b(V-b)} \tag{6}$$

where

$$a = \frac{0.45724 R^2 T_C^2}{P_C}\left(1 + \kappa\left(1 - \sqrt{T_r}\right)\right)^2 \tag{7}$$

$$\kappa = 0.37464 + 1.54226\omega - 0.26992\omega^2 \tag{8}$$

$$b = \frac{0.07780 RT_C}{P_C} \tag{9}$$

Substituting values of T_C, P_C, V_C, and ω in Eqs. (7), (8), and (9) yields:

$$b = \frac{0.07780 \times 8.314 \times 430.8}{78.8 \times 10^5} = 3.54 \times 10^{-5}\, \text{m}^3/\text{mol} \tag{10}$$

$$\kappa = 0.37464 + 1.54226 \times 0.251 - 0.26992 \times 0.251^2 = 0.7447 \tag{11}$$

$$a = \frac{0.45724 \times 8.314^2 \times 430.8^2}{78.8 \times 10^5}\left(1 + 0.7447\left(1 - \sqrt{\frac{T}{430.8}}\right)\right)^2$$

$$= 0.7444 \times \left(1 + 0.7447\left(1 - \sqrt{\frac{T}{430.8}}\right)\right)^2 \text{Jm}^3/\text{mol}^2 \tag{12}$$

$$
\frac{da}{dT} = 0.7444 \times 2 \times \left(1 + 0.7447\left(1 - \sqrt{\frac{T}{430.8}}\right)\right) \times 0.7447 \times (-1) \times \left(\frac{1}{2}\right)
$$

$$
\times \sqrt{\frac{1}{430.8 \times T}} \tag{13}
$$

$$
= 9.5798 \times 10^{-4} - \frac{0.0466}{\sqrt{T}} \ \mathrm{J\,m^3/K\,mol^2}
$$

In view of Eq. (6), to evaluate Eq. (5), we need to evaluate the integral, $\int_{V_1}^{V_2} \frac{a(\omega,T_r) - T(da/dT)}{V(V+b) + b(V-b)} dV$, with the appropriate limits of integration. Because the numerator in the integrand is independent of V, we can take it out of the integral, leaving the following integral for evaluation:

$$
\int_{V_1}^{V_2} \frac{1}{V(V+b) + b(V-b)} dV = \int_{V_1}^{V_2} \frac{1}{V^2 + 2Vb - b^2} dV
$$

$$
= \int_{V_1}^{V_2} \frac{1}{V^2 + 2Vb + b^2 - 2b^2} dV
$$

$$
= \int_{V_1}^{V_2} \frac{1}{(V+b)^2 - 2b^2} dV
$$

$$
= \frac{1}{2\sqrt{2}b} \int_{V_1}^{V_2} \frac{(V+b+\sqrt{2}b) - (V+b-\sqrt{2}b)}{(V+b-\sqrt{2}b)(V+b+\sqrt{2}b)} dV \tag{14}
$$

$$
= \frac{1}{2\sqrt{2}b} \int_{V_1}^{V_2} \frac{1}{(V+b-\sqrt{2}b)} - \frac{1}{(V+b+\sqrt{2}b)} dV
$$

$$
= \frac{1}{2\sqrt{2}b} \ln \left|\frac{V+b-\sqrt{2}b}{V+b+\sqrt{2}b}\right|_{V_1}^{V_2}
$$

Using Eq. (14), we can simplify Eq. (5) as follows:

$$U_f - U_i = 0 = \int\limits_{V_i}^{V^*} \left[T\left(\frac{\partial P}{\partial T}\right)_V - P \right]_{T=T_i} dV + \int\limits_{T_i}^{T_f} C_V|_{V=V^*} dT + \int\limits_{V^*}^{V_f} \left[T\left(\frac{\partial P}{\partial T}\right)_V - P \right]_{T=T_f} dV$$

$$= \frac{a(\omega, T_{i,r}) - T_i(da/dT)_{T=T_i}}{2\sqrt{2}b} \ln \left| \frac{V+b-\sqrt{2}b}{V+b+\sqrt{2}b} \right|_{V_i}^{V^*\to\infty} + \int\limits_{T_i}^{T_f} C_V|_{V=V^*\to\infty} dT +$$

$$\frac{a(\omega, T_{f,r}) - T_f(da/dT)_{T=T_f}}{2\sqrt{2}b} \ln \left| \frac{V+b-\sqrt{2}b}{V+b+\sqrt{2}b} \right|_{V^*\to\infty}^{V_f=2V_i}$$

$$= \frac{a(\omega, T_{i,r}) - T_i(da/dT)_{T=T_i}}{2\sqrt{2}b} \left(\lim_{V^*\to\infty} \ln \left| \frac{1 + \left(\frac{b-\sqrt{2}b}{V^*}\right)}{1 + \left(\frac{b+\sqrt{2}b}{V^*}\right)} \right| - \ln \left| \frac{V_i+b-\sqrt{2}b}{V_i+b+\sqrt{2}b} \right| \right) +$$

$$\int\limits_{T_i}^{T_f} C_V|_{V=V^*\to\infty} dT$$

$$+ \frac{a(\omega, T_{f,r}) - T_f(da/dT)_{T=T_f}}{2\sqrt{2}b} \left(\ln \left| \frac{2V_i+b-\sqrt{2}b}{2V_i+b+\sqrt{2}b} \right| - \lim_{V^*\to\infty} \ln \left| \frac{1 + \left(\frac{b-\sqrt{2}b}{V^*}\right)}{1 + \left(\frac{b+\sqrt{2}b}{V^*}\right)} \right| \right)$$

$$= \frac{a(\omega, T_{i,r}) - T_i(da/dT)_{T=T_i}}{2\sqrt{2}b} \left(\ln 1 - \ln \left| \frac{V_i+b-\sqrt{2}b}{V_i+b+\sqrt{2}b} \right| \right)$$

$$+ \int\limits_{T_i}^{T_f} C_V|_{V=V^*\to\infty} dT + \frac{a(\omega, T_{f,r}) - T_f(da/dT)_{T=T_f}}{2\sqrt{2}b} \left(\ln \left| \frac{2V_i+b-\sqrt{2}b}{2V_i+b+\sqrt{2}b} \right| - \ln 1 \right)$$

$$= \frac{a(\omega, T_{f,r}) - T_f(da/dT)_{T=T_f}}{2\sqrt{2}b} \left(\ln \left| \frac{2V_i+b-\sqrt{2}b}{2V_i+b+\sqrt{2}b} \right| \right) -$$

$$\frac{a(\omega, T_{i,r}) - T_i(da/dT)_{T=T_i}}{2\sqrt{2}b} \left(\ln \left| \frac{V_i+b-\sqrt{2}b}{V_i+b+\sqrt{2}b} \right| \right) + \int\limits_{T_i}^{T_f} C_V|_{V=V^*\to\infty} dT$$

$$(15)$$

Using the given values of the parameters in Eq. (15), we obtain an equation that we can solve for V_i.

Specifically:

$$100 \times 10^5 = \frac{8.314 \times 520}{(V_i - 3.54 \times 10^{-5})} - \frac{0.7444 \times \left(1 + 0.7447\left(1 - \sqrt{\frac{520}{430.8}}\right)\right)^2}{V_i(V_i + 3.54 \times 10^{-5}) + 3.54 \times 10^{-5} \times (V_i - 3.54 \times 10^{-5})}$$

$$\Rightarrow 10^7 = \frac{4323.28}{(V_i - 3.54 \times 10^{-5})} - \frac{0.639}{V_i^2 + V_i \times 7.08 \times 10^{-5} - 1.253 \times 10^{-9}}$$

$$\Rightarrow V_i = 3.2125 \times 10^{-4} \, \text{m}^3/\text{mol}$$

$$(16)$$

Note that in carrying out the integration in Eq. (5), shown in Eq. (15), we did not encounter any singularity, and therefore, did not have to add or subtract any term to remove the singularity, as we discussed in Part I.

The last piece in calculating the change in the molar internal energy of the system involves evaluating the temperature integral. Because the temperature integral is carried out in the attenuated state ($V^* \to \infty$), we can use the ideal gas heat capacity to compute the integral as shown below:

$$\int_{T_i}^{T_f} C_V|_{V=V^* \to \infty} dT$$

$$= \int_{T_i}^{T_f} C_V^0 dT = \int_{T_i}^{T_f} (C_P^0 - R) dT$$

$$= \int_{T_i}^{T_f} (23.852 + 6.699 \times 10^{-2} T - 4.961 \times 10^{-5} T^2 + 1.328 \times 10^{-8} T^3 - 8.314) dT$$

$$= \left[15.538 \times T + 6.699 \times 10^{-2} \left(\frac{T^2}{2} \right) - 4.961 \times 10^{-5} \left(\frac{T^3}{3} \right) + 1.328 \times 10^{-8} \left(\frac{T^4}{4} \right) \right]_{T_i}^{T_f}$$

$$(17)$$

Combining Eqs. (15) and (17), and substituting values of a, da/dT, b, T_i, and V_i, yields an equation that can be solved for T_f:

$$U_f - U_i = 0 = \frac{0.7444 \times \left(1 + 0.7447\left(1 - \sqrt{\frac{T_f}{430.8}}\right)\right)^2 - T_f\left(9.5798 \times 10^{-4} - \frac{0.0466}{\sqrt{T_f}}\right)}{2\sqrt{2} \times 3.54 \times 10^{-5}}$$

$$\left(\ln\left|\frac{2 \times 3.2125 \times 10^{-4} + 3.54 \times 10^{-5}(1 - \sqrt{2})}{2 \times 3.2125 \times 10^{-4} + 3.54 \times 10^{-5}(1 + \sqrt{2})}\right|\right)$$

$$-\frac{0.7444 \times \left(1 + 0.7447\left(1 - \sqrt{\frac{520}{430.8}}\right)\right)^2 - 520\left(9.5798 \times 10^{-4} - \frac{0.0466}{\sqrt{520}}\right)}{2\sqrt{2} \times 3.54 \times 10^{-5}}$$

$$\left(\ln\left|\frac{3.2125 \times 10^{-4} + 3.54 \times 10^{-5}(1 - \sqrt{2})}{3.2125 \times 10^{-4} + 3.54 \times 10^{-5}(1 + \sqrt{2})}\right|\right)$$

$$+ 15.538 \times T_f + 6.699 \times 10^{-2}\left(\frac{T_f^2}{2}\right) - 4.961 \times 10^{-5}\left(\frac{T_f^3}{3}\right) + 1.328 \times 10^{-8}\left(\frac{T_f^4}{4}\right)$$

$$- 15.538 \times 520 - 6.699 \times 10^{-2}\left(\frac{520^2}{2}\right)$$

$$+ 4.961 \times 10^{-5}\left(\frac{520^3}{3}\right) - 1.328 \times 10^{-8}\left(\frac{520^4}{4}\right)$$

$$= \left(7434.61 \times \left(1 + 0.7447\left(1 - \sqrt{\frac{T_f}{430.8}}\right)\right)^2 - 9.5677 \times T_f + 465.41 \times \sqrt{T_f}\right)$$

$$\times (-0.1480)$$

$$- 12020.07 \times (-0.2826) + 15.538 \times T_f + 0.0335 \times T_f^2 - 1.654 \times 10^{-5} \times T_f^3$$

$$+ 3.32 \times 10^{-9} \times T_f^4 - 15054.37$$

$$= -1100.32 \times \left(1 + 0.7447\left(1 - \sqrt{\frac{T_f}{430.8}}\right)\right)^2 - 68.88 \times \sqrt{T_f} + 16.954 \times T_f$$

$$+ 0.0335 \times T_f^2 - 1.654 \times 10^{-5} \times T_f^3 + 3.32 \times 10^{-9} \times T_f^4 - 11657.5$$

$$\tag{18}$$

Equation (18) can be solved for T_f using a nonlinear equation solver. Solving it using the goal seek function in MS Excel yields $T_f = 479.35$ K. Using this value of the final temperature and final volume = 2 (initial volume), we can obtain the final pressure from the equation of state as follows:

$$P = \frac{8.314 \times 479.3478}{\left(2 \times 3.2125 \times 10^{-4} - 3.54 \times 10^{-5}\right)}$$

$$- \frac{0.7444 \times \left(1 + 0.7447\left(1 - \sqrt{\frac{479.3478}{430.8}}\right)\right)^2}{2 \times 3.2125 \times 10^{-4}\left(2 \times 3.2125 \times 10^{-4} + 3.54 \times 10^{-5}\right) + 3.54 \times 10^{-5} \times \left(2 \times 3.2125 \times 10^{-4} - 3.54 \times 10^{-5}\right)}$$

$$= 50.66\,\text{bar}$$

$$(19)$$

Therefore, the final temperature and pressure of the gas are 479.35 K and 50.66 bar, respectively. Note that the final temperature of the gas modeled using the Peng-Robinson equation of state is lower than that for an ideal gas. This is because, in this case, the molar internal energy of the gas increased due to an increase in the volume. However, because the system (the gas) is isolated, the molar internal energy of the system cannot increase. Therefore, to maintain constant molar internal energy, the gas temperature had to decrease so that it offsets the increase in internal energy due to the increase in volume.

4. **Steps to avoid**

   1. Setting $\Delta \underline{S} = 0$ – Although the system is an isolated system undergoing an adiabatic expansion, the expansion is not reversible in this case, and therefore, the process is not isentropic.
   2. Using T and P as the two independent intensive variables when using a pressure-explicit equation of state – Although there is nothing fundamentally incorrect in using T and P as the two independent intensive variables, it is not advisable to use them in this case because this will lead to a significantly more complicated mathematical derivation.

5. **Other solution strategies**

We could also use the departure-function approach presented in Part I. In addition, other suitable equations of state may be used.

Problem 10

Adapted from Problem 8.15 in Tester and Modell

Oxygen gas at 150 K and 30 bar is to be pressurized to 100 bar in a compressor with an efficiency of 80% (based on an isentropic process). If the flow rate is 10 kg/s, what would be the power required? Carry out your calculation using the Peng-Robinson equation of state, and use both the attenuated state approach and the departure function approach. The following information is available: $T_c = 154.6K$, $P_c = 50.46$ bar, and $\omega = 0.021$. Also assume that C_P is 29.3 J/mol K, independent of temperature.

Solution to Problem 10

Solution Strategy

To solve this problem, we begin with:

$$\delta W = \frac{\delta W_{isen}}{0.80}$$

where δW is the real, differential work done by the compressor on the gas, which includes pressurizing the δn moles and the flow work associated with the δn moles, δW_{isen} is the differential work done by the compressor on the gas in an isentropic process, and 0.80 is the efficiency of the actual compressor relative to the isentropic process. Recall that this is not the Carnot efficiency introduced in Part I!

We recognize that δW is also the differential amount of work that must be supplied to the compressor so that it can compress δn moles of gas in this open system. Therefore, dividing both sides of the equation above by dt yields:

$$\frac{\delta W}{dt} = \frac{1}{0.80} \frac{\delta W_{isen}}{dt}$$

$$\dot{W} = \frac{1}{0.80} \dot{W}_{isen} \tag{1}$$

Based on Eq. (1), it is clear that more power needs to be supplied to the compressor in the real process than in the isentropic process. This makes physical sense because the isentropic process is reversible. Therefore, we can first calculate the power required in the isentropic process and then divide this value by 0.80 to evaluate the actual power required.

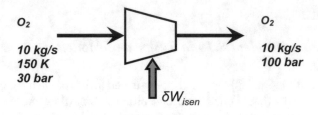

Fig. 1

In Fig. 1, $\delta Q = 0$, because an isentropic process is equivalent to a reversible, adiabatic process (see below). Carrying out an entropy balance on the gas in the compressor yields:

$$d\underline{S} = 0 = \frac{\delta Q}{T} + (S_{in} - S_{out})\delta n + d\underline{\sigma} \text{ (Steady state)} \tag{2}$$

To derive Eq. (2), we used the fact that at steady state, $\delta n_{in} = \delta n_{out} = \delta n$. If the process is adiabatic ($\delta Q = 0$) and reversible ($d\underline{\sigma} = 0$), Eq. (2) reduces to:

$$(S_{in} - S_{out})\delta n = 0$$

Because $\delta n \neq 0$, the equation above shows that:

$$S_{in} = S_{out} \text{ (For an isentropic process)}$$

Next, let us carry out a First Law of Thermodynamics analysis of the gas in the compressor:

$$d\underline{U} = \delta Q + \delta W + H_{in}\delta n_{in} - H_{out}\delta n_{out} \tag{3}$$

$$d\underline{U} = 0 \text{ (Steady state)}$$

$$\delta Q = 0 \quad \text{(Adiabatic)}$$

$$\delta n_{in} = \delta n_{out} = \delta n \quad \text{(Steady state)}$$

Therefore,

$$\delta W_{isen} + (H_{in} - H_{out})\delta n = 0$$

where we have highlighted the fact that the differential work corresponds to that in an isentropic process! Dividing the last equation by dt and rearranging yields:

$$0 = \dot{W}_{isen} + (H_{in} - H_{out})\dot{n}$$
$$\dot{W}_{isen} = (H_{out} - H_{in})\dot{n} = (H_2 - H_1)\dot{n} \tag{3a}$$

Derivations

We are given an ideal gas heat capacity, and therefore, we need to carry out the temperature change when the gas is ideal. We are also given an equation of state that describes the gas. With this given information, as discussed in Part I, we can use the "attenuated state approach" or the "departure function approach" to determine the change in any derived intensive property of the gas.

Essentially, as shown in Eq. (3a), we need to calculate $(H_2 - H_1)$, so that we can evaluate $\dot{W}_{isen}$ and then $\dot{W}$ using Eq. (1). In order to calculate $(H_2 - H_1)$, we must also use the fact that $(S_2 - S_1) = 0$. The "attenuated state approach" or the "departure function approach" can be used to derive expressions for $(S_2 - S_1)$ and $(H_2 - H_1)$. We begin with the attenuated state approach.

Attenuated-State Approach

Because we are given a pressure-explicit EOS, we choose T and V as the two independent intensive variables. In the (V-T) phase diagram shown in Fig. 2, we see that:

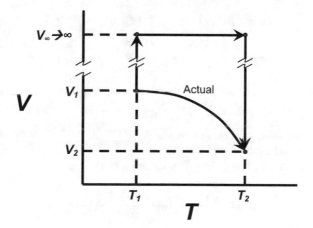

Fig. 2

(i) $V_2 < V_1$, because the gas is compressed, and (ii) $T_2 > T_1$, because work is done on the gas without heat transfer, and as a result, the internal energy of the gas increases.

Let us first derive an expression for $(S_2 - S_1)$, where $S\,(T,V)$. The differential of S can be written as follows:

$$dS = \left(\frac{\partial S}{\partial T}\right)_V dT + \left(\frac{\partial S}{\partial V}\right)_T dV$$

In Part I, we saw that the fist partial derivative in the last equation can be related to the heat capacity at constant volume and the second partial derivative can be related to a partial derivative that can be calculated using a pressure-explicit EOS. Specifically:

$$dS = \frac{C_V}{T} dT + \left(\frac{\partial P}{\partial T}\right)_V dV \tag{4}$$

Working with C_V, although we are given C_P, is no problem when we deal with an ideal gas. Indeed, as shown in Fig. 2, we will be changing the temperature when the gas behaves ideally, for which the heat capacities are related by the following equation derived in Part I:

$$C_V = C_P - R \tag{5}$$

As shown in Fig. 2, $(S_2 - S_1)$ can be calculated using an isothermal step followed by an isochoric step and then completed with a second isothermal step. Specifically:

$$S_2 - S_1 = \underbrace{\int_{V_1}^{V_\infty} \left(\frac{\partial P}{\partial T}\right)_V \bigg|_{T_1} dV}_{\text{constant } T} + \underbrace{\int_{T_1}^{T_2} \frac{C_V^0}{T} dT}_{\text{constant } V} + \underbrace{\int_{V_\infty}^{V_2} \left(\frac{\partial P}{\partial T}\right)_V \bigg|_{T_2} dV}_{\text{constant } T} \tag{6}$$

As discussed in Part I, it is generally good practice to add and subtract $\int_{V_1}^{V_\infty} \left(\frac{\partial P^{\text{ideal}}}{\partial T}\right)_V \bigg|_{T_1} dV$ as well as $\int_{V_2}^{V_\infty} \left(\frac{\partial P^{\text{ideal}}}{\partial T}\right)_V \bigg|_{T_2} dV$ to handle the two $\ln(V_\infty \to \infty)$ singularities. Nevertheless, in this solution, we will work directly with Eq. (6), without adding and subtracting integrals.

As discussed in Part I, the Peng-Robinson EOS is given by:

$$P = \frac{RT}{V - b} - \frac{a(\omega, T_r)}{V(V + b) + b(V - b)} \tag{7}$$

where

$$a(\omega, T_r) = a_c \alpha(\omega, T_r) \tag{8}$$

$$a_c = \frac{0.45724R^2T_c^2}{P_c} \tag{9}$$

$$\alpha(\omega, T_r) = \left[1 + \kappa\left(1 - \sqrt{T_r}\right)\right]^2 \tag{10}$$

$$\kappa = 0.37464 + 1.54226\omega - 0.26992\omega^2 \tag{11}$$

$$b = \frac{0.07780RT_c}{P_c} \tag{12}$$

$$\left(\frac{\partial P}{\partial T}\right)_V = \frac{R}{V-b} - \frac{1}{V(V+b)+b(V-b)} \underbrace{\frac{da(,T_r)}{dT}}$$

$a'(T)$ is a full
derivative, because
"a" is not a function
of P or V

where

$$\frac{da}{dT} = a'(T) = a_c\frac{d\alpha}{dT} = a_c\frac{d}{dT}\left(\left[1 + \kappa\left(1 - \sqrt{T_r}\right)\right]^2\right)$$

It then follows that:

$$a'(T) = a_c 2\left[1 + \kappa\left(1 - \left(\frac{T}{T_C}\right)^{1/2}\right)\right]\left[-\kappa\frac{1}{2}\left(\frac{1}{T_C}\frac{1}{T}\right)^{1/2}\right]$$

$$= \frac{-a_c\kappa\left[1 + \kappa\left(1 - \left(\frac{T}{T_C}\right)^{1/2}\right)\right]}{\sqrt{TT_C}}$$

Let us next derive a general expression for $\int_{V_i}^{V_f}\left(\frac{\partial P}{\partial T}\right)_V\Big|_{T_0} dV$, which we will need to use twice in Eq. (6):

$$\int_{V_i}^{V_f} \left(\frac{\partial P}{\partial T}\right)_V\Big|_{T_0} dV = R\int_{V_i}^{V_f} \frac{dV}{V-b} - a'(T)\underbrace{\int_{V_i}^{V_f} \frac{dV}{V^2 + 2bV - b^2}}$$

Using Maple to find the integral

$$= R\ln\left(\frac{V_f - b}{V_i - b}\right) - \frac{a'(T)}{2b\sqrt{2}}\left\{\ln\left(\frac{V_f + b(1-\sqrt{2})}{V_f + b(1+\sqrt{2})}\right) - \ln\left(\frac{V_i + b(1-\sqrt{2})}{V_i + b(1+\sqrt{2})}\right)\right\}$$

$$(13)$$

In addition, we know that:

$$\int_{T_1}^{T_2} \frac{C_V^0}{T} dT = C_V^0 \ln\left(\frac{T_2}{T_1}\right) \tag{14}$$

Using Eqs. (13) and (14) in Eq. (6) yields:

$$S_2 - S_1 = \int_{V_1}^{V_\infty}\left(\frac{\partial P}{\partial T}\right)_V\Big|_{T_1} dV + \int_{T_1}^{T_2}\frac{C_V^0}{T}dT + \int_{V_\infty}^{V_2}\left(\frac{\partial P}{\partial T}\right)_V\Big|_{T_2} dV$$

$$= R\ln\left(\frac{V_\infty - b}{V_1 - b}\right) - \frac{a'(T_1)}{2b\sqrt{2}}\left\{\ln\left(\frac{V_\infty + b(1-\sqrt{2})}{V_\infty + b(1+\sqrt{2})}\right) - \ln\left(\frac{V_1 + b(1-\sqrt{2})}{V_1 + b(1+\sqrt{2})}\right)\right\}$$

$$+ C_V^0 \ln\left(\frac{T_2}{T_1}\right)$$

$$+ R\ln\left(\frac{V_2 - b}{V_\infty - b}\right) - \frac{a'(T_2)}{2b\sqrt{2}}\left\{\ln\left(\frac{V_2 + b(1-\sqrt{2})}{V_2 + b(1+\sqrt{2})}\right) - \ln\left(\frac{V_\infty + b(1-\sqrt{2})}{V_\infty + b(1+\sqrt{2})}\right)\right\}$$

$$(15)$$

Rearranging Eq. (15) and recognizing that $S_2 - S_1 = 0$, we obtain:

$$0 = C_V^0 \ln\left(\frac{T_2}{T_1}\right) + R\ln\left(\frac{V_2 - b}{V_1 - b}\right) + R\ln\left(\frac{\cancel{V_\infty - b}}{\cancel{V_\infty - b}}\right)$$

$$- \frac{a'(T_1)}{2b\sqrt{2}}\left\{\ln\left(\frac{V_\infty + b(1-\sqrt{2})}{V_\infty + b(1+\sqrt{2})}\right) - \ln\left(\frac{V_1 + b(1-\sqrt{2})}{V_1 + b(1+\sqrt{2})}\right)\right\}$$

$$- \frac{a'(T_2)}{2b\sqrt{2}}\left\{\ln\left(\frac{V_2 + b(1-\sqrt{2})}{V_2 + b(1+\sqrt{2})}\right) - \ln\left(\frac{V_\infty + b(1-\sqrt{2})}{V_\infty + b(1+\sqrt{2})}\right)\right\}$$

Recall that $a'(T)$ refers to the derivative of $a(\omega, T_r)$ with respect to T.
Because $V_\infty \to \infty$, we can further simplify the last equation as follows:

$$0 = C_V^0 \ln\left(\frac{T_2}{T_1}\right) + R \ln\left(\frac{V_2 - b}{V_1 - b}\right)$$

$$-\frac{a'(T_1)}{2b\sqrt{2}}\left\{\ln\left(\frac{V_\infty + b\left(1 - \sqrt{2}\right)}{V_\infty + b\left(1 + \sqrt{2}\right)}\right) - \ln\left(\frac{V_1 + b\left(1 - \sqrt{2}\right)}{V_1 + b\left(1 + \sqrt{2}\right)}\right)\right\}$$

$$-\frac{a'(T_2)}{2b\sqrt{2}}\left\{\ln\left(\frac{V_2 + b\left(1 - \sqrt{2}\right)}{V_2 + b\left(1 + \sqrt{2}\right)}\right) - \ln\left(\frac{V_\infty + b\left(1 - \sqrt{2}\right)}{V_\infty + b\left(1 + \sqrt{2}\right)}\right)\right\}$$

or

$$0 = C_V^0 \ln\left(\frac{T_2}{T_1}\right) + R\ln\left(\frac{V_2 - b}{V_1 - b}\right) - \frac{a'(T_1)}{2b\sqrt{2}}\ln\left(\frac{V_1 + b(1 - \sqrt{2})}{V_1 + b(1 + \sqrt{2})}\right)$$

$$-\frac{a'(T_2)}{2b\sqrt{2}}\ln\left(\frac{V_2 + b(1 - \sqrt{2})}{V_2 + b(1 + \sqrt{2})}\right) \tag{16}$$

We know that $T_1 = 150\,\mathrm{K}$, and we can calculate V_1 by numerically solving Eq. (7) with $T_1 = 150\,\mathrm{K}$ and $P_1 = 30 \times 10^5\,\mathrm{Pa}$. Specifically, we solved Eq. (7) using Maple, and V_1 was found to be $2.936 \times 10^{-4}\,\mathrm{m^3/mol}$. Therefore, in Eq. (16), T_2 and V_2 are the only two unknowns. Because we know that $P_2 = 100 \times 10^5\,\mathrm{Pa}$, T_2 and V_2 are also the only two unknowns in Eq. (7), the Peng-Robinson equation of state. Consequently, T_2 and V_2 can be calculated numerically by simultaneously solving Eq. (7) and Eq. (16) using Maple. This yields:

$$T_2 = 217.7\,\mathrm{K}$$

$$V_2 = 1.361 \times 10^{-4}\,\mathrm{m^3/mol}$$

Now that we have (T_1, V_1) and (T_2, V_2), we can calculate $(H_2 - H_1)$ by expanding dH in terms of dT and dV. Recall that $(H_2 - H_1)$ is the key quantity that we are after.

Like we did for the entropy, in order to utilize the pressure-explicit Peng-Robinson EOS, let us expand dH in terms of dT and dV. Specifically:

$$dH = \left(\frac{\partial H}{\partial T}\right)_V dT + \left(\frac{\partial H}{\partial V}\right)_T dV$$

In Part I, we saw that:

$$\left(\frac{\partial H}{\partial T}\right)_V = C_V + V\left(\frac{\partial P}{\partial T}\right)_V$$

$$\left(\frac{\partial H}{\partial V}\right)_T = T\left(\frac{\partial P}{\partial T}\right)_V + V\left(\frac{\partial P}{\partial V}\right)_T$$

Using the last two results in the expression for dH above, we obtain:

$$dH = \left[C_V + V\left(\frac{\partial P}{\partial T}\right)_V\right]dT + \left[T\left(\frac{\partial P}{\partial T}\right)_V + V\left(\frac{\partial P}{\partial V}\right)_T\right]dV$$

We again proceed with the integration of the last equation along the isothermal (at T_1) $\rightarrow$ isochoric (at V_∞) $\rightarrow$ isothermal (at T_2) path indicated in the (V-T) phase diagram shown in Fig. 2. This yields:

$$H_2 - H_1 = \underbrace{\int_{V_1}^{V_\infty} \left[T\left(\frac{\partial P}{\partial T}\right)_V + V\left(\frac{\partial P}{\partial V}\right)_T\right]\Bigg|_{T_1} dV}_{dT=0} + \underbrace{\int_{T_1}^{T_2} \left[C_V^0 + V\left(\frac{\partial P^{Ideal}}{\partial T}\right)_V\right]\Bigg|_{V_\infty} dT}_{dV=0}$$

$$+ \underbrace{\int_{V_\infty}^{V_2} \left[T\left(\frac{\partial P}{\partial T}\right)_V + V\left(\frac{\partial P}{\partial V}\right)_T\right]\Bigg|_{T_2} dV}_{dT=0}$$

or

$$H_2 - H_1 = T_1 \int_{V_1}^{V_\infty} \left(\frac{\partial P}{\partial T}\right)_V\Bigg|_{T_1} dV + \int_{V_1}^{V_\infty} V\left(\frac{\partial P}{\partial V}\right)_T\Bigg|_{T_1} dV + C_V^0(T_2 - T_1) + \int_{T_1}^{T_2} V\left(\frac{\partial}{\partial T}\left(\frac{RT}{V}\right)\right)_V\Bigg|_{V_\infty} dT$$

$$+ T_2 \int_{V_\infty}^{V_2} \left(\frac{\partial P}{\partial T}\right)_V\Bigg|_{T_2} dV + \int_{V_\infty}^{V_2} V\left(\frac{\partial P}{\partial V}\right)_T\Bigg|_{T_2} dV$$

Carrying out the temperature integration in the last equation yields:

$$H_2 - H_1 = T_1 \int_{V_1}^{V_\infty} \left(\frac{\partial P}{\partial T}\right)_V\Bigg|_{T_1} dV + \int_{V_1}^{V_\infty} V\left(\frac{\partial P}{\partial V}\right)_T\Bigg|_{T_1} dV + C_V^0(T_2 - T_1) + R(T_2 - T_1)$$

$$+ T_2 \int_{V_\infty}^{V_2} \left(\frac{\partial P}{\partial T}\right)_V\Bigg|_{T_2} dV + \int_{V_\infty}^{V_2} V\left(\frac{\partial P}{\partial V}\right)_T\Bigg|_{T_2} dV$$

$$\tag{17}$$

Using Eq. (13), the general expression for $\int_{V_i}^{V_f} \left(\frac{\partial P}{\partial T}\right)_V \Big|_{T_0} dV$, in the two pertinent integrals in Eq. (17), we obtain:

$$
T_1 \int_{V_1}^{V_\infty} \left(\frac{\partial P}{\partial T}\right)_V \Big|_{T_1} dV + T_2 \int_{V_\infty}^{V_2} \left(\frac{\partial P}{\partial T}\right)_V \Big|_{T_2} dV
$$

$$
= T_1 R \ln\left(\frac{V_\infty - b}{V_1 - b}\right) - T_1 \frac{a'(T_1)}{2b\sqrt{2}} \left\{ \ln\left(\frac{V_\infty + b(1 - \sqrt{2})}{V_\infty + b(1 + \sqrt{2})}\right) - \ln\left(\frac{V_1 + b(1 - \sqrt{2})}{V_1 + b(1 + \sqrt{2})}\right) \right\}
$$

$$
\underbrace{\phantom{\ln\left(\frac{V_\infty + b(1 - \sqrt{2})}{V_\infty + b(1 + \sqrt{2})}\right)}}_{0,\, as\ V_\infty \text{ goes to infinity}}
$$

$$
+ T_2 R \ln\left(\frac{V_2 - b}{V_\infty - b}\right) - T_2 \frac{a'(T_2)}{2b\sqrt{2}} \left\{ \ln\left(\frac{V_2 + b(1 - \sqrt{2})}{V_2 + b(1 + \sqrt{2})}\right) - \ln\left(\frac{V_\infty + b(1 - \sqrt{2})}{V_\infty + b(1 + \sqrt{2})}\right) \right\}
$$

$$
\underbrace{\phantom{\ln\left(\frac{V_\infty + b(1 - \sqrt{2})}{V_\infty + b(1 + \sqrt{2})}\right)}}_{0,\, as\ V_\infty \text{ goes to infinity}}
$$

$$
= T_1 R \ln\left(\frac{V_\infty - b}{V_1 - b}\right) + T_1 \frac{a'(T_1)}{2b\sqrt{2}} \ln\left(\frac{V_1 + b(1 - \sqrt{2})}{V_1 + b(1 + \sqrt{2})}\right)
$$

$$
+ T_2 R \ln\left(\frac{V_2 - b}{V_\infty - b}\right) + T_2 \frac{a'(T_2)}{2b\sqrt{2}} \ln\left(\frac{V_2 + b(1 - \sqrt{2})}{V_2 + b(1 + \sqrt{2})}\right)
$$

$$
\tag{18}
$$

Before we can proceed further, we need to find the general expression for:

$$
\int_{V_i}^{V_f} V \left(\frac{\partial P}{\partial V}\right)_T \Big|_{T_0} dV
$$

which we will then use with the appropriate integration limits in Eq. (17). Based on Eq. (7), we can directly take the partial derivative of P with respect to V at constant temperature. This yields:

$$
\left(\frac{\partial P}{\partial V}\right)_T = -\frac{RT}{(V - b)^2} + \frac{a(T)(2V + 2b)}{\left(V^2 + 2bV - b^2\right)^2}
$$

Then,

$$
\int_{V_i}^{V_f} V \left(\frac{\partial P}{\partial V}\right)_T \Big|_{T_0} dV = -RT_0 \int_{V_i}^{V_f} \frac{V}{(V - b)^2} dV + a(T_0) \int_{V_i}^{V_f} \frac{V(2V + 2b)}{\left(V^2 + 2bV - b^2\right)^2} dV
$$

Using Maple, we can simplify the last result as follows:

$$
\begin{aligned}
\int_{V_i}^{V_f} V\left(\frac{\partial P}{\partial V}\right)_T\bigg|_{T_0} dV = {} & RT_0\left[\frac{b}{V_f - b} - \frac{b}{V_i - b}\right] - RT_0 \ln\left(\frac{V_f - b}{V_i - b}\right) \\
& -a(T_0)\left[\frac{V_f}{V_f{}^2 + 2bV_f - b^2} - \frac{V_i}{V_i{}^2 + 2bV_i - b^2}\right] \\
& +\frac{a'(T_0)}{2b\sqrt{2}} \ln\left\{\left(\frac{V_f + b(1 - \sqrt{2})}{V_f + b(1 + \sqrt{2})}\right)\left(\frac{V_i + b(1 + \sqrt{2})}{V_i + b(1 - \sqrt{2})}\right)\right\}
\end{aligned} \tag{19}
$$

We can then use Eq. (19) for the two relevant dV integrals in Eq. (17). This yields:

$$
\begin{aligned}
\int_{V_1}^{V_\infty} & V\left(\frac{\partial P}{\partial V}\right)_T\bigg|_{T_1} dV + \int_{V_\infty}^{V_2} V\left(\frac{\partial P}{\partial V}\right)_T\bigg|_{T_2} dV \\
= {} & RT_1\left[\frac{b}{V_\infty - b} - \frac{b}{V_1 - b}\right] - RT_1 \ln\left(\frac{V_\infty - b}{V_1 - b}\right) \\
& -a(T_1)\left[\frac{V_\infty}{V_\infty{}^2 + 2bV_\infty - b^2} - \frac{V_1}{V_1{}^2 + 2bV_1 - b^2}\right] \\
& +\frac{a'(T_1)}{2b\sqrt{2}} \ln\left\{\left(\frac{V_\infty + b(1 - \sqrt{2})}{V_\infty + b(1 + \sqrt{2})}\right)\left(\frac{V_1 + b(1 + \sqrt{2})}{V_1 + b(1 - \sqrt{2})}\right)\right\} \\
& +RT_2\left[\frac{b}{V_2 - b} - \frac{b}{V_\infty - b}\right] - RT_2 \ln\left(\frac{V_2 - b}{V_\infty - b}\right) \\
& -a(T_2)\left[\frac{V_2}{V_2{}^2 + 2bV_2 - b^2} - \frac{V_\infty}{V_\infty{}^2 + 2bV_\infty - b^2}\right] \\
& +\frac{a'(T_2)}{2b\sqrt{2}} \ln\left\{\left(\frac{V_2 + b(1 - \sqrt{2})}{V_2 + b(1 + \sqrt{2})}\right)\left(\frac{V_\infty + b(1 + \sqrt{2})}{V_\infty + b(1 - \sqrt{2})}\right)\right\}
\end{aligned} \tag{20}
$$

Substituting Eqs. (19) and (20) in Eq. (17), we obtain:

$$H_2 - H_1 = RT_1 \left[\frac{b}{V_\infty - b} - \frac{b}{V_1 - b} \right] - RT_1 \ln\left(\frac{V_\infty - b}{V_1 - b} \right)$$

$$-a(T_1) \left[\frac{V_\infty}{V_\infty{}^2 + 2bV_\infty - b^2} - \frac{V_1}{V_1{}^2 + 2bV_1 - b^2} \right]$$

$$+\frac{a'(T_1)}{2b\sqrt{2}} \ln\left\{ \left(\frac{V_\infty + b(1 - \sqrt{2})}{V_\infty + b(1 + \sqrt{2})} \right) \left(\frac{V_1 + b(1 + \sqrt{2})}{V_1 + b(1 - \sqrt{2})} \right) \right\}$$

$$+RT_2 \left[\frac{b}{V_2 - b} - \frac{b}{V_\infty - b} \right] - RT_2 \ln\left(\frac{V_2 - b}{V_\infty - b} \right)$$

$$-a(T_2) \left[\frac{V_2}{V_2{}^2 + 2bV_2 - b^2} - \frac{V_\infty}{V_\infty{}^2 + 2bV_\infty - b^2} \right]$$

$$+\frac{a'(T_2)}{2b\sqrt{2}} \ln\left\{ \left(\frac{V_2 + b(1 - \sqrt{2})}{V_2 + b(1 + \sqrt{2})} \right) \left(\frac{V_\infty + b(1 + \sqrt{2})}{V_\infty + b(1 - \sqrt{2})} \right) \right\}$$

$$+C_V^0(T_2 - T_1) + R(T_2 - T_1)$$

$$+T_1 R \ln\left(\frac{V_\infty - b}{V_1 - b} \right) + T_1 \frac{a'(T_1)}{2b\sqrt{2}} \ln\left(\frac{V_1 + b(1 - \sqrt{2})}{V_1 + b(1 + \sqrt{2})} \right)$$

$$+T_2 R \ln\left(\frac{V_2 - b}{V_\infty - b} \right) + T_2 \frac{a'(T_2)}{2b\sqrt{2}} \ln\left(\frac{V_2 + b(1 - \sqrt{2})}{V_2 + b(1 + \sqrt{2})} \right)$$

Combining terms in the last equation and letting V_∞ go to infinity, we obtain:

$$H_2 - H_1 = RT_2 \frac{b}{V_2 - b} - RT_1 \frac{b}{V_1 - b} + a(T_1) \left[\frac{V_1}{V_1{}^2 + 2bV_1 - b^2} \right]$$

$$-a(T_2) \left[\frac{V_2}{V_2{}^2 + 2bV_2 - b^2} \right]$$

$$-\frac{a'(T_1) - T_1 a'(T_1)}{2b\sqrt{2}} \ln\left(\frac{V_1 + b(1 - \sqrt{2})}{V_1 + b(1 + \sqrt{2})} \right) \qquad (21)$$

$$+\frac{a(T_2) - T_2 a'(T_2)}{2b\sqrt{2}} \ln\left(\frac{V_2 + b(1 - \sqrt{2})}{V_2 + b(1 + \sqrt{2})} \right)$$

$$+C_P^0(T_2 - T_1)$$

Equation (21) is the final expression for $H_2 - H_1$. Substituting T_1, V_1, T_2, and V_2, we find that:

$$\Delta H = H_2 - H_1 = 1336\,\mathrm{J/mol}$$

$$\dot{W}_{isen} = (H_2 - H_1)\dot{n} = (1336\,\text{J/mol})(10\,\text{kg/s} \cdot 1000\,\text{g/kg} \cdot \text{mol/32}\,\text{g})$$
$$= 417500\,\text{J/s}$$

$$\dot{W} = \frac{\dot{W}_{isen}}{0.80} = \frac{417500\,\text{J/s}}{0.80} = 522\,\text{kW}$$

Next, we will pursue the departure function approach.

Departure Function Approach

As discussed in Part I, it is convenient to first analyze the Helmholtz free energy departure function, DA, because we chose T and V as the two independent intensive variables. Again, recall that this choice was motivated because we have access to the Peng-Robinson pressure-explicit EOS. The (V-T) phase diagram in Fig. 3 shows the three steps that we need to follow to implement the departure function approach.

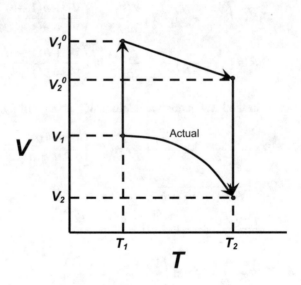

Fig. 3

Following the derivation in Part I, it follows that:

$$A(T, V) - A^0\left(T, V^0\right) = -\int_{V_\infty}^{V} \left[P - \frac{RT}{V}\right] dV + RT \ln\left(\frac{V^0}{V}\right) \tag{22}$$

In addition, in Part I, we showed that:

$$S(T,V) - S^0(T,V^0) = \left(\frac{\partial}{\partial T}\left[\int_{V_\infty}^{V}\left[P - \frac{RT}{V}\right]dV\right]\right)_V - R\ln\left(\frac{V^0}{V}\right) \qquad (23)$$

$$U(T,V) - U^0(T,V^0) = \left[A(T,V) - A^0(T,V^0)\right] + T\left[S(T,V) - S^0(T,V^0)\right] \quad (24)$$

$$H(T,V) - H^0(T,V^0) = \left[U(T,V) - U^0(T,V^0)\right] + (PV - RT) \qquad (25)$$

Carrying out the integrals in Eqs. (23) and (24), one can show that:

$$S(T,V) - S^0(T,V^0) = -R\ln\frac{V^0}{V-b} - \frac{a'(T)}{2\sqrt{2}b}\ln\left[\frac{V + b(1 - \sqrt{2})}{V + b(1 + \sqrt{2})}\right] \qquad (26)$$

and

$$H(T,V) - H^0(T,V^0) = \frac{a(T)}{2\sqrt{2}b}\ln\left[\frac{V + b(1 - \sqrt{2})}{V + b(1 + \sqrt{2})}\right]$$
$$- \frac{Ta'(T)}{2\sqrt{2}b}\ln\left[\frac{V + b(1 - \sqrt{2})}{V + b(1 + \sqrt{2})}\right] + PV - RT \qquad (27)$$

Using Eq. (26), let us next calculate $S_2 - S_1 = S(T_2, V_2) - S(T_1, V_1)$. Specifically:

$$S(T_2, V_2) - S(T_1, V_1) = \underbrace{\left[S(T_2, V_2) - S^0(T_2, V_2^0)\right]}_{(1)}$$
$$+ \underbrace{\left[S^0(T_2, V_2^0) - S^0(T_1, V_1^0)\right]}_{(2)} \qquad (28)$$
$$+ \underbrace{\left[S^0(T_1, V_1^0) - S(T_1, V_1)\right]}_{(3)}$$

The first term (1) and the third term (3) are departure functions at T_2 and T_1, respectively. The second term involves a temperature and volume change when the gas behaves ideally. Using Eq. (26) as needed, we obtain:

$$\underbrace{\left[S(T_2, V_2) - S^0(T_2, V_2^0)\right]}_{(1)} = -R\ln\frac{V_2^0}{V_2 - b} - \frac{a'(T_2)}{2\sqrt{2}b}\ln\left[\frac{V_2 + b(1 - \sqrt{2})}{V_2 + b(1 + \sqrt{2})}\right]$$

and

$$\underbrace{\left[S^0\left(T_1, V_1^0\right) - S\left(T_1, V_1\right)\right]}_{(3)} = -R\ln\frac{V_1^0}{V_1 - b} - \frac{a'(T_1)}{2\sqrt{2}b}\ln\left[\frac{V_1 + b\left(1 - \sqrt{2}\right)}{V_1 + b\left(1 + \sqrt{2}\right)}\right]$$

For the ideal gas contribution (2), we obtain:

$$\underbrace{\left[S^0\left(T_2, V_2^0\right) - S^0\left(T_1, V_1^0\right)\right]}_{(2)} = C_V^0\ln\left(\frac{T_2}{T_1}\right) + R\ln\left(\frac{V_2^0}{V_1^0}\right)$$

Substituting the three last equations in Eq. (28), we obtain:

$$S(T_2, V_2) - S(T_1, V_1) = C_V^0\ln\left(\frac{T_2}{T_1}\right) + R\ln\frac{V_2 - b}{V_1 - b}$$
$$+ \frac{a'(T_1)}{2\sqrt{2}b}\ln\left[\frac{V_1 + b\left(1 - \sqrt{2}\right)}{V_1 + b\left(1 + \sqrt{2}\right)}\right] - \frac{a'(T_2)}{2\sqrt{2}b}\ln\left[\frac{V_2 + b\left(1 - \sqrt{2}\right)}{V_2 + b\left(1 + \sqrt{2}\right)}\right]$$
$$(29)$$

Because $S_2 - S_1 = S(T_2, V_2) - S(T_1, V_1) = 0$, Eq. (29) yields:

$$C_V^0\ln\left(\frac{T_2}{T_1}\right) + R\ln\frac{V_2 - b}{V_1 - b} + \frac{a'(T_1)}{2\sqrt{2}b}\ln\left[\frac{V_1 + b\left(1 - \sqrt{2}\right)}{V_1 + b\left(1 + \sqrt{2}\right)}\right]$$
$$- \frac{a'(T_2)}{2\sqrt{2}b}\ln\left[\frac{V_2 + b\left(1 - \sqrt{2}\right)}{V_2 + b\left(1 + \sqrt{2}\right)}\right]$$
$$= 0$$

Note that the last equation is identical to Eq. (16) that we derived above using the attenuated state approach!

Next, let us calculate $H_2 - H_1 = H(T_2, V_2) - H(T_1, V_1)$. Specifically:

$$H(T_2, V_2) - H(T_1, V_1) = \underbrace{\left[H(T_2, V_2) - H^0\left(T_2, V_2^0\right)\right]}_{(1)}$$
$$+ \underbrace{\left[H^0\left(T_2, V_2^0\right) - H^0\left(T_1, V_1^0\right)\right]}_{(2)} \qquad (30)$$
$$+ \underbrace{\left[H^0\left(T_1, V_1^0\right) - H(T_1, V_1)\right]}_{(3)}$$

Like in the entropy case, the first term (1) and the third term (3) are departure functions at T_2 and T_1, respectively. The second term involves a temperature and volume change when the gas behaves ideally. The three contributions in Eq. (30) are given by:

$$\underbrace{\left[H(T_2, V_2) - H^0(T_2, V_2^0)\right]}_{(1)} = \frac{a(T_2)}{2\sqrt{2}b} \ln\left[\frac{V_2 + b(1 - \sqrt{2})}{V_2 + b(1 + \sqrt{2})}\right]$$

$$-\frac{T_2 a'(T_2)}{2\sqrt{2}b} \ln\left[\frac{V_2 + b(1 - \sqrt{2})}{V_2 + b(1 + \sqrt{2})}\right]$$

$$+P_2 V_2 - RT_2$$

$$\underbrace{\left[H^0(T_1, V_1^0) - H(T_1, V_1)\right]}_{(3)} = \frac{a(T_1)}{2\sqrt{2}b} \ln\left[\frac{V_1 + b(1 - \sqrt{2})}{V_1 + b(1 + \sqrt{2})}\right]$$

$$-\frac{T_1 a'(T_1)}{2\sqrt{2}b} \ln\left[\frac{V_1 + b(1 - \sqrt{2})}{V_1 + b(1 + \sqrt{2})}\right]$$

$$+P_1 V_1 - RT_1$$

$$\underbrace{\left[H^0(T_2, V_2^0) - H^0(T_1, V_1^0)\right]}_{(2)} = C_P^0(T_2 - T_1)$$

Substituting the last three equations in Eq. (30) yields:

$$H(T_2, V_2) - H(T_1, V_1) = C_P^0(T_2 - T_1) + \underbrace{\left[\frac{RT_2}{V_2 - b} - \frac{a(T_2)}{V_2^2 + 2bV_2 - b^2}\right]}_{P_2} V_2 - RT_2$$

$$-\underbrace{\left[\frac{RT_1}{V_1 - b} - \frac{a(T_1)}{V_1^2 + 2bV_1 - b^2}\right]}_{P_1} V_2 + RT_1$$

$$-\frac{a'(T_1) - T_1 a'(T_1)}{2b\sqrt{2}} \ln\left(\frac{V_1 + b(1 - \sqrt{2})}{V_1 + b(1 + \sqrt{2})}\right)$$

$$+\frac{a(T_2) - T_2 a'(T_2)}{2b\sqrt{2}} \ln\left(\frac{V_2 + b(1 - \sqrt{2})}{V_2 + b(1 + \sqrt{2})}\right)$$

Collecting terms, we obtain:

$$H(T_2, V_2) - H(T_1, V_1) = RT_2 \frac{b}{V_2 - b} - RT_1 \frac{b}{V_1 - b} + a(T_1) \left[\frac{V_1}{V_1{}^2 + 2bV_1 - b^2} \right]$$

$$-a(T_2) \left[\frac{V_2}{V_2{}^2 + 2bV_2 - b^2} \right] - \frac{a'(T_1) - T_1 a'(T_1)}{2b\sqrt{2}} \ln \left(\frac{V_1 + b(1 - \sqrt{2})}{V_1 + b(1 + \sqrt{2})} \right)$$

$$+ \frac{a(T_2) - T_2 a'(T_2)}{2b\sqrt{2}} \ln \left(\frac{V_2 + b(1 - \sqrt{2})}{V_2 + b(1 + \sqrt{2})} \right) + C_P^0(T_2 - T_1)$$

Note that the last equation is identical to Eq. (21) that we derived above using the attenuated state approach!

Solved Problems for Part II

Problem 11

Problem 9.24 in Tester and Modell

The enthalpy of mixing of a ternary solution containing components 1, 2, and 3 is given by:

$$\Delta H_{123} = [100x_1x_2 + 50(x_1x_3 + x_2x_3)]/[x_1 + 2x_2 + 0.5x_3]$$

where ΔH_{123} is measured in J/(mole of solution) and x_i is the mole fraction of component i for $i = 1$, 2, and 3.

(a) Determine the partial molar enthalpy of mixing of component 2 at $x_1 = 0.2$ and $x_2 = 0.5$. Express your result in J/(mole of component 2).

(b) A blender is used to mix two binary solutions to make an equimolar ternary solution of composition $x_1 = x_2 = x_3$. The compositions of the two binary solutions are ($x_1 = 0.333$, $x_2 = 0.667$) and ($x_1 = 0.333$, $x_3 = 0.667$), respectively. If the mixing is carried out isothermally and isobarically as a continuous process operating at steady state, determine the heating load required. Express your result in J/(mole of ternary product).

Solution to Problem 11

Solution strategy

Two general methods can be used to solve this problem:

1. Working with the extensive form of the enthalpy of mixing
2. Working with the intensive form of the enthalpy of mixing

Of course, the two methods are equally valid, but use different sets of variables to calculate the same quantity, which is the partial molar enthalpy of mixing with respect to component 2.

Part (a)

Method 1

This method involves correctly converting the mole fractions to extensive moles. To implement this method, we first multiply the given intensive enthalpy of mixing expression by N to obtain the extensive form and then take the partial derivative with respect to N_2. Specifically, we first multiply ΔH_{123} given in the Problem Statement by N and then we apply the partial molar operator $\left(\dfrac{\partial}{\partial N_2}\right)_{T,P,N_1,N_3}$, that is:

$$\Delta H_{123} = \frac{[100x_1x_2 + 50(x_1x_3 + x_2x_3)]}{[x_1 + 2x_2 + 0.5x_3]} \qquad (1)$$

$$\Delta \underline{H}_{123} = N\Delta H_{123} = \frac{N[100x_1x_2 + 50(x_1x_3 + x_2x_3)]}{[x_1 + 2x_2 + 0.5x_3]}$$

$$= \frac{N\left[100\dfrac{N x_1 x_2}{N^2} + 50\left(\dfrac{N^2 x_1 x_3 + N^2 x_2 x_3}{N^2}\right)\right]}{\left[\dfrac{Nx_1 + 2Nx_2 + 0.5Nx_3}{N}\right]}$$

$$= \frac{100N_1N_2 + 50N_1N_3 + 50N_2N_3}{N_1 + 2N_2 + 0.5N_3}$$

$$\Delta \overline{H}_{123,2} = \left(\frac{\partial \Delta \underline{H}_{123}}{\partial N_2}\right)_{T,P,N_1,N_3}$$

$$= \left(\frac{\partial}{\partial N_2}\left(\frac{100N_1N_2 + 50N_1N_3 + 50N_2N_3}{N_1 + 2N_2 + 0.5N_3}\right)\right)_{T,P,N_1,N_3}$$

Simplifying the last equation, we obtain:

$$\Delta \overline{H}_{123,2} = \frac{100x_1{}^2 + 25(1 - x_1 - x_2)^2}{[x_1 + 2x_2 + 0.5(1 - x_1 - x_2)]^2}$$

At $x_1 = 0.2$ and $x_2 = 0.5$, we obtain:

$$\Delta \overline{H}_{123,2} = \frac{100 \times 0.2^2 + 25(1 - 0.2 - 0.5)^2}{[0.2 + 2 \times 0.5 + 0.5(1 - 0.2 - 0.5)]^2} = 3.42 \, \text{J}/(\text{mol of component 2})$$

Therefore, the partial molar enthalpy of mixing of component 2 at $x_1 = 0.2$ and $x_2 = 0.5$ is 3.42 J/(mol of component 2).

Method 2

In this method, we use an expression for $\Delta \overline{H}_{123,2}$ discussed in Part I, that is:

$$\Delta \overline{H}_{123,2} = \Delta H_{123} - x_1\left(\frac{\partial \Delta H_{123}}{\partial x_1}\right)_{T,P,x_3} - x_3\left(\frac{\partial \Delta H_{123}}{\partial x_1}\right)_{T,P,x_1} \qquad (2)$$

To calculate the partial derivatives in Eq. (2), it is first convenient to replace x_2 by $(1 - x_1 - x_3)$ in Eq. (1). Specifically:

$$\Delta H_{123} = \frac{[100x_1(1 - x_1 - x_3) + 50(x_1x_3 + (1 - x_1 - x_3)x_3)]}{[x_1 + 2(1 - x_1 - x_3) + 0.5x_3]}$$
$$= \frac{100x_1 - 100x_1^2 - 100x_1x_3 + 50x_3 - 50x_3^2}{2 - x_1 - 1.5x_3} \tag{3}$$

We then need to calculate the two partial derivatives in Eq. (2) using Eq. (3). Specifically:

$$\left(\frac{\partial \Delta H_{123}}{\partial x_1} \right)_{T,P,x_3} = \frac{200 - 300x_3 + 100x_1^2 + 300x_1x_3 - 400x_1 + 100x_3^2}{(2 - x_1 - 1.5x_3)^2} \tag{4}$$

and

$$\left(\frac{\partial \Delta H_{123}}{\partial x_3} \right)_{T,P,x_1} = \frac{100 - 50x_1^2 + 100x_1x_3 - 100x_1 - 200x_3 + 75x_3^2}{(2 - x_1 - 1.5x_3)^2} \tag{5}$$

Substituting Eqs. (4) and (5) in Eq. (2) yields:

$$\Delta \overline{H}_{123,2} = \Delta H_{123} - x_1 \frac{200 - 300x_3 + 100x_1^2 + 300x_1x_3 - 400x_1 + 100x_3^2}{(2 - x_1 - 1.5x_3)^2}$$
$$- x_3 \frac{100 - 50x_1^2 + 100x_1x_3 - 100x_1 - 200x_3 + 75x_3^2}{(2 - x_1 - 1.5x_3)^2} \tag{6}$$

At $x_1 = 0.2$, $x_2 = 0.5$, and $x_3 = 1 - x_1 - x_2 = 0.3$, Eq. (6) yields:

$$\Delta \overline{H}_{123,2} = 3.42 \, \text{J}/(\text{mol of component 2})$$

As expected, this is the same answer that we obtained using Method 1.

Part (b)

To solve this part of the problem, we first recall our discussions in Part I to evaluate the thermodynamic behavior of a given system (the blender, in this case). A diagram of the system in question is shown in Fig. 1.

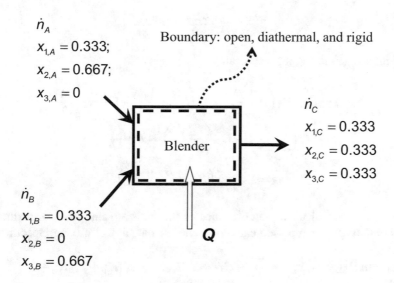

Fig. 1

We first write the First Law of Thermodynamics for the blender (chosen as the system), which is surrounded by a boundary that is open, diathermal, and rigid (see the dashed boundary in Fig. 1). Specifically:

$$dU = \delta W + \delta Q + \sum H_{in} \delta n_{in} - \sum H_{out} \delta n_{out} \tag{7}$$

Because the boundary is rigid ($\delta W = 0$) and the blender operates at steady state ($\frac{dU}{dt} = 0$), Eq. (7) reduces to:

$$0 = \dot{Q} + H_A \frac{dN_A}{dt} + H_B \frac{dN_B}{dt} + H_C \frac{dN_C}{dt} \tag{8}$$

Although Eqs. (7) and (8) look very similar to those that we encountered many times in Part I, it is imperative to recognize that each stream that appears in these equations involves a mixture.

Given the steady-state operation of the blender, we know that there is no mole accumulation in the blender. Carrying out a mass balance for the overall system, it follows that:

$$\frac{dN_A}{dt} + \frac{dN_B}{dt} = -\frac{dN_C}{dt}$$

Similarly, carrying out a mole balance for each of the components in the system (i.e., components 1, 2, and 3), it follows that:

$$2\frac{dN_A}{dt} = 2\frac{dN_B}{dt} = -\frac{dN_C}{dt} \tag{9}$$

Using Eq. (9) in Eq. (8), we obtain:

$$0 = \dot{Q} + H_A\left(-\frac{1}{2}\frac{dN_C}{dt}\right) + H_B\left(-\frac{1}{2}\frac{dN_C}{dt}\right) + H_C\frac{dN_C}{dt}$$

$$\dot{Q} = H_A\left(\frac{1}{2}\frac{dN_C}{dt}\right) + H_B\left(\frac{1}{2}\frac{dN_C}{dt}\right) - H_C\frac{dN_C}{dt}$$

$$\dot{Q} = -\frac{dN_C}{dt}\left(H_C - \frac{H_A + H_B}{2}\right) \tag{10}$$

Next, we can calculate the enthalpies of the three streams (each a mixture) in terms of their pure component enthalpies and the enthalpy of mixing. Specifically:

$$H_A = \Delta H_{123}(x_{A,1}, x_{A,2}) + \sum_i x_{i,A}H_i = \Delta H_{123,A} + x_{A,1}H_1 + x_{A,2}H_2 + x_{A,3}H_3$$

$$H_B = \Delta H_{123}(x_{B,1}, x_{B,2}) + \sum_i x_{i,B}H_i = \Delta H_{123,B} + x_{B,1}H_1 + x_{B,2}H_2 + x_{B,3}H_3$$

$$H_C = \Delta H_{123}(x_{C,1}, x_{C,2}) + \sum_i x_{i,C}H_i = \Delta H_{123,C} + x_{C,1}H_1 + x_{C,2}H_2 + x_{C,3}H_3$$

$$\tag{11}$$

Combining Eqs. (10) and (11), we obtain:

$$\dot{Q} = -\frac{dN_C}{dt}\begin{pmatrix} \Delta H_{123,C} + x_{C,1}H_1 + x_{C,2}H_2 + x_{C,3}H_3 \\ -\dfrac{\Delta H_{123,A}}{2} - \dfrac{x_{A,1}H_1}{2} - \dfrac{x_{A,2}H_2}{2} - \dfrac{x_{A,3}H_3}{2} \\ -\dfrac{\Delta H_{123,B}}{2} - \dfrac{x_{B,1}H_1}{2} - \dfrac{x_{B,2}H_2}{2} - \dfrac{x_{B,3}H_3}{2} \end{pmatrix}$$

$$\frac{\dot{Q}}{-dN_C/dt} = \begin{pmatrix} \Delta H_{123,C} - \dfrac{\Delta H_{123,A}}{2} - \dfrac{\Delta H_{123,B}}{2} + x_{C,1}H_1 - \dfrac{x_{A,1}H_1}{2} - \dfrac{x_{B,1}H_1}{2} \\ + x_{C,2}H_2 - \dfrac{x_{A,2}H_2}{2} - \dfrac{x_{B,2}H_2}{2} + x_{C,3}H_3 - \dfrac{x_{A,3}H_3}{2} - \dfrac{x_{B,3}H_3}{2} \end{pmatrix}$$

$$\frac{\dot{Q}}{(-dN_C/dt)} = \Delta H_{123,C} + \left(x_{C,1} - \frac{x_{A,1} + x_{B,1}}{2}\right)H_1 + \left(x_{C,2} - \frac{x_{A,2} + x_{B,2}}{2}\right)H_2$$

$$+ \left(x_{C,3} - \frac{x_{A,3} + x_{B,3}}{2}\right)H_3 - \frac{\Delta H_{123,A} + \Delta H_{123,B}}{2} \tag{12}$$

Using Eq. (1), we can calculate the enthalpy of mixing for each of the three species:

$$\Delta H_{123,A}(x_{A,1}, x_{A,2}) = 19.0476 \, \text{J/mol}$$
$$\Delta H_{123,B}(x_{B,1}, x_{B,2}) = 13.3333 \, \text{J/mol} \tag{13}$$
$$\Delta H_{123,C}(x_{C,1}, x_{C,2}) = 16.6667 \, \text{J/mol}$$

Combining the results in Eq. (13) with Eq. (12), we obtain the heat duty required per mole of ternary product:

$$\text{Heat/mole of ternary product} = \frac{\dot{Q}}{(-dN_C/dt)} = 4.05 \, \text{J/(mol of C)} \tag{14}$$

Problem 12

Problem 9.2 in Tester and Modell

We are faced with the problem of diluting a 90 wt % H_2SO_4 solution with water in the following manner. A tank contains 500 kg of pure water at 298 K; it is equipped with a cooling device to remove any heat of mixing. This cooling device operates with a boiling refrigerant reflux condenser system to maintain the temperature at 298 K. Because of the peculiarities of the system, the rate of heat transfer (W/m^2) must be constant. We wish to add 1500 kg of acid solution (at a variable rate) in 1 h. The acid is initially at 298 K. Enthalpy data are provided in the table below.

(a) Plot the heat of solution (kJ/kg solution) versus weight fraction H_2SO_4 with the reference states as pure water and pure H_2SO_4, liquid, at 298 K.
(b) What is the total heat transferred in the dilution process described?
(c) Derive a differential equation to express the mass flow of 90 wt % acid, kg/min, as a function of the acid concentration in the solution.
(d) Using the result from Part (c), determine the mass flow of 90 wt % acid when the overall tank liquid is 64.5 wt % acid.

$x_{H_2SO_4}$	$(\bar{H} - H^0)_{H_2O}$ (J/Mol)	$(\bar{H} - H^0)_{H_2SO_4}$ (J/Mol)
0.00	0	0
0.05	−183	17,290
0.10	−1228	32,360
0.15	−2428	38,980
0.20	−4187	46,850
0.25	−6071	53,090
0.30	−7997	58,490
0.35	−10,340	63,350
0.40	−12,810	67,660
0.45	−16,250	72,180
0.50	−20,310	76,660
0.55	−23,990	79,720
0.60	−26,380	81,770
0.65	−28,010	82,650
0.70	−29,350	83,360
0.75	−30,480	83,850
0.80	−31,360	84,150
0.85	−32,240	84,300
0.90	−32,950	84,380
0.95	−33,700	84,460
1.00	−34,440	84,570

Solution to Problem 12

Solution Strategy

For this solution, the subscript "w" denotes water and the subscript "a" denotes sulfuric acid.

Part (a)

To begin solving this problem, we would like to derive an expression for the heat (or enthalpy) of mixing of the solution on a mass basis as a function of the weight fraction of sulfuric acid. In general, as discussed in Part II, the enthalpy of mixing of a solution is given by:

$$\Delta H_{mix} = H - \sum_i x_i \overline{H}_i^+ \tag{1}$$

In Eq. (1), H is the weighted sum of the partial molar enthalpies $\overline{H}_i$, that is:

$$H = \sum_i x_i \overline{H}_i.$$

x_i is the mole fraction of component i, and $\overline{H}_i^+$ is the partial molar enthalpy of component i in the reference state (+). Because the solution consists of sulfuric acid and water, Eq. (1) combined with the last equation yields:

$$\Delta H_{mix} = \left(x_a \overline{H}_a + x_w \overline{H}_w \right) - \left(x_a \overline{H}_a^+ + x_w \overline{H}_w^+ \right)$$
$$= x_a \left(\overline{H}_a - \overline{H}_a^+ \right) + x_w \left(\overline{H}_w - \overline{H}_w^+ \right) \tag{2}$$

In Eq. (2), the reference state of water is pure water, $\overline{H}_w^+ = H_w^0$, and the reference state of sulfuric acid is pure sulfuric acid. Comparing the reference state for sulfuric acid in Eq. (2) with the data provided in the table in the Problem Statement, we recognize that the table entries for sulfuric acid are defined with respect to an infinite-dilution reference state, that is, $\overline{H}_a^+ \neq H_a^0$! Therefore, we cannot simply use the table for this column. Instead, we can define both sulfuric acid enthalpies in terms of this reference state:

$$x_a \left(\overline{H}_a - \overline{H}_a^+ \right) = x_a \left(\left(\overline{H}_a - \overline{H}_a^0 \right) - \left(\overline{H}_a^+ - \overline{H}_a^0 \right) \right) \tag{3}$$

Using Eq. (3) in Eq. (2) yields:

$$\Delta H_{mix} = x_a \left(\left(\overline{H}_a - \overline{H}_a^0 \right) - \left(\overline{H}_a^+ - \overline{H}_a^0 \right) \right) + x_w \left(\overline{H}_w - \overline{H}_w^+ \right) \qquad (4)$$

The mole fractions are related by:

$$x_w = 1 - x_a,$$

Using the last equation in Eq. (4), we obtain:

$$\Delta H_{mix} = x_a \left(\left(\overline{H}_a - \overline{H}_a^0 \right) - \left(\overline{H}_a^+ - \overline{H}_a^0 \right) \right) + (1 - x_a) \left(\overline{H}_w - \overline{H}_w^+ \right) \qquad (5)$$

Equation (5) is written on a molar basis. However, we need to express everything on a mass basis. The simplest way to do this is to recognize that the molecular weight of 1 mole of solution is given by:

$$M_{sol} = x_a M_a + x_w M_w = x_a \left(0.098 \frac{kg}{mol} \right) + (1 - x_a) \left(0.018 \frac{kg}{mol} \right) \qquad (6)$$

and to divide Eq. (5) by M_{sol} given in Eq. (6). This yields:

$$\Delta \widetilde{H}_{mix} = \frac{1}{M_{sol}} \left[x_a \left(\left(\overline{H}_a - \overline{H}_a^0 \right) - \left(\overline{H}_a^+ - \overline{H}_a^0 \right) \right) + (1 - x_a) \left(\overline{H}_w - \overline{H}_w^+ \right) \right] \qquad (7)$$

where $\Delta \widetilde{H}_{mix}$ is the enthalpy of mixing on a mass basis. Finally, we can determine $\overline{H}_a^+ - \overline{H}_a^0$ from the table in the Problem Statement. We recognize that $\overline{H}_a^+$ is $\overline{H}$ when $x_a = 1.00$, and therefore, the table shows that:

$$\overline{H}_a^+ - \overline{H}_a^0 = 8.457 \times 10^4 \text{ J/mol}$$

Using the last result in Eq. (7) yields:

$$\Delta \widetilde{H}_{mix} = \frac{1}{M_{sol}}$$
$$\times \left[x_a \left(\left(\overline{H}_a - \overline{H}_a^0 \right) - \left(8.457 \times 10^4 \text{ J/mol} \right) \right) + (1 - x_a) \left(\overline{H}_w - \overline{H}_w^+ \right) \right] \qquad (8)$$

To create the required plot, we also need to determine the weight fraction corresponding to each value of x_a. Using Eq. (6), this can be readily done as follows:

$$w_a = \frac{0.098 x_a}{0.098 x_a + 0.018(1 - x_a)} = \frac{0.098 x_a}{0.08 x_a + 0.018} \qquad (9)$$

and

$$w_w = 1 - w_a \tag{10}$$

Alternatively, we can work on a term-by-term basis and divide each term by its appropriate molar weight and then multiply the appropriate weight fractions, which are given in Eqs. (9) and (10). After some simple algebra, we obtain:

$$\Delta \tilde{H}_{mix} = \frac{w_a \left(\left(\overline{H}_a - \overline{H}_a^0 \right) - \left(\overline{H}_a^+ - \overline{H}_a^0 \right) \right)}{0.098 \text{ kg/mol}} + \frac{(1 - w_a) \left(\overline{H}_w - \overline{H}_w^+ \right)}{0.018 \text{ kg/mol}} \tag{11}$$

where the tilde indicates the mass basis of the enthalpy of mixing. Analyzing the data in the table using Eq. (11) and plotting them in a spreadsheet, we obtain the following data table and plot (Table 1 and Fig. 1 below). Note that the numerical answers obtained using either Eq. (8) or Eq. (11) will be identical.

Table 1 Data to calculate the heat of mixing in Part (a)

x_a	$\left(\overline{H} - H^0 \right)_w$	$\left(\overline{H} - \overline{H}^0 \right)_a$	w_a	$\Delta \tilde{H}_{mix}$
	J/Mol	J/Mol		kJ/kg
0	0	0	0.000	0.00
0.05	−183	17,290	0.223	−160.81
0.10	−1228	32,360	0.377	−243.32
0.15	2428	38,980	0.490	−296.74
0.20	−4187	46,850	0.576	−320.40
0.25	−6071	53,090	0.645	−326.93
0.30	−7997	58,490	0.700	−319.57
0.35	−10,340	63,350	0.746	−307.57
0.40	−12,810	67,660	0.784	−289.00
0.45	−16,250	72,180	0.817	−268.76
0.50	−20,310	76,660	0.845	−243.28
0.55	−23,990	79,720	0.869	−217.15
0.60	−26,380	81,770	0.891	−185.33
0.65	−28,010	82,650	0.910	−157.88
0.70	−29,350	83,360	0.927	−130.43
0.75	−30,180	83,850	0.942	−104.62
0.80	−31,360	84,150	0.956	−80.59
0.85	−32,240	84,300	0.969	−58.90
0.90	−32,950	84,380	0.980	−38.51
0.95	−33,700	84,460	0.990	−19.04
1.00	−34,440	84,570	1.000	0.00

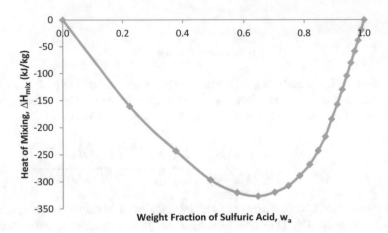

Fig. 1

Part (b)

Next, we examine the process of adding sulfuric acid to the water, considering the complete solution to be our system. This system is clearly *open*, but it is also *diathermal* and *movable*, because the volume of the system is changing, although the pressure is (effectively) constant. Because we want to calculate heat transfer in the system, we begin with the First Law of Thermodynamics for an open system given by:

$$dU = \delta Q + \delta W + \sum_{in} H_{in}\delta n_{in} - \sum_{out} H_{out}\delta n_{out} \tag{12}$$

Because there is only one inlet stream and no outlet streams, it follows that $\delta n_{in} = dN$, and $\delta n_{out} = 0$.

The work term δW in Eq. (12) can be written as $-Pd\underline{V}$. However, because the pressure is constant, it follows that:

$$\delta W = -Pd\underline{V} = -d(P\underline{V})$$

Combining these results, we can express Eq. (12) as follows:

$$dU + d(P\underline{V}) = \delta Q + H_{in}dN \tag{13}$$

The left-hand side of Eq. (13) is equal to the differential enthalpy, and therefore, Eq. (13) can be expressed as follows:

$$dH = \delta Q + H_{in}dN \tag{14}$$

Integrating Eq. (14) from the initial (i) to the final (f) states yields:

$$\underline{H}_f - \underline{H}_i = Q + H_{in}(N_f - N_i) \tag{15}$$

Expressing the left-hand side of Eq. (15) on a molar basis we obtain:

$$N_f H_f - N_i H_i = Q + H_{in}(N_f - N_i) \tag{16}$$

Examining the terms in Eq. (16), we recognize that $N_i = N_{wi}$ is the number of moles of water initially charged and $H_i = H_w$ is the enthalpy of pure water. Similarly, $N_f = N_{in} + N_{wi}$ is the final number of moles of solution. The expression for H_f is related to the enthalpy of mixing in the final state. Specifically:

$$\Delta H_{mix,f} = H_f - \left(x_{wf}\overline{H}_w^+ + x_{af}\overline{H}_a^+\right) \tag{17a}$$

which we can rewrite as follows:

$$H_f = \Delta H_{mix,f} + \left(x_{wf}\overline{H}_w^+ + x_{af}\overline{II}_a^+\right) \tag{17b}$$

We can also describe the enthalpy of the incoming solution using an expression similar to Eq. (17b), that is:

$$H_{in} = \Delta H_{mix,in} + \left(x_{w,in}\overline{H}_w^+ + x_{a,in}\overline{H}_a^+\right) \tag{18}$$

In Eq. (17b) and Eq. (18), we have defined H_f and H_{in} relative to a pair of reference states $\overline{H}_w^+$ and $\overline{H}_a^+$. For convenience, we choose these reference states to be the pure component reference states at temperature T and pressure P of the solution. As a result, we have $\overline{H}_w^+ = H_w^0$ and $\overline{H}_a^+ = H_a^0$, where $H_w^0 \equiv$ pure component enthalpy of water and $H_a^0 \equiv$ pure component enthalpy of the acid. This allows us to rewrite Eqs. (17b) and (18) as follows:

$$H_f = \Delta H_{mix,f} + \left(x_{wf}H_w^0 + x_{af}H_u^0\right) \tag{19}$$

$$H_{in} = \Delta H_{mix,in} + \left(x_{w,in}H_w^0 + x_{a,in}H_a^0\right) \tag{20}$$

Using Eqs. (19) and (20) in Eq. (16), we obtain:

$$N_f\left[\Delta H_{mix,f} + \left(x_{wf}H_w^0 + x_{af}H_a^0\right)\right] - N_{wi}H_w^0$$
$$= Q + N_{in}\left[\Delta H_{mix,in} + \left(x_{w,in}H_w^0 + x_{a,in}H_a^0\right)\right]$$

which we can rearrange as an expression for Q, that is:

$$Q = N_f\left[\Delta H_{mix,f} + \left(x_{wf}H_w^0 + x_{af}H_a^0\right)\right] - N_{wi}H_w^0$$
$$- N_{in}\left[\Delta H_{mix,in} + \left(x_{w,in}H_w^0 + x_{a,in}H_a^0\right)\right]$$

$$= N_f\Delta H_{mix,f} + \left(N_f x_{wf} - N_{wi} - N_{in}x_{w,in}\right)H_w^0 + \left(N_f x_{af} - N_{in}x_{a,in}\right)H_a^0$$
$$- N_{in}\Delta H_{mix,in} \tag{21}$$

Equation (21) indicates that the coefficients of H_w^0 and H_a^0 are both equal to zero. This important result follows because we selected the pure component reference states for all our calculations. Indeed:

$$\left(N_f x_{wf} - N_{wi} - N_{in}x_{w,in}\right) = N_{wf} - N_{wi} - N_{w,in} = 0$$
$$\left(N_f x_{af} - N_{in}x_{a,in}\right) = N_{af} - N_{a,in} = 0$$

and therefore, Eq. (21) reduces to:

$$Q = N_f\Delta H_{mix,f} - N_{in}\Delta H_{mix,in} \tag{22}$$

Before we can determine the heat transfer using Eq. (22) and Table 1, we recognize that we again need to convert to a mass basis. Doing so yields:

$$Q = N_f M_{sol,f}\frac{\Delta H_{mix,f}}{M_{sol,f}} - N_{in}M_{sol,in}\frac{\Delta H_{mix,in}}{M_{sol,in}} = m_f\Delta\tilde{H}_{mix,f} - m_{in}\Delta\tilde{H}_{mix,in} \tag{23}$$

Using the data provided in the Problem Statement, we see that in Eq. (23), $m_w = 500$ kg, $m_{in} = 1500$ kg, and $m_f = m_w + m_{in} = 2000$ kg. To determine $\Delta\tilde{H}_{mix,in}$, we read off the plot in Fig. 1 at $w_a = 0.9$. Specifically, we can fit the data with a fourth-order polynomial and then use the resulting equation to calculate $\Delta\tilde{H}_{mix,in}(w_a = 0.9)$. This yields:

$$\Delta\tilde{H}_{mix,in} \approx -175 \text{ kJ/kg}$$

In order to calculate $\Delta\tilde{H}_{mix,f}$, we have to determine the weight fraction of acid in the final solution, which is given by:

$$w_a = \frac{0.90(1500 \text{ kg})}{1500 \text{ kg} + 500 \text{ kg}} = 0.675$$

from which we can find from Fig. 1 that:

$$\Delta\tilde{H}_{mix,f} \approx -325 \text{ kJ/kg}$$

Alternatively, we can obtain estimates via linear interpolation of the relevant values in Table 1, finding that:

$$\Delta\tilde{H}_{mix,in} = -185.33 \text{ kJ/kg} + \frac{0.9 - 0.891}{0.910 - 0.891}$$
$$\times (-157.88 \text{ kJ/kg} + 185.33 \text{ kJ/kg})$$
$$= -172.33 \text{ kJ/kg} \tag{24}$$

$$\Delta\tilde{H}_{mix,f} = -326.93 \text{ kJ/kg} + \frac{0.675 - 0.645}{0.700 - 0.645}$$
$$\times (-319.57 \text{ kJ/kg} + 326.93 \text{ kJ/kg})$$
$$= -322.92 \text{ kJ/kg} \tag{25}$$

Substitution of the results in Eqs. (24) and (25) in Eq. (23) yields:

$$Q = (2000 \text{ kg})\left(-325\frac{\text{kJ}}{\text{kg}}\right) - (1500 \text{ kg})\left(-175\frac{\text{kJ}}{\text{kg}}\right) = -388 \text{ MJ}$$

$$Q = (2000 \text{ kg})\left(-322.92\frac{\text{kJ}}{\text{kg}}\right) - (1500 \text{ kg})\left(-172.33\frac{\text{kJ}}{\text{kg}}\right) = -387 \text{ MJ}$$

Part (c)

The most logical way to begin our derivation of the differential equation describing the mass flow as a function of acid concentration is to let $m \equiv$ total mass of the liquid in the tank and use the differential relationship in Eq. (14) given by:

$$d\underline{H} = \delta Q + H_{in}dN \tag{26}$$

Because the Problem Statement requests a mass flow rate, rather than a molar flow rate, it is necessary to describe the respective amounts of water and acid in the tank in terms of mass (m) and mass fractions (w), rather than in terms of moles (N) and mole fractions (x). With this need in mind, Eq. (26) can be expressed as follows:

$$d\underline{H} = \delta Q + \tilde{H}_{in}dm \tag{27}$$

where $\tilde{H}_{in}$ represents the enthalpy per kg of solution added to the tank. According to Postulate I, $\underline{H}$ for this binary ($n = 2$) system containing water and acid depends on $n + 2 = 2 + 2 = 4$ independent variables. Because we need to work in mass units, we will choose $\underline{H}$ to be a function of T, P, m_w, and m_a. In terms of these four variables, it follows that:

$$dH = \left(\frac{\partial H}{\partial T}\right)_{P,m_a,m_w} dT + \left(\frac{\partial H}{\partial P}\right)_{T,m_a,m_w} dP + \tilde{\tilde{H}}_w dm_w + \tilde{\tilde{H}}_a dm_a \qquad (28)$$

In Eq. (28), we recognize that $\tilde{\tilde{H}}_w$ represents the change in the enthalpy of the solution resulting from the addition of 1 kg of water at constant T and P. Note that $\tilde{\tilde{H}}_w$ is not equivalent to $\overline{H}_w \equiv$ partial molar enthalpy of water in the solution. Setting $dT = dP = 0$ in Eq. (28), we obtain:

$$dH = \tilde{\tilde{H}}_w dm_w + \tilde{\tilde{H}}_a dm_a \qquad (29)$$

Equation (20) taken on a mass basis becomes:

$$\tilde{H}_{in} = \Delta \tilde{H}_{mix,in} + \left(w_{w,in}\tilde{H}_w^0 + w_{a,in}\tilde{H}_a^0\right) \qquad (30)$$

We can now equate Eq. (27) to Eq. (29) to obtain:

$$\delta Q + \tilde{H}_{in} dm = \tilde{\tilde{H}}_w dm_w + \tilde{\tilde{H}}_a dm_a \qquad (31)$$

Substituting Eq. (30) in Eq. (31) yields:

$$\delta Q + \left(\Delta \tilde{H}_{mix,in}\right) dm + \left(w_{w,in}\tilde{H}_w^0 + w_{a,in}\tilde{H}_a^0\right) dm = \tilde{\tilde{H}}_w dm_w + \tilde{\tilde{H}}_a dm_a \qquad (32)$$

Writing the species balance separately for the water and the acid, we obtain:

$$w_{a,in} dm = dm_a \qquad (33)$$

and

$$w_{w,in} dm = dm_w \qquad (34)$$

Equations (33) and (34) allow us to express Eq. (32) as follows:

$$\tilde{\tilde{H}}_w w_{w,in} dm + \tilde{\tilde{H}}_a w_{a,in} dm = \delta Q + \Delta \tilde{H}_{mix,in} dm + \tilde{H}_a^0 w_{a,in} dm + \tilde{H}_w^0 w_{w,in} dm \qquad (35)$$

We can rearrange Eq. (35) to obtain:

$$\frac{\delta Q}{dm} = \left(\tilde{\tilde{H}}_w - \tilde{H}_w^0\right) w_{w,in} + \left(\tilde{\tilde{H}}_a - \tilde{H}_a^0\right) w_{a,in} - \Delta \tilde{H}_{mix,in} \qquad (36)$$

We can rearrange Eq. (36) and take the derivative of both sides with respect to time to obtain:

$$\frac{dm}{dt} = \left(\left(\tilde{\bar{H}}_w - \tilde{H}_w^0 \right) w_{w,in} + \left(\tilde{\bar{H}}_a - \tilde{H}_a^0 \right) w_{a,in} - \Delta \tilde{H}_{mix,in} \right)^{-1} \frac{\delta Q}{dt} \qquad (37)$$

Part (d)

After deriving the differential equation in Eq. (37), we would like to evaluate it numerically for a specific solution composition. To this end, we only need to evaluate the individual terms in Eq. (37). The simplest term is $\delta Q/dt$, and because it is constant, we obtain:

$$\frac{\delta Q}{dt} = \frac{\Delta Q}{\Delta t} = -\frac{387 \text{ MJ}}{1 \text{ h}} = \frac{387000 \text{ kJ}}{(60 \text{ min})(60 \text{ s})} = -107.5 \text{ kW}$$

Based on our knowledge of the incoming solution, we know that $w_{w,\,in} = 0.1$ and $w_{a,\,in} = 0.9$. In addition, from Eq. (24), we know that $\Delta \tilde{H}_{mix,in} = -172$ kJ/kg. Next, we need to evaluate the enthalpy changes $\tilde{\bar{H}}_w - \tilde{H}_w^0$ and $\tilde{\bar{H}}_a - \tilde{H}_a^0$, which should be determined at a weight fraction of $w_a = 0.645$. If we convert mass fractions to mole fractions, the two required enthalpy changes can be calculated from the table provided in the Problem Statement and from the results in Part (a), respectively. We can utilize the table by using the relation:

$$\left(\tilde{\bar{H}}_w - \tilde{H}_w^0 \right) = \frac{\bar{H}_w - H_w^0}{18 \text{ g/mol}} \qquad (38)$$

followed by converting w_a to x_a as follows:

$$x_a = \frac{\left(\frac{w_a}{98 \text{ g/mol}} \right)}{\left(\frac{w_a}{98 \text{ g/mol}} \right) + \left(\frac{1 - w_a}{18 \text{ g/mol}} \right)} \qquad (39)$$

Substituting $w_a = 0.645$ in Eq. (39) yields $x_a = 0.25$. From the table in the Problem Statement, it follows that:

$$\left(\bar{H} - H^0 \right)_{w_{x_a} = 0.25} = -6.07 \text{ kJ/mol}$$

Substituting the last result in Eq. (38) yields $\left(\tilde{\bar{H}}_w - \tilde{H}_w^0 \right) = -337$ kJ/kg. To evaluate $\left(\tilde{\bar{H}}_a - \tilde{H}_a^0 \right)$, we need to revisit Part (a), including using the data provided in the Problem Statement. We can begin with:

$$\left(\tilde{\bar{H}}_a - \tilde{H}_a^0 \right) = \frac{\bar{H}_a - H_a^0}{98 \text{ g/mol}} \qquad (40)$$

and then determine $\left(\bar{H}_a - H_a^0 \right)$. To this end, examining Eq. (3), we obtain:

$$\left(\overline{H}_a - \overline{H}_a^+\right) = \left(\overline{H}_a - \overline{H}_a^0\right) - \left(\overline{H}_a^+ - \overline{H}_a^0\right)$$

We recognize that the left-hand side of Eq. (3) is equivalent to $\left(\overline{H}_a - H_a^0\right)$. From Part (a), we know that $\left(\overline{H}_a^+ - \overline{H}_a^0\right) = 84.57$ kJ/mol. For $x_a = 0.25$, we can use the table in the Problem Statement to find that $\left(\overline{H}_a - \overline{H}_a^0\right) = 53.09$ kJ/mol. Substituting the above values in Eq. (3) and equating to $\left(\overline{H}_a - H_a^0\right)$ yields:

$$\left(\overline{H}_a - H_a^0\right) = 53.09 \text{ kJ/mol} - 84.57 \text{ kJ/mol} = -31.48 \text{ kJ/mol}$$

Next, we can convert the last result to a mass basis by using Eq. (40). This yields:

$$\left(\widetilde{\overline{H}}_a - \widetilde{H}_a^0\right) = \frac{-31.48 \text{ kJ/mol}}{98 \text{ g/mol}} = -321 \text{ kJ/mol}$$

Finally, we can substitute the mass basis enthalpies above in Eq. (37) to obtain:

$$\frac{dm}{dt} = \frac{(-107.5 \text{ kJ/s})}{(-337 \text{ kJ/kg})(0.1) + (-321 \text{ kJ/kg})(0.9) - 172 \text{ kJ/kg}} = 0.714 \text{ kg/s}$$
$$= 42.8 \text{ kg/min}$$

Problem 13

Problem 9.23 in Tester and Modell

A stream composed of a binary mixture of benzene and cyclohexane that contains 0.5 mole fraction benzene is separated into two product streams. One stream contains a 0.98 mole fraction of benzene and the second contains a 0.90 mole fraction of cyclohexane. The separation process is performed at a constant temperature of 300 K and a constant pressure of 1 bar under steady-state conditions.

You can assume that the activity coefficient for benzene, γ_B, is given by:

$$\gamma_B = \exp\left[0.56(1 - x_B)^2\right]$$

where x_B is the mole fraction of benzene in the mixture. What is the minimum work required to perform the separation?

Solution to Problem 13

Solution Strategy

This problem deals with separating a mixture stream of a given composition into two mixed streams of different compositions. The only information given to us about the process is that it is carried out under isothermal and isobaric conditions. Using this information, we are asked to calculate the minimum work required to perform this separation process.

Because we have been asked to calculate work, it is clear that we will have to carry out a First Law of Thermodynamics analysis of the system. Further, because we have been asked to calculate the minimum work, it is very likely that we will also have to use the Second Law of Thermodynamics to relate the heat interaction in terms of the properties of the inlet and the outlet streams. In addition, we are dealing with mixtures, and therefore, we anticipate that mixing free energies will be useful. In summary, the key concepts to be used in the solution of this problem include:

1. First Law of Thermodynamics analysis
2. Second Law of Thermodynamics analysis
3. Mixing energies for a binary mixture using activity coefficient models

Selection of System and Boundaries

As discussed in Part I, we begin solving the problem by defining our system and its boundaries. As shown in Fig. 1, our system consists of a black box in which the separation process is carried out. Next, we proceed to determine the characteristics of the boundary of the system (see Fig. 1).

Fig. 1

1. Permeable or impermeable? The mixture stream to be separated enters our system (denoted as Black Box in Fig. 1). Therefore, the system boundary is permeable, and the system is open. The conditions of the inlet stream are given by:

$$P_{in} = 1 \text{ bar} = 10^5 \text{ Pa}$$

$$T_{in} = 300 \text{ K}$$

$$\dot{n}_{in} = ?$$

The conditions of the two outlet streams are given by:

$$P_{in} = 1 \text{ bar} = 10^5 \text{ Pa}$$

$$T_{in} = 300 \text{ K}$$

$$\dot{n}_{out,1} = ?$$

$$x_{B,out,1} = 0.98$$

$$\dot{n}_{out,2} = ?$$

$$x_{B,out,2} = 0.10$$

2. Rigid or Movable? There are no movable parts in the system boundary, and therefore, it is rigid. Due to the rigidity of the boundary, there is no P-$\underline{V}$ work interaction between the system and its environment. However, we have been told that external work is done to perform this operation. Therefore:

$$\delta \dot{W} \neq 0 = ?$$

3. Adiabatic or Diathermal? The Problem Statement provides no information about this property of the boundary, and therefore, we can assume that there is some heat interaction between the system and the surroundings. Moreover, we are told that the process is carried out isothermally, which provides an additional reason to assume that heat is supplied to, taken away from, taken away from the system in order to maintain the process isothermal. Therefore,

$$\delta \dot{Q} \neq 0 = ?$$

Next, we can use the information above to write the First Law of Thermodynamics for our open system. Specifically:

$$\frac{dE}{dt} = \frac{dU}{dt} = \delta \dot{Q} + \delta \dot{W} + H_{in}\delta \dot{n}_{in} - H_{out,1}\delta \dot{n}_{out,1} - H_{out,2}\delta \dot{n}_{out,2} \tag{1}$$

In Eq. (1), we assumed that kinetic energy effects and potential energy effects are negligible, such that $\underline{E} = \underline{U}$. Further, assuming that the process is carried out at steady state, the first and second terms in Eq. (1) are both zero. We can then integrate Eq. (1) to obtain:

$$\dot{Q} + \dot{W} + H_{in}\dot{n}_{in} - H_{out,1}\dot{n}_{out,1} - H_{out,2}\dot{n}_{out,2} = 0 \tag{2}$$

where in the equations above, the dots in Q, W, n_{in}, $n_{out,1}$, and $n_{out,2}$ denote time derivatives. We recognize that the flow rates of the different inlet and outlet streams are not independent of each other, because they have to satisfy the mass balance for each of the two species comprising the system. Below, we seek to relate the different flow rates. The species mass balance for benzene can be written as follows:

$$x_{B,in}\dot{n}_{in} = x_{B,out,1}\dot{n}_{out,1} + x_{B,out,2}\dot{n}_{out,2}$$
$$\Rightarrow 0.50\dot{n}_{in} = 0.98\dot{n}_{out,1} + 0.10\dot{n}_{out,2} \tag{3}$$
$$\Rightarrow \dot{n}_{in} = 1.96\dot{n}_{out,1} + 0.20\dot{n}_{out,2}$$

Similarly, the species mass balance for cyclohexane yields:

$$x_{C,in}\dot{n}_{in} = x_{C,out,1}\dot{n}_{out,1} + x_{C,out,2}\dot{n}_{out,2}$$
$$\Rightarrow 0.50\dot{n}_{in} = 0.02\dot{n}_{out,1} + 0.90\dot{n}_{out,2} \tag{4}$$
$$\Rightarrow \dot{n}_{in} = 0.04\dot{n}_{out,1} + 1.80\dot{n}_{out,2}$$

Substituting Eq. (3) in Eq. (4) yields:

$$1.96\dot{n}_{out,1} + 0.20\dot{n}_{out,2} = 0.04\dot{n}_{out,1} + 1.80\dot{n}_{out,2}$$

$$\Rightarrow 1.92\dot{n}_{out,1} = 1.60\dot{n}_{out,2} \tag{5}$$

$$\Rightarrow \dot{n}_{out,1} = 5/6\dot{n}_{out,2}$$

Substituting Eq. (5) in Eq. (3) yields:

$$\dot{n}_{in} = 1.96(5/6)\dot{n}_{out,2} + 0.20\dot{n}_{out,2} = (11/6)\dot{n}_{out,2} \tag{6}$$

Using Eqs. (3), (5), and (6), we can relate the two outlet flow rates to the inlet flow rate as follows:

$$\dot{n}_{out,2} = (6/11)\dot{n}_{in} \tag{7}$$

$$\dot{n}_{out,1} = (5/11)\dot{n}_{in} \tag{8}$$

Substituting Eqs. (7) and (8) in Eq. (2) yields:

$$\dot{Q} + \dot{W} + H_{in}\dot{n}_{in} - H_{out,1}(5/11)\dot{n}_{in} - H_{out,2}(6/11)\dot{n}_{in} = 0$$
$$\Rightarrow \dot{W} = -\dot{Q} + \dot{n}_{in}(H_{out,1}(5/11) + H_{out,2}(6/11) - H_{in}) \tag{9}$$

Our next task is to calculate the heat interaction between the system and its environment that would result in the smallest value of $\dot{W}$. The work requirement would be minimum when the operation is carried out reversibly such that the cumulative entropy change of the system and the surroundings would amount to 0. Therefore, we write an entropy balance on the system when the process is carried out reversibly so that we can relate the heat interaction to the entropy changes of the streams. Specifically:

$$\frac{dS}{dt} = S_{in}\delta\dot{n}_{in} - S_{out,1}\delta\dot{n}_{out,1} - S_{out,2}\delta\dot{n}_{out,2} + \frac{\delta\dot{Q}}{T} + S_{generation}$$

$$\Rightarrow S_{in}\dot{n}_{in} - S_{out,1}(5/11)\dot{n}_{in} - S_{out,2}(6/11)\dot{n}_{in} + \frac{\dot{Q}}{T} = 0$$

$$\Rightarrow \dot{Q} = T\dot{n}_{in}((5/11)S_{out,1} + (6/11)S_{out,2} - S_{in}) \tag{10}$$

where again we have assumed that the process is carried out reversibly and is at steady state. In addition, we have used Eqs. (7) and (8) to eliminate the outlet flow rates in terms of the inlet flow rate.

Substituting Eq. (10) in Eq. (9) yields:

$$\dot{W} = -T\dot{n}_{in}((5/11)S_{out,1} + (6/11)S_{out,2} - S_{in}) + \dot{n}_{in}(H_{out,1}(5/11) + H_{out,2}(6/11) - H_{in})$$
$$\Rightarrow \dot{W} = \dot{n}_{in}((5/11)(H_{out,1} - TS_{out,1}) + (6/11)(H_{out,2} - TS_{out,2}) - (H_{in} - TS_{in}))$$
$$\Rightarrow \dot{W} = \dot{n}_{in}((5/11)G_{out,1} + (6/11)G_{out,2} - G_{in})$$

$$(11)$$

Equation (11) shows that the minimum work required to perform the separation process is given by the change in the Gibbs free energy of the streams. Accordingly, the last step required to solve this problem involves calculating the Gibbs free energies of the different streams. This can be done using the information about the activity coefficient model provided in the Problem Statement. In Part II, we showed that the excess Gibbs free energy can be related to the activity coefficient model using the following relationship:

$$\underline{G}^{EX} = RT\sum_i N_i \ln\gamma_i$$
$$\Rightarrow d\underline{G}^{EX} = RT\sum_i (N_i d\ln\gamma_i + \ln\gamma_i dN_i)$$

$$(12)$$

From Postulate 1, we know that any extensive property of a simple system can be written in terms of the masses (moles) of the different species and two other independent variables. If we take T and P as the two independent variables (recall that the Problem Statement indicates that T and P are held constant), we can write $\underline{G}^{EX}$ as follows:

$$\underline{G}^{EX} = \underline{G}^{EX}(T,P,N_B,N_C)$$
$$\Rightarrow d\underline{G}^{EX} = \left(\frac{\partial \underline{G}^{EX}}{\partial T}\right)_{P,N_B,N_C} dT + \left(\frac{\partial \underline{G}^{EX}}{\partial P}\right)_{T,N_B,N_C} dP + \sum_i \left(\frac{\partial \underline{G}^{EX}}{\partial N_i}\right)_{T,P,N_{j\neq i}} dN_i$$
$$\Rightarrow d\underline{G}^{EX} = \left(\frac{\partial \underline{G}^{EX}}{\partial T}\right)_{P,N_B,N_C} dT + \left(\frac{\partial \underline{G}^{EX}}{\partial P}\right)_{T,N_B,N_C} dP + RT\sum_i \ln\gamma_i dN_i$$

$$(13)$$

Equating Eqs. (13) and (12), at constant temperature and pressure, yields:

$$d\underline{G}^{EX} = RT\sum_i \ln\gamma_i dN_i - RT\sum_i (N_i d\ln\gamma_i + \ln\gamma_i dN_i)$$
$$\Rightarrow RT\sum_i N_i d\ln\gamma_i = 0$$
$$\Rightarrow RTN\sum_i x_i d\ln\gamma_i = 0$$
$$\Rightarrow \sum_i x_i d\ln\gamma_i = 0$$

$$(14)$$

Substituting the activity model given in the Problem Statement in Eq. (14) yields:

$$x_B d \ln \gamma_B + x_C d \ln \gamma_C = 0$$
$$\Rightarrow x_B d \ln \left(\exp \left[0.56(1 - x_B)^2 \right] \right) + (1 - x_B) d \ln \gamma_C = 0$$
$$\Rightarrow x_B \cdot 0.56 \cdot 2(1 - x_B)(-1) dx_B + (1 - x_B) d \ln \gamma_C = 0$$
$$\Rightarrow d \ln \gamma_C = 1.12 x_B dx_B$$

$$\Rightarrow \int_{\ln \gamma_C = 0}^{\ln \gamma_C} d \ln \gamma_C = \int_{x_B = 0}^{x_B} 1.12 x_B dx_B \qquad (15)$$

$$\Rightarrow \ln \gamma_C - 0 = 0.56 \left(x_B^2 - 0 \right)$$
$$\Rightarrow \gamma_C = \exp \left[0.56(1 - x_C)^2 \right]$$

Substituting Eq. (15) in Eq. (12) yields:

$$\underline{G}^{EX} = RT(N_B \ln \gamma_B + N_C \ln \gamma_C) = NRT \left(x_B 0.56(1 - x_B)^2 + x_C 0.56(1 - x_C)^2 \right)$$
$$= 0.56 NRT \left(x_C^2 x_B + x_B^2 x_C \right)$$
$$= 0.56 NRT x_B x_C (x_C + x_B)$$
$$= 0.56 NRT x_B x_C$$

$$(16)$$

Using the definition of $\underline{G}^{EX}$, we can find the Gibbs free energy of any stream as follows:

$$\underline{G}^{EX} = \Delta \underline{G}^{EX} = \Delta \underline{G} - \Delta \underline{G}^{ID} = \underline{G} - \underline{G}^0 - \Delta \underline{G}^{ID}$$

$$\underline{G} = \underline{G}^0 + \Delta \underline{G}^{ID} + \underline{G}^{EX} = \sum_i \left(N_i G_i^0 + N_i RT \ln x_i \right) + 0.56 NRT x_B x_C$$

$$= NRT \left(x_B \left(G_B^0 / RT \right) + x_C \left(G_C^0 / RT \right) + x_B \ln x_B + x_C \ln x_C + 0.56 x_B x_C \right)$$

$$(17)$$

where G_B^0 and G_C^0 are the molar Gibbs free energies of pure benzene and cyclohexane at $T = 300$ K and $P = 1$ bar, respectively.

Substituting Eq. (17) in Eq. (11) yields:

$$\dot{W} = \dot{n}_{in}RT \begin{pmatrix} (5/11)\left(\dfrac{x_{B,out,1}\cancel{(G_B^0/RT)}+x_{C,out,1}\cancel{(G_C^0/RT)}+x_{B,out,1}\ln x_{B,out,1}+x_{C,out,1}\ln x_{C,out,1}}{+0.56x_{B,out,1}x_{C,out,1}}\right) \\[2mm] +(6/11)\left(\dfrac{x_{B,out,2}\cancel{(G_B^0/RT)}+x_{C,out,2}\cancel{(G_C^0/RT)}+x_{B,out,2}\ln x_{B,out,2}+x_{C,out,2}\ln x_{C,out,2}}{+0.56x_{B,out,2}x_{C,out,2}}\right) \\[2mm] -\left(\dfrac{x_{B,in}\cancel{(G_B^0/RT)}+x_{C,in}\cancel{(G_C^0/RT)}+x_{B,in}\ln x_{B,in}+x_{C,in}\ln x_{C,in}}{+0.56x_{B,in}x_{C,in}}\right) \end{pmatrix}$$

$$= \dot{n}_{in}RT\begin{pmatrix} (5/11)(0.98\ln 0.98 + 0.02\ln 0.02 + 0.56\cdot 0.98\cdot 0.02) \\ +(6/11)(0.10\ln 0.10 + 0.90\ln 0.90 + 0.56\cdot 0.10\cdot 0.90) \\ -(0.50\ln 0.50 + 0.50\ln 0.50 + 0.56\cdot 0.50\cdot 0.50) \end{pmatrix}$$

$$= \dot{n}_{in}RT((5/11)(-0.087) + (6/11)(-0.275) - (-0.553))$$
$$= 0.364\dot{n}_{in}RT = 907\dot{n}_{in}$$

$$(18)$$

Note that in Eq. (18), all the standard-state contributions cancelled out due to the mass balance constraints on the system (see Eqs. (3)–(8)). From Eq. (18), we conclude that the minimum work required to separate the stream is 907 J/mole flow rate of the inlet stream.

Problem 14

Problem 15.4 in Tester and Modell

In a binary solution of two components, the eutectic point is the lowest freezing point of the mixture. It is less than the freezing point of either pure component.

Assume that the liquid phase forms an ideal solution and that all solid phases are pure components (i.e., no mixed crystals form). The vapor phase forms an ideal gas mixture. Some data that may be of use are given below. Neglect any pressure effects. Assume that ΔH values do not vary with temperature.

	Nitrogen	Oxygen
Freezing point (K)	63.3	54.4
$\Delta H_{vaporization}$ (J/mol)	6000	7490
$\Delta H_{sublimation}$ (J/mol)	6720	7940
ΔH_{fusion} (J/mol)	721	447

Estimate the eutectic point (i.e., composition and temperature) for a liquid-air mixture (O_2 and N_2).

Solution to Problem 14

Solution Strategy

To solve this problem, we will utilize the general strategy presented in Part II to solve phase equilibria problems:

1. Draw the system and describe the boundaries
2. Use the Gibbs Phase Rule to determine how much information is required to find a unique solution based on the given boundaries
3. Draw a phase diagram to better understand the problem
4. Decide whether the integral approach or the differential approach is more appropriate to calculate the chemical potential equalities
5. Use the approach selected in 4 above to solve the problem

Draw the System and Describe the Boundaries

As per the Problem Statement, the system boundaries are diathermal, movable, and permeable. The only constraint mentioned in the Problem Statement is that both solid phases consist of pure component solids (rather than of a solid mixture). However, this does not prevent the pure component solid phases from coexisting with each other. Using this information, we can draw the diagram shown in Fig. 1.

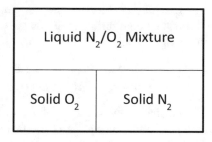

The existence of all three phases, as we show below using the Gibbs Phase Rule, can only occur at one temperature and composition for a given pressure, which is referred to as the Eutectic Point.

Use the Gibbs Phase Rule

At the Eutectic Point, there are two components in three phases (pure solid O_2, pure solid N_2, and liquid mixture containing both O_2 and N_2). Because two of the solid phases are pure components – indicating that there is a barrier preventing O_2 from entering the solid N_2 phase and a barrier preventing N_2 from entering the solid O_2 phase – the system is not simple, and therefore, not all the conditions of thermodynamic equilibria apply. It then follows that we must use the Gibbs Phase Rule analysis for composite systems presented in Part II.

We begin by determining the number of independent intensive variables needed to describe each phase (a simple system). From the Corollary to Postulate I, it follows that we need $(n + 1)$ independent intensive variables to define each phase:

Liquid Phase (α): $n = 2$ (O_2 and N_2), $n + 1 = 2 + 1 = 3$ variables
Solid O_2 Phase (β): $n = 1$ (O_2 only), $n + 1 = 1 + 1 = 2$ variables
Solid N_2 Phase (γ): $n = 1$ (N_2 only), $n + 1 = 1 + 1 = 2$ variables

Total number of independent intensive variables needed to characterize each phase $= 3 + 2 + 2 = 7$ variables.

After we determine the total number of intensive variables needed to describe the three phases, we need to determine the number of constraints relating these intensive variables by identifying the conditions of thermodynamic equilibria that must be satisfied. We know that there are no adiabatic or rigid internal barriers, and therefore, thermal and mechanical equilibria are satisfied between the three phases. Specifically:

Thermal Equilibrium (T.E.): $T^\alpha = T^\beta = T^\gamma$ ($\pi - 1 = 3 - 1 = 2$ constraints)
Mechanical Equilibrium (M.E.): $P^\alpha = P^\beta = P^\gamma$ ($\pi - 1 = 3 - 1 = 2$ constraints)

The composite system has internal barriers that are impermeable to certain species. Therefore, the diffusional equilibria conditions are different than those in a simple three-phase system. Specifically:

Diffusional Equilibrium (D.E.) for O_2: $\mu_{O_2}^{\alpha} = \mu_{O_2}^{\beta}$ (1 constraint)
Diffusional Equilibrium (D.E.) for N_2: $\mu_{N_2}^{\alpha} = \mu_{N_2}^{\gamma}$ (1 constraint)

Total number of constraints that apply $= 2 + 2 + 1 + 1 = 6$.

The variance for a composite system is equal to the total number of independent intensive variables needed to characterize all the phases (seven in this case) minus the total number of constraints that apply (six in this case). Therefore, for the system under consideration, it follows that:

$$L = 7 - 6 = 1 \tag{a}$$

Therefore, at the eutectic point (three phases), the system is monovariant, indicating that only one variable needs to be specified to find a unique solution. If we specify the pressure, there exists only one temperature and liquid phase composition at which three phases are simultaneously present.

At other liquid-solid equilibrium conditions, only one of the solid phases will be present. Using the same procedure as above, we find that five variables are needed to characterize each phase independently (three for the liquid phase and two for the solid phase). We also recognize that there are three constraints on the system (one for T.E., one for M.E., and one for D.E.). Therefore, the variance is given by:

$$L = 5 - 3 = 2 \tag{b}$$

Therefore, the Gibbs Phase Rule indicates that the system is divariant, indicating that if we specify the pressure, another variable would have to be specified to find a unique solution.

Draw a Schematic Phase Diagram

To better understand the problem, we can draw a sketch of the phase diagram that represents our problem (see Fig. 2). We know that there are two liquid-solid equilibrium lines, whose temperatures vary with composition (see the left and the right curves in Fig. 2). At the pure component mole fractions (0 and 1), the liquid-solid equilibrium temperatures are equal to the pure component temperatures (63.3 K and 54.4 K). The temperatures decrease along the left and the right lines as the system evolves from the pure components to the eutectic composition due to the interactions between the two components.

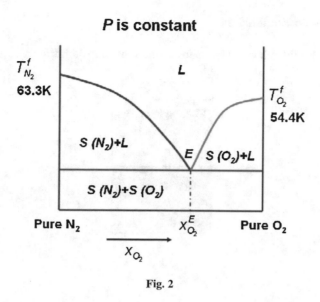

P is constant

Fig. 2

Decide Whether the Integral Approach or the Differential Approach Is More Appropriate

To decide which approach to use, it is useful to examine the information given in the Problem Statement. In order to use the integral approach, we would need the vapor pressures of the different phases in order to calculate the values of the fugacities. Instead, we are given several different enthalpies of phase change, which can be used in the differential approach. We may be concerned that no information is provided about the volume changes associated with the phase changes. However, these changes are related to pressure effects which the Problem Statement asks us to neglect. Accordingly, we will utilize the differential approach.

Solve the Problem

To solve this problem, we need to find the point at which the liquid mixture is in equilibrium with both pure solid phases. To find this special point (the Eutectic), we will first calculate the two curves that describe each individual component in the liquid mixture being in equilibrium with that component in the corresponding solid phase (these correspond to the left and right curves shown in Fig. 2). Subsequently, we will calculate the point at which the two curves intersect, that is, the Eutectic point. At that intersection, both liquid components will be in equilibrium with their solid phases. Note that because the equilibrium curves are not necessarily continuous and differentiable, setting a derivate equal to zero and solving could lead to erroneous results!

N₂ Liquid/Solid Equilibrium

Figure 3 shows the relevant phases in thermodynamic equilibrium:

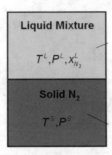

Fig. 3

At equilibrium:

$$T^L = T^S = T^f \tag{1}$$

$$P^L = P^S = P \tag{2}$$

$$\mu_{N_2}^L\left(T^L, P^L, x_{N_2}^L\right) = \mu_{N_2}^S\left(T^S, P^S\right)$$

$$=> \widehat{f}_{N_2}^L\left(T^f, P, x_{N_2}^L\right) = f_{N_2}^S\left(T^S, P\right) \tag{3}$$

Therefore:

$$d\ln\widehat{f}_{N_2}^L\left(T^f, P, x_{N_2}^L\right) = d\ln f_{N_2}^S\left(T^S, P\right) \tag{4}$$

Expanding each fugacity term in Eq. (4) yields:

$$\left(\frac{\partial \ln\widehat{f}_{N_2}^L}{\partial T}\right)_{P, x_{N_2}^L} dT + \left(\frac{\partial \ln\widehat{f}_{N_2}^L}{\partial P}\right)_{T, x_{N_2}^L} dP + \left(\frac{\partial \ln\widehat{f}_{N_2}^L}{\partial x_{N_2}}\right)_{T, P} dx_{N_2}$$

$$= \left(\frac{\partial \ln f_{N_2}^S}{\partial T}\right)_P dT + \left(\frac{\partial \ln f_{N_2}^S}{\partial P}\right)_T dP \tag{5}$$

At constant pressure, Eq. (5) becomes:

$$\left(\frac{\partial \ln \hat{f}_{N_2}^L}{\partial T}\right)_{P,x_{N_2}^L} dT + \left(\frac{\partial \ln \hat{f}_{N_2}^L}{\partial x_{N_2}}\right)_{T,P} dx_{N_2} = \left(\frac{\partial \ln f_{N_2}^S}{\partial T}\right)_P dT \quad (6)$$

Because

$$\left(\frac{\partial \ln \hat{f}_{N_2}^L}{\partial T}\right)_{P,x_{N_2}^L} = -\left[\frac{\overline{H}_{N_2}^L - H_{N_2}^0(T)}{RT^2}\right]$$

and

$$\left(\frac{\partial \ln f_{N_2}^S}{\partial T}\right)_P = -\left[\frac{H_{N_2}^S - H_{N_2}^0(T)}{RT^2}\right]$$

it follows that:

$$-\left[\frac{\overline{H}_{N_2}^L - H_{N_2}^0(T)}{RT^2}\right] dT + \left(\frac{\partial \ln \hat{f}_{N_2}^L}{\partial x_{N_2}}\right)_{T,P} dx_{N_2} = -\left[\frac{H_{N_2}^S - H_{N_2}^0(T)}{RT^2}\right] dT \quad (7)$$

We are told that the liquid phase forms an ideal solution, and therefore:

$$\overline{H}_{N_2}^L = H_{N_2}^L(T,P) \quad (8)$$

$$\hat{f}_{N_2}^L = x_{N_2} f_{N_2}^L \quad (9)$$

$$\Longrightarrow \ln \hat{f}_{N_2}^L = \ln x_{N_2} + \ln f_{N_2}^L$$

$$\longrightarrow d \ln \hat{f}_{N_2}^L = d \ln x_{N_2} + d \ln f_{N_2}^l$$

$$\Longrightarrow \left(\frac{\partial \left(\ln \hat{f}_{N_2}^L\right)}{\partial x_{N_2}}\right)_{T,P} = \left(\frac{\partial (\ln x_{N_2})}{\partial x_{N_2}}\right)_{T,P} + \left(\frac{\partial \left(\ln f_{N_2}^L\right)}{\partial x_{N_2}}\right)_{T,P} = \frac{1}{x_{N_2}} \quad (10)$$

Substituting Eqs. (8) and (10) in Eq. (7) yields:

$$-\left[\frac{\overline{H}_{N_2}^L - H_{N_2}^0(T)}{RT^2}\right]dT + \frac{1}{x_{N_2}}dx_{N_2} = -\left[\frac{H_{N_2}^S - H_{N_2}^0(T)}{RT^2}\right]dT \qquad (11)$$

Rearranging Eq. (11) yields:

$$\Rightarrow \frac{1}{x_{N_2}}dx_{N_2} = \left[\frac{\overline{H}_{N_2}^L - H_{N_2}^S}{RT^2}\right]dT \qquad (12)$$

$$\Rightarrow \frac{1}{x_{N_2}}dx_{N_2} = \frac{\Delta H_{N_2}^f}{RT^2}dT \qquad (13)$$

$$\Rightarrow d\ln x_{N_2} = -\frac{\Delta H_{N_2}^f}{R}d\left(\frac{1}{T}\right) \qquad (14)$$

Because $\Delta H_{N_2}^f$ is not a function of temperature, we can integrate Eq. (14) from the freezing point of pure N_2 ($x_{N_2} = 1$) to the freezing temperature of a liquid mixture with N_2 composition (x_{N_2}). This yields:

$$\int_1^{x_{N_2}} d\ln x_{N_2} = -\frac{\Delta H_{N_2}^f}{R}\int_{T_{N_2}^f}^{T^f} d\left(\frac{1}{T}\right)$$

$$\Rightarrow \ln x_{N_2} = -\frac{\Delta H_{N_2}^f}{R}\left(\frac{1}{T^f} - \frac{1}{T_{N_2}^f}\right)$$

$$\Rightarrow \frac{1}{T^f} = -\frac{R}{\Delta H_{N_2}^f}\ln x_{N_2} + \frac{1}{T_{N_2}^f} \qquad (15)$$

Because $\ln x_{N_2} < 0$, we have $\frac{1}{T^f} > \frac{1}{T_{N_2}^f}$ or $T_{N_2}^f > T^f$.

O_2 Liquid/Solid Equilibrium

Following a similar approach to model this solid/liquid equilibrium, we obtain:

$$\frac{1}{T^f} = -\frac{R}{\Delta H_{O_2}^f}\ln x_{O_2} + \frac{1}{T_{O_2}^f} \qquad (16)$$

As before, we have $T_{O_2}^f > T^f$.

At the eutectic point, the two equilibrium curves intersect, and we have:

$$T^f = T_E, x_{N_2} = x_{N_2}^E, x_{O_2} = x_{O_2}^E = 1 - x_{N_2}^E \tag{17}$$

Using Eqs. (15) and (16), it follows that:

$$-\frac{R}{\Delta H_{N_2}^f} \ln x_{N_2} + \frac{1}{T_{N_2}^f} = -\frac{R}{\Delta H_{O_2}^f} \ln x_{O_2} + \frac{1}{T_{O_2}^f}$$

$$=> -\frac{R}{\Delta H_{N_2}^f} \ln x_{N_2} + \frac{1}{T_{N_2}^f} = -\frac{R}{\Delta H_{O_2}^f} \ln (1 - x_{N_2}) + \frac{1}{T_{O_2}^f}$$

$$=> \left(\frac{1}{T_{N_2}^f} - \frac{1}{T_{O_2}^f}\right) = R\left[\frac{\ln x_{N_2}}{\Delta H_{N_2}^f} - \frac{\ln (1 - x_{N_2})}{\Delta H_{O_2}^f}\right] \tag{18}$$

Using the values given in the Problem Statement in Eq. (18), we obtain:

$$\left(\frac{1}{63.3} - \frac{1}{54.4}\right) = 8.314 \left[\frac{\ln x_{N_2}^E}{721} - \frac{\ln \left(1 - x_{N_2}^E\right)}{447}\right] \tag{19}$$

Solving Eq. (19), we obtain:

$$\boxed{x_{N_2}^E = 0.373 \text{ and } x_{O_2}^E = 0.627}$$

Using these two mole fraction values in Eqs. (15) or (16), we can calculate T_E:

$$\frac{1}{T_E} = \frac{1}{T^f} = -\frac{8.314}{721} \ln 0.373 + \frac{1}{63.3}$$

or

$$\boxed{T_E - 36.8K}$$

Problem 15

Problem 15.13 in Tester and Modell

The *International Critical Tables*, Vol. III, p. 313, lists the boiling points for mixtures of acetaldehyde and ethyl alcohol at various pressures. The data are summarized below for mixtures containing 80 mole % ethyl alcohol in the liquid.

From these data, determine the molar enthalpy of vaporization of ethyl alcohol at 320.7 K, from a liquid mixture containing 80% ethyl alcohol. [That is, what is $\left(\overline{H}^V - \overline{H}^L\right)$ alcohol?]

T (K)	P (N/m^2)	Mole fraction ethyl alcohol in vapor
331.3	9.319×10^4	0.318
320.7	5.306×10^4	0.385
299.9	1.027×10^4	0.330

Note: The vapor phase may be considered to be an ideal gas mixture, but the liquid phase is a non-ideal solution.

Solution to Problem 15

Solution Strategy

This problem asks us to evaluate partial molar enthalpies using phase equilibria data. From our knowledge of phase equilibria, we know that at thermodynamic equilibrium, the chemical potentials (and therefore the fugacities) of each of the components in each of the phases are equal. Furthermore, we know that the temperature derivatives of ln of the fugacities can be expressed in terms of the partial molar enthalpies. In addition, we know that the pressure derivatives of ln of the fugacities can be expressed in terms of the partial molar volumes. Therefore, it is convenient to solve this problem by writing the phase equilibria criteria in terms of the fugacities and then use the differential approach to phase equilibria to calculate the difference between the partial molar enthalpies of ethyl alcohol in the vapor phase and in the liquid phase.

Solution Using the Differential Approach to Phase Equilibria

For the binary mixture in liquid/vapor equilibrium, the conditions of thermodynamic equilibrium are given by:

Thermal equilibrium:

$$T^V = T^L = T \tag{1}$$

Mechanical equilibrium:

$$P^V = P^L = P \tag{2}$$

Diffusional equilibrium:

$$\mu_E^V = \mu_E^L \quad \Rightarrow \hat{f}_E^V = \hat{f}_E^L \tag{3}$$

$$\mu_A^V = \mu_A^L \quad \Rightarrow \hat{f}_A^V = \hat{f}_A^L \tag{4}$$

where in Eqs. (1), (2), (3), and (4), the subscripts A and E denote acetaldehyde and ethyl alcohol, respectively, and the superscripts V and L denote the vapor and liquid phases, respectively. Because the problem asks us to evaluate the difference between the partial molar enthalpies in the vapor and the liquid phases for ethyl alcohol only, we only need to use Eqs. (1), (2), and (3). In order to relate Eq. (3) to the partial molar enthalpy, we take its total differential and simplify it using the information given in the Problem Statement. Specifically, we will eliminate any dependence on x_E because the Problem Statement tells us that the liquid mole fraction is held constant at 0.8. It then follows that:

$$\hat{f}_E^V = \hat{f}_E^L \Rightarrow \ln \hat{f}_E^V = \ln \hat{f}_E^L \Rightarrow d\ln \hat{f}_E^V = d\ln \hat{f}_E^L$$

$$\Rightarrow \left(\frac{\partial}{\partial T} \ln \hat{f}_E^V\right)_{P,y} dT + \left(\frac{\partial}{\partial P} \ln \hat{f}_E^V\right)_{y,T} dP + \left(\frac{\partial}{\partial y_E} \ln \hat{f}_E^V\right)_{T,P} dy_E = \left(\frac{\partial}{\partial T} \ln \hat{f}_E^L\right)_{P,x} dT$$

$$+ \left(\frac{\partial}{\partial P} \ln \hat{f}_E^L\right)_{x,T} dP + \left(\frac{\partial}{\partial x_E} \ln \hat{f}_E^L\right)_{T,P} dx_E$$

$$\Rightarrow \frac{-\Delta \overline{H}_E^V}{RT^2} dT + \frac{\overline{V}_E^V}{RT} dP + \left(\frac{\partial}{\partial y_E} \ln (y_E P)\right)_{T,P} dy_E = \frac{-\Delta \overline{H}_E^L}{RT^2} ldT + \frac{\overline{V}_E^L}{RT} dP$$

$$\text{Ideal Gas Mixture}$$

$$\Rightarrow \frac{-\left(\Delta \overline{H}_E^V - \Delta \overline{H}_E^L\right)}{RT^2} dT + \frac{\left(\overline{V}_E^V - \overline{V}_E^L\right)}{RT} dP + \frac{P}{y_E P} dy_E = 0$$

However, because the same reference state is used for the two phases, it follows that:

$$\Delta \overline{H}_E^V - \Delta \overline{H}_E^L = \overline{H}_E^V - \overline{H}_E^L.$$

In addition, because the pressure is not high, it is safe to assume that $\overline{V}_E^V \gg \overline{V}_E^L$ and $\overline{V}_E^V - \overline{V}_E^L \approx \overline{V}_E^V$. Accordingly, the last equation becomes:

$$\Rightarrow \frac{-\left(\bar{H}_E^V - \bar{H}_E^L\right)}{RT^2} dT + \frac{\overline{V}_E^V}{RT} dP + \frac{dy_E}{y_E} = 0 \qquad \text{Ideal Gas Mixture}$$

$$\Rightarrow \frac{-\left(\bar{H}_E^V - \bar{H}_E^L\right)}{RT^2} dT + \frac{\cancel{R}T/P}{\cancel{R}T} dP + \frac{dy_E}{y_E} = 0 \qquad \text{Ideal Gas Mixture}$$

$$\Rightarrow \frac{-\left(\bar{H}_E^V - \bar{H}_E^L\right)}{RT^2} dT + \frac{dP}{P} + \frac{dy_E}{y_E} = 0$$

$$\Rightarrow \frac{\left(\bar{H}_E^V - \bar{H}_E^L\right)}{R} d\left(\frac{1}{T}\right) + d\ln P + d\ln y_E = 0$$

$$\Rightarrow \frac{\left(\bar{H}_E^V - \bar{H}_E^L\right)}{R} d\left(\frac{1}{T}\right) + d\ln\left(y_E P\right) = 0 \Rightarrow \frac{d\ln\left(y_E P\right)}{d(1/T)} = -\frac{\left(\bar{H}_E^V - \bar{H}_E^L\right)}{R} \qquad (5)$$

Using the data given in the Problem Statement, we can obtain a polynomial fit for $\ln(y_E P)$ in terms of $(1/T)$ and then evaluate its derivative at the required condition $(T = 320.7\,\text{K})$ to obtain $\left(\overline{H}_E^V - \overline{H}_E^L\right)$. From the data provided, the highest-order polynomial fit that we can obtain is a quadratic fit (we have only three data points). Although it may seem inappropriate to fit a quadratic equation when we have only three data points, it should be noted that the three points are quite close to each other (299.9–331.3 K), and therefore, fitting a quadratic equation is a reasonable approximation. Note that this is widely done in numerical schemes. For example, in finite differences, this approximation is used to obtain the three-point finite difference formula for the second derivative. Similarly, in numerical integration, this approximation is the basis for Simpson's rule. The main difference between these algorithms and the approximation that we use in this problem is that our points are not equidistant.

Performing a quadratic fit between $\ln(y_E P)$ in terms of $(1/T)$ (see Fig. 1), we obtain:

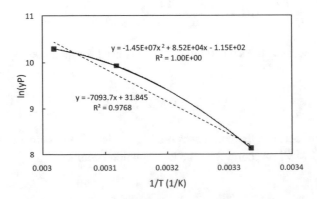

Fig. 1

$$\ln(y_E P) = -1.45 \times 10^7 \times (1/T)^2 + 8.52 \times 10^4 \times (1/T) - 1.15 \times 10^2 \quad (6)$$

Differentiating Eq. (6) with respect to $1/T$ and evaluating it at $T = 320.7$ K yields:

$$\frac{\partial \ln(y_E P)}{\partial(1/T)} = -1.45 \times 10^7 \times 2 \times (1/T) + 8.52 \times 10^4$$

$$\Rightarrow \frac{\partial \ln(y_E P)}{\partial(1/T)}\bigg|_{T=320.7} = -2.90 \times 10^7 \times 2 \times (1/320.7) + 8.52 \times 10^4 = -5227.19$$

$$\quad (7)$$

Substituting Eq. (7) in Eq. (5) yields:

$$\frac{d \ln(y_E P)}{d(1/T)} = -\frac{\left(\overline{H}_E^V - \overline{H}_E^L\right)}{R} = -5227.19$$

$$\Rightarrow \left(\overline{H}_E^V - \overline{H}_E^L\right) = 5227.19 \times 8.314 = 43458.68 \,\text{J/mol} = 43.46 \,\text{kJ/mol} \quad (8)$$

Instead of fitting a quadratic, if we had assumed that $\left(\overline{H}_E^V - \overline{H}_E^L\right)$ was a constant and that $\ln(y_E P)$ was a linear function of $(1/T)$, we would have obtained the following value for $\left(\overline{H}_E^V - \overline{H}_E^L\right)$:

$$\ln(y_E P) = -7093.7 \times (1/T) + 31.845$$

$$\Rightarrow \frac{d \ln(y_E P)}{d(1/T)} = -7093.7$$

$$\Rightarrow -\frac{\left(\overline{H}_E^V - \overline{H}_E^L\right)}{R} = -7093.7 \quad (9)$$

$$\Rightarrow \left(\overline{H}_E^V - \overline{H}_E^L\right) = 7093.7 \times 8.314 = 58977.02 \,\text{J/mol}$$

$$\rightarrow \left(\overline{H}_E^V - \overline{H}_E^L\right) = 58.98 \,\text{kJ/mol}$$

which is more than 30% higher than the value obtained using Eq. (8)! However, in the absence of error bars on the data points and additional data points, it is difficult to evaluate the statistical significance of this difference.

Problem 16

Problem 16.7 in Tester and Modell

Michael K. Jones, a close relative of Rocky and Rochelle and an avid inventor, claims to be able to produce diamonds from β-graphite at room temperature by a process involving the application of 37 kbar pressure. In view of the data shown below, are his claims to be taken seriously?

$$\text{Specific gravity, } \beta - \text{graphite} = 2.26$$

$$\text{Specific gravity, diamond} = 3.51$$

$$C_{\beta\text{-graphite}} = C_{\text{diamond}} \, \Delta G^0_{298} = 2870 \, \text{J/g-atom}$$

Both solids are incompressible and no solid solutions are formed.

Solution to Problem 16

Solution Strategy

The strategy to solve this problem involves the following two steps:

1. Derive a thermodynamic criterion to determine whether or not the process in question is physically possible
2. Use the given information to determine if the actual process satisfies the criterion in 1

To carry out Step 1, we will use the techniques discussed in Part I. We recall that two thermodynamic laws (the First and the Second) govern all open systems:

$$d\underline{U} = \delta W + \delta Q + \sum H_{in}\delta n_{in} - \sum H_{out}\delta n_{out} \tag{1}$$

and

$$d\underline{S}_{universe} \geq 0 \tag{2}$$

For the process under consideration, we can draw the diagram shown in Fig. 1 to represent the inlets and outlets of the system. Using this diagram, we can rewrite Eq. (1) as follows:

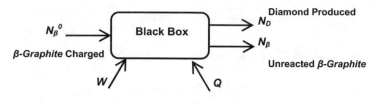

Fig. 1

$$dU = \delta W + \delta Q + H_\beta^{in} \delta n_\beta^{in} - H_D^{out} \delta n_D^{out} - H_\beta^{out} \delta n_\beta^{out} = 0 \qquad (3)$$

where the differential change in the internal energy in Eq. (1) was set to zero because the system is at steady state. By integrating and cancelling out the enthalpies of the unreacted graphite, we obtain:

$$W + Q + H_\beta N_\beta^0 - H_D N_D - H_\beta N_\beta = 0$$
$$\Rightarrow W + Q + H_\beta N_\beta^0 - H_D N_D - H_\beta \left(N_\beta^0 - N_D \right) = 0 \text{ (From a stoichiometric balance)}$$
$$\Rightarrow W + Q - H_D N_D + H_\beta N_D = 0$$
$$\Rightarrow -Q = W + \left(H_\beta - H_D \right) N_D$$

$$(4)$$

After obtaining the last expression in Eq. (4) based on the First Law of Thermodynamics, we can begin working through the Second Law of Thermodynamics. To this end, we can expand Eq. (2) by splitting the entropy into two terms as follows:

$$\Delta S_{universe} = \Delta S_{process} + \Delta S_{env} \qquad (5)$$

We can then separately evaluate the two entropy terms in Eq. (5). For the entropy term associated with the process, we obtain:

$$\Delta S_{process} = S^{out} - S^{in} = S_D N_D + S_\beta N_\beta - S_\beta N_\beta^0$$
$$= S_D N_D + S_\beta \left(N_\beta^0 - N_D \right) - S_\beta N_\beta^0 \text{ (From a stoichiometric balance)} \qquad (6)$$
$$= N_D \left(S_D - S_\beta \right)$$

For the entropy term associated with the environment, we can show that:

$$\Delta S_{env} = -\frac{Q}{T_{env}} \qquad (7)$$

where T_{env} is the temperature of the environment. Combining Eqs. (5), (6), and (7), we obtain:

$$\Delta \underline{S}_{universe} = N_D(S_D - S_\beta) - \frac{Q}{T_{env}} \geq 0 \tag{8}$$

By rearranging Eq. (8), we obtain:

$$Q \leq T_{env} N_D (S_D - S_\beta) \tag{9}$$

Combining Eqs. (4) and (9), we obtain:

$$W \geq (H_D - H_\beta)N_D - T_D(S_D - S_\beta)N_D \tag{10}$$

If we assume that there is no shaft work in the system and use the definition for the Gibbs free energy, we can express Eq. (9) as follows:

$$0 \geq (G_D - G_\beta)N_D$$
$$\Rightarrow \boxed{0 \geq \Delta G(T,P)} \tag{11}$$

where T is the system temperature and $P = 37$ kbar is the system pressure. Equation (11) provides us with a criterion that must be satisfied for the process to be possible.

Next, we must take the information provided in the Problem Statement and evaluate the reaction free energy at the system temperature and pressure. This can be done as follows:

$$\left(\frac{\partial G_i}{\partial P}\right)_{T,N} = V_i$$

$$\Rightarrow \left(\frac{\partial \Delta G}{\partial P}\right)_{T,N} = \Delta V$$

$$\Rightarrow \int_{\Delta G^0(T, 1\,\text{bar})}^{\Delta G^0(T, 37\,\text{kbar})} d\Delta G = \int_{1\,\text{bar}}^{37\,\text{kbar}} \Delta V dP$$

$$\Rightarrow \Delta G(T, 37\,\text{kbar}) - \Delta G(T, 1\,\text{bar}) = \int_{1\,\text{bar}}^{37\,\text{kbar}} (V_D - V_\beta) dP$$

$$\Rightarrow \Delta G(T, 37\,\text{kbar}) - \Delta G^0_{298} = \int_{1\,\text{bar}}^{37\,\text{kbar}} (V_D - V_\beta) dP \tag{12}$$

$$\Rightarrow \Delta G(T, 37\,\text{kbar}) = \int_{1\,\text{bar}}^{37\,\text{kbar}} (V_D - V_\beta) dP + 2870\,\text{J/mol}$$

Because the volumes can be approximated as being independent of pressure, the integral in Eq. (12) can be readily evaluated. Specifically:

$$\Delta G(T, 37\,\text{kbar}) = (V_D - V_\beta)(37\,\text{kbar} - 1\,\text{bar}) + 2870\,\text{J/mol}$$

$$= \left(\frac{12 \times 10^{-6}}{1 \times 3.51} - \frac{12 \times 10^{-6}}{1 \times 2.26}\right)(37 \times 10^8 - 1 \times 10^5) + 2870 \quad (13)$$

$$= -4126.\,\text{J/mol} < 0$$

Because the free energy of the reaction is negative, Eq. (11) is satisfied, demonstrating that the process is possible!

Other Possible Solution Strategies

It is possible that considering an equilibrium reaction, one would decide to solve the problem by evaluating the equilibrium constant and showing that the conversion of the reaction is nonzero. If we proceed along those lines, we will realize that there is some sort of discrepancy between the equilibrium constant calculated using the standard Gibbs free energy of reaction and calculated using the fugacity of the reacting species. If we define our reference states as pure solids at the system temperature and 1 bar pressure, we can write the equilibrium constant as follows:

$$K = \exp\left(-\frac{\Delta G(T, 1\,\text{bar})}{RT}\right) = \exp\left(-\frac{2870}{8.314 \times 298}\right) = 0.314 \quad (14)$$

However, if we calculate the equilibrium constant based on its definition in terms of the fugacities of the reacting species, we obtain:

$$K = \left(\frac{f_\beta(T, 37\,\text{kbar})}{f_\beta(T, 1\,\text{bar})}\right)^{-1} \left(\frac{f_D(T, 37\,\text{kbar})}{f_D(T, 1\,\text{bar})}\right)$$

$$= \left(\frac{f_\beta(T, 1\,\text{bar})\exp\displaystyle\int_{1\,\text{bar}}^{37\,\text{kbar}}(V_\beta/RT)\,dP}{f_\beta(T, 1\,\text{bar})}\right)^{-1} \left(\frac{f_D(T, 1\,\text{bar})\exp\displaystyle\int_{1\,\text{bar}}^{37\,\text{kbar}}(V_D/RT)\,dP}{f_D(T, 1\,\text{bar})}\right)$$

$$K = \exp\left((V_D - V_\beta)(37\,\text{kbar} - 1\,\text{bar})/RT\right)$$

$$= \exp\left(\left(\frac{12 \times 10^{-6}}{1 \times 3.51} - \frac{12 \times 10^{-6}}{1 \times 2.26}\right)(37 \times 10^8 - 1 \times 10^5)/(8.314 \times 298)\right)$$

$$= \exp\left(\left(\frac{12 \times 10^{-6}}{1 \times 3.51} - \frac{12 \times 10^{-6}}{1 \times 2.26}\right)(37 \times 10^8 - 1 \times 10^5)/(8.314 \times 298)\right)$$

$$= 0.059$$

$$(15)$$

which is different from the value obtained in Eq. (14). To explain this difference, let us invoke the Gibbs Phase Rule to see if we indeed have two phases at the given T and P condition. In the last analysis, we have assumed that we have two pure-component phases ($n + 1 = 2$ per phase, total four independent intensive variables), two conditions of equilibrium (thermal and mechanical), and one chemical reaction. This results in a variance which is equal to 4–2–$1 = 1$. Therefore, once the temperature is specified, equilibrium exists at a unique pressure. In other words, we are not allowed to independently specify both temperature and pressure. This is the reason behind the discrepancy in the values obtained for the equilibrium constant in Eqs. (13) and (14) using two seemingly valid approaches to calculate it. In light of the fact that equilibrium need not exist at a pressure of 37 kbar, it is clear that we cannot use Eq. (14) to calculate the equilibrium constant because it inherently assumes that there exists equilibrium at 37 kbar. We can calculate the system pressure using the following equation:

$$K = \exp\left(-\frac{\Delta G(T, 1\,\text{bar})}{RT}\right) = 0.314 = \left(\frac{f_\beta(T, P)}{f_\beta(T, 1\,\text{bar})}\right)^{-1}\left(\frac{f_D(T, P)}{f_D(T, 1\,\text{bar})}\right)$$

$$\Rightarrow 0.314 = \exp\left((V_D - V_\beta)(P - 1\,\text{bar})/RT\right)$$

$$\Rightarrow \ln 0.314 = \left(\frac{12 \times 10^{-6}}{1 \times 3.51} - \frac{12 \times 10^{-6}}{1 \times 2.26}\right)(P - 1 \times 10^5)/(8.314 \times 298)$$

$$\Rightarrow P = 15.18\,\text{kbar}$$

$$(16)$$

This pressure corresponds to a standard molar Gibbs free energy of reaction which is equal to 0. At any pressure lower or higher than this pressure, the reaction will proceed to completion, and there would exist only one component and one phase. In other words, there would be no chemical equilibrium between the two components, and application of the Gibbs Phase Rule will yields a variance of 2 (if n is equal to 2, π is also equal to 2, while if n is equal to 1, π is also equal to 1). Note that, in both cases, r is equal to 0.

Problem 17

Problem 16.11 in Tester and Modell

We do not seem to be able to locate any data for the standard Gibbs free energy of reaction, nor for the standard enthalpy of reaction, ΔH^0, for the following chemical reaction:

$$D_{(g)} = A_{(g)} + B_{(g)}$$

However, one of our research students has conducted a few experiments as described below. A constant volume pressure vessel was immersed in a constant-temperature bath. The vessel was first evacuated and then filled with pure gaseous D at 298 K and 1 bar. After heating the closed vessel to 473 K, reaction occurred to form A and B. When no further pressure change was noted, the gauge read 3 bar. Further heating to 523 K caused a pressure increase to 3.30 bar.

With these data, can one calculate $\Delta G^0_{523\,K}$ and ΔH^0? If so, provide numerical values. If not, describe what additional data are required. You may assume that the vapor phase forms an ideal gas mixture and that ΔH^0 is not a function of temperature.

Solution to Problem 17

Solution Strategy

This problem provides information about the equilibrium temperatures and pressures of a gas-phase chemical reaction and asks us to evaluate the standard molar Gibbs free energy of reaction using the data provided. Let us begin by summarizing all the experiments which were carried out in a constant volume vessel having volume $\underline{V}$:

Experiment 1

$$P_1 = 1\,\text{bar}$$
$$T_1 = 298\,\text{K}$$
$$\underline{V}_1 = \underline{V}$$
$$N_{1,A} = 0$$
$$N_{1,B} = 0$$
$$N_{1,D} = ? = \frac{P_1 \underline{V}}{RT_1}$$

Experiment 2

$$P_2 = 3 \, \text{bar}$$

$$T_2 = 473 \, \text{K}$$

$$\underline{V}_2 = \underline{V}$$

$$N_{2,A} = ?$$

$$N_{2,B} = ?$$

$$N_{2,D} = ?$$

In Experiment 2, the moles of the three gaseous species are related to each other based on the stoichiometry of the chemical reaction. The relationship is reported in the Stoichiometric Table below.

	D	A	B
Initial moles	$N_{1,D}$	0	0
Final moles	$N_{2,D} = N_{1,D} - N_{2,A}$	$N_{2,A}$	$N_{2,B} = N_{2,A}$
Final mole fraction	$(N_{1,D} - N_{2,A})/(N_{1,D} + N_{2,A})$	$N_{2,A}/(N_{1,D} + N_{2,A})$	$N_{2,A}/(N_{1,D} + N_{2,A})$

From the Stoichiometric table, we can deduce that the total number of moles of gas in the reactor vessel at the end of Experiment 2 is given by $N_{2,A} + N_{1,D}$, which can be related to P_2 and T_2 using the ideal gas law as follows:

$$N_{2,A} + N_{1,D} = \frac{P_2 V}{RT_2} \tag{1}$$

Finally, we can relate the standard molar Gibbs free energy of reaction to the measurable experimental properties using the equilibrium constant for the chemical reaction as follows:

$$K = \left(\frac{\hat{f}_D}{f_D^0}\right)^{-1} \left(\frac{\hat{f}_A}{f_A^0}\right) \left(\frac{\hat{f}_B}{f_B^0}\right)$$

$$= \left(\frac{y_{2,D}\hat{\phi}_{2,D}P_2}{P^0}\right)^{-1} \left(\frac{y_{2,A}\hat{\phi}_{2,A}P_2}{P^0}\right) \left(\frac{y_{2,B}\hat{\phi}_{2,B}P_2}{P^0}\right)$$

$$= \left(\frac{y_{2,A}y_{2,B}P_2}{y_{2,D}P^0}\right) \qquad \text{ideal gas mixture; } \hat{\phi}_i = 1$$

$$= \left(\frac{N_{2,A}N_{2,B}(N_{1,D} + N_{2,A})P_2}{(N_{1,D} + N_{2,A})(N_{1,D} - N_{2,A})P^0}\right) \tag{2}$$

$$= \left(\frac{N_{2,A}^2 P_2}{(N_{1,D} + N_{2,A})(N_{1,D} - N_{2,A})P^0}\right)$$

$$= \exp\left(-\frac{\Delta G^0(T_2, P^0)}{RT_2}\right)$$

Substituting the values of $N_{1,D}$ and $N_{2,A}$ in terms of $\underline{V}$ using Eq. (1) yields:

$$\left(\frac{N_{2,A}^2 P_2}{(N_{1,D} + N_{2,A})(N_{1,D} - N_{2,A})P^0}\right) = \exp\left(-\frac{\Delta G^0(T_2, P^0)}{RT_2}\right)$$

$$\Rightarrow \left(\frac{\left(\frac{P_2\underline{V}}{RT_2} - \frac{P_1\underline{V}}{RT_1}\right)^2 P_2}{\frac{P_2\underline{V}}{RT_2}\left(2\frac{P_1\underline{V}}{RT_1} - \frac{P_2\underline{V}}{RT_2}\right)P^0}\right) = \exp\left(-\frac{\Delta G^0(T_2, P^0)}{RT_2}\right)$$

$$\Rightarrow \left(\frac{T_2(P_2/T_2 - P_1/T_1)^2}{(2P_1/T_1 - P_2/T_2)P^0}\right) = \exp\left(-\frac{\Delta G^0(T_2, P^0)}{RT_2}\right) \tag{3}$$

$$\Rightarrow \left(\frac{473(3/473 - 1/298)^2}{(2/298 - 3/473)1}\right) = \exp\left(-\frac{\Delta G^0(T_2, P^0)}{RT_2}\right)$$

$$\Rightarrow 11.4739 = \exp\left(-\frac{\Delta G^0(T_2, P^0)}{8.314 \times 473}\right)$$

$$\Rightarrow \Delta G^0(T_2, P^0) = -9583.2748\,\text{J/mol}$$

Experiment 3

$$P_3 = 3.30\,\text{bar}$$

$$T_2 = 523\,\text{K}$$

$$\underline{V}_3 = \underline{V}$$

$$N_{3,A} = ?$$

$$N_{3,B} = ?$$

$$N_{3,D} = ?$$

An analysis identical to that done for Experiment 2 would lead to the following Stoichiometric Table:

	D	A	B
Initial moles	$N_{1,D}$	0	0
Final moles	$N_{3,D} = N_{1,D} - N_{3,A}$	$N_{3,A}$	$N_{3,\,B} = N_{3,A}$
Final mole fraction	$(N_{1,D} - N_{3,A})/(N_{1,D} + N_{3,A})$	$N_{3,A}/(N_{1,D} + N_{3,A})$	$N_{3,\,A}/(N_{1,D} + N_{3,A})$

Similar to Experiment 2, the total number of moles of gas in the reactor vessel at the end of Experiment 3 is given by $N_{3,\,A} + N_{1,\,D}$, which can be related to P_3 and T_3 as follows:

$$N_{3,A} + N_{1,D} = \frac{P_3 \underline{V}}{RT_3} \tag{4}$$

In addition, we can relate the standard molar Gibbs free energy of reaction for Experiment 3 to the measurable experimental properties using the equilibrium constant as follows:

$$
\begin{aligned}
K &= \left(\frac{y_{3,D}\hat{\phi}_{3,D}\cancel{P_3}}{\cancel{P^0}}\right)^{-1}\left(\frac{y_{3,A}\hat{\phi}_{3,A}\cancel{P_3}}{\cancel{P^0}}\right)\left(\frac{y_{3,B}\hat{\phi}_{3,B}P_3}{P^0}\right) \\
&= \left(\frac{y_{3,A}y_{3,B}P_3}{y_{3,D}P^0}\right) = \left(\frac{N_{3,A}^2 P_3}{(N_{1,D} + N_{3,A})(N_{1,D} - N_{3,A})P^0}\right) \\
&= \exp\left(-\frac{\Delta G^0\left(T_3, P^0\right)}{RT_3}\right)
\end{aligned}
\tag{5}
$$

Substituting the values of $N_{1,D}$ and $N_{3,A}$ in terms of $\underline{V}$ yields:

$$\left(\frac{N_{3,A}^2 P_3}{(N_{1,D}+N_{3,A})(N_{1,D}-N_{3,A})P^0}\right) = \exp\left(-\frac{\Delta G^0\left(T_3, P^0\right)}{RT_3}\right)$$

$$\Rightarrow \left(\frac{\left(\frac{P_3\underline{V}}{RT_3}-\frac{P_1\underline{V}}{RT_1}\right)^2 P_2}{\frac{P_3\underline{V}}{RT_3}\left(2\frac{P_1\underline{V}}{RT_1}-\frac{P_3\underline{V}}{RT_3}\right)P^0}\right) = \exp\left(-\frac{\Delta G^0\left(T_3, P^0\right)}{RT_3}\right)$$

$$\Rightarrow \left(\frac{T_3(P_3/T_3 - P_1/T_1)^2}{(2P_1/T_1 - P_3/T_3)P^0}\right) = \exp\left(-\frac{\Delta G^0\left(T_3, P^0\right)}{RT_3}\right) \tag{6}$$

$$\Rightarrow \left(\frac{523(3.3/523 - 1/298)^2}{(2/298 - 3.3/523)1}\right) = \exp\left(-\frac{\Delta G^0\left(T_3, P^0\right)}{RT_3}\right)$$

$$\Rightarrow 11.3627 = \exp\left(-\frac{\Delta G^0\left(T_3, P^0\right)}{8.314 \times 523}\right)$$

$$\Rightarrow \Delta G^0\left(523\,\mathrm{K}, P^0\right) = -10567.6256\,\mathrm{J/mol}$$

We can then calculate ΔH^0 from the standard molar Gibbs free energy of reaction using the Gibbs-Helmholtz equation. Specifically:

$$\left(\frac{\partial\left(\Delta G^0/T\right)}{\partial T}\right)_P = -\frac{\Delta H^0}{T^2}$$

$$\Rightarrow \left(\frac{\partial\left(\Delta G^0/T\right)}{\partial(1/T)}\right)_P = \Delta H^0 \tag{7}$$

The Problem Statement indicates that the standard molar enthalpy of reaction is independent of temperature. Therefore, we can calculate the partial derivative in Eq. (7) using the two data points for ΔG^0 as follows:

$$\left(\frac{\partial\left(\Delta G^0/T\right)}{\partial(1/T)}\right)_P = \Delta H^0$$

$$\Rightarrow \left(\frac{\Delta G^0\left(T_3, P^0\right)/T_3 - \Delta G^0\left(T_2, P^0\right)/T_2}{(1/T_3) - (1/T_2)}\right) = \Delta H^0$$

$$\Rightarrow \left(\frac{(-10567.6256/523) - (-9583.2748/473)}{(1/523) - (1/473)}\right) = \Delta H^0 \tag{8}$$

$$\Rightarrow \Delta H^0 = -271.3162\,\mathrm{J/mol}$$

Problem 18

Problem 9.3 in Tester and Modell

In an experiment, a mixture of helium and ammonia was prepared as follows. As shown in the figure below, separate supply manifolds are available for the helium and the ammonia gases. The aluminum mixing tank is first evacuated to a very low pressure. Helium gas is then admitted very rapidly until the tank pressure is at 2 bar. The helium supply valve is then closed.

Ten minutes later, the ammonia supply valve is opened to allow ammonia gas to flow rapidly into the tank. The valve is closed when the tank pressure reaches 3 bar.

Data:

1. Assume that the gases behave ideally. The heat capacities of helium and ammonia may be considered to be constants with the following values:

$$C_P(\text{He}) = 20.9 \text{ J/mol K}$$
$$C_P(\text{NH}_3) = 35.6 \text{ J/mol K}$$

2. Assume that the He-NH$_3$ gas mixture is ideal.
3. The mixing tank dimensions are 0.3 m (inside diameter), 0.3 m tall.
4. The initial wall temperature of the mixing tank is 310 K and may be assumed to remain constant.
5. It may be assumed that the heat transfer from the gas to the mixing tank walls occurs by natural convection. For purposes of computation, assume that the heat transfer coefficient has a constant value of 15 W/m^{2}K.
6. The helium gas in its manifold is at 310 K and 10 bar, and the ammonia gas in its manifold is at 310 K and 5 bar.

With only this description and the given data, use your engineering reasoning to answer the following questions:

(a) What is the amount of He gas in the mixing tank and its temperature when the tank pressure reaches 2 bar?
(b) What is the temperature and the pressure of the He gas in the mixing tank 10 min later?
(c) What is the temperature and composition of the He-NH$_3$ gas mixture in the mixing tank when the tank pressure reaches 3 bar?

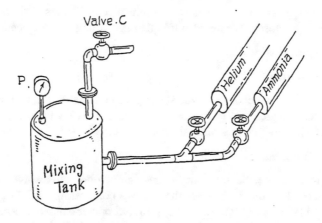

Solution to Problem 18

Solution Strategy

To setup this problem, before solving it, we will use the four-step strategy discussed in Part I when we utilized the First Law of Thermodynamics for closed and open systems. In this problem, we will also need to do this for a one component gas (in Parts (a) and (b)), as well as for an ideal binary gas mixture in Part (c). The four steps include:

1. Draw the pertinent problem configuration
2. Summarize the given information
3. Identify critical issues
4. Make physically reasonable approximations

1. **Draw the pertinent problem configuration**

   A figure depicting the tank and the gas supply tubes is provided in the Problem Statement. To solve the three parts of this problem, we will treat the contents of the tank at any given time as the system to be studied (see 3 below).

2. **Summarize the given information**

   This problem consists of three parts, with a different process occurring in each part. For Part (a), the helium valve is opened, and helium flows into the tank until the tank pressure reaches 2 bars. Once this pressure is reached, the helium valve is closed. We are also told that this helium filling process occurs very rapidly. For Part (b), we are asked to examine how the temperature of helium in the tank changes in 10 min. During this 10-min period, no additional gases are added to the tank. Finally, for Part (c), the ammonia valve is opened, and ammonia is allowed to flow into the helium-containing tank until the tank pressure reaches 3 bars. Subsequently, the ammonia valve is closed. We are also told that this filling process occurs rapidly.

3. Identify critical issues

Defining the System

As stated above, we choose as our system the contents of the tank at any time. This allows us to neglect any Pd$\underline{V}$-type work (i.e., $\delta W = 0$), which will simplify our First Law of Thermodynamics analysis and equations.

Defining the Boundaries

For all three parts, the boundaries are rigid (solid, unmovable tank walls). However, depending on the process that we are considering, some of the system boundaries will change. In Part (a), the system has open boundaries because helium flows into the tank. The boundaries are also considered adiabatic because we are told that this filling process occurs very rapidly, and therefore, we can assume that there is no time for heat transfer to occur between helium and the tank walls. In Part (b), the system has closed boundaries. Unlike in Part (a), however, we are specifically told that there is heat transfer between helium and the tank walls. Finally, in Part (c), the boundaries are open because ammonia flows into the tank. The boundaries are similar to those in Part (a), including being adiabatic.

Mixing of Gases

In this problem, helium and ammonia are considered to be ideal gases. We will assume that helium and ammonia will be well-mixed within the tank. As a result, each gas will occupy the entire tank volume and will have uniform temperatures and pressures.

4. Make physically reasonable approximations

Ideal Gas Behavior

The problem states that helium and ammonia behave ideally. Therefore, helium and ammonia will be assumed to have constant C_V and C_P values. Nevertheless, we note that the two gases have different C_V (and therefore C_P) values!

Adiabatic process for Parts (a) and (c)

As stated above, we take our system to be the contents of the tank at any time. For Parts (a) and (c), we are told that the filling occurs rapidly. Consequently, we will assume that there is not enough time for heat transfer between the gas and the tank walls to occur (i.e., adiabatic filling process).

Heat Transfer by Convection for Part (b)

For Part (b), we are told that heat transfer from helium in the tank to the tank walls occurs by natural convection. Because the tank is kept at a constant temperature of 310 K, we will also assume that any heat transferred from helium to the tank walls will be immediately transferred to the surroundings, so that the tank walls are kept at a constant temperature.

Part (a)

To solve this part of the problem, we can make the following assumptions:

- Helium is an ideal gas of constant C_V and $C_P = C_V + R$.
- Helium is well-mixed and occupies the entire aluminum tank. As a result, $d\underline{V} = 0$, and $\delta W = 0$ (no Pd$\underline{V}$-type work).
- This is a fast filling process, implying that there is no heat transfer. That is, $\delta Q = 0$ (adiabatic operation).
- The initial (i) tank pressure is assumed to be zero ($P_i = 0$). That is, $N_i = 0$ for helium (The aluminum tank is initially empty).

We begin by carrying out a First Law of Thermodynamics analysis of the tank contents at any time. This system is simple, open, rigid, and adiabatic. As a result,

$$
\begin{aligned}
&d\underline{U} = \delta Q + \delta W + H_{in}\delta n_{in} - H_{out}\delta n_{out} \\
&\delta Q = 0 \\
&\delta W = 0 \\
&\delta n_{out} = 0 \\
&\delta n_{in} = dN \\
&H_{in} = H_{in}(T_{in}) = H_{in}(310K) = \text{constant}
\end{aligned}
\tag{1}
$$

Therefore, the top expression in Eq. (1) can be rewritten as follows:

$$
d\underline{U} = H_{in}dN
\tag{2}
$$

We can integrate Eq. (2) directly from (N_i, $\underline{U}_i$) to (N_f, $\underline{U}_f$). Note that $N_i = 0$, and therefore, $\underline{U}_i = 0$. Upon integration, we obtain:

$$
\begin{aligned}
\underline{U}_f &= H_{in}N_f \\
U_f N_f &= H_{in}N_f
\end{aligned}
$$

$$
U_f = H_{in}
\tag{3}
$$

Because $\frac{dU}{dT} = C_V$ and $\frac{dH}{dT} = C_P$, upon integration, we can write:

$$
\begin{aligned}
U_f &= U_0 + C_V(T_f - T_0) \\
H_{in} &= H_0 + C_P(T_{in} - T_0)
\end{aligned}
$$

Using the expressions above for U_f and H_{in} in Eq. (3) yields:

$$U_f = H_{in}$$
$$U_0 + C_V T_f - C_V T_0 = H_0 + C_P T_{in} - C_P T_0$$
$$C_V T_f = C_P T_{in} + (H_0 - (C_P - C_V)T_0 - U_0) \tag{4}$$
$$C_V T_f = C_P T_{in} + (H_0 - H_0)$$
$$C_V T_f = C_P T_{in}$$

Equation (4) shows that:

$$T_f = \frac{C_P T_{in}}{C_V} \tag{5}$$

Denoting $T_f \equiv T_1$ and using the data provided in the Problem Statement, we obtain:

$$T_f = T_1 = (310\,\text{K}) \left[\frac{20.9}{20.9 - 8.314} \right] = 514.8\,\text{K}$$

$$T_1 = 514.8\,\text{K} \tag{6}$$

Because helium is treated as an ideal gas, we know that it is described by the ideal gas EOS. Therefore:

$$N_1 = \frac{P_1 V}{R T_1} \tag{7}$$

Using the required values in Eq. (7), we find that:

$$P_1 = 2\,\text{bar} = 2 \times 10^5\,\text{Pa}$$
$$T_1 = 514.8\,\text{K}$$
$$R = 8.314\,\text{J/mol K}$$
$$V = \frac{\pi D^2 L}{4} = \frac{3.14(0.3)^2 0.3}{4} = 2.12 \times 10^{-2} E^3$$
$$N_1 = \frac{(2 \times 10^5\,\text{Pa})(2.12 \times 10^{-2}\,\text{m}^3)}{(8.314)(514.8\,\text{K})} = 0.991\,\text{mol}$$

$$\boxed{N_1 = 0.991\,\text{mol}} \tag{8}$$

Part (b)

After a 10-min waiting period, during which helium in the aluminum tank exchanges heat with the tank walls via a natural convection process, helium cools down due to the heat transfer to the vessel walls. We will assume that the wall temperature, T_w, does not change. Then, if we consider the helium gas as a well-mixed, closed, simple system that exchanges heat, a First Law of Thermodynamics analysis yields:

$$d\underline{U} = \delta Q + \delta W = \delta Q \tag{9}$$

where δQ refers to the heat absorbed by the gas, and $\delta W = 0$ because of the rigid tank walls. We also know that:

$$d\underline{U} = N_1 C_V dT \tag{10}$$

According to the Problem Statement, we know that heat transfer between helium and the tank walls occurs via convection, that is:

$$\delta Q = -h_Q A (T - T_w) dt \tag{11}$$

In Eq. (11), h_Q denotes the heat transfer coefficient given in the Problem Statement, and A denotes the surface area of the tank. A negative sign is necessary to account for the fact that helium loses heat (it is at a temperature of 515 K, which is higher than the tank wall temperature of 310 K).

Using Eqs. (10) and (11) in Eq. (9) yields:

$$N_1 C_V dT = -h_Q A (T - T_w) dt \tag{12}$$

Because T_w and h_Q are constant, Eq. (12) can be integrated as follows:

$$\frac{dT}{(T - T_w)} = -\left(\frac{h_Q A}{N_1 C_V}\right) dt$$

$$\int_{T_1}^{T_2} \frac{dT}{(T - T_w)} = -\left(\frac{h_Q A}{N_1 C_V}\right) \int_{0}^{t=600 \text{ sec}} dt$$

$$\ln\left(\frac{T_2 - T_w}{T_1 - T_w}\right) = -\left(\frac{h_Q A}{N_1 C_V}\right) t$$

or

$$T_2 = T_w + (T_1 - T_w) \exp\left[-\left(\frac{h_Q A}{N_1 C_V}\right) t\right] \tag{13}$$

Using the information given below:

$$A = \pi D L + 2\left(\frac{\pi D^2}{4}\right) = (3.14)(0.3\,\text{m})(0.3\ \text{m}) + \frac{2(3.14)(0.3\,\text{m})^2}{4} = 0.424\text{m}^2$$

$N_1 = 0.991$ mol

$C_V = (20.9 - 8.314)$ J/mol K $= 12.586$ J/mol K

$h_Q = 15\,\text{W/m}^2\text{K}$

$t = 10$ min $= 600$ sec

$T_1 = 514.8\,\text{K}$ (From Part (a))

in Eq. (13), we obtain:

$$T_2 = 310\,\text{K} + (514.8 - 310)\,\text{K}\exp\left[\frac{(-1)(15)(0.424)(600)}{(0.991)(12.586)}\right]$$

$$T_2 = 310\,\text{K} + (204.8)\exp(-305)$$

or

$$\boxed{T_2 \cong 310\,\text{K}} \tag{14}$$

Equation (14) shows that in 10 min, helium cools down to the wall temperature. Further, the helium pressure changes to $P_2 = \frac{N_1 R T_2}{V}$. However, $P_1 = \frac{N_1 R T_1}{V}$, so that:

$$\frac{P_1}{T_1} = \frac{P_2}{T_2}$$

$$P_2 = \frac{P_1 T_2}{T_1} = (2\,\text{bar})\left(\frac{310}{514.8}\right)$$

or

$$\boxed{P_2 = 1.2\text{bar}} \tag{15}$$

Part (c)

This part of the problem deals with the addition of ammonia to the tank already filled with helium. We can make the same assumptions that we made in Part (a):

- Helium and ammonia behave ideally
- There is rapid addition of ammonia, so that $\delta Q = 0$
- $\delta n_{out} = 0$, $\delta n_{in} = dN$
- We will analyze the well-mixed case. This implies that $\delta W = 0$

We first note that inside the aluminum tank, we have a binary mixture of ideal gases (helium and ammonia), and we are told that the gas mixture itself is also ideal. As in Part (a), we carry out a First Law of Thermodynamics analysis of the binary gas mixture at any time. The system is simple, open, adiabatic, and rigid. Accordingly:

$$dU = \delta Q + \delta W + H_{in}\delta n_{in} - H_{out}\delta n_{out} \tag{16}$$

In Eq. (16), we have:

$$H_{in} = H_{in}(T_{in}) = H_{in}(310K) = \text{constant} \tag{17}$$

Like in Part (a), Eq. (16) reduces to:

$$dU = H_{in}\delta n_{in} \tag{18}$$

As we did in Part (a), we can integrate Eq. (18) directly from state i to state f, which yields:

$$U_f - U_i = H_{in}N_{in} \tag{19}$$

where $N_{in} = N_{NH3}$ (the total number of moles of ammonia that entered into the aluminum tank).

We can next relate U to H, P, and V as follows:

$$\begin{aligned} U_f &= H_f - P_f V_f \\ U_i &= H_i - P_i V_i \\ V_f &= V_i = V \end{aligned} \tag{20}$$

Using the three equations in Eq. (20), we can write that:

$$U_f - U_i = (H_f - H_i) - (P_f - P_i)V \tag{21}$$

Because the final (f) state corresponds to an ideal binary mixture of helium and ammonia in the tank and the initial (i) state corresponds to pure helium in the tank, the following enthalpy expressions can be written down:

$$\begin{aligned} H_f &= N_{He}^f H_{He}^f + N_{NH_3}^f H_{NH_3}^f \\ H_i &= \text{pure helium} = N_{He}^i H_{He}^i \\ H_{in} &= H_{NH_3}^i \end{aligned}$$

Recall that $N_{He}^f = N_{He}^i = N_{He}$ and $N_{NH_3}^f \equiv N_{NH_3}$. Hence, using the three expressions above in Eq. (21), rearranging, and using Eq. (19) where in denotes ammonia, we obtain:

$$\underline{U}_f - \underline{U}_i = \left(N_{He}H_{He}{}^f + N_{NH_3}H_{NH_3}{}^f - P_f\underline{V}\right) - \left(N_{He}H_{He}{}^i - P_i\underline{V}\right)$$

$$= H_{NH_3}N_{NH_3} \qquad (22)$$

Rearranging Eq. (22), we obtain:

$$N_{He}\left(H_{He}{}^f - H_{He}{}^i\right) + N_{NH_3}\left(H_{NH_3}{}^f - H_{NH_3}{}^i\right) - \underline{V}\left(P_f - P_i\right) = 0 \qquad (23)$$

Using the fact that helium and ammonia are considered here as ideal gases, we can express the molar enthalpy changes of each gas in Eq. (23) as follows:

$$H_{He}{}^f - H_{He}{}^i = C_P{}^{He}\left(T_f - T_0\right) - C_P{}^{He}(T_2 - T_0) = C_P{}^{He}\left(T_f - T_2\right) \qquad (24)$$

$$H_{NH_3}{}^f - H_{NH_3}{}^i = C_P{}^{NH_3}\left(T_f - T_0\right) - C_P{}^{NH_3}(T_{in} - T_0) = C_P{}^{NH_3}\left(T_f - T_{in}\right) \qquad (25)$$

Using Eqs. (24) and (25) in Eq. (23), we obtain:

$$N_{He}\left[C_P{}^{He}\left(T_f - T_2\right)\right] + N_{NH_3}\left[C_P{}^{NH_3}\left(T_f - T_{in}\right)\right] - \underline{V}\left(P_f - P_i\right) = 0 \qquad (26)$$

where $N_{He} = N_1 = 0.991$ mol, $T_2 = 310$ K, $T_{in} = 310$ K, $\underline{V} = 2.12 \times 10^{-2}$ m^3, $P_f = 3$ bar, and $P_2 = 1.2$ bar.

Equation (26) relates the two unknowns, N_{NH3} and T_f. In order to solve for them, we need a second equation relating them that we can then solve simultaneously with Eq. (26). Because ammonia is treated here as an ideal gas, the second equation is the ideal gas EOS relating T_f and N_{NH3}. Specifically:

$$P_f\underline{V} = N_{total}RT_f$$

$$N_{total} = \frac{P_f\underline{V}}{RT_f} = \frac{\left(3 \times 10^5\right)\left(2.12 \times 10^{-2}\right)}{(8.314)T_f}$$

$$N_{NH_3} = N_{total} - N_1 = \left(\frac{765}{T_f} - 0.991\right)$$

$$N_{NH_3} = \left(\frac{765}{T_f} - 0.991\right) \qquad (27)$$

Using Eq. (27) in Eq. (26), we obtain:

$$(0.991)(20.9)(T_f - 310) + \left(\frac{765}{T_f} - 0.991\right)(35.6)(T_f - 310)$$
$$-(2.12 \times 10^{-2})(3 - 1.2)(10^5) = 0$$
$$-14.57 T_f^2 + 27935 T_f - 8442540 = 0$$

or

$$T_f^2 - 1917 T_f + 579447 = 0 \qquad (28)$$

Solving the quadratic equation above, we find that that there are two roots, that is:

$$T_f = \left(\frac{1917 \pm 1165}{2}\right) K$$
$$T_f^- = 376 \, K$$
$$T_f^+ = 1541 \, K$$

To determine which of the two T_f values obtained above is the correct one, we use these values in Eq. (27). This yields:

$$N_{NH_3}{}^- = \frac{765}{376} - 0.991 = 1.044 \, mol$$
$$N_{NH_3}{}^+ = \frac{765}{1541} - 0.991 < 0!$$

Clearly, using $T_f = 1541$ K results in a negative value for N_{NH3}, which is not physical. Therefore, the correct final temperature of the helium-ammonia mixture is 376 K. This implies that 1.044 mol of NH_3 were added to the tank during the final operation and, therefore, that ammonia makes up 51.3% of the final gas mixture in the tank. To summarize, we obtain:

$$\boxed{T_f = 376 \, K, \quad N_{NH_3} = 1.044 \, mol}$$

Note that the final helium-ammonia mixture temperature is higher than its initial temperature ($T_f > T_2$). This reflects ammonia entering the tank, which originally contained helium only. Because of the well-mixing assumption made, there is no compression of the existing helium gas (i.e., the stratified layer model discussed in Part I does not apply). Instead, the increase in temperature of the helium-ammonia mixture results from the enthalpy of ammonia flowing in. An examination of Eq. (19) reveals that the internal energy of the helium-ammonia mixture (and therefore, its temperature) increases as NH_3 enters the system. Therefore, as expected, $T_f > T_2$.

Problem 19

A creative Chemical Engineering graduate student is trying his luck in the business world to supplement his monthly stipend. He plans to produce freshwater for the crew of a nuclear submarine by inserting a semipermeable membrane, which allows solely passage of freshwater, in the hull of the submarine, and uses the pressure difference between the outside (the sea) and the inside of the submarine to provide a driving force to produce freshwater by reverse osmosis. The student believes that the US Navy will be interested in his idea.

As proof of concept, the engineer would like to calculate what the minimum depth of the submarine, h_{min}, has to be for the first drop of freshwater to pass through the membrane. You are asked to utilize your expertise in classical thermodynamics to answer the following question: Derive an explicit expression for h_{min}, and then use it to obtain a numerical value for the minimum submarine depth. For this purpose, you can make the following assumptions:

1. Seawater can be considered as a binary water-NaCl solution.
2. NaCl in seawater is not dissociated.
3. The NaCl concentration in seawater is uniform and has a value of 5 wt%.

The mass density of seawater, ρ_{SW}, and the partial specific volume of water, $\overline{V}_w$, are independent of pressure, and they have constant values of:

$$\rho_{SW} = 1.145 \text{ g/cm}^3$$
$$\overline{V}_w = 0.994 \text{ cm}^3/\text{g}$$

4. The seawater temperature is uniform and has a value of 25 °C.
5. The temperature in the hull of the submarine is maintained at a constant value of 25 °C.
6. The pressure in the hull of the submarine is maintained at a constant value of 1 atm.
7. The graph of water chemical potential versus weight fraction of NaCl, given below, can be utilized.

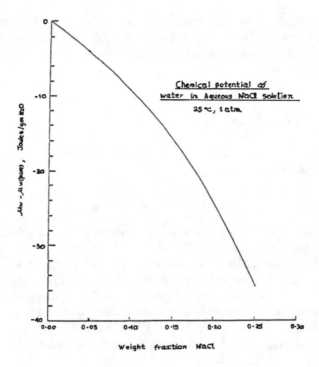

Chemical potential of water in Aqueous NaCl solution 25°C, 1 atm.

(y-axis: $\mu_w - \mu_{w(pure)}$, Joules/gm H2O)

(x-axis: Weight fraction NaCl)

Solution to Problem 19

Solution Strategy

This is an interesting problem which explores the possibility of obtaining freshwater in the hull of a nuclear submarine from seawater through reverse osmosis. Because as the submarine descends there is a pressure difference (higher pressure outside the submarine, P, relative to that inside the submarine, $P_0 = 1$ atm), the equality of the chemical potentials (or fugacities) at the given T_0 will be the required condition.

Before we begin to solve this problem, we would like to take a moment to think about what is actually going on. We know that when pure water is separated from an aqueous solution of a solute by a semipermeable membrane (one that is impermeable to the solute), water flows from the pure water side to the solution side. This is known as osmosis, which is the result of the higher water chemical potential of pure water. However, in this problem, this would mean that freshwater is flowing from the hull of the submarine into the sea, which is not very helpful.

Fortunately, because of the presence of hydrostatic pressure, the pressure outside the submarine is much larger than that in the hull of the submarine. When this pressure difference is large enough to overcompensate for the concentration difference, we can have water flowing in the reverse direction, that is, from the sea into the hull of the submarine. This process is known as reverse osmosis. Therefore, if we need to calculate the submarine depth at which the first drop of freshwater flows from the sea into the hull of the submarine, we simply need to calculate the depth at which the pressure difference *just* balances out the concentration difference to allow reverse osmosis to occur.

Calculation of Phase Equilibria

With the above discussion in mind, we can write the required phase equilibria condition as follows:

$$\mu_w(T_0, P_0) = \mu_w(T_0, P, x_s) \tag{1}$$

where $T_0 = 25\ °C$, $P > P_0 = 1$ atm, and x_s refers to the sea salt concentration (or equivalently, $x_w = 1 - x_s$).

Because the Problem Statement provides graphical information about the water chemical potential, it is convenient to continue using Eq. (1) without switching to water fugacities, as reflected in the integral or the differential approaches to phase equilibria that we discussed in Part II. In order to utilize the given graph of chemical potential (which is given at $T_0 = 25\ °C$ and $P = 1$ atm) at $T_0 = 25\ °C$ and $P > 1$ atm, we need to express $\mu_w(T_0, P, x_s)$ in terms of $\mu_w(T_0, P_0, x_s)$. This can be readily done because, as discussed in Part II:

$$\left(\frac{\partial \mu_i}{\partial P}\right)_{T,x} = \overline{V}_i \tag{2}$$

Integrating Eq. (2) for $i = w$, at $T = T_0$ and $x = x_s$, from P_0 to P, including rearranging, we obtain:

$$\left(\frac{\partial \mu_w}{\partial P}\right)_{T_0, x_s} = \overline{V}_w$$

$$\left(\frac{\partial \mu_w}{\partial P}\right)_{T_0, x_s} dP = \overline{V}_w dP$$

$$\int_{P_0}^{P} \left(\frac{\partial \mu_w}{\partial P}\right)_{T_0, x_s} dP = \mu_w(T_0, P, x_s) - \mu_w(T_0, P_0, x_s) = \int_{P_0}^{P} \overline{V}_w dP$$

or

$$\mu_w(T_0, P, x_s) = \mu_w(T_0, P_0, x_s) + \int_{P_0}^{P} \overline{V}_w dP \tag{3}$$

Using Eq. (3) in Eq. (1), we find that:

$$\mu_w(T_0, P_0, x_s) + \int_{P_0}^{P} \overline{V}_w dP = \mu_w(T_0, P_0)$$

or

$$\mu_w(T_0, P_0, x_s) - \mu_w(T_0, P_0) = -\int_{P_0}^{P} \overline{V}_w dP \qquad (4)$$

Note that the left-hand side of Eq. (4) is given by the chemical potential graph (see Fig. 1), because $T_0 = 25\ ^\circ C$ and $P_0 = 1$ atm for any salt concentration. In particular, for $x_s = 5$ wt% $= 0.05$, we read:

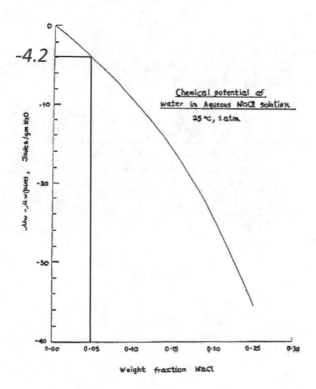

Fig. 1

$$\Delta\mu_w = \mu_w\left(25^\circ\text{C}, 1\,\text{atm}, 0.05\right) - \mu_w\left(25^\circ\text{C}, 1\ \text{atm}\right) = -4.2\,\text{Joule}/\text{gH}_2\text{O}$$

Because $\overline{V}_w$ is constant, Eq. (4) shows that:

$$\Delta\mu_w = -\overline{V}_w(P - P_0)$$

or that:

$$P - P_0 = -\frac{\Delta\mu_w}{\overline{V}_w} \tag{5}$$

Because of the hydrostatic pressure, in Eq. (5), $P - P_0 = \rho_{sw}gh$, which yields:

$$h = \frac{-\dfrac{\Delta\mu_w}{\overline{V}_w}}{\rho_{sw}g} \tag{6}$$

Using all the information given in the Problem Statement, that is, $\Delta\mu_w = -4\,\text{J}/\text{gH}_2\text{O} = -4\,\text{Nm}/\text{gH}_2\text{O}$, $\overline{V}_w = 0.994\,\text{cm}^3/\text{gH}_2\text{O}$, $\rho_{sw} = 1.145\,\text{gH}_2\text{O}/\text{cm}^3$, and $g \cong 9.8\,\text{m/sec}^2$, in Eq. (6), we find that:

$$h = \frac{\dfrac{4\,\text{N}\cancel{\text{m}}/\cancel{g}}{0.994 \times 10^{-6}\,\text{m}^{\cancel{3}/2}/\cancel{g}}}{\left(\dfrac{1.145 \times 10^{-3}\,\text{kg}}{10^{-6}\,\text{m}^{\cancel{3}/2}}\right)\left(10\,\dfrac{\text{m}}{\text{sec}^2}\right)} = \frac{\left(\dfrac{40}{0.994}\right)10^5\,\dfrac{\text{N}}{\text{m}^2}}{1.145 \times 10^4\,\dfrac{\text{N}}{\text{m}^3}}$$

$$h = \frac{\left(\dfrac{40}{0.994}\right) \times 10^5}{1.145 \times 10^4}\,\text{m}$$

or

$$\boxed{h \cong 351\,\text{m}} \tag{7}$$

Problem 20

As discussed in Part II, the azeotropic point in a binary mixture of components A and B coexisting in liquid and vapor phases results in special properties. For example, at the azeotropic point, the relative volatility is unity because the liquid and the vapor compositions are identical. In addition, a plot of the logarithm of the azeotropic pressure, P_{az}, versus $1/T$ is typically linear for many real systems.

Starting with the phase equilibria criteria for this liquid-vapor equilibrium azeotropic system, develop an explicit expression to relate $P_{az} = f(T)$. Describe under what conditions you would expect a plot of $\ln(P_{az})$ to be linear in $1/T$.

Solution to Problem 20

Solution Strategy

First, we recall that at the azeotrope, the Gibbs Phase Rule requires that:

$$L = n + 2 - \pi - r - s = 1$$

where $n = 2$, $\pi = 2$, $r = 0$, and $s = 1$ (recall that at the azeotrope, the liquid composition is equal to the vapor composition, which is a constraint, such that $s = 1$). This indicates that as discussed in Part II, the system is indeed monovariant.

Because $L = 1$, we know that if we choose T as the intensive variable, at the azeotrope, it should be possible to express P_{az} as a unique function of T. That is, $P_{az} = f(T)$!

Conditions of Phase Equilibria

Next, we utilize the three conditions of phase equilibria, that is:

$$(i) \quad T_V = T_L \equiv T \qquad \qquad \text{(Thermal)}$$
$$(ii) \quad P_V = P_L \equiv P \qquad \qquad \text{(Mechanical)}$$
$$(iii) \quad \hat{f}_A^{\,V} = \hat{f}_A^{\,L} \text{ and } \hat{f}_B^{\,V} = \hat{f}_B^{\,L} \qquad \text{(Diffusional)}$$

Choosing the Differential Approach to Phase Equilibria

Because we need to calculate how $P_{az} = f(T)$, it is convenient to utilize the differential approach to phase equilibria. As discussed in Part II, we write:

$$-\left(\frac{\overline{H}_A^V - \overline{H}_A^L}{RT^2}\right)dT + \left(\frac{\overline{V}_A^V - \overline{V}_A^L}{RT}\right)dP + \left(\frac{\partial \ln \widehat{f}_A}{\partial y_A}\right)_{T,P}^V dy_A - \left(\frac{\partial \ln \widehat{f}_A}{\partial x_A}\right)_{T,P}^L dx_A = 0$$

$$(1)$$

and

$$-\left(\frac{\overline{H}_B^V - \overline{H}_B^L}{RT^2}\right)dT + \left(\frac{\overline{V}_B^V - \overline{V}_B^L}{RT}\right)dP + \left(\frac{\partial \ln \widehat{f}_B}{\partial y_A}\right)_{T,P}^V dy_A - \left(\frac{\partial \ln \widehat{f}_B}{\partial x_A}\right)_{T,P}^L dx_A = 0$$

$$(2)$$

Next, we multiply Eq. (1) by x_A and Eq. (2) by x_B, which yields:

$$-x_A\left(\frac{\overline{H}_A^V - \overline{H}_A^L}{RT^2}\right)dT + x_A\left(\frac{\overline{V}_A^V - \overline{V}_A^L}{RT}\right)dP + x_A\left(\frac{\partial \ln \widehat{f}_A}{\partial y_A}\right)_{T,P}^V dy_A$$
$$-x_A\left(\frac{\partial \ln \widehat{f}_A}{\partial x_A}\right)_{T,P}^L dx_A = 0$$

$$(3)$$

and

$$-x_B\left(\frac{\overline{H}_B^V - \overline{H}_B^L}{RT^2}\right)dT + x_B\left(\frac{\overline{V}_B^V - \overline{V}_B^L}{RT}\right)dP + x_B\left(\frac{\partial \ln \widehat{f}_B}{\partial y_A}\right)_{T,P}^V dy_A$$
$$-x_B\left(\frac{\partial \ln \widehat{f}_B}{\partial x_A}\right)_{T,P}^L dx_A = 0$$

$$(4)$$

Adding up Eqs. (3) and (4) yields:

$$-\left(\frac{x_A\left(\overline{H}_A^V - \overline{H}_A^L\right) + x_B\left(\overline{H}_B^V - \overline{H}_B^L\right)}{RT^2}\right)dT$$
$$+\left(\frac{x_A\left(\overline{V}_A^V - \overline{V}_A^L\right) + x_B\left(\overline{V}_B^V - \overline{V}_B^L\right)}{RT}\right)dP$$
$$+\left[x_A\left(\frac{\partial \ln \widehat{f}_A}{\partial y_A}\right)_{T,P}^V + x_B\left(\frac{\partial \ln \widehat{f}_B}{\partial y_A}\right)_{T,P}^V\right]dy_A$$
$$-\left[x_A\left(\frac{\partial \ln \widehat{f}_A}{\partial x_A}\right)_{T,P}^L + x_B\left(\frac{\partial \ln \widehat{f}_B}{\partial x_A}\right)_{T,P}^L\right]dx_A = 0$$

$$(5)$$

Because $L = 1$, we recognize that we should be able to get rid of the dy_A and dx_A terms in Eq. (5) to obtain a unique relation between dT and dP. To do that, we use the Gibbs-Duhem equation in each phase.

Solving Equation (5)

- In the liquid phase, at constant T and P, the Gibbs-Duhem Equation applies. Specifically:

$$x_A d\left(\ln \widehat{f}_A^{\,L}\right)_{T,P} + x_B d\left(\ln \widehat{f}_B^{\,L}\right)_{T,P} = 0$$

or

$$x_A \left(\frac{\partial \ln \widehat{f}_A^{\,L}}{\partial x_A}\right)_{T,P} + x_B \left(\frac{\partial \ln \widehat{f}_B^{\,L}}{\partial x_A}\right)_{T,P} = 0 \tag{6}$$

- In the gas phase, at constant T and P, the Gibbs-Duhem Equation applies. Specifically:

$$y_A d\left(\ln \widehat{f}_A^{\,V}\right)_{T,P} + y_B d\left(\ln \widehat{f}_B^{\,V}\right)_{T,P} = 0$$

$$y_A \left(\frac{\partial \ln \widehat{f}_A^{\,V}}{\partial y_A}\right)_{T,P} + y_B \left(\frac{\partial \ln \widehat{f}_B^{\,V}}{\partial y_A}\right)_{T,P} = 0$$

$$\left(\frac{\partial \ln \widehat{f}_B^{\,V}}{\partial y_A}\right)_{T,P} = -\frac{y_A}{y_B}\left(\frac{\partial \ln \widehat{f}_A^{\,V}}{\partial y_A}\right)_{T,P}$$

or

$$\left(\frac{\partial \ln \widehat{f}_B^{\,V}}{\partial y_A}\right)_{T,P} = -\left(\frac{y_A}{1 - y_A}\right)\left(\frac{\partial \ln \widehat{f}_A^{\,V}}{\partial y_A}\right)_{T,P} \tag{7}$$

Equation (6) shows that the term in the square brackets multiplying dx_A in Eq. (5) is zero!

Equation (7) shows that the term in the square brackets multiplying dy_A in Eq. (5) can be expressed as follows:

$$x_A \left(\frac{\partial \ln \widehat{f}_A^{\,V}}{\partial y_A}\right)_{T,P} + x_B \left(\frac{\partial \ln \widehat{f}_B^{\,V}}{\partial y_A}\right)_{T,P} = x_A \left(\frac{\partial \ln \widehat{f}_A^{\,V}}{\partial y_A}\right)_{T,P}$$

$$+ x_B \left(-\left(\frac{y_A}{1-y_A}\right)\left(\frac{\partial \ln \widehat{f}_A^{\,V}}{\partial y_A}\right)_{T,P}\right)$$

$$x_A \left(\frac{\partial \ln \widehat{f}_A^{\,V}}{\partial y_A}\right)_{T,P} + x_B \left(\frac{\partial \ln \widehat{f}_B^{\,V}}{\partial y_A}\right)_{T,P} = \left[x_A - \frac{x_B y_A}{1-y_A}\right]\left(\frac{\partial \ln \widehat{f}_A^{\,V}}{\partial y_A}\right)_{T,P}$$

$$x_A \left(\frac{\partial \ln \widehat{f}_A^{\,V}}{\partial y_A}\right)_{T,P} + x_B \left(\frac{\partial \ln \widehat{f}_B^{\,V}}{\partial y_A}\right)_{T,P} = \left[x_A - \frac{(1-x_A) y_A}{1-y_A}\right]\left(\frac{\partial \ln \widehat{f}_A^{\,V}}{\partial y_A}\right)_{T,P}$$

$$x_A \left(\frac{\partial \ln \widehat{f}_A^{\,V}}{\partial y_A}\right)_{T,P} + x_B \left(\frac{\partial \ln \widehat{f}_B^{\,V}}{\partial y_A}\right)_{T,P} = x_A \left[1 - \frac{y_A(1-x_A)}{x_A(1-y_A)}\right]\left(\frac{\partial \ln \widehat{f}_A^{\,V}}{\partial y_A}\right)_{T,P}$$

However, at the azeotrope, $x_A = y_A$. Therefore, the last term in the square brackets in the last equation is equal to zero! Accordingly, as expected because $L = 1$, at the azeotrope, Eq. (5) reduces to a relation between dP and dT!

Next, in Eq. (5), we can simplify the numerators of the terms multiplying dT and dP. The vaporization of N_A moles of component A and N_B moles of component B involves the following changes in enthalpy and volume:

$$\underline{H}_L = N_A \overline{H}_A^{\,L} + N_B \overline{H}_B^{\,L} \Rightarrow H_L = x_A \overline{H}_A^{\,L} + x_B \overline{H}_B^{\,L}$$
$$\underline{H}_V = N_A \overline{H}_A^{\,V} + N_B \overline{H}_B^{\,V} \Rightarrow H_V = x_A \overline{H}_A^{\,V} + x_B \overline{H}_B^{\,V}$$

Therefore, we can express the mixture molar enthalpy of vaporization as follows:

$$\boxed{(H_V - H_L) = x_A \left(\overline{H}_A^{\,V} - \overline{H}_A^{\,L}\right) + x_B \left(\overline{H}_B^{\,V} - \overline{H}_B^{\,L}\right) = \Delta H_{vap}^{\,mix}} \qquad (8)$$

Similarly, we can express the mixture molar volume of vaporization as follows:

$$\underline{V}_L = N_A \overline{V}_A^{\,L} + N_B \overline{V}_B^{\,L} \Rightarrow V_L = x_A \overline{V}_A^{\,L} + x_B \overline{V}_B^{\,L}$$
$$\underline{V}_V = N_A \overline{V}_A^{\,V} + N_B \overline{V}_B^{\,V} \Rightarrow V_V = x_A \overline{V}_A^{\,V} + x_B \overline{V}_B^{\,V}$$

$$\boxed{(V_V - V_L) = x_A \left(\overline{V}_A^{\,V} - \overline{V}_A^{\,L}\right) + x_B \left(\overline{V}_B^{\,V} - \overline{V}_B^{\,L}\right) = \Delta V_{vap}^{\,mix}} \qquad (9)$$

Using Eqs. (8) and (9) in Eq. (5), with the dy_A and dx_A terms set equal to zero, yields:

$$\boxed{-\left(\frac{\Delta H_{vap}^{\ mix}}{RT^2}\right)dT + \left(\frac{\Delta V_{vap}^{\ mix}}{RT}\right)dP = 0}$$

or

$$\boxed{\left(\frac{dP}{dT}\right)_{L/V,azeotrope} = \left(\frac{\Delta H_{vap}^{\ mix}}{T\Delta V_{vap}^{\ mix}}\right)} \qquad (10)$$

Equation (10) is the Clapeyron equation for the [L/V] equilibrium at the azeotrope.

If we make the following assumptions:

(1) $\Delta H_{vap}^{\ mix} \approx$ constant

(2) $V_V >> V_L$

$$\Delta V_{vap}^{\ mix} \approx V_V = x_A \overline{V}_A^{\ V} + x_B \overline{V}_B^{\ V}$$

$$\overline{V}_A^{\ V} \approx V_A^{\ V} = RT/P \qquad \text{(Ideal)}$$

$$\overline{V}_B^{\ V} \approx V_B^{\ V} = RT/P \qquad \text{(Ideal)}$$

$$\boxed{V_V = (x_A + x_B)(RT/P) = RT/P}$$

Using the last result in Eq. (10), including rearranging, we obtain:

$$\left(\frac{dP}{dT}\right)_{L/V,azeotrope} = \frac{\Delta H_{vap}^{\ mix}}{T(RT/P)}$$

$$\left(\frac{dP}{P}\right)_{L/V,azeotrope} = \frac{\Delta H_{vap}^{\ mix}}{R}\frac{dT}{T^2}$$

$$\boxed{\left(\frac{d\ln P}{d(1/T)}\right)_{L/V,azeotrope} = -\frac{\Delta H_{vap}^{\ mix}}{R}} \qquad (11)$$

Equation (11) is the Clausius-Clapeyron equation at the azeotrope. Because $\Delta H_{vap}^{\ mix} \approx$ constant, Eq. (11) shows that P_{az} is indeed a linear function of $1/T$!

Solved Problems for Part III

Problem 21

Problem 21.1

A molecule has the following three energy levels and degeneracies:

Level	Energy	Degeneracy
1	0	1
2	ε	1
3	2ε	γ

(a) Write an expression for the molecular partition function, q, as a function of ε, γ, and the temperature, T.

(b) Write an expression for the average molecular energy, $<\varepsilon>$, as a function of ε, γ, and T.

(c) For $\varepsilon/k_BT = 1$, and $\gamma = 1$, compute the populations or probabilities of occupancy, p_1, p_2, and p_3, of the three levels.

(d) Find the temperature, T^*, at which $p_1 = p_3$. Express your result in terms of ε and γ.

(e) Compute p_1, p_2, and p_3 at the condition given in (d). Express your result in terms of γ.

(f) Explain what happens to the results in (d) and (e) when $\gamma = 1$.

Problem 21.2

Consider a system of independent, distinguishable particles that have only two quantum states with energies 0 and ε. Calculate the molecular heat capacity of such a system, and show that C_V/k_B plotted against $\beta\varepsilon$ passes through a maximum at $\beta\varepsilon = 2.40$, which corresponds to the solution of the equation, $\beta\varepsilon/2 = \coth(\beta\varepsilon/2)$.

Problem 21.3

The Canonical partition function of a monoatomic van der Waals gas is given by the following expression:

$$Q(N, \underline{V}, T) = \frac{1}{N!} \left(\frac{2\pi m k_B T}{h^2} \right)^{3N/2} (\underline{V} - Nb)^N \exp\left(aN^2/\underline{V}k_B T \right)$$

where a and b are the van der Waals constants.

(a) Derive an expression for the energy of a monoatomic van der Waals gas. Compare your result with that for a monoatomic ideal gas.
(b) Calculate the heat capacity, C_V, of a monoatomic van der Waals gas. Compare your result with that of a monoatomic ideal gas.
(c) Derive an expression for the pressure of a monoatomic van der Waals gas.

Solution to Problem 21

Solution to Problem 21.1

Part (a)

The molecular partition function is given by:

$$q = \sum_{j=1}^{3} g_j \exp\left(-\frac{\epsilon_j}{k_B T}\right) \tag{1}$$

In Eq. (1), j represents energy levels and not energy states, because we have already considered the existence of distinct states having the same energy by including the g_j term which indicates degeneracy. A table specifying different energy levels and their respective degeneracies is provided in the Problem Statement. Using the information provided in the Table in Eq. (1) yields:

$$q = 1e^0 + 1e^{-\epsilon/k_B T} + \gamma e^{-2\epsilon/k_B T}$$

$$q = 1 + 1e^{-\epsilon/k_B T} + \gamma e^{-2\epsilon/k_B T} \tag{2}$$

Part (b)

The average molecular energy is given by:

$$< \epsilon >= \sum_{j=1}^{3} p_j \epsilon_j \tag{3}$$

where p_j, the probability of occurrence of energy level, ϵ_j, is given by:

$$p_j = \frac{g_j \exp\left(-\frac{\epsilon_j}{k_B T}\right)}{q} \tag{4}$$

In Eq. (4), the molecular partition function, q, is given by Eq. (2). Using Eqs. (4) and (2), it follows that:

$$p_1 = \frac{1 \exp(0)}{q} = \frac{1}{q}$$

(5)

$$p_2 = \frac{1 \exp\left(\frac{-\epsilon}{k_B T}\right)}{q} = \frac{\exp\left(\frac{-\epsilon}{k_B T}\right)}{q}$$

(6)

$$p_3 = \frac{\gamma \exp\left(\frac{-2\epsilon}{k_B T}\right)}{q}$$

(7)

Substituting p_1, p_2, and p_3 from Eqs. (5), (6), and (7), respectively, as well as q from Eq. (2), in Eq. (3) then yields:

$$< \epsilon > = p_1 0 + p_2 \epsilon + p_3 (2\epsilon)$$

$$< \epsilon > = \frac{1}{q}\left[\epsilon \exp\left(\frac{-\epsilon}{k_B T}\right) + 2\epsilon\gamma \exp\left(\frac{-\epsilon}{k_B T}\right)\right]$$

$$< \epsilon > = \frac{\epsilon \exp\left(\frac{-\epsilon}{k_B T}\right) + 2\epsilon\gamma \exp\left(\frac{-\epsilon}{k_B T}\right)}{1 + 1 e^{-\epsilon/k_B T} + \gamma e^{-2\epsilon/k_B T}}$$

(8)

Equation (8) can also be obtained from the relation:

$$< \epsilon > = k_B T^2 \left(\frac{\partial \ln q}{\partial T}\right)_{\underline{V}}$$

(9)

Indeed, starting from Eq. (2) and recognizing that q does not depend on $\underline{V}$ in this case, it follows that:

$$\left(\frac{\partial \ln q}{\partial T}\right)_{\underline{V}} = \left(\frac{d \ln q}{dT}\right) = \frac{d}{dT}\left(\ln\left(1 + 1 e^{-\epsilon/k_B T} + \gamma e^{-2\epsilon/k_B T}\right)\right)$$

$$\left(\frac{\partial \ln q}{\partial T}\right)_{\underline{V}} = \frac{0 + \left(\frac{\epsilon}{k_B T^2}\right)\exp\left(-\epsilon/k_B T\right) + \gamma\left(\frac{2\epsilon}{k_B T^2}\right)\exp\left(-2\epsilon/k_B T\right)}{1 + 1 \exp\left(-\epsilon/k_B T\right) + \gamma \exp\left(-2\epsilon/k_B T\right)}$$

(10)

Substituting the expression in Eq. (10) in Eq. (9) yields:

$$< \epsilon > = \frac{\epsilon \exp\left(-\epsilon/k_B T\right) + 2\gamma\epsilon \exp\left(-2\epsilon/k_B T\right)}{1 + 1 \exp\left(-\epsilon/k_B T\right) + \gamma \exp\left(-2\epsilon/k_B T\right)}$$

(11)

which is identical to the result derived in Eq. (8).

Part (c)

When $\epsilon/k_BT = 1$ and $\gamma = 1$, the molecular partition function q can be calculated using Eq. (2) as follows:

$$q = 1 + e^{-1} + e^{-2} = 1 + 0.3678 + 0.1354 = 1.5032 \tag{12}$$

Using Eq. (12) in Eqs. (5), (6), and (7), it follows that:

$$p_1 = \frac{1}{q} = 0.6652$$

$$p_2 = \frac{e^{-1}}{q} = 0.2447$$

$$p_3 = \frac{e^{-2}}{q} = 0.0981 \tag{13}$$

Part (d)

We are asked to find the temperature, T^*, at which $p_1 = p_3$. Using Eqs. (5) and (7), we obtain:

$$p_1 = p_3 \quad \rightarrow \quad \frac{1}{q} = \frac{\gamma \exp\left(-\frac{2\epsilon}{k_BT^*}\right)}{q}$$

$$\exp\left(-\frac{2\epsilon}{k_BT^*}\right) = \frac{1}{\gamma}$$

$$-\frac{2\epsilon}{k_BT^*} = \ln\left(\frac{1}{\gamma}\right) = -\ln\gamma$$

$$T^* = \frac{2}{\ln\gamma}\left(\frac{\epsilon}{k_B}\right) \tag{14}$$

Clearly, to obtain a finite value of T^*, γ has to be greater than 1.

Part (e)

At the conditions given in Part (d), we have:

$$k_BT^* = \frac{2\epsilon}{\ln\gamma} \tag{15}$$

From Eq. (2), it follows that:

$$q(T^*) = 1 + 1e^{-\epsilon/k_BT^*} + \gamma e^{-2\epsilon/k_BT^*}$$

$$q(T^*) = 1 + e^{-\frac{\ln \gamma}{2}} + \gamma e^{-\ln \gamma} = 1 + \frac{1}{\sqrt{\gamma}} + \frac{\gamma}{\gamma}$$

$$q(T^*) = 2 + \frac{1}{\sqrt{\gamma}} \tag{16}$$

With $q(T^*)$ given by Eq. (16), we can use Eqs. (5), (6), and (7) to determine p_1, p_2, and p_3. Specifically:

$$p_1^* = \frac{1}{q} = \frac{1}{2 + 1/\sqrt{\gamma}}$$

$$p_1^* = \frac{\sqrt{\gamma}}{1 + 2\sqrt{\gamma}} \tag{17}$$

$$p_2^* = \frac{\exp\left(\frac{-\epsilon}{k_B T^*}\right)}{q} = \frac{\exp\left(\frac{-\epsilon}{k_B \frac{2}{\ln \gamma}\left(\frac{\epsilon}{k_B}\right)}\right)}{q}$$

$$= \frac{\exp\left(-\frac{\ln \gamma}{2}\right)}{q} = \frac{\frac{1}{\sqrt{\gamma}}}{2 + \frac{1}{\sqrt{\gamma}}}$$

$$p_2^* = \frac{1}{1 + 2\sqrt{\gamma}} \tag{18}$$

$$p_3^* = \frac{\gamma \exp\left(\frac{-2\epsilon}{k_B T^*}\right)}{q} = \frac{\gamma \exp\left(\frac{-2\epsilon}{k_B \frac{2}{\ln \gamma}\left(\frac{\epsilon}{k_B}\right)}\right)}{q} \tag{19}$$

$$= \frac{\gamma \exp\left(\ln \gamma\right)}{q} = \frac{\gamma/\gamma}{2 + \frac{1}{\sqrt{\gamma}}}$$

$$p_3^* = \frac{\sqrt{\gamma}}{1 + 2\sqrt{\gamma}} \tag{20}$$

As expected, at the conditions in Part (d), $p_1 = p_3$.

Part (f)

When $\gamma = 1$, Eq. (14) reveals that:

$$T^* = \frac{2}{\ln 1}\left(\frac{\epsilon}{k_B}\right) = \infty \tag{21}$$

In this case, the three levels become equally probable, with a probability of 1/3 each. This can be verified from Eqs. (17), (18), and (19) when $\gamma = 1$

Solution to Problem 21.2

The particle partition function q is given by:

$$q = \sum_j \exp(-\beta\epsilon_j) = \exp(0) + \exp(-\beta\epsilon) = 1 + \exp(-\beta\epsilon) \qquad (22)$$

The average energy of a particle is given by:

$$<\epsilon> = \sum_j \frac{\epsilon_j \exp(-\beta\epsilon_j)}{q} = 0 + \frac{\epsilon \exp(-\beta\epsilon)}{q} \qquad (23)$$

Because the particles are independent, it follows that:

$$<\underline{E}> = N <\epsilon> = \frac{N\epsilon \exp(-\beta\epsilon)}{1 + \exp(-\beta\epsilon)} \qquad (24)$$

We know that:

$$C_V = \frac{1}{N}\left(\frac{\partial <\underline{E}>}{\partial T}\right)_{N,\underline{V}} \qquad (25)$$

Next, substituting Eq. (24) for $\underline{E}$ in Eq. (25), we obtain:

$$C_V = \frac{1}{N}\left(\frac{\partial}{\partial T}\left[\frac{N\epsilon \exp(-\beta\epsilon)}{1 + \exp(-\beta\epsilon)}\right]_{N,\underline{V}}\right) = \frac{\partial}{\partial T}\left[\frac{\epsilon \exp(-\beta\epsilon)}{1 + \exp(-\beta\epsilon)}\right] \qquad (26)$$

Recalling that $\beta = 1/(k_B T)$ and that $\frac{d}{dT} = -\frac{1}{k_B T^2}\frac{d}{d\beta}$, it follows that Eq. (26) can be expressed as follows:

$$C_V = -\frac{1}{k_B T^2}\frac{d}{d\beta}\left[\frac{\epsilon \exp(-\beta\epsilon)}{1 + \exp(-\beta\epsilon)}\right] \qquad (27)$$

$$C_V = -\frac{1}{k_B T^2}$$

$$\times \left[\frac{(1 + \exp{(-\beta\epsilon)})\exp{(-\beta\epsilon)}(-\epsilon) - \exp{(-\beta\epsilon)}(\exp{(-\beta\epsilon)})(-\epsilon)}{(1 + \exp{(-\beta\epsilon)})^2} \right] \quad (28)$$

$$C_V = -\frac{1}{k_B T^2}\left[\frac{-\epsilon\exp{(-\beta\epsilon)} - \epsilon\exp{(-2\beta\epsilon)} + \epsilon\exp{(-2\beta\epsilon)}}{(1 + \exp{(-\beta\epsilon)})^2} \right]$$

$$= \frac{\epsilon^2\exp{(-\beta\epsilon)}}{k_B T^2(1 + \exp{(-\beta\epsilon)})^2} \quad (29)$$

In order to express C_V in terms of β, we write T in Eq. (29) as $T = \frac{1}{k_B\beta}$. This yields:

$$C_V = \frac{k_B(\beta\epsilon)^2\exp{(-\beta\epsilon)}}{(1 + \exp{(-\beta\epsilon)})^2} \quad (30)$$

To find the maximum value of $\frac{C_V}{k_B}$, let $\beta\epsilon = x$ in Eq. (30). This yields:

$$\frac{C_V}{k_B} = \frac{x^2\exp{(-x)}}{(1 + \exp{(-x)})^2} \quad (31)$$

At the maximum value of $\frac{C_V}{k_B}$, we know that:

$$\frac{d(C_V/k_B)}{dx} = 0 \quad (32)$$

Therefore, differentiating Eq. (31) with respect to x yields:

$$\frac{d(C_V/k_B)}{dx} = \frac{2x\exp{(-x)}}{(1 + \exp{(-x)})^2} - \frac{x^2\exp{(-x)}}{(1 + \exp{(-x)})^2}$$

$$+ \frac{(x^2\exp{(-x)})(2\exp{(-x)})}{(1 + \exp{(-x)})^3} \quad (33)$$

$$\frac{d(C_V/k_B)}{dx} = \frac{\exp{(-x)}x}{(1 + \exp{(-x)})^2}\left[2 - x + \frac{2x\exp{(-x)}}{1 + \exp{(-x)}} \right] \quad (34)$$

Substituting Eq. (34) in Eq. (32) and equating to zero yields:

$$2 - x + \frac{2x\exp{(-x)}}{1 + \exp{(-x)}} = 0 \quad \rightarrow \quad x - \frac{2x\exp{(-x)}}{1 + \exp{(-x)}} = 2 \quad (35)$$

The solution of Eq. (35) yields the value of x when $\frac{C_V}{k_B}$ attains its maximum value. Equation (35) can be written in a more compact form after some manipulation. Specifically:

$$x\frac{1 + \exp(-x)2\exp(-x)}{1 + \exp(-x)} = 2 \tag{36}$$

$$x\frac{1 - \exp(-x)}{1 + \exp(-x)} = 2 \quad \rightarrow \quad \frac{x}{2} = \frac{1 + \exp(-x)}{1 - \exp(-x)} \tag{37}$$

$$\frac{x}{2} = \frac{\exp(-x/2)}{\exp(-x/2)}\left(\frac{\exp(x/2) + \exp(-x/2)}{\exp(x/2) - \exp(-x/2)}\right) \tag{38}$$

Cancelling the equal terms in Eq. (38), and recognizing that the remaining term on the right-hand side of the equal sign is equal to $\coth(x/2)$, we obtain:

$$x/2 = \coth(x/2) \tag{39}$$

The solution of Eq. (39) is $x = \beta\epsilon = 2.40$.

Remark

A rigorous analysis to obtain the value of x at which C_V/k_B attains its maximum value should include checking the second-order derivative condition, in addition to the first-order derivative condition. An analytical differentiation of Eq. (34) to obtain the second differential may be quite messy. In such cases, one way out is to plot the function C_V/k_B numerically. Figure 1 shows the plot of C_V/k_B as a function of x. One can clearly see that $x = 2.40$ corresponds to the maximum of the function.

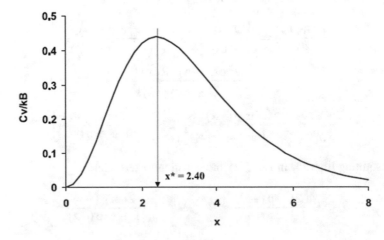

Fig. 1

Solution to Problem 21.3

Part (a)

We know that:

$$< \underline{E} >= k_B T^2 \left(\frac{\partial \ln Q}{\partial T} \right)_{N,\underline{V}} \tag{40}$$

The calculation of $< \underline{E} >$ involves finding $\left(\frac{\partial \ln Q}{\partial T} \right)_{N,\underline{V}}$ given the expression for Q in the Problem Statement. It is first useful to isolate the T-dependent terms in Q. Specifically:

$$Q(N, \underline{V}, T) = \frac{(\underline{V} - Nb)^N}{N!} \left(\frac{2\pi m k_B}{h^2} \right)^{3N/2} \exp \left(\frac{aN^2}{\underline{V} k_B T} \right) T^{3N/2} \tag{41}$$

Taking the logarithm of Eq. (41) yields:

$$\ln (Q) = \frac{3N}{2} \ln T + \frac{aN^2}{\underline{V} k_B T} + \text{Terms which do not contain } T \tag{42}$$

Taking the derivative of $\ln Q$ with respect to T, at constant N and $\underline{V}$, yields:

$$\left(\frac{\partial \ln Q}{\partial T} \right)_{N,\underline{V}} = \frac{3N}{2T} - \frac{aN^2}{\underline{V} k_B T^2} \tag{43}$$

Substituting Eq. (43) in Eq. (40) yields the desired result:

$$< \underline{E} >= k_B T^2 \left(\frac{3N}{2T} - \frac{aN^2}{\underline{V} k_B T^2} \right) \tag{44}$$

$$< \underline{E} >= \frac{3}{2} N k_B T - \frac{aN^2}{\underline{V}} \tag{45}$$

In Eq. (45), the van der Waals parameter a represents attractive interactions between the atoms and has a positive value. For $a > 0$, we see that the energy of a monoatomic van der Walls gas is lower than that of an ideal gas ($\frac{3}{2} N k_B T$) by an amount, $\frac{aN^2}{\underline{V}}$, which reflects the pairwise attractions between the atoms in the van der Waals gas, which are absent in the monatomic ideal gas.

Part (b)

After calculating $< E >$, we can calculate C_V by recalling that:

$$U = < E >$$ (46)

$$C_V = \left(\frac{\partial U}{\partial T}\right)_V = \frac{1}{N}\left(\frac{\partial \underline{U}}{\partial T}\right)_{\underline{V},N} = \frac{1}{N}\left(\frac{\partial < E >}{\partial T}\right)_{\underline{V},N}$$ (47)

Substituting Eq. (45) in Eq. (47) yields:

$$C_V = \frac{1}{N}\left(\frac{\partial}{\partial T}\left[\frac{3}{2}Nk_BT - \frac{aN^2}{\underline{V}}\right]\right)_{\underline{V},N}$$ (48)

$$C_V = \frac{3}{2}k_B$$ (49)

Equation (49) shows that C_V is the same for the monoatomic van der Waals gas and for a monoatomic ideal gas. This reflects the fact that the van der Waals parameter, a, is not a function of temperature!

Part (c)

The gas pressure can be calculated from the partition function as follows:

$$P = k_BT\left(\frac{\partial \ln Q}{\partial \underline{V}}\right)_{N,T}$$ (50)

This time, we will isolate the terms in Q that depend on $\underline{V}$, then take the logarithm, and finally differentiate with respect to $\underline{V}$ at constant N and T. Specifically:

$$Q(N, \underline{V}, T) = \frac{1}{N!}\left(\frac{2\pi mk_BT}{h^2}\right)^{3N/2}(\underline{V} - Nb)^N \exp\left(aN^2/\underline{V}k_BT\right)$$

$$Q(N, \underline{V}, T) = \frac{1}{N!}\left(\frac{2\pi mk_BT}{h^2}\right)^{3N/2}(\underline{V} - Nb)^N \exp\left(aN^2/\underline{V}k_BT\right)$$

$$Q(N, \underline{V}, T) = \frac{1}{N!}\left(\frac{2\pi mk_BT}{h^2}\right)^{3N/2}(\underline{V} - Nb)^N \exp\left(aN^2/\underline{V}k_BT\right)$$ (51)

$$\ln Q = N \ln(\underline{V} - Nb) + \frac{aN^2}{\underline{V}k_BT} + \text{Terms which do not contain } \underline{V}$$ (52)

$$\left(\frac{\partial \ln Q}{\partial \underline{V}}\right)_{N,T} = \frac{N}{\underline{V} - Nb} - \frac{aN^2}{\underline{V}^2k_BT}$$ (53)

Substituting Eq. (53) in Eq. (50) then yields the desired expression for P. Specifically:

$$P = \frac{Nk_BT}{V - Nb} - \frac{aN^2}{V^2} \tag{54}$$

Equation (54) is, of course, the celebrated van der Waals equation of state first introduced in Part I, which we have now derived molecularly using statistical mechanics!

Problem 22

Problem 22.1

In analogy with the characteristic vibrational and rotational temperatures, θ_{vib} and θ_{rot}, respectively, one can define a characteristic electronic temperature by:

$$\theta_{elec,j} = \varepsilon_{ej}/k_B$$

where ε_{ej} is the energy of the j th excited electronic state relative to the ground state.

(a) If one defines the ground electronic state to be the zero of energy, derive an expression for the electronic partition function, q_{elec}, expressed in terms of $\theta_{elec,\,j}$.

(b) The first ($j = 1$) and the second ($j = 2$) excited electronic states of O(g) lie 158.2 cm^{-1} and 226.5 cm^{-1} above the ground electronic state ($j = 0$). Given the degeneracies, $g_{e0} = 5$, $g_{e1} = 3$, and $g_{e2} = 1$, calculate the values of $\theta_{elec,1}$, $\theta_{elec,2}$, and q_{elec} (ignoring any higher excited electronic states) for O(g) at 5000 K.

(c) Calculate the fraction of O(g) atoms in the ground, first, and second electronic states at 5000 K. What is the fraction of O(g) atoms in all the remaining excited electronic states?

Note: When energies are given in cm^{-1} (which is typical in the case of electronic states, it is convenient to use $k_B = 0.69509$ cm^{-1} K^{-1}).

Problem 22.2

Molecular nitrogen is heated in an electronic arc. The spectroscopically determined relative populations of the excited vibrational states are listed below:

n	0	1	2	3	4	$\ldots$
f_n/f_0	1.000	0.200	0.040	0.008	0.002	$\ldots$

You are asked to test if the spectroscopic data provided above corresponds to molecular nitrogen being in thermodynamic equilibrium with respect to vibrational energy, as well as to determine the actual temperature at which this equilibrium would be established.

Problem 22.3

For $O_2(g)$, the following spectroscopic data is available:

$$Mass = 53.15 \times 10^{-27} \text{ kg}$$

$$Bond\ Length = 1.21 \times 10^{-10} \text{ m}$$

$$v = 1567 \text{ cm}^{-1}$$

$$D_0 = 118 \text{ kcal mol}^{-1}$$

(a) What is the fraction of $O_2(g)$ in the ground translational state when $T = 298$ K and $V = 1000$ cm^3?

(b) What is the fraction of $O_2(g)$ in the ground rotational state when $T = 298$ K?

(c) What is the fraction of $O_2(g)$ in the ground vibrational state when $T = 298$ K?

(d) Compute the enthalpy per molecule of $O_2(g)$ at 298 K and 1 atm.

Solution to Problem 22

Solution to Problem 22.1

Part (a)

As discussed in Part III, if we define the ground electronic state ($j = 0$) to be the zero of energy (such that $\epsilon_{e0}=0$), the electronic partition function can be written as follows:

$$q_{elec} = \sum_{j=0}^{\infty} g_{ej} \exp\left(\frac{-\epsilon_{ej}}{k_B T}\right) = \sum_{j=0}^{\infty} g_{ej} \exp\left(\frac{-\theta_{elec,j}}{T}\right) \tag{1}$$

Note that j in Eq. (1) represents energy levels. Applying Eq. (1) to the given case:

$$q_{elec} = g_{e0} \exp(0) + \sum_{j=1}^{\infty} g_{ej} \exp\left(\frac{-\theta_{elec,j}}{T}\right) \tag{2}$$

$$\boxed{q_{elec} = g_{e0} + \sum_{j=1}^{\infty} g_{ej} \exp\left(\frac{-\theta_{elec,j}}{T}\right)} \tag{3}$$

Part (b)

Using the definition of $\theta_{elec,j}$, we can compute $\theta_{elec,1}$ and $\theta_{elec,2}$ as follows:

$$\Theta_{elec,1} = \frac{\epsilon_{e1}}{k_B} = \frac{158.2 \ cm^{-1}}{0.69509 \ cm^{-1}K^{-1}} = 227.6 \ K$$

$$\Theta_{elec,2} = \frac{\epsilon_{e2}}{k_B} = \frac{226.5 \ cm^{-1}}{0.69509 \ cm^{-1}K^{-1}} = 325.8 \ K$$

Ignoring excited electronic states with energy levels $j > 2$, we can calculate q_{elec} using Eq. (3) and the data given in the Problem Statement, as follows:

$$q_{elec} = g_{e0} + g_{e1} \ exp \left(\frac{-\Theta_{elec,1}}{T} \right) + g_{e2} \ exp \left(\frac{-\Theta_{elec,2}}{T} \right)$$

$$q_{elec} = 5 + 3 \ exp \left(\frac{-227.6}{5000} \right) + 1 \ exp \left(\frac{-325.8}{5000} \right) \tag{4}$$

$$\boxed{q_{elec} = 8.8034} \tag{5}$$

Part (c)

The various fractions can be calculated using the following expression:

$$f_j = \frac{g_{ej} \exp \left(-\frac{\epsilon_j}{k_B T} \right)}{q_{elec}} \tag{6}$$

Recall that j in Eq. (6) denotes energy levels. Because each energy level can be g_{ej} degenerate, the g_{ej} factor appears in the expression for f_j. Using Eq. (6) for $j = 0$, 1, and 2, we obtain:

$$f_0 = \frac{g_{e0}}{q_{elec}} = \frac{5}{8.8034} = 0.5679 \tag{7}$$

$$f_1 = \frac{g_{e1} \exp \left(-\frac{\epsilon_1}{k_B T} \right)}{q_{elec}} = \frac{3 \exp \left(-\frac{227.6}{5000} \right)}{8.8034} = 0.3256 \tag{8}$$

$$f_2 = \frac{g_{e2} \exp \left(-\frac{\epsilon_2}{k_B T} \right)}{q_{elec}} = \frac{1 \exp \left(-\frac{325.8}{5000} \right)}{8.8034} = 0.1064 \tag{9}$$

Because higher electronic states were ignored in the calculation of q_{elec}, the fraction of O(g) in the higher states will be 0. However, if the contributions of the higher-energy states are included in the calculation of q_{elec}, the fraction will turn out to be a very small positive quantity.

Solution to Problem 22.2

At thermal equilibrium, the population of state n is given by:

$$f_n = \frac{1 * \exp\left(-\beta\left[n + \frac{1}{2}\right]h\nu\right)}{q_{vib}} \tag{10}$$

The term "1" in the numerator of Eq. (10) is due to the fact that the vibrational energy levels are nondegenerate. For $n = 0$, Eq. (10) yields the population of the ground vibrational state:

$$f_0 = \frac{\exp\left(-\beta\frac{h\nu}{2}\right)}{q_{vib}} \tag{11}$$

Dividing Eq. (10) by Eq. (11) yields:

$$\boxed{\frac{f_n}{f_0} = \exp\left(-\beta n h\nu\right)} \tag{12}$$

According to Eq. (12), if nitrogen is in thermodynamic equilibrium with respect to vibrational energy, then a plot of $\ln\left(\frac{f_n}{f_0}\right)$ versus n will be a straight line passing through the origin with a slope of $(-\beta n h\nu)$.

Let us examine the spectroscopic data given in the Problem Statement, and plot the values of $\ln\left(\frac{f_n}{f_0}\right)$ versus n. The table below summarizes the results:

n	0	1	2	3	4
f_n/f_0	1.000	0.200	0.040	0.008	0.002
$\ln(f_n/f_0)$	0.000	-1.609	-3.218	-4.828	-6.221

The best fit to the data is a straight line passing through the origin with a slope $- 1.5804$ (see Fig. 1). From Eq. (12), the slope is given by:

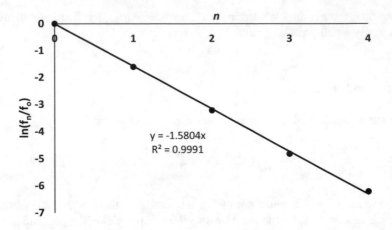

Fig. 1

$$-\beta n h\nu = -\frac{h\nu}{k_B T} = -\frac{hc\nu}{k_B T} = -1.5804$$

$$T = \frac{hc\widetilde{v}}{1.5804 k_B} \qquad (13)$$

Substituting

$$h = 6.626 \times 10^{-34} \ Js,$$

$$c = 2.998 \times 10^{10} \ cms^{-1}$$

$$k_B = 1.381 \times 10^{-23} \ J \ K^{-1}$$

$$\widetilde{v} = 2330 \ cm^{-1}$$

in Eq. (13), we obtain:

$$\boxed{T = 2121 \ K}$$

Solution to Problem 22.3

The fraction of $O_2(g)$ in a specific state j is equivalent to the probability of finding $O_2(g)$ in that state and is given by:

$$f_j = p_j = \frac{\exp(-\beta \epsilon_j)}{q} \tag{14}$$

Equation (14) shows that in order to calculate p_j, one has to determine ϵ_j and q.

Contrary to Eq. (6), g_j does not appear in the numerator of Eq. (14). This is because f_j in Eq. (14) represents the fraction in a specific energy state, while f_j in Eq. (6) represents the fraction in a specific energy level.

Part (a)

As discussed in Part III, the ground-state energy for translational motion can be obtained from the expression:

$$\epsilon_{n_x,n_y,n_z} = \frac{h^2 \left(n_x^2 + n_y^2 + n_z^2 \right)}{8m\underline{V}^{2/3}}, \qquad \text{with} \qquad n_x = n_y = n_z = 1 \tag{15}$$

That is:

$$\epsilon_{111} = \frac{3h^2}{8m\underline{V}^{2/3}} \tag{16}$$

The translational partition function is given by:

$$q_{trans} = \left(\frac{2\pi m k_B T}{h^2} \right)^{3/2} \underline{V} \tag{17}$$

Using the given values of m and $\underline{V}$ in Eq. (16) yields:

$$\epsilon_{111} = \frac{3\left(6.626 \times 10^{-34} \text{ Js}\right)^2}{8\left(53.15 \times 10^{-27} \text{kg}\right)\left(1000 \text{ cm}^3\right)^{\frac{2}{3}}}$$

$$\boxed{\epsilon_{111} = 3.097 \times 10^{-40} \text{ J}}$$

Next, we calculate q_{trans} using the values of $h = 6.626 \times 10^{-34}$ Js, $m = 53.15 \times 10^{-27}$ kg, $k_B = 1.381 \times 10^{-23}$ J K^{-1}, $T = 298$ K, and $\underline{V} = 1000$ cm^3 in Eq. (17). This yields:

$$\boxed{q_{trans} = 1.75 \times 10^{29}}$$

From Eq. (14), the fraction of molecular oxygen in the ground translational state is given by:

$$f_{111}^{trans} = \frac{\exp\left(-\beta\epsilon_j\right)}{q_{trans}} \tag{18}$$

Substituting the values of $\epsilon_{111} = 3.097 \times 10^{-40}$ J and $q_{trans} = 1.75 \times 10^{29}$ in Eq. (18) yields:

$$\boxed{f_{111}^{trans} = 5.7143 \times 10^{-30}}$$

Because $k_B T = 4.115 \times 10^{-25} \gg \epsilon_{111}$, most translational states are accessible. As a result, the probability of occupancy of any one of the available translational states is the same and is given by $\frac{1}{q_{trans}}$.

Part (b)

Similar to our approach in Part (a), we will first determine the ground-state rotational energy and the rotational partition function and then use Eq. (14) to calculate the fraction of $O_2(g)$ in the ground-rotational state. The energy levels of a rotational state are given by:

$$\epsilon_J = \frac{h^2 J(J+1)}{2I} \tag{19}$$

The ground-state rotational energy corresponds to $J = 0$ in Eq. (19), such that:

$$\epsilon_0 = 0 \tag{20}$$

The rotational partition function of $O_2(g)$ is given by:

$$q_{rot} = \frac{T}{\sigma\Theta_{rot}} \tag{21}$$

where in the case of $O_2(g)$, the symmetry number, $\sigma = 2$, and where:

$$\Theta_{rot} = \frac{h}{8\pi^2 I k_B} \tag{22}$$

In Eq. (22), the moment of inertia, I, is given by:

$$I = \mu d^2 \tag{23}$$

where μ is the reduced mass of the molecule and is given by:

$$\mu = \frac{m(O)m(O)}{m(O) + m(O)} = \frac{m(O)}{2} = \frac{m(O_2)}{4} \tag{24}$$

Using the value of $m(O_2) = 53.15 \times 10^{-27}$ kg in Eq. (24) yields:

$$\mu = \frac{m(O_2)}{4} = 13.28 \times 10^{-27} \text{kg} \tag{25}$$

Using the given value of the bond length, d, and $\mu = 13.28 \times 10^{-27}$ kg in Eq. (23) yields:

$$I = 1.945 \times 10^{-46} \text{kgm}^2$$

Using this value of I in Eq. (22), Θ_{rot} is given by:

$$\Theta_{rot} = 2.07 \ K \tag{26}$$

Recall that Eq. (22) was derived under the assumption that $\Theta_{rot} \ll T$. Indeed, at $T = 298$ K, we see that $\Theta_{rot} \ll T$. We can now use Θ_{rot} in Eq. (26), and $\sigma = 2$, to calculate q_{rot} in Eq. (21). Specifically,

$$q_{rot} = \frac{298}{2(2.07)} = 71.98$$

We can next calculate the fraction of $O_2(g)$ in the ground rotational state using Eq. (14), which yields:

$$f_0^{rot} = \frac{\exp(-\beta\epsilon_0)}{q_{rot}}$$

Substituting the values of $\epsilon_0 = 0$ from Eq. (20), and $q_{rot} = 71.98$ from Eq. (26), in the last result yields:

$$f_0^{rot} = \frac{\exp(-\beta(0))}{71.98} = \frac{1}{71.98}$$

$$\boxed{f_0^{rot} = 0.0139}$$

Part (c)

We will follow the same approach as in Parts (a) and (b) to determine the fraction of $O_2(g)$ in the ground vibrational state. That is, we will first calculate the ground-state energy, the vibrational partition function, and then use Eq. (14) to obtain the desired quantity.

The vibrational energies are given by:

$$\epsilon_n = \left(n + \frac{1}{2}\right)h\nu = \left(n + \frac{1}{2}\right)hc\tilde{\nu} \tag{27}$$

The ground-state vibrational energy corresponds to n = 0, and hence:

$$\epsilon_0 = \frac{1}{2}hc\tilde{\nu} \tag{28}$$

The vibrational partition function is given by:

$$q_{vib} = \frac{\exp\left(-\Theta_{vib}/2T\right)}{1 - \exp\left(-\Theta_{vib}/T\right)} \tag{29}$$

where

$$\Theta_{vib} = \frac{hc\tilde{\nu}}{k_B} \tag{30}$$

Now that we derived expressions for ϵ_0 and q_{vib}, let us compute the values of these quantities. Using the values of $h = 6.626 \times 10^{-34}$ Js, $c = 2.998 \times 10^{10}$ cm s^{-1}, $k_B = 1.381 \times 10^{-23}$ J, and $\tilde{\nu} = 1567$ cm^{-1} in Eqs. (28) and (30), ϵ_0 and Θ_{vib} can be calculated as follows:

$$\epsilon_0 = 1.5564 \times 10^{-20} J \tag{31}$$

$$\Theta_{vib} = 2254\ K \tag{32}$$

Because $\Theta_{vib} \gg T$, we anticipate that few vibrational states will be accessible besides the ground vibrational state. That is, we can expect to observe a high fraction of $O_2(g)$ in the ground vibrational state. Let us see if our calculation reflects that. Using Θ_{vib} from Eq. (32) in Eq. (29), we find:

$$q_{vib} = 0.2279 \tag{33}$$

We can now calculate the fraction of $O_2(g)$ in the ground vibrational state by substituting the values of $\epsilon_0 = 1.5564 \times 10^{-20} J$ from Eq. (31) and $q_{vib} = 0.2279$ from Eq. (33) in Eq. (14). This yields:

$$\boxed{f_0^{vib} = \frac{\exp\left(-\beta\epsilon_0\right)}{q_{vib}} = 0.99956} \tag{34}$$

As expected, Eq. (34) clearly shows that $O_2(g)$ is essentially in the ground vibrational state.

Part (d)

Calculating the molecular enthalpy for $O_2(g)$ requires that we calculate the molecular internal energy of $O_2(g)$ using the expression derived in Part III for diatomic molecules and then use the thermodynamic relation:

$$H = U + PV \tag{35}$$

where

$$U = k_B T \left\{ \frac{3}{2} + \frac{2}{2} + \left[\frac{\Theta_{vib}}{2T} + \frac{\frac{\Theta_{vib}}{T}}{\exp\left(\frac{\Theta_{vib}}{T}\right) - 1} \right] \right\} - D_e \tag{36}$$

In Eq. (35), because $T = 298$ K and $P = 1$ atm, oxygen can be modeled as an ideal gas, for which:

$$PV = k_B T \tag{37}$$

Using Eqs. (36) and (37) in Eq. (35), we obtain:

$$H = k_B T \left\{ \frac{3}{2} + \frac{2}{2} + 1 + \left[\frac{\Theta_{vib}}{2T} + \frac{\frac{\Theta_{vib}}{T}}{\exp\left(\frac{\Theta_{vib}}{T}\right) - 1} \right] \right\} - D_e \tag{38}$$

Next, we will calculate the values of each of the terms in Eq. (38) separately and then will substitute them in Eq. (38) to determine H. From the calculation carried out in Part (c), we obtain:

$$\Theta_{vib} = 2254 \ K \tag{39}$$

$$D_e = D_0 + \frac{1}{2} h\nu = D_0 + \frac{1}{2} hc\tilde{\nu} \tag{40}$$

$$D_e = \frac{118 \times 4.184 \ \text{Jmol}^{-1}}{6.023 \times 10^{23} \text{mol}^{-1}} + \frac{1}{2} \left(6.626 \times 10^{-34} \text{Js} \right) \left(2.998 \times 10^{10} \text{cm s}^{-1} \right) \\ \times \left(1567 \ \text{cm}^{-1} \right) \tag{41}$$

In Eq. (41), we note that we converted the basis of the given value of D_0 from moles to molecules by dividing by Avogadro's number. Use of Eq. (41) yields:

$$D_e = 8.5097 \times 10^{-19} \text{J} \tag{42}$$

We are now ready to compute H in Eq. (39) using Θ_{vib} from Eq. (39) and D_e from Eq. (42). Substituting the values of $\Theta_{vib} = 2254K$, $D_e = 8.5097 \times 10^{-19}$J, and $T = 298$ K in Eq. (38), we obtain:

$$\boxed{H = -8.2099 \times 10^{-19}\text{J}}$$

Note: There is no need to be concerned about the negative value of H. This is because of the negative contribution by D_e to H and the fact that D_e has a negative contribution because of the choice of reference state for the electronic energy levels.

Problem 23

Problem 23.1

Consider a system of N identical, but distinguishable, particles, each having two energy levels with energies 0 and $\varepsilon > 0$, respectively. The upper level is g-fold degenerate and the lower level is nondegenerate. The total energy of the system is $\underline{E}$.

(a) Use the Micro-Canonical ensemble to calculate the entropy of the system. Express your result in terms of g, N, and the occupation numbers of the upper and lower levels, n_+ and n_0, respectively, where $n_+ + n_0 = N$ and $\underline{E} = n_+\varepsilon$. Note that your result corresponds to the entropy fundamental equation, $\underline{S} = \underline{S}\,(\underline{E}, N)$, where in the present case, there is no explicit dependence on the system volume, $\underline{V}$.

(b) Use the result in Part (a) to derive an expression for the temperature of the system.

(c) Use the result in Part (b) to derive expressions for the occupation numbers n_0 and n_+. Show that the same expressions can be derived much more readily in the context of the Canonical ensemble.

Problem 23.2

Consider a vessel containing a gas consisting of N independent and indistinguishable atoms at pressure, P, and temperature, T. The walls of the vessel have n adsorbing sites, each of which can only adsorb one atom. Let $-\varepsilon$ be the energy of an adsorbed atom.

(a) Use the Grand-Canonical ensemble to derive an expression for the fugacity, $\lambda = \exp(\beta\mu)$. Express your result in terms of T, P, and the necessary atomic constants.

(b) Derive an expression for the average number of atoms, $<N>$, that adsorb from the gas onto the walls of the vessel. Express your result in terms of n, ε, T, P, and the necessary atomic constants. Discuss the low- and high-pressure and temperature behaviors of $<N>$, and provide a physical interpretation of these behaviors.

Problem 23.3

(a) Calculate the vibrational partition function of a diatomic molecule treating the vibrational mode classically.

(b) Calculate the average vibrational energy of a diatomic molecule treating the vibrational mode classically.

(c) Compare your results in Parts (a) and (b) with the corresponding quantum mechanical results derived in Part III, and determine the temperature conditions at which the classical and the quantum mechanical results become identical.

Solution to Problem 23

Solution to Problem 23.1

In this problem, we are asked to find the entropy, $\underline{S}$, of a system of N identical, but distinguishable, particles using the Micro-Canonical ensemble, where the independent variables are $\underline{E}$, $\underline{V}$, and N. Knowing $\underline{S}$ as a function of $\underline{E}$, $\underline{V}$, and N, T can be evaluated using concepts from classical thermodynamics discussed in Part I. Lastly, we are asked to compare the results obtained using the Micro-Canonical ensemble to the results obtained using the Canonical ensemble, where the independent variables are T, $\underline{V}$, and N.

Part (a): Calculating W and $\underline{S}$

We begin with Boltzmann's celebrated expression for the entropy in the context of the Micro-Canonical ensemble. Specifically:

$$\underline{S} = k_B \ln(W) \tag{1}$$

In Eq. (1), k_B is the Boltzmann constant, and W is the number of distinct states of the system for a given $\underline{E}$, $\underline{V}$, and N, that is, the degeneracy corresponding to $\underline{E}$, $\underline{V}$, and N. Note that W is not the degeneracy, g, of an energy level. Recall that according to the Problem Statement, we have:

$$\underline{E} = n_+\varepsilon \tag{2}$$

Therefore, we can also regard W as the number of distinct ways in which we can arrange N particles such that the total energy is always $\underline{E}$. An examination of Eq. (2) shows that only the number of particles in the upper energy level, n_+, contributes to the total energy. Calculation of W, therefore, involves counting the distinct number of ways in which n_+ distinguishable particles out of the available N particles can occupy the upper energy level, ε, including recognizing that the n_+ particles can be assigned to any one of the g available degenerate states. Accordingly, W can be expressed as follows:

$$W = \left(\frac{N!}{n_+!(N - n_+)!}\right)(g^{n_+}) \tag{3}$$

In Eq. (3), the first quantity in parentheses represents the number of distinct ways in which n_+ distinguishable particles can occupy the upper energy level, and the second quantity in parentheses represents the number of distinct ways of assigning the n_+ particles to g degenerate states.

Using Eq. (3) in Eq. (1), we obtain:

$$\underline{S} = k_B \ln \left[\left(\frac{N!}{n_+!(N - n_+)!} \right) (g^{n_+}) \right] \tag{4}$$

Rearranging Eq. (4) yields:

$$\underline{S} = k_B \{ \ln N! - \ln n_+! - \ln (N - n_+)! + n_+ \ln g \} \tag{5}$$

Because N >> 1, we can use Stirling's approximation, that is:

$$\ln y! = y \ln y - y, \text{ for } y \gg 1 \tag{6}$$

Using Eq. (6) in Eq. (5), we obtain:

$$\underline{S} = k_B \{ N \ln N - N - n_+ \ln n_+ + n_+ - (N - n_+) \ln (N - n_+) + N - n_+ + n_+ \ln g \}$$

Rearranging and simplifying the last expression yields:

$$\underline{S} = -k_B \left\{ n_+ \ln \frac{n_+}{N} + (N - n_+) \ln \frac{(N - n_+)}{N} - n_+ \ln g \right\} \tag{7}$$

For convenience, let us define the fraction of molecules in the upper level as follows:

$$x = \frac{n_+}{N} = \frac{E/\varepsilon}{N} = \frac{E}{\varepsilon N} \tag{8}$$

Recall that $\underline{E} = n_+\varepsilon$ (see the Problem Statement). Substituting Eq. (8) in Eq. (7) yields:

$$\underline{S} = -k_B N \{ x \ln x + (1 - x) \ln (1 - x) - x \ln g \} \tag{9}$$

Equation (9) can also be expressed as follows.

$$\underline{S} = k_B N \left\{ \ln \frac{1}{1 - x} + x \ln \frac{1 - x}{x} + x \ln g \right\} \tag{10}$$

Part (b): Obtaining T

Note that because $x = E/N\varepsilon$, Eq. (9), or Eq. (10), expresses $\underline{S}$ as a function of $\underline{E}$ and N (the volume $\underline{V}$ does not appear explicitly in Eq. (9), although it could affect the value of ε which is assumed to be a constant). Therefore, Eq. (9), or Eq. (10), is essentially the entropy fundamental equation corresponding to this system. As discussed in Part I, we can calculate the temperature recalling that $\underline{S}$ and $\underline{E}$ are related as follows:

$$\left(\frac{\partial \underline{S}}{\partial \underline{E}}\right)_N = \frac{1}{T} \tag{11}$$

Because $\underline{E} = N\varepsilon x$, Eq. (11) can be expressed as follows:

$$\left(\frac{\partial \underline{S}}{\partial \underline{E}}\right)_N = \frac{1}{N\varepsilon}\left(\frac{\partial \underline{S}}{\partial x}\right)_N \tag{12}$$

Differentiating Eq. (9) (or Eq. (10)) with respect to x, at constant N, yields:

$$\left(\frac{\partial \underline{S}}{\partial x}\right)_N = -kN\left\{\ln x + \frac{x}{x} - \ln(1-x) - \frac{1-x}{1-x} - \ln g\right\}$$

$$\left(\frac{\partial \underline{S}}{\partial x}\right)_N = kN\{\ln(1-x) + \ln g - \ln x\} = kN\ln\frac{g(1-x)}{x} \tag{13}$$

Using Eq. (3) in Eq. (12) yields:

$$\left(\frac{\partial \underline{S}}{\partial \underline{E}}\right)_N = \frac{1}{N\varepsilon}kN\ln\frac{g(1-x)}{x} = \frac{k}{\varepsilon}\ln\frac{g(1-x)}{x} \tag{14}$$

Using Eqs. (11) and (14), we can now calculate the temperature which is given by:

$$T = \frac{\varepsilon}{k}\left[\ln\frac{g(1-x)}{x}\right]^{-1}, x = \frac{n_+}{N} = \frac{E}{N\varepsilon} \tag{15}$$

Part (c): Determining n_+ and n_o

For large values of N, $\frac{n_+}{N}$ represents the probablility of finding a particle in the upper energy level (p_+). In that case, rearranging Eq. (15) for x yields:

$$x = \frac{n_+}{N} = p_+ = \frac{g\,exp\left(\frac{-\varepsilon}{kT}\right)}{1 + g\,exp\left(\frac{-\varepsilon}{kT}\right)} = \frac{g\,exp\left(-\varepsilon\beta\right)}{1 + g\,exp\left(-\varepsilon\beta\right)} \tag{16}$$

Equation (16) can be readily derived using the molecular partition function in the context of the Canonical ensemble, where q is given by:

$$q = exp\left(-\beta\varepsilon_o\right) + g\exp\left(-\beta\varepsilon\right) = 1 + g\exp\left(-\beta\varepsilon\right)$$

recalling that $\varepsilon_o = 0$. The fraction of particles in the lower-energy level, or p_o, is given by:

$$p_o = \frac{1}{1 + g\exp\left(-\beta\varepsilon\right)} \tag{17}$$

Similarly, p_+ is given by:

$$p_+ = \frac{g\exp\left(-\varepsilon\beta\right)}{1 + g\exp\left(-\varepsilon\beta\right)} \tag{18}$$

Upon comparison, Eq. (18) (obtained using the Canonical ensemble) is identical to Eq. (16) (obtained using the Micro-Canonical ensemble).

We can next calculate n_o and n_+ using Eqs. (17) and (18), respectively. Specifically:

$$n_o = \frac{N}{1 + g\exp\left(-\beta\varepsilon\right)} \tag{19}$$

$$n_+ = \frac{Ng\exp\left(-\varepsilon\beta\right)}{1 + g\exp\left(-\varepsilon\beta\right)} \tag{20}$$

Solution to Problem 23.2

This problem asks us to use the Grand-Canonical ensemble to study this system. The Grand-Canonical ensemble is typically used to study systems in equilibrium. The system described in this problem contains two simple systems: the gas in the three-dimensional region and the two-dimensional adsorption sites. We will focus on each of these systems separately. Then, we can relate the two systems by imposing the condition of equality of the chemical potentials. In Part (a), we are asked to derive an expression for λ, or the fugacity, in terms of P and T, and in Part (b), we are asked to derive an expression for $<N>$, the average number of atoms adsorbed onto a wall. Since this is an equilibrium situation, where atoms are in equilibrium between the gas phase and the adsorbed phase, the use of the Grand-Canonical ensemble is most appropriate. Recall that the Grand-Canonical ensemble was discussed in detail in Part III.

Part (a): Calculate the Fugacity, λ, Using the Grand-Canonical Ensemble

Because the gas atoms can adsorb onto the walls of the vessel, the number of gas atoms in the three-dimensional gas region does not remain constant. As discussed in Part III, the Grand-Canonical partition function is given by:

$$\Xi(\underline{V}, T, \mu) = \sum_{N=0}^{\infty} Q(N, \underline{V}, T)\lambda^N, \text{ where } \lambda = \exp(\beta\mu) \tag{21}$$

For a collection of N independent, indistinguishable atoms, we know that:

$$Q(N, \underline{V}, T) = \frac{[q(\underline{V}, T)]^N}{N!} \tag{22}$$

where q is the atomic partition function.
Substituting Eq. (22) in Eq. (21) yields:

$$\Xi(\underline{V}, T, \mu) = \sum_{N=0}^{\infty} \frac{[q(\underline{V}, T)\lambda]^N}{N!} \tag{23}$$

where $\sum_{N=0}^{\infty} \frac{[q(\underline{V}, T)\lambda]^N}{N!} = e^{q\lambda}$. Using the last result in Eq. (23), we obtain:

$$\Xi(\underline{V}, T, \mu) = \exp(q\lambda) \tag{24}$$

In Part III, we showed that:

$$P\underline{V} = k_B T \ln \Xi \tag{25}$$

Combining Eq. (25) with Eq. (24) yields:

$$\lambda = \frac{P\underline{V}}{k_B T q} \tag{26}$$

For a monoatomic ideal gas, we know that:

$$q = \left(\frac{2\pi m k_B T}{h^2}\right)^{3/2} \underline{V} g_{el} \tag{27}$$

Using Eq. (27) in Eq. (26), we obtain:

$$\lambda = P(k_B T)^{-5/2} \left(\frac{h^2}{2\pi m}\right)^{3/2} \frac{1}{g_{el}} \tag{28}$$

where λ is not a function of $\underline{V}$.

Part (b): Calculate $<N>$

We are asked to calculate the average number of atoms in the adsorbed state. Naturally, the focus is the adsorbing wall, which we regard as being in equilibrium with the gas. Accordingly, μ and T are the same for both the gas and the adsorbed atoms. In fact, each adsorbing site acts independently, is separately in equilibrium with the gas, and has a Grand-Canonical partition function derived in Part III. Specifically:

$$\Xi_{site} = \sum_{N=0}^{1} Q(N, \underline{V}, T)\lambda^N \tag{29}$$

and

$$Q(N, \underline{V}, T) = \sum_{j} e^{-\beta \underline{E}_j}$$

where the summation is over all possible states corresponding to N and $\underline{V}$. In our case, there exists only one energy level for a given N and $\underline{V}$, that is, $-N\varepsilon$. Therefore, Eq. (29) becomes:

$$\Xi_{site} = \sum_{N=0}^{1} \exp\ (\beta N \varepsilon)\lambda^N \tag{30}$$

In other words:

$$\Xi_{site} = e^0 \lambda^0 + e^{\beta \varepsilon} \lambda = 1 + \lambda e^{\beta \varepsilon} \tag{31}$$

The average number of adsorbed atoms per site was derived in Part III. Specifically:

$$< N>_{site} = kT\left(\frac{\partial \ln \Xi_{site}}{\partial \mu}\right)_{T,\underline{V}} = kT\left[\frac{\partial}{\partial \mu}\ \ln\ (1 + e^{\beta \varepsilon} e^{\beta \mu})\right]_{T,\underline{V}} \tag{32}$$

$$< N>_{site} = kT\left(\frac{\beta e^{\beta \varepsilon} e^{\beta \mu}}{1 + e^{\beta \varepsilon} e^{\beta \mu}}\right) = \frac{e^{\beta \varepsilon} e^{\beta \mu}}{1 + e^{\beta \varepsilon} e^{\beta \mu}} = \frac{1}{1 + e^{-\beta \varepsilon} e^{-\beta \mu}} = \frac{1}{1 + \lambda^{-1} e^{-\beta \varepsilon}} \tag{33}$$

It then follows that for n adsorbing sites, the average number of adsorbing atoms is given by:

$$< N >=< N>_{site}n = \frac{n}{1 + \lambda^{-1}e^{-\beta\varepsilon}} \tag{34}$$

At this stage, we impose the condition of thermodynamic equilibrium between the gas and the adsorbing wall. Accordingly, using the expression for λ (see Eq. (28)) in Eq. (34), we obtain the desired expression for the average number of adsorbed atoms as a function of T and P. Specifically:

$$< N >= \frac{n}{1 + \left[\left(\frac{2\pi m}{h^2}\right)^{3/2}g_{el}\right]P^{-1}(kT)^{5/2}e^{-\beta\varepsilon}} \tag{35}$$

At fixed temperature, $< N >$ varies with pressure in a manner that is expected intuitively. At high pressures, $P \gg 1$, $P^{-1} \ll 1$, and $< N > \sim n$. Physically, at high pressures, the gas density is high and the atoms frequently approach and adsorb onto the adsorbing sites, so that $< N > \sim n$. At low pressures, $P \ll 1$, $P^{-1} \gg 1$, and $< N > \sim 0$. The effect of temperature is also apparent. At high temperatures, T $\rightarrow\infty$, $e^{-\beta\varepsilon} \sim 1$, and $(kT)^{5/2} \gg 1$, so that $< N > \sim 0$, because the adsorbed atoms are easily released from the adsorbing sites. At low temperatures, T $\rightarrow 0$, $e^{-\beta\varepsilon} \sim 0$, and $(kT)^{5/2} \rightarrow 0$, so that $< N > \sim n$. Clearly, at high temperatures, a high pressure is required to keep the adsorbing sites filled, while only a modest pressure is required at low temperatures.

With respect to the temperature dependence, the atoms can be thought of as possessing kinetic energy in the three-dimensional phase and zero kinetic energy in the two-dimensional phase. When adsorbed onto the two-dimensional wall, an atom has no kinetic energy but gains $-\varepsilon$ of energy through its interaction with the wall. When the temperature is very low, the kinetic energy of an atom is so small relative to ε, that the atom prefers to gain ε of energy by adsorbing onto the wall. On the other hand, when the temperature is high, the atom has far more kinetic energy than ε and, therefore, prefers to access the three-dimensional gas phase.

Solution to Problem 23.3

In this problem, the main goal is to compare the classical and quantum mechanical representations of q and <ε>. At some temperature T, the classical and the quantum mechanical values approach each other. Although we are asked to calculate the vibrational contribution, similar calculations could be carried out for the translational or the rotational contributions. In Part III, we discussed the vibrational partition function in detail and provided an introductory exposition to the classical partition function.

Part(a): Calculate the Classical Vibrational Partition Function

We have already calculated the quantum mechanical vibrational contribution to the partition function for a diatomic molecule. Let us recall the results:

(i) The energy levels of the harmonic oscillator are quantized according to:

$$\varepsilon_n = \left(n + \frac{1}{2}\right)h\nu, n = 0, 1, 2 \ldots \tag{36}$$

(ii) Results presented in Part III indicate that:

$$q_{vib}^{QM} = \frac{exp\left(-\theta_{vib}/2T\right)}{1 - exp\left(-\theta_{vib}/T\right)}, \theta_{vib} = h\nu/k \tag{37}$$

$$< \varepsilon_{vib}>^{QM} = kT\left\{\frac{\theta_{vib}}{2T} + \frac{\frac{\theta_{vib}}{T} exp\left(\frac{-\theta_{vib}}{T}\right)}{1 - exp\left(\frac{-\theta_{vib}}{T}\right)}\right\} \tag{38}$$

where QM stands for quantum mechanics.

In this problem, we are asked to calculate q_{vib} and $<\varepsilon_{vib}>$ treating the vibrational modes classically. We therefore consider the energy of each possible vibrational state as a continuous variable. We consider the classical Hamiltonian description of the energy of a diatomic molecule in terms of its position, $\vec{q}$, and momentum, $\vec{p}$. Here, we envision a diatomic molecule as made up of two atoms connected by a Hookean spring. In this case, we have two degrees of freedom: the displacement, x, and the momentum associated with that displacement, p. We can write the classical Hamiltonian describing the motion of the two atoms and their interactions (through the stretch of the spring) as follows (see Part III):

$$H = \frac{p^2}{2m} + \frac{1}{2}k_s x^2$$

where m is the reduced mass of the diatomic molecule and k_s is the spring constant. As shown in Part III, the spring constant is given by

$$k_s = 4\pi^2 \nu^2 m \tag{39}$$

where ν is the characteristic vibrational frequency of the system.

We can also use a result derived in Part III to calculate q_{vib}^C, where C stands for classical. Specifically:

$$q_{vib}^c = \frac{1}{h} \int_{-\infty}^{\infty} \int_{-\infty}^{\infty} e^{-\beta H(p,x)} dp dx \tag{40}$$

Using the expression for H above in Eq. (40) and separating the integrals over p and x, we obtain:

$$q_{vib}^c = \frac{1}{h} \left(\int_{-\infty}^{\infty} e^{-\frac{\beta p^2}{2m}} dp \right) \left(\int_{-\infty}^{\infty} e^{-\frac{\beta}{2} k_s x^2} dx \right) \tag{41}$$

where the two integrals in Eq. (41) are Gaussian integrals. In Part III, we showed that:

$$\int_{-\infty}^{\infty} e^{-a^2 y^2} dy = \frac{\sqrt{\pi}}{a} \tag{42}$$

Using Eq. (42) in Eq. (41), we obtain:

$$q_{vib}^c = \frac{1}{h} \left(\frac{2\pi m}{\beta} \right)^{1/2} \left(\frac{2\pi}{\beta k_s} \right)^{1/2} = \frac{2\pi m^{1/2}}{h\beta k_s^{1/2}} \tag{43}$$

or

$$q_{vib}^c = \left(\frac{2\pi m^{1/2}}{h k_s^{1/2}} \right) k_B T \tag{44}$$

Part (b): Calculate the Average Vibrational Energy

To calculate the average vibrational energy, we use the following equation derived in Part III:

$$<\varepsilon_{vib}>^c = k_B T^2 \left(\frac{\partial \ln q_{vib}^c}{\partial T} \right)_{\underline{V}} \tag{45}$$

Using Eq. (44) in Eq. (45) yields:

$$<\varepsilon_{vib}>^c = k_B T^2 \left(\frac{\partial}{\partial T} (\ln T + \text{terms that do not contain } T) \right)_{\underline{V}} \tag{46}$$

$$<\varepsilon_{vib}>^c = k_B T \tag{47}$$

Part (c): Compare the Classical and the Quantum Mechanical Limits

An examination of Eq. (37) shows that if $T \gg \theta_{vib}$, we obtain the following result (see Part III):

$$\lim T \gg \theta_{vib}, q_{vib}^{QM} = \frac{1 - \theta_{vib}/2T}{1 - (1 + (\theta_{vib}/T))} = \frac{T}{\theta_{vib}} = \frac{T}{\frac{h\nu}{k_B}} = \frac{k_B T}{h\nu} \qquad (48)$$

According to Eq. (44):

$$q_{vib}^c = \left(\frac{2\pi m^{1/2}}{h k_s^{1/2}} \right) k_B T$$

Using the expression for k_s, as well as Eq. (56), in Eq. (61) yields:

$$q_{vib}^c = \frac{k_B T}{h\nu} \qquad (49)$$

A comparison of Eq. (66) and Eq. (65) shows that when $T \gg \theta_{vib}$, $q_{vib}^c = q_{vib}^{QM}$.

Next, let us consider the average energy in the quantum mechanical representation. Recall that for the QM case:

$$<\varepsilon_{vib}>^{QM} = kT \left\{ \frac{\theta_{vib}}{2T} + \frac{\frac{\theta_{vib}}{T} \exp\left(\frac{-\theta_{vib}}{T} \right)}{1 - \exp\left(\frac{-\theta_{vib}}{T} \right)} \right\} \qquad (50)$$

In the limit $T \gg \theta_{vib}$, Eq. (67) reduces to:

$$<\varepsilon_{vib}>^{QM} = kT \left\{ \frac{\theta_{vib}}{2T} + \frac{\frac{\theta_{vib}}{T} \left(1 - \frac{\theta_{vib}}{T} \right)}{1 - 1 + \frac{\theta_{vib}}{T}} \right\} = k_B T \qquad (51)$$

A comparison of Eq. (51) and Eq. (47) shows that in the limit $T \gg \theta_{vib}$, $<\varepsilon_{vib}>^c = <\varepsilon_{vib}>^{QM}$.

Problem 24

Adapted from *Molecular Driving Forces - Statistical Thermodynamics in Chemistry and Biology* by Ken A. Dill and Sarina Bromberg, Garland Science, Taylor & Francis Group, New York and London (2003). Hereafter, we will abbreviate the names of the authors as D&B.

Problem 24.1

(a) One mole of a molecular system can occupy any one of the four energy states below:

_____ $E_3 = 11$ kCal/mol

$E_1 = 8$ kCal/mol _____   _____ $E_2 = 8$ kCal/mol

_____ $E_0 = 3$ kCal/mol

1. Calculate U at $T = 300$ K.
2. Calculate the probability that a given snapshot of the system will have an energy of 3 kCal/mol at 300 K.
3. If the energy of each state is increased by 2 kCal/mol, what is the probability that a given snapshot of the system will have an energy of 3 kCal/mol at 300 K?
4. Calculate U as T gets very large.
5. Calculate U as T gets very small.

(b) A four-bead chain can adopt several conformations that may be grouped as shown in the energy ladder below:

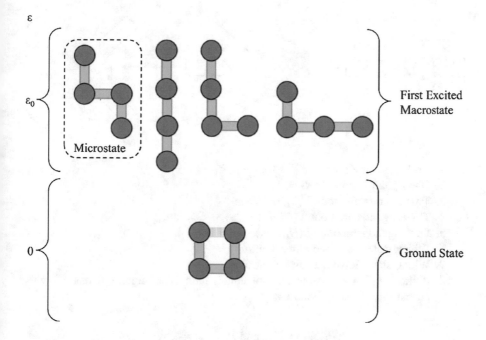

The distance between the chain ends is 1 lattice unit in the compact conformation, 3 lattice units in the extended conformation, and $\sqrt{5}$ lattice units in each of the other three chain conformations.

1. Calculate the average end-to-end distance of the chain (in lattice units) as a function of temperature.
2. Calculate the maximum value of the average end-to-end distance of the chain (in lattice units). At which temperature will this be realized?
3. Calculate the temperature at which the average end-to-end distance of the chain is equal to one half of its maximum value.

Problem 24.2

(a) The protein below has four distinguishable binding sites (α, β, γ, and δ) for the ligand L. Find the protein equilibrium binding population for a case where the ligand-protein association and dissociation constants are equal. Specifically, calculate the number of distinct arrangements, W, and the entropy (in units of k_B) when:

1. No ligands are bound ($N_L = 0$).
2. One ligand is bound ($N_L = 1$).
3. Two ligands are bound ($N_L = 2$).
4. Three ligands are bound ($N_L = 3$).
5. Four ligands are bound ($N_L = 4$).
6. Which states have the highest entropy?
7. Which states have the lowest entropy?
8. If the binding constants are not equal, namely, if ligand L has a higher probability of being bound, say:

$$P_{bound} = 75\% \text{ and } P_{unbound} = 25\%$$

Calculate the probability distribution for all the available states.

(b) Consider a zipper that has N links. Each link can be in two states, where state 1 means that the zipper is closed and has energy of zero (the zipper ground state) and state 2 means that the zipper is open and has energy ε (the zipper excited state). The zipper can only unzip from the left end, and the ith link cannot open unless all the links to its left $(1, 2, 3 \ldots i-1)$ are already open.

HINT: $\sum_{n=0}^{k} ar^n = a\frac{1-r^{k+1}}{1-r}$, when $r < 1$.

1. Think carefully about what microstates are allowed, and derive an explicit expression for the Canonical partition function, Q, for the zipper.
2. If the average energy of the zipper is $<E>$, find the average number of open links, $<i>$, in the low-temperature limit, $\varepsilon/k_BT \gg 1$. (Do not leave $<E>$ in your final expression, but instead, calculate it.)

Problem 24.3

(a) Given the intermolecular potential, $\varphi_2(r)$, as a function of intermolecular separation, r, shown below, derive an expression for the second virial coefficient, $B_2(T)$. Express your result solely in terms of ε, σ, $R\sigma$, and k_BT.

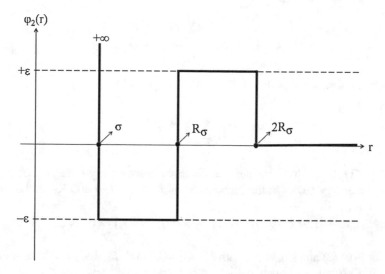

(b) Consider an ideal gas of molecules which possess permanent dipole moments, $\vec{\mu}$, in an external electric field, $\vec{\varepsilon}$. It is known that the potential energy of a single dipolar molecule in an external electric field is $u = -\mu\varepsilon\cos\theta$, where θ is the angle between the vectors $\vec{\mu}$ and $\vec{\varepsilon}$.

The Hamiltonian for a single molecule possessing a permanent dipole moment and interacting with an external electric field can be expressed as the sum of translational, rotational, and potential energy contributions. The contribution to the Hamiltonian related to rotational energy can be modeled using a rigid rotor.

1. Write the Hamiltonian expression for a single molecule possessing a permanent dipole moment in an external electric field.
2. Derive the classical partition function for the single gas molecule in Part (a).
3. Calculate the additional contribution to the ideal gas internal energy resulting from the dipole-electric field interactions.

Solution to Problem 24

Solution to Problem 24.1

(a1) Because the number of moles, the temperature, and the system volume are constant, we can use the Canonical ensemble to solve this problem. Working on a per mole basis, underbars will not be carried in the solution of this problem. It is most convenient to compute the average energy, $\langle E \rangle$, or internal energy, U, as follows:

$$U = \langle E \rangle = \frac{\sum_{i=0}^{3} E_i e^{-E_i/RT}}{Q} \tag{1}$$

where the Canonical partition function, Q, is given by:

$$Q = \sum_{i=0}^{3} e^{-E_i/RT} \tag{2}$$

In Eqs. (1) and (2), we consider four distinguishable energy states, E_0, E_1, E_2, and E_3 (see the energy ladder in the Problem Statement). At $T = 300$ K, it follows that:

$$RT|_{T=300\,K} = (1.98717 \times 10^{-3}\,kCal/(molK))(300\,K) = 0.5962\,kCal/mol \tag{3}$$

Using Eqs. (2) and (3), along with $E_0 = 3$ kCal/mol, $E_1 = E_2 = 8$ kCal/mol, and $E_3 = 11$ kCal/mol, we obtain:

$$Q = e^{-(3)/(0.5962)} + 2e^{-(8)/(0.5962)} + e^{-(11)/(0.5962)} = 0.006527 \tag{4}$$

In addition, the numerator in Eq. (1) is given by:

$$\sum_{i=0}^{3} E_i e^{E_i/RT} = \left[(3)e^{-(3)/(0.5962)} + 2(8)e^{-(8)/(0.5962)} + (11)e^{-(11)/(0.5962)} \right] kCal/mol$$

$$\sum_{i=0}^{3} E_i e^{E_i/RT} = 0.019596\,kCal/mol \tag{5}$$

Using Eqs. (5) and (4) in Eq. (1) yields the desired result:

$$\boxed{U = \frac{0.019532\,kCal/mol}{0.006506} = 3.00229\,kCal/mol} \tag{6}$$

(a2) The probability that the system will be in the ground state (0) having an energy $E_0 = 3$ kCal/mol is given by:

$$p_0 = \frac{e^{-E_0/RT}}{Q} \tag{7}$$

Using $E_0 = 3$ kCal/mol, along with Eqs. (3) and (4), in Eq. (7) yields:

$$p_0 = \frac{e^{-3/0.5962}}{0.006506} = 0.99954, \text{ or } 99.954\%, \text{ a very high probability!} \tag{8}$$

(a3) If each energy is increased by 2 kCal/mol, then, the possible energy states will be $E_0 = 5\text{kCal/mol}$, $E_1 = E_2 = 10\text{kCal/mol}$, and $E_3 = 13\text{kCal/mol}$. Therefore, the lowest possible energy state that the system can attain is now 5 kCal/mol. As a result, the system can never have $E = 3$ kCal/mol, and hence P ($E = 3$ kCal/mol) = 0!

(a4) As $T \to \infty$, the term $e^{-E_i/RT}$ tends to 1. It then follows that:

$$p_i = \frac{e^{-E_i/RT}}{Q} = \frac{1}{Q} = \frac{1}{4} \tag{9}$$

Equation (9) indicates that the four available energy states are likely to be equally populated. As a result, the ensemble-averaged energy, U, is equal to the average of E_0, E_1, E_2, and E_3. That is:

$$U = \langle E \rangle = \frac{[3 + (2)(8) + 11]\,\text{kCal/mol}}{4} = \left(\frac{30}{4}\right)\text{kCal/mol} = 7.5\,\text{kCal/mol} \tag{10}$$

(a5) As $T \to 0$, the term $e^{-E_i/RT}$ tends to zero. However, it tends to zero more rapidly for the higher-energy states than for the lower-energy states. As a result, the only energy state that is likely to be populated is the ground state (0) having energy $E_0 = 3$ kCal/mol. In other words, as $T \to 0$, the ensemble-averaged energy is equal to the energy of the ground state. In mathematical terms, one obtains:

$$U|_{T\to 0} = \langle E \rangle|_{T\to 0} = \lim_{T\to 0} \frac{\sum_{i=0}^{3} E_i e^{-E_i/RT}}{\sum_{i=0}^{3} e^{-E_i/RT}} = \lim_{T\to 0} \frac{E_0 e^{-E_0/RT}}{e^{-E_0/RT}} = E_0 = 3\,\text{kCal/mol} \tag{11}$$

(b1) As can be seen from the energy ladder in the Problem Statement, there are five conformations of the four-bead chain: the conformation with the bead-bead contact is taken to have energy $\varepsilon = 0$ (the ground state). The other four conformations have no bead-bead contacts, and all have energy $\varepsilon = \varepsilon_0$, where ε_0 is a constant. The average end-to-end distance of the four-bead chain is given by:

$$\langle d \rangle = \sum_{i=1}^{5} d_i p_i \tag{12}$$

where according to the Problem Statement (in lattice units):

$$d_1 = 1$$
$$d_2 = d_3 = d_4 = \sqrt{5}$$
$$d_5 = 3 \tag{13}$$
$$\varepsilon_1 = 0$$
$$\varepsilon_2 = \varepsilon_3 = \varepsilon_4 = \varepsilon_5 = \varepsilon_0$$

In addition, the various probabilities are given by:

$$p_1 = \frac{e^{-\varepsilon_1/k_BT}}{q} = \frac{1}{q}, \quad p_2 = p_3 = p_4 = p_5 = \frac{e^{-\varepsilon_0/k_BT}}{q}, \quad and \quad q = \sum_{i=1}^{5} e^{-\varepsilon_i/k_BT}$$

$$= 1 + 4e^{-\varepsilon_0/k_BT} \tag{14}$$

Using Eqs. (13) and (14) in Eq. (12) yields:

$$\langle d \rangle = 1 \left(\frac{1}{q} \right) + \frac{3\left(\sqrt{5} \right) e^{-\varepsilon_0/k_BT}}{q} + \frac{3 e^{-\varepsilon_0/k_BT}}{q} \tag{15}$$

Using q in Eq. (14) in Eq. (15) yields:

$$\boxed{\langle d \rangle = \frac{1 + 9.71 e^{-\varepsilon_0/k_BT}}{1 + 4 e^{-\varepsilon_0/k_BT}}} \tag{16}$$

(b2) The maximum value of the end-to-end distance of the chain is obtained when $T \to \infty$ and $e^{-\varepsilon_0/k_BT} \to 1$. In that case (in lattice units):

$$\boxed{\lim_{T\to\infty} \langle d \rangle = \langle d \rangle_{max} = \frac{1 + 9.71}{1 + 4} = \frac{10.71}{5} \cong 2.14} \tag{17}$$

(b3) We are asked to calculate at what temperature the end-to-end distance of the chain is equal to half of $\langle d \rangle_{max}$ given in Eq. (17). We therefore require that $\langle d \rangle$ in Eq. (16) be equal to $\frac{\langle d \rangle_{max}}{2}$. Specifically:

$$\frac{1 + 9.71 e^{-\varepsilon_0/k_BT}}{1 + 4 e^{-\varepsilon_0/k_BT}} = \frac{\langle d \rangle_{max}}{2} = 1.07 \tag{18}$$

$$1 + 9.71 e^{-\varepsilon_0/k_BT} = 1.07 \left(1 + 4 e^{-\varepsilon_0/k_BT} \right)$$

$$(9.71 - 1.07(4))e^{-\varepsilon_0/k_BT} = 5.43e^{-\varepsilon_0/k_BT} = (1.07 - 1) = 0.07$$

$$e^{-\varepsilon_0/k_BT} = \frac{0.07}{5.43} = 0.01305$$

$$-\frac{\varepsilon_0}{k_BT} = \ln(0.01305) = -4.339$$

$$\boxed{T = 0.2305\left(\frac{\varepsilon_0}{k_B}\right) = 0.2305T_0} \tag{19}$$

Solution to Problem 24.2

(a1) $N_L = 0 \implies W = \frac{4!}{0!4!} = 1$, only one arrangement is possible.

$$S = k_B \ln W = k_B \ln(1) = 0$$

(a2) $N_L = 1 \implies W = \frac{4!}{1!3!} = 4$, four ways to arrange on $\alpha, \beta, \gamma,$ or δ.

$$S = k_B \ln W = k_B \ln(4)$$

(a3) $N_L = 2 \implies W = \frac{4!}{2!2!} = 6$, six ways to arrange: $\alpha\beta, \alpha\gamma, \alpha\delta, \beta\gamma, \beta\delta,$ or $\gamma\delta$.

$$S = k_B \ln W = k_B \ln(6)$$

(a4) $N_L = 3 \implies W = \frac{4!}{3!1!} = 4$, four ways to arrange: the vacant spot is on $\alpha, \beta, \gamma,$ or δ.

$$S = k_B \ln W = k_B \ln(4)$$

(a5) $N_L = 4 \implies W = \frac{4!}{4!0!} = 1$, only one arrangement is possible.

$$S = k_B \ln W = k_B \ln(1) = 0$$

(a6) The states of highest entropy have $N_L = 2$.
(a7) The states of lowest entropy have $N_L = 0$ or $N_L = 4$.

(a8) The probability distribution associated with binding N_L ligands with probability p and not binding $(M-N_L)$ ligands with probability $(1-p)$ onto M sites on the protein is given by:

$$p(N_L, M) = \frac{M!}{N_L!(M-N_L)!} p^{N_L}(1-p)^{M-N_L} \tag{20}$$

Using Eq. (20), it follows that:

(i)   For $N_L = 0$, $p(0,4) = \frac{4!}{0!4!}(0.75)^0(0.25)^4 = 3.91 \times 10^{-3}$.

(ii) For $N_L = 1$, $p(1,4) = \frac{4!}{1!3!}(0.75)^1(0.25)^3 = 4.69 \times 10^{-2}$.

(iii) For $N_L = 2$, $p(2,4) = \frac{4!}{2!2!}(0.75)^2(0.25)^2 = 0.211$.

(iv) For $N_L = 3$, $p(3,4) = \frac{4!}{3!1!}(0.75)^3(0.25)^1 = 0.422$.

(v)   For $N_L = 4$, $p(4,4) = \frac{4!}{4!0!}(0.75)^4(0.25)^0 = 0.316$.

The resulting probability distribution is plotted in Fig. 1.

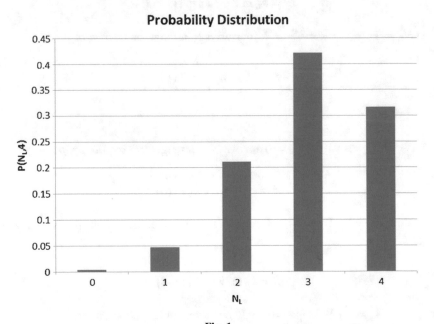

Probability Distribution

Fig. 1

(b1) The possible states of the zipper are determined by the open link number i. Each state of the zipper has an energy of $i\varepsilon$, and the Canonical partition function can be obtained by summing e^{-E_i/k_BT} for every possible state. Specifically:

$$Q = \sum_{i=0}^{N} e^{-i\varepsilon/k_BT} = e^0 + e^{-\varepsilon/k_BT} + e^{-2\varepsilon/k_BT} + \ldots + e^{-N\varepsilon/k_BT} \qquad (21)$$

Equation (21) is a finite geometric series (recall that $e^{-\varepsilon/k_BT} < 1$) and can be summed according to the **HINT** given in the Problem Statement ($\sum_{n=0}^{k} ar^n = a\frac{1-r^{k+1}}{1-r}$, for $r < 1$). Specifically:

$$Q = \frac{1 - e^{-(N+1)\varepsilon/k_BT}}{1 - e^{-\varepsilon/k_BT}} \qquad (22)$$

(b2) The average energy of the zipper can be obtained by differentiating $\ln Q$ with respect to $\beta = 1/k_BT$. Specifically, from Eq. (22), it follows that:

$$\ln Q = \ln\left(1 - e^{-(N+1)\beta\varepsilon}\right) - \ln\left(1 - e^{-\beta\varepsilon}\right) \qquad (23)$$

$$\langle E \rangle = -\left(\frac{\partial \ln Q}{\partial \beta}\right) \qquad (24)$$

$$\langle E \rangle = -\left(\frac{1}{1 - e^{-(N+1)\beta\varepsilon}}\frac{\partial}{\partial\beta}\left(1 - e^{-(N+1)\beta\varepsilon}\right) - \frac{1}{1 - e^{-\beta\varepsilon}}\frac{\partial}{\partial\beta}\left(1 - e^{-\beta\varepsilon}\right)\right)$$

$$= -\left[\frac{(N+1)\varepsilon e^{-(N+1)\beta\varepsilon}}{1 - e^{-(N+1)\beta\varepsilon}} - \frac{\varepsilon e^{-\beta\varepsilon}}{1 - e^{-\beta\varepsilon}}\right]$$

For $e^{-\beta\varepsilon} \ll 1$, the two denominators will both approach 1. Hence, we obtain:

$$\langle E \rangle = -\left[(N+1)\varepsilon e^{-(N+1)\beta\varepsilon} - \varepsilon e^{-\beta\varepsilon}\right]$$

$$= -\left[(N+1)e^{-N\beta\varepsilon} - 1\right]\varepsilon e^{-\beta\varepsilon}$$

Again, using the fact that $e^{-\beta\varepsilon} \ll 1$ (the term in brackets goes to -1) yields:

$$\langle E \rangle \cong \varepsilon e^{-\beta\varepsilon} \qquad (25)$$

Because $\langle E \rangle = \varepsilon\langle i \rangle$, it follows that:

$$\boxed{\varepsilon\langle i \rangle = \varepsilon e^{-\beta\varepsilon} \quad => \quad \langle i \rangle = e^{-\beta\varepsilon}} \tag{26}$$

Solution to Problem 24.3

(a) By inspection of the figure in the Problem Statement, it follows that:

$$\varphi(r) = \begin{cases} \infty, & for\ r < \sigma & (I) \\ -\varepsilon, & for\ R_\sigma > r \geq \sigma & (II) \\ \varepsilon, & for\ 2R_\sigma > r \geq R_\sigma & (III) \\ 0, & for\ r \geq 2R_\sigma & (IV) \end{cases} \tag{27}$$

As discussed in Part III, the second virial coefficient is related to the intermolecular potential as follows:

$$B_2(T) = -2\pi \int_0^\infty \left[e^{-\beta\varphi_2(r)} - 1 \right] r^2 dr \tag{28}$$

Using $B_2(T) = B_2^I(T) + B_2^{II}(T) + B_2^{III}(T) + B_2^{IV}(T)$, we find:

$$B_2^I(T) = -2\pi \int_0^\sigma [e^{-\infty} - 1] r^2 dr = 2\pi \int_0^\sigma r^2 dr = \frac{2}{3}\pi\sigma^3 \tag{29}$$

$$B_2^{II}(T) = -2\pi \int_\sigma^{R_\sigma} \left[e^{-\beta(-\varepsilon)} - 1 \right] r^2 dr = \frac{2}{3}\pi\left(1 - e^{\beta\varepsilon}\right)\left((R_\sigma)^3 - \sigma^3\right) \tag{30}$$

$$B_2^{III}(T) = -2\pi \int_{R_\sigma}^{2R_\sigma} \left[e^{-\beta\varepsilon} - 1 \right] r^2 dr = \frac{2}{3}\pi(R_\sigma)^3 \left(1 - e^{-\beta\varepsilon}\right)(8 - 1)$$

$$= \frac{14}{3}\pi(R_\sigma)^3 \left(1 - e^{-\beta\varepsilon}\right) \tag{31}$$

$$B_2^{IV}(T) = -2\pi \int_{2R_\sigma}^\infty [e^0 - 1] r^2 dr = -2\pi \int_{2R_\sigma}^\infty 0 r^2 dr = 0 \tag{32}$$

Adding up Eqs. (29)–(32) yields the desired result:

$$\boxed{B_2(T) = \frac{2}{3}\pi\sigma^3 \left\{ 1 + \left(1 - e^{\beta\varepsilon}\right)\left(R_\sigma^3 - 1\right) + 7R_\sigma^3 \left(1 - e^{-\beta\varepsilon}\right) \right\}} \tag{33}$$

(b) The solution to this problem involves (i) deriving an expression for the Hamiltonian of a dipolar molecule interacting with an external electric field, (ii) calculating the partition function of the dipolar molecule, and (iii) calculating the additional contribution to the ideal gas internal energy resulting from the dipole-electric field interactions.

In this problem, the kinetic energy is not affected by the presence of the dipoles in the molecules. In addition, while the dipolar molecules do not interact with each other by assumption (ideal gas), they interact with the external electric field, $\vec{\varepsilon}$, according to the interaction given in the Problem Statement ($u = -\mu\varepsilon\cos\Theta$) (see Fig. 2).

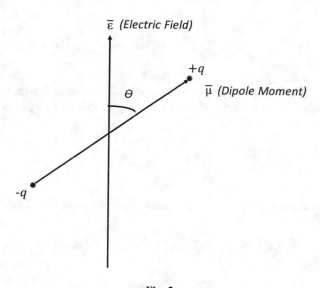

Fig. 2

Figure 1 shows only the in-plane projection, where the out-of-plane angle, ϕ, is not shown.

The five degrees of freedom of one dipolar molecule include:

- Translational: (x, y, z, p_x, p_y, p_z) – 3 translations. The limits for x, y, and z are $(-\infty$ to $\infty)$.
- Rotational: $\Theta, \phi, P_\Theta, P_\phi$ – 2 rotations. The limits for Θ and ϕ are $(0$ to $\pi)$ and $(0$ to $2\pi)$, respectively.
- Because the **dipole is fixed**, no vibrations are possible.

The Classical Hamiltonian for the Rigid Rotor is given by:

$$H = \left(\frac{p_x^2 + p_y^2 + p_z^2}{2m} \right) + \frac{1}{2I} \left(p_\Theta^2 + \frac{p_\phi^2}{\sin^2 \Theta} \right) - \mu \varepsilon \cos \Theta \qquad (34)$$

where the first two terms in Eq. (34) were discussed in Part III and the third term is new and corresponds to the potential energy of the dipole in an external electric field.

The classical partition function of the dipolar molecule possessing five degrees of freedom (see above) is then given by:

$$q = \frac{V}{h^5} \int_{-\infty}^{\infty} dp_x \int_{-\infty}^{\infty} dp_y \int_{-\infty}^{\infty} dp_z \int_{-\infty}^{\infty} dp_\Theta \int_{-\infty}^{\infty} dp_\phi \int_0^{2\pi} d\phi \int_0^{\pi} d\Theta e^{-H/k_B T} \qquad (35)$$

$$q = \frac{V}{\Lambda^3} \frac{2\pi}{h^2} \int_{-\infty}^{\infty} e^{-P_\Theta^2 / 2Ik_B T} dp_\Theta \int_{-\infty}^{\infty} e^{-P_\phi^2 / 2Ik_B T \sin^2 \Theta} dp_\phi \int_0^{\pi} e^{\mu \varepsilon \cos \Theta / k_B T} d\Theta \qquad (36)$$

The integrals over p_Θ and p_ϕ in Eq. (36) are both Gaussian integrals that can be evaluated using the expression presented in Part III, which is repeated below for completeness:

$$\int_{-\infty}^{\infty} e^{-a^2 x^2} dx = \frac{\sqrt{\pi}}{a} \qquad (37)$$

Carrying out the two Gaussian integrals in Eq. (36) using Eq. (37) yields:

$$q = \left(\frac{V}{\Lambda^3} \right) \left(\frac{8\pi^2 Ik_B T}{h^2} \right) \int_0^{\pi} \frac{1}{2} e^{\mu \varepsilon \cos \Theta / k_B T} \sin \Theta d\Theta \qquad (38)$$

As shown in Part III, in Eq. (38), the first two terms in parentheses correspond to the translational and the rotational contributions to the molecular partition function of a rigid rotor, respectively. In other words, we can rewrite Eq. (38) as follows:

$$q = q_{tran} q_{rot} \int_0^{\pi} \frac{1}{2} e^{\mu \varepsilon \cos \Theta / k_B T} \sin \Theta d\Theta \qquad (39)$$

An examination of Eq. (39) reveals that the additional contribution of the μ-ε interaction to the molecular partition function, that is, of $q_{dipole-\varepsilon}$ in Eq. (39), corresponds to:

$$q_{dipole-\varepsilon} = \int_0^{\pi} \frac{1}{2} e^{\mu \varepsilon \cos \Theta / k_B T} \sin \Theta d\Theta \qquad (40)$$

To calculate the integral in Eq. (40), it is convenient to change variables as follows:

$$x = \cos \Theta, dx = - \sin \Theta d\Theta \tag{41}$$

Using Eq. (41) in Eq. (40) yields:

$$q_{dipole-\varepsilon} = \int_{1}^{-1} \frac{1}{2} e^{\frac{\mu \varepsilon x}{k_B T}} (-dx) = -\frac{1}{2} \int_{1}^{-1} e^{\frac{\mu \varepsilon x}{k_B T}} dx \tag{42}$$

$$q_{dipole-\varepsilon} = \frac{1}{2} \int_{-1}^{1} e^{\frac{\mu \varepsilon x}{k_B T}} dx \tag{43}$$

$$q_{dipole-\varepsilon} = \frac{1}{2} \frac{k_B T}{\mu \varepsilon} \left(e^{\frac{\mu \varepsilon x}{k_B T}} \right)_{-1}^{1} = \frac{k_B T}{\mu \varepsilon} \left(\frac{e^{\frac{\mu \varepsilon}{k_B T}} - e^{-\frac{\mu \varepsilon}{k_B T}}}{2} \right) \tag{44}$$

$$q_{dipole-\varepsilon} = \frac{k_B T}{\mu \varepsilon} \sinh \left(\frac{\mu \varepsilon}{k_B T} \right) \tag{45}$$

Accordingly, the expression for the total partition function of a dipolar molecule interacting with an external electric field is given by:

$$q = \left(\frac{V}{\Lambda^3} \right) \left(\frac{8\pi^2 I k_B T}{h^2} \right) \frac{k_B T}{\mu \varepsilon} \sinh \left(\frac{\mu \varepsilon}{k_B T} \right) \tag{46}$$

(c) The additional contribution (i.e., excluding the translations and the rotations) to the total internal energy of an ideal gas of N dipolar molecules, each interacting with an external electric field, is given by:

$$\underline{U}_{dipole-\varepsilon}(N, \underline{V}, T, \varepsilon) = N k_B T^2 \left[\frac{\partial \ln q_{dipole-\varepsilon}}{\partial T} \right]_{N, \underline{V}, \varepsilon} \tag{47}$$

Using Eq. (46) in Eq. (47), including taking the temperature partial derivative, we obtain the desired result:

$$\underline{U}_{dipole-\varepsilon}(N, \underline{V}, T, \varepsilon) = N k_B T^2 \left[\frac{\partial \left(\ln \frac{k_B T}{\mu \varepsilon} \right)}{\partial T} + \frac{\partial \ln \left(\sinh \left(\frac{\mu \varepsilon}{k_B T} \right) \right)}{\partial T} \right]_{N, \underline{V}, \varepsilon} \tag{48}$$

$$\underline{U}_{dipole-\varepsilon}(N, \underline{V}, T, \varepsilon) = N k_B T^2 \left[\frac{k_B/\mu\varepsilon}{k_B T/\mu\varepsilon} - \frac{\mu\varepsilon}{k_B T^2} \frac{\cosh\left(\frac{\mu\varepsilon}{k_B T}\right)}{\sinh\left(\frac{\mu\varepsilon}{k_B T}\right)} \right] \tag{49}$$

$$\boxed{\underline{U}_{dipole-\varepsilon}(N, \underline{V}, T, \varepsilon) = N k_B T \left[1 - \frac{\mu\varepsilon}{k_B T} \cosh\left(\frac{\mu\varepsilon}{k_B T}\right) \right]} \tag{50}$$

Problem 25

Problem 25.1

(a) The energies and degeneracies of the two lowest electronic levels of atomic iodine are listed below:

Level	Energy (cm^{-1})	Degeneracy
1	0.0	4
2	7603.2	2

Calculate the temperature at which 2% of the iodine atoms will be in the excited electronic state.

NOTE: $1 \text{ cm}^{-1} = 1.986 \times 10^{-23}$ J, and $k_B = 0.69509 \text{ cm}^{-1} \text{ K}^{-1}$.

(b) Two monoatomic ideal gases consisting of N_1 atoms and N_2 atoms, respectively, are mixed in a vessel of volume, $\underline{V}$, at temperature, T. It is known that the masses of the atoms in each gas are m_1 and m_2, respectively.

(i) Calculate the Canonical partition function of the binary gas mixture.
(ii) Calculate the energy and pressure of the binary gas mixture.

Problem 25.2 (Adapted from D&B)

Statistical mechanics can be used to predict the dependence of protein folding on temperature. Consider a polypeptide chain consisting of six beads having the energy ladder shown below, where the energy increment of the unfolded states, ε_0, is positive. You are asked to answer the following questions:

(a) What is the degeneracy of each energy state?
(b) What is the Canonical partition function of the polypeptide chain?
(c) What are the probabilities of observing the polypeptide chain in each of the three states as a function of temperature? Discuss the low-temperature and the high-temperature limits.

Problem 25.3

(a) Use the Grand-Canonical ensemble to calculate the standard-state chemical potential, $\mu_0(T)$, of Ar (g) at 298 K. It is known that (i) the standard-state pressure, P_0, equals 1 bar, (ii) the mass of argon is 0.03995 kg/mol, and (iii) the first electronic state of argon is nondegenerate. How does your result compare with the experimentally measured value of -39.97 kJ/mol.

(b) It has been suggested that the triangular potential:

$$\Phi(r) = \begin{cases} \infty, & \text{for } r \leq \sigma_0 \\ -\epsilon\left(\dfrac{r - \sigma_1}{\sigma_0 - \sigma_1}\right), & \text{for } \sigma_0 < r \leq \sigma_1 \\ 0, & \text{for } r > \sigma_1 \end{cases}$$

may provide adequate second viral coefficients, $B_2(T)$. Use the model for $\Phi(r)$ above to derive an expression for $B_2(T)$ in the high-temperature limit (to linear order in $\epsilon\beta$) in terms of the model parameters σ_0, σ_1, and ϵ.

Solution to Problem 25

Solution to Problem 25.1

(a) The probability that the iodine atoms will be in excited electronic (e) state i is given by:

$$P_{ei} = \frac{g_{ei} \exp\left(-\frac{\epsilon_i}{k_B T}\right)}{q_e} \tag{1}$$

where q_e, the electronic partition function, is given by:

$$q_e = \sum_i g_{ei} \exp\left(-\frac{\epsilon_i}{k_B T}\right) \tag{2}$$

and g_{ei} is the degeneracy of the electronic energy level i.

Assuming that only the ground ($i = 1$) and the first ($i = 2$) excited electronic states contribute significantly to q_e, that is, assuming that the energy of the second ($i = 3$) and the higher ($i > 3$) excited electronic states are extremely high, q_e in Eq. (2) is given by:

$$q_e = g_{e1} \exp\left(-\frac{\epsilon_1}{k_B T}\right) + g_{e2} \exp\left(-\frac{\epsilon_2}{k_B T}\right) \tag{3}$$

Using the values of g_{e1}, g_{e2}, ϵ_1, and ϵ_2 given in the Table in the Problem Statement, Eq. (3) yields:

$$q_e = 4 + 2x, \text{ where } x = \exp\left(-\frac{7603.2}{0.69509T}\right) \tag{4}$$

where in Eq. (4), T has units of Kelvin. We are asked to find the temperature at which 2% of the iodine atoms will be in the excited ($i = 2$) electronic state. From Eqs. (1) and (4), it follows that:

$$P_{e2} = \frac{g_{e2} \exp\left(-\frac{\epsilon_2}{k_B T}\right)}{q_e} = \frac{2x}{4 + 2x} = 0.02 \tag{5}$$

Clearly, x can be determined from Eq. (5). Once x is known, T can be calculated using Eq. (4). Rearranging Eq. (5) yields:

$$0.08 + 0.04x = 2x \rightarrow 0.08 = 1.96x \qquad (6)$$

$$x = 0.040816 \qquad (7)$$

Using Eq. (4), T can be calculated from x as follows:

$$T = -\frac{10938.44}{\ln(x)} = -\frac{10938.44}{\ln(0.040816)} \qquad (8)$$

$$T = 3419.67 \text{ K} \qquad (9)$$

(bi) Because the atoms in a monoatomic ideal gas are independent and atoms of type 1 and 2 are distinguishable, the Canonical partition function of a mixture of monoatomic ideal gases can be written as the product of the Canonical partition functions of each gas, that is:

$$Q(N_1, N_2, \underline{V}, T) = Q_1(N_1, \underline{V}, T)Q_2(N_2, \underline{V}, T) \qquad (10)$$

Because in each gas the atoms are independent and indistinguishable, it follows that:

$$Q_1(N_1, \underline{V}, T) = \frac{[q_1(\underline{V}, T)]^{N_1}}{N_1!} \qquad (11)$$

and

$$Q_2(N_2, \underline{V}, T) = \frac{[q_2(\underline{V}, T)]^{N_2}}{N_2!} \qquad (12)$$

where q_1 and q_2 are the atomic partition functions of gases 1 and 2, respectively. For a monoatomic ideal gas, these are given by:

$$q_1(\underline{V}, T) = \left(\frac{2\pi m_1 k_B T}{h^2}\right)^{3/2} \underline{V} g_{e1,1} \qquad (13)$$

· where $g_{e1,1}$ is the degeneracy of the ground electronic state for gas 1. Similarly:

$$q_2(\underline{V}, T) = \left(\frac{2\pi m_2 k_B T}{h^2}\right)^{3/2} \underline{V} g_{e1,2} \qquad (14)$$

where $g_{e1,2}$ is the degeneracy of the ground electronic state for gas 2.
Substituting Eqs. (11) through (14) in Eq. (10) yields the desired result:

$$Q(N_1, N_2, \underline{V}, T) = \frac{1}{N_1! N_2!} \left[\left(\frac{2\pi m_1 k_B T}{h^2} \right)^{3/2} \underline{V} g_{el,1} \right]^{N_1} \left[\left(\frac{2\pi m_2 k_B T}{h^2} \right)^{3/2} \underline{V} g_{el,2} \right]^{N_2}$$

$$(15)$$

(bii) To calculate the energy of the gas mixture, we use the expression presented in Part III, that is:

$$< \underline{E} >= \underline{U} = k_B T^2 \left(\frac{\partial \ln Q}{\partial T} \right)_{N_1, N_2, \underline{V}}$$

$$(16)$$

It is convenient to first write down the expression for $\ln Q$ where we isolate the terms that depend explicitly on T. From Eq. (15), it follows that:

$$\ln Q = \frac{3N_1}{2} \ln T + \frac{3N_2}{2} \ln T + \text{Terms that do not depend on } T \qquad (17)$$

Differentiating Eq. (17) with respect to T, at constant N_1, N_2, and $\underline{V}$, yields:

$$\left(\frac{\partial \ln Q}{\partial T} \right)_{N_1, N_2, \underline{V}} = \frac{3N_1}{2T} + \frac{3N_2}{2T}$$

$$(18)$$

Using Eq. (18) in Eq. (16) yields the desired result:

$$< \underline{E} >= \underline{U} = \frac{3}{2}(N_1 + N_2) k_B T$$

$$(19)$$

To calculate the pressure of the gas mixture, we use the expression presented in Part III, that is:

$$P = k_B T \left(\frac{\partial \ln Q}{\partial \underline{V}} \right)_{N_1, N_2, T}$$

$$(20)$$

Again, it is convenient to isolate the terms that depend explicitly on $\underline{V}$ in the expression of $\ln Q$ before differentiation with respect to $\underline{V}$. Specifically:

$$\ln Q = N_1 \ln \underline{V} + N_2 \ln \underline{V} + \text{Terms that do not depend on } \underline{V} \qquad (21)$$

Differentiating Eq. (21) with respect to $\underline{V}$, at constant N_1, N_2, and T, yields:

$$\left(\frac{\partial \ln Q}{\partial \underline{V}}\right)_{N_1,N_2,T} = \frac{N_1}{\underline{V}} + \frac{N_2}{\underline{V}} \tag{22}$$

Using Eq. (22) in Eq. (20) yields the desired result:

$$P = \frac{k_B T(N_1 + N_2)}{\underline{V}} \tag{23}$$

Equations (19) and (23) indicate that a gas mixture of the two monoatomic ideal gases behaves like an ideal gas consisting of $(N_1 + N_2)$ atoms.

Problem Solution 25.2

(a) The degeneracy corresponds to the number of microstates in each energy level. Examination of Fig. 1 in the Problem Statement shows that:

$$E_0 = 0, \quad g_0 = 4 \tag{24}$$

$$E_1 = \epsilon_0, \quad g_1 = 11 \tag{25}$$

$$E_2 = 2\epsilon_0, \quad g_2 = 21 \tag{26}$$

(b) The partition function of the polypeptide chain, corresponding to the energy ladder in Fig. 1 in the Problem Statement, is given by:

$$q = \sum_i g_i \exp\left(-\beta E_i\right) \tag{27}$$

$$q = g_0 \exp\left(-\beta E_0\right) + g_1 \exp\left(-\beta E_1\right) + g_2 \exp\left(-\beta E_2\right) \tag{28}$$

Using the information from Fig. 1 and the Problem Statement, as well as using the degeneracies found in Part (a), yields:

$$q = 4 \exp\left(-\beta(0)\right) + 11 \exp\left(-\beta(\epsilon_0)\right) + 21 \exp\left(-\beta(2\epsilon_0)\right) \tag{29}$$

$$q = 4 + 11 \exp\left(-\beta\epsilon_0\right) + 21 \exp\left(-2\beta\epsilon_0\right) \tag{30}$$

or

$$q = 4 + 11 \exp\left(-\epsilon_0/k_B T\right) + 21 \exp\left(-2\epsilon_0/k_B T\right) \tag{31}$$

(c) The probabilities of finding the polypeptide chain in the folded state (0) and in the two unfolded states (1 and 2) are given by:

$$p_0(\text{folded state}) = \frac{g_0 \exp\left(-0/k_B T\right)}{q} = 4/q \tag{32}$$

$$p_1(\text{partially unfolded state}) = \frac{g_1 \exp\left(-\epsilon_0/k_B T\right)}{q} = 11 \exp\left(-\epsilon_0/k_B T\right)/q \tag{33}$$

$$p_2(\text{fully unfolded state}) = \frac{g_2 \exp\left(-2\epsilon_0/k_B T\right)}{q} = 21 \exp\left(-2\epsilon_0/k_B T\right)/q \tag{34}$$

Because each energy level has a different degeneracy ($g_0 \neq g_1 \neq g_2$), the energy level with the largest degeneracy is more likely to be observed at higher temperatures. At high temperature ($T \to \infty$),

$$\lim_{T \to \infty} q = 4 + 11 + 21 = 36 \tag{35}$$

$$\lim_{T \to \infty} p_0 = 4/36 \approx 0.1111 \tag{36}$$

$$\lim_{T \to \infty} p_1 = 11/36 \approx 0.3056 \tag{37}$$

$$\lim_{T \to \infty} p_2 = 21/36 \approx 0.5833 \tag{38}$$

As expected, $p_2 > p_1 > p_0$. We can see that as T increases, the unfolded states (denatured polypeptide) are more likely to be observed, as seen in nature.

At low temperatures, the lowest-energy levels are more likely, because at low temperatures ($T \to 0$), it follows that:

$$\lim_{T \to 0} q = 4 \tag{39}$$

$$\lim_{T \to 0} p_0 = 4/4 = 1 \tag{40}$$

$$\lim_{T \to 0} p_1 = 0/4 = 0 \tag{41}$$

$$\lim_{T \to 0} p_2 = 4/4 = 0 \tag{42}$$

As expected, $p_0 = 1$ and $p_1 = p_2 = 0$ as $T \to 0$.

Solution to Problem 25.3

(a) We are asked to calculate the chemical potential, μ, using the Grand-Canonical ensemble. The Grand-Canonical ensemble partition function is given by:

$$\Xi = \sum_{N=0}^{\infty} Q(N, \underline{V}, T) \exp(N\mu/k_B T) \tag{43}$$

where $Q(N, \underline{V}, T)$ is the Canonical partition function. We anticipate that at $P_0 = 1$ bar and $T = 298$ K, argon behaves like an ideal gas. In that case, the Canonical partition function is given by:

$$Q(N, \underline{V}, T) = \frac{[q(\underline{V}, T)]^N}{N!} \tag{44}$$

where the atomic partition function, q, is given by:

$$q(\underline{V}, T) = \left(\frac{2\pi m k_B T}{h^2}\right)^{3/2} \underline{V} g_{el} \tag{45}$$

We are asked to calculate μ when $P_0 = 1$ bar and $T = 298$ K. Therefore, if we could express Ξ in terms of P_0, we could use Eq. (43) to calculate μ as a function of P_0. Recall that Ξ and P are related by:

$$\Xi = \exp(P\underline{V}/k_B T) \tag{46}$$

Using Eq. (46) in Eq. (43) and substituting Eqs. (44) and (45) yields:

$$\exp(P\underline{V}/k_B T) = \sum_{N=0}^{\infty} \frac{[q(\underline{V}, T)]^N}{N!} \lambda^N, \text{where} \quad \lambda = \exp(\mu/k_B T) \tag{47}$$

Recall that:

$$e^x = \sum_{N=0}^{\infty} \frac{x^N}{N!} \tag{48}$$

In view of Eq. (48), Eq. (47) can be written as follows:

$$\exp(P\underline{V}/k_B T) = \exp(q(\underline{V}, T)\lambda) \tag{49}$$

Substituting Eq. (45) for $q(\underline{V}, T)$ in Eq. (49), and subsequently taking the natural logarithm of both sides of the resulting equation, yields:

$$\frac{P\underline{V}}{k_B T} = \left(\frac{2\pi m k_B T}{h^2}\right)^{3/2} \underline{V} g_{el} \lambda \tag{50}$$

Rearranging Eq. (50) yields:

$$\lambda = \frac{P}{g_{el}} (k_B T)^{-5/2} \left(\frac{h^2}{2\pi m}\right)^{3/2} \tag{51}$$

Recalling that $\lambda = \exp(\mu/k_B T)$, we obtain the desired expression:

$$\mu = k_B T \ln \left[\frac{P}{g_{el}} (k_B T)^{-5/2} \left(\frac{h^2}{2\pi m}\right)^{3/2}\right] \tag{52}$$

Note that the expression for μ in Eq. (52) is the same as the expression obtained in Part III using the Canonical ensemble approach.

Equation (52) enables calculation of the chemical potential at a given P and T. Note that the standard-state chemical potential is simply the chemical potential at the reference pressure $P = P_0$, that is:

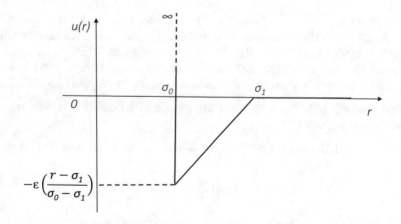

Fig. 1

$$\mu_0(T, P_0) = k_B T \ln \left[\frac{P_0}{g_{el}} (k_B T)^{-5/2} \left(\frac{h^2}{2\pi m}\right)^{3/2}\right] \tag{53}$$

Next, let us calculate the chemical potential of argon at $P = P_0 = 1$ bar and $T = 298$ K. The values of the various terms that appear in Eq. (53) are as follows:

$$g_{el} = 1 \text{ (nondegenerate first electronic ground state)}$$

$$h = 6.626 \times 10^{-34} \text{ Js}$$

$$m = 0.03995 \text{kg/mol} = 6.633 \times 10^{-26} \text{ kg/molecule}$$

$$k_B = 1.381 \times 10^{-23} \text{ J/K}$$

$$T = 298 \text{ K}$$

$$P_0 = 1 \text{ bar} = 1.0 \times 10^5 \text{ N/m}^2 \tag{54}$$

Substituting the values in Eq. (54) in Eq. (53) yields:

$$\mu_0(298 \text{ K}, 1 \text{ bar}) = -6.636 \times 10^{-23} \text{ kJ/molecule} \tag{55}$$

We next need to convert the value in Eq. (55) from a per molecule basis to a per mole basis and then compare the resulting predicted value with the experimental value given in the Problem Statement. To this end, we simply multiply the result in Eq. (55) by Avogadro's number. This yields:

$$\mu_0(298 \text{ K}, 1 \text{ bar}) = -39.97 \text{ kJ/mol} \tag{56}$$

It turns out that $\mu_0^{\text{experiment}}(1 \text{ bar}, 298 \text{ K}) = -39.97 \text{ kJ/mol}$. It then follows that the statistical mechanical prediction is in remarkable agreement with the experimental result. It is important to note that the calculated $\mu(1 \text{ bar}, 298 \text{ K})$ is based on the assumption that the ground electronic state of the argon atom is zero. If this is not the case, additional contributions may appear in the expression for $\mu_0(1 \text{ bar}, 298 \text{ K})$.

(b) This problem involves calculating the second virial coefficient from a given interaction potential.

We begin with the expression for $B_2(T)$ in terms of $u(r)$ presented in Part III:

$$B_2(T) = -2\pi \int_0^\infty \left[e^{-\beta u(r)} - 1 \right] r^2 dr \tag{57}$$

The given interaction potential, $u(r)$, is plotted in Fig. 1 below:

Because $u(r)$ shows distinct behaviors in three regions (see Fig. 1), we break the integration range in Eq. (57) into three regions as follows:

$$B_2(T) = -2\pi \left[\int_0^{\sigma_0} \left[e^{-\beta u(r)} - 1 \right] r^2 dr + \int_{\sigma_0}^{\sigma_1} \left[e^{-\beta u(r)} - 1 \right] r^2 dr + \int_{\sigma_1}^\infty \left[e^{-\beta u(r)} - 1 \right] r^2 dr \right]$$

$$B_2(T) = -2\pi [I_1 + I_2 + I_3] \tag{58}$$

We will next calculate I_1, I_2, and I_3 as follows:

$$I_1 = \int_0^{\sigma_0} [e^{-\infty} - 1] r^2 dr = \int_0^{\sigma_0} [-1] r^2 dr = -\frac{\sigma_0^3}{3} \tag{59}$$

$$I_2 = \int_{\sigma_0}^{\sigma_1} \left[e^{\beta \varepsilon \frac{r-\sigma_1}{\sigma_0 - \sigma_1}} - 1 \right] r^2 dr \tag{60}$$

$$I_3 = \int_{\sigma_1}^{\infty} \left[e^{-\beta(0)} - 1 \right] r^2 dr = \int_0^{\sigma_0} [0] r^2 dr = 0 \tag{61}$$

Using Eqs. (59), (60), and (61) in Eq. (58) yields:

$$B_2(T) = \frac{2\pi\sigma_0^3}{3} - 2\pi \int_{\sigma_0}^{\sigma_1} \left[e^{\beta \varepsilon \frac{r-\sigma_1}{\sigma_0 - \sigma_1}} - 1 \right] r^2 dr \tag{62}$$

At high temperatures, $\beta\varepsilon \ll 1$, and we can expand the exponential term in Eq. (62) as follows:

$$e^{\beta \varepsilon \frac{r-\sigma_1}{\sigma_0 - \sigma_1}} - 1 = 1 + \beta\varepsilon \frac{r - \sigma_1}{\sigma_0 - \sigma_1} - 1 + O\left((r - \sigma_0)^2\right) \approx \frac{\beta\varepsilon r}{\sigma_0 - \sigma_1} - \frac{\beta\varepsilon\sigma_1}{\sigma_0 - \sigma_1} \tag{63}$$

Using Eq. (63) in Eq. (60) yields:

$$I_2 = \int_{\sigma_0}^{\sigma_1} \frac{\beta\varepsilon r}{\sigma_0 - \sigma_1} r^2 dr - \int_{\sigma_0}^{\sigma_1} \frac{\beta\varepsilon\sigma_1}{\sigma_0 - \sigma_1} r^2 dr \tag{64}$$

$$I_2 = \frac{\beta\varepsilon}{\sigma_0 - \sigma_1} \left(\frac{\sigma_1^4 - \sigma_0^4}{4} \right) - \frac{\beta\varepsilon\sigma_1}{\sigma_0 - \sigma_1} \left(\frac{\sigma_1^3 - \sigma_0^3}{3} \right) \tag{65}$$

$$I_2 = \frac{\beta\varepsilon}{\sigma_0 - \sigma_1} \left(\frac{(\sigma_1^2 - \sigma_0^2)(\sigma_1^2 + \sigma_0^2)}{4} \right) - \frac{\beta\varepsilon\sigma_1}{\sigma_0 - \sigma_1}$$
$$\times \left(\frac{(\sigma_1^2 + \sigma_0\sigma_1 + \sigma_0^2)(\sigma_1 - \sigma_0)}{3} \right) \tag{66}$$

or

$$I_2 = -\beta\varepsilon \left(\frac{(\sigma_1 + \sigma_0)(\sigma_1^2 + \sigma_0^2)}{4} \right) + \beta\varepsilon\sigma_1 \left(\frac{(\sigma_1^2 + \sigma_0\sigma_1 + \sigma_0^2)}{3} \right) \tag{67}$$

Combining the two terms in Eq. (67) yields:

$$I_2 = \beta\varepsilon \left(\frac{(\sigma_1^3 + \sigma_0^2\sigma_1 + \sigma_1^2\sigma_0 - 3\sigma_0^3)}{12} \right) \tag{68}$$

Equation (68) can be further simplified as follows:

$$I_2 = \beta\varepsilon\left(\frac{(\sigma_1{}^2 + 2\sigma_0\sigma_1 + 3\sigma_0{}^2)(\sigma_1 - \sigma_0)}{12}\right) \tag{69}$$

Using Eqs. (59), (61), and (69) in Eq. (58) yields the desired result:

$$B_2(T) = \frac{2\pi\sigma_0^3}{3} - \pi\beta\varepsilon\left(\frac{(\sigma_1{}^2 + 2\sigma_0\sigma_1 + 3\sigma_0{}^2)(\sigma_1 - \sigma_0)}{6}\right) \tag{70}$$

The first term in Eq. (70) is due to the hard-sphere repulsive part of the interaction potential, $u(r)$, and, therefore, has a positive contribution to B_2. As Fig. 1 shows, $\varepsilon > 0$ corresponds to an attractive interaction and, consequently, has a negative contribution to B_2. On the other hand, $\varepsilon < 0$ corresponds to a repulsive interaction and, consequently, has a negative contribution to B_2.

Printed in the United States
by Baker & Taylor Publisher Services